AF785144

Fate of Antioxidants in Gut and Interaction of Gut Metabolites and Gut Microbiota

Fate of Antioxidants in Gut and Interaction of Gut Metabolites and Gut Microbiota

Editor

Baojun Xu

MDPI

Basel • Beijing • Wuhan • Barcelona • Belgrade • Novi Sad • Cluj • Manchester

Editor
Baojun Xu
Beijing Normal University-
Hong Kong Baptist University
United International College
Zhuhai,
China

Editorial Office
MDPI
St. Alban-Anlage 66
4052 Basel, Switzerland

This is a reprint of articles from the Special Issue published online in the open access journal *Antioxidants* (ISSN 2076-3921) (available at: https://www.mdpi.com/journal/antioxidants/special_issues/Antioxidants_Gut).

For citation purposes, cite each article independently as indicated on the article page online and as indicated below:

Lastname, A.A.; Lastname, B.B. Article Title. *Journal Name* **Year**, *Volume Number*, Page Range.

ISBN 978-3-0365-8532-1 (Hbk)
ISBN 978-3-0365-8533-8 (PDF)
doi.org/10.3390/books978-3-0365-8533-8

Contents

About the Editor

Baojun Xu

Dr. Baojun Xu is a Chair Professor in Beijing Normal University–Hong Kong Baptist University United International College (UIC, a full English-teaching college in China), Fellow of the Royal Society of Chemistry, Zhuhai Scholar Distinguished Professor, Department Head of the Department of Life Sciences, Program Director of Food Science and Technology Program, author of over 300 peer-reviewed papers. Dr. Xu received a Ph.D. in Food Science from Chungnam National University, South Korea. He conducted postdoctoral research work in North Dakota State University (NDSU), Purdue University, and Gerald P. Murphy Cancer Foundation in USA during 2005-2009. He did short-term visiting research in NDSU in 2012, and University Georgia in 2014, followed by visiting research during his sabbatical leave (7 months) in Pennsylvania State University in USA in 2016. Dr. Xu is serving as Associate Editor-in-Chief of *Food Science and Human Wellness*, Associate Editor of *Food Research International*, Associate Editor of *Food Frontiers*, and the editorial board member of around 10 international journals. He received the inaugural President's Award for Outstanding Research of UIC in 2016, President's Award for Outstanding Service of UIC in 2020. Dr. Xu has been listed in the world's top 2% scientists by Stanford University in 2020, 2021, 2022, and has been listed as one of the best scientists in the world in the field of Biology and Biochemistry at Research.com in 2023.

Preface

The interactions between antioxidants and gut microbiota significantly impact the bioavailability of antioxidants by boosting the diversity of gut microbiota and the generation of bioactive metabolites by the microbiome. Thus, the potent health advantages of dietary antioxidants may be explained by their active, reciprocal interactions or cascades between commensal gut bacteria, food antioxidants, and the metabolites they generate. However, there is still a significant gap in our knowledge about the final fate of antioxidants in the gut, how antioxidant metabolites affect the gut microbiota, and how gut microbiota affects the metabolism of antioxidants. Thus, this Special Issue discusses the latest knowledge on the fate of antioxidants in the gut and the role antioxidant gut metabolites play in reducing oxidative stress in various gut and metabolic diseases. This Special Issue is a balanced collection of original scientific research papers and authoritative reviews on in vitro or in vivo studies relating to any of the following topics: the fate of antioxidants in gut; antioxidative activities of phytochemicals in the digestive system; molecular mechanisms of phytochemical antioxidants in maintaining gut health; and interactions of antioxidant metabolites and gut microbiota contributed by invited experts in the field as well as a careful selection of submissions based on a general call for papers.

This Special Issue includes seven research papers and four reviews on recent findings in the field. Tang et al. report that Bacillus amyloliquefaciens SC06 can induce AKT–FOXO signaling-pathway-mediated autophagy to reduce oxidative-stress-induced apoptosis and cell damage in IPEC-J2 Cells. Tan and co-workers observed that quercetin supplementation in male C57BL/6J mice fed with a high-fat diet significantly increased the population of Akkermansia and decreased the Firmicutes/Bacteroidetes ratio in feces. In another study, Kang et al. reported that the oral administration of selenium- and zinc-enriched *Lactobacillus plantarum* SeZi significantly enhanced the concentrations of selenium and zinc in blood samples of mice. The consumption of *Dendropanax morbifera* leaf (DML) extracts containing high concentrations of chlorogenic acid and rutin showed concentration-dependent liver protection in animal models. The oral administration of Polymethoxyflavone (PMF)-rich fraction from Ougan (*Citrus reticulata* cv. Suavissima) fruit enhanced the population of Lactobacillus and Bifidobacterium in the gut microbiota. In another study by Gu et al, lipids such as coconut oil, sunflower oil, and beef tallow were reported to promote the bioaccessibility of both anthocyanins and phenolic acids from black carrot during intestinal digestion. Time-trend differential pulse voltammetry kinetic data by Olivas-Aguirre et al. revealed concurrent epithelial permeability and biotransformation of bioactive compounds from small berries. In a review, Ballway et al. have efficiently described the mechanisms of gut dysbiosis, leaky gut, endotoxemia, and fatty liver disease with a specific focus on the alcohol-associated pathways. Makarewicz et al. reviewed the bidirectional relationship between polyphenols and the gut microbiome in detail. Koudoufio et al. documented that the interaction of dietary polyphenols and micro-ecology promotes multiple physiological functions in humans. In addition, Singh et al. discuss in detail the effects of polyphenols on inflammation in the brain–liver–gut axis.

Overall, this Special Issue provides up-to-date findings related to the fate of antioxidants in the gut, interactions of antioxidant metabolites and gut microbiota and molecular mechanisms of phytochemical antioxidants in maintaining gut health and managing various diseases. We would like to take this opportunity to thank Prof. Stanley Omaye, the Editor-in-Chief of *Antioxidants*, for giving us this great opportunity to publish this Special Issue "Fate of Antioxidants in Gut and Interaction of Gut Metabolites and Gut Microbiota". We also appreciate the contributions from subject matter experts and congratulate them for their significant findings and hard work. We also owe our peer reviewers a huge debt of gratitude for their insightful comments that helped to improve the quality of manuscripts published in this issue.

Baojun Xu
Editor

antioxidants

MDPI

Article

First-Pass Metabolism of Polyphenols from Selected Berries: A High-Throughput Bioanalytical Approach

Francisco J. Olivas-Aguirre [1,*], Sandra Mendoza [2], Emilio Alvarez-Parrilla [3], Gustavo A. Gonzalez-Aguilar [4], Monica A. Villegas-Ochoa [4], Jael T.J. Quintero-Vargas [1] and Abraham Wall-Medrano [3,*]

1 Departamento de Ciencias de la Salud, Universidad de Sonora (Campus Cajeme), Blvd Bordo Nuevo s/n, Ejido Providencia, Cd, Obregón 85199, Mexico; jael.quintero@unison.mx

2 Departamento de Investigación y Posgrado en Alimentos (PROPAC), Facultad de Química, Universidad Autónoma de Querétaro, Cerro de las Campanas s/n, Santiago de Querétaro 76010, Mexico; smendoza@uaq.mx

3 Instituto de Ciencias Biomédicas, Universidad Autónoma de Ciudad Juárez, Anillo Envolvente del PRONAF y Estocolmo s/n, Ciudad Juárez 32310, Mexico; ealvarez@uacj.mx

4 Centro de Investigación en Alimentación y Desarrollo, A.C. (CIAD), Carretera a Ejido La Victoria, Km. 0.6, Hermosillo 83304, Mexico; gustavo@ciad.mx (G.A.G.-A.); mvillegas@ciad.mx (M.A.V.-O.)

* Correspondence: francisco.olivas@unison.mx (F.J.O.-A.); awall@uacj.mx (A.W.-M.)

Received: 29 March 2020; Accepted: 11 April 2020; Published: 13 April 2020

Abstract: Small berries are rich in polyphenols whose first-pass metabolism may alter their ultimate physiological effects. The antioxidant capacity and polyphenol profile of three freeze-dried berries (blackberry, raspberry, Red Globe grape) were measured and their apparent permeability (Papp) and first-pass biotransformation were tracked with an ex vivo bioanalytical system [everted gut sac (rat) + three detection methods: spectrophotometry, HPLC-ESI-QTOF-MS, differential pulse voltammetry (DPV)]. Total polyphenol (ratio 0.07-0.14-1.0) and molecular diversity (anthocyanins > flavan-3-ols), antioxidant capacity (DPPH, FRAP), anodic current *maxima* and Papp (efflux > uptake) were in the following order: blackberry > raspberry > Red Globe grape. Epicatechin, pelargonidin & cyanin (all), callistephin (raspberry/blackberry), catechin (grape), cyanidin glycosides (blackberry) and their derived metabolites [quinic acid, epicatechin, cyanidin/malvidin glucosides, and chlorogenic/caffeic acids] were fruit-specific and concentration-dependent. Time-trend DPV kinetic data revealed concurrent epithelial permeability & biotransformation processes. Regular permeability and high-biotransformation of berry polyphenols suggest fruit-specific health effects apparently at the intestinal level.

Keywords: anthocyanins; berries; polyphenols; bioaccesibility; differential pulse voltammetry; first-pass metabolism; HPLC-ESI-QTOF-MS; apparent permeability

1. Introduction

Regular consumption of small berries has been associated with several health benefits. Epidemiological studies and controlled clinical trials indicate that their acute-chronic consumption exerts synergistic and independent effects on lowering several pathophysiological markers including hyperglycemia, hyperinsulinemia, dyslipidemia, pro-inflammatory cytokines, hypertensive factors and oxidative stressors [1]. In fact, there is also an inverse association between berry consumption and many risk factors for cardiovascular disease and type-2 diabetes [2]; most, if not all, of these health benefits, are related to the amount and phytochemical diversity present in each berry, from which those with antioxidant activity (e.g., polyphenols) have been the most studied [3]. However, the heterogeneity in physiological response after their intake can hinder their beneficial effects in specific subpopulations [1,4].

To exert their physiological effects, berry polyphenols must be present in sufficient amounts in raw or prepared foods, be both bioaccessible (the fraction released from the food matrix during gastrointestinal digestion) and bioavailable (the fraction that reaches systemic circulation as the parent compound or a metabolite). The bioaccessibility of polyphenols is closely related to their physicochemical structure, the food matrix that contains them and the presence of anti-nutritional factors that could interfere with their release ability and intestinal absorption [5]. In this sense, the bioaccessibility of polyphenols from berry fruits is higher compared to other fruits, due to a concomitant effect between their natural higher level [6] and their low content of non-digestible carbohydrates and protein, both associated with an efficient gastrointestinal (GI) delivery that enhances their bioavailability [2,7].

On the other hand, the absorption, pharmacokinetics and systemic metabolism of polyphenols and their and biotransformation by the GI microbiota, have been extensively studied in the last decade [2,5]; however, the first-pass metabolism (a.k.a. pre-systemic metabolism) of bioaccessible polyphenols also modifies their ultimate health effects [1]. Polyphenols present in plant foods are commonly biotransformed (e.g., conjugation, de-glycation) before and during their pre-systemic passage, a phenomenon that involves several brush-border enzymes and a tightly-regulated influx/efflux interchange that sustain their cellular homeostasis [8,9]. These events are not normally considered when evaluating the pharmacokinetics of polyphenols, partially due to the absence of high sensitivity methods [10,11] to record such metabolic changes.

The evaluation of bioaccessible and bioavailable fractions of polyphenols has been proposed recently as a "quality" parameter in berry breeding programs. However, studies reporting the extent to which bioaccessible polyphenols are bio-transformed during their first-pass metabolism and colonic fermentation are still very scarce [2,9]. In this study, we used a high-throughput ex vivo bioanalytical system [everted gut sac (rat) + three detection methods: spectrophotometry, HPLC-ESI-QTOF-MS, differential pulse voltammetry (DPV)] to evaluate the apparent permeability (P_{app}) and enteral biotransformation of polyphenols from three berry fruits with graded levels of polyphenols (blackberry > raspberry > Red Globe grape); to the best of our knowledge, this ex vivo bioanalytical approach is reported for the first time.

2. Materials and Methods

2.1. Chemicals and Standards

Pure (≥ 93%) chemical standards were purchased from Cayman Chemicals (Ann Arbor, MI, USA). 2,2-Diphenyl-1-picrylhydrazyl (DPPH), 2,2′-azobis(2-amidinopropane) dihydrochloride (AAPH), 6-hydroxy-2,5,7,8-tetramethylchroman-2-carboxylic acid (Trolox), fluorescein, Folin–Ciocalteu (FC) phenol reagent, ACS-grade salts & acids and all enzymes and chemicals used for the *in vitro* digestion and *ex vivo* apparent permeability assays, were purchased from Sigma-Aldrich Fluka (St. Louis, MO, USA). Analytical and HPLC-MS grade solvents were obtained from JT-Baker (Avantor Performance Materials S.A. de C.V., Ecatepec de Morelos, Estado de Mexico, Mexico); sodium pentobarbital (Pisabental®) was acquired from PISA Agropecuaria (Guadalajara, Jalisco, Mexico).

2.2. Samples and Extracts

Fully ripe Red Globe grape (*Vitis vinifera* L., 18° Brix, pH 4), raspberry (*Rubus idaeus L*, 10° Brix, pH 3) and blackberry (*Rubus* spp., 10° Brix, pH 3) were purchased locally (Ciudad Juarez, Chihuahua, Mexico; 31°44′22″N, 106°29′13″O), transported immediately under cooling conditions (2–4 °C), frozen (−80 °C), freeze-dried [−42 °C, 48 h; light-protected vessels (Labconco™ Freezone 6, Labconco Co., Kansas City, MO, USA)], grounded to a fine powder (≤0.40 µm) and kept at −20 °C until use. Organic extracts (80% methanol) from all three freeze-dried samples (1:20 w/v; three batches per sample) were obtained by ultra-sonication (10 min; Fisher Scientific FS220H, Thermo Fisher Scientific, Waltham,

MA, USA), centrifugation (4 °C, 15 min, 1650× *g*; Eppendorff® centrifuge, mod. AG 5810R, Hamburg, Germany) and rotoevaporation (40 °C; Büchi® R-114 evaporator, Büchi Labortechnik AG, Flawil, Switzerland). Freeze-dried samples were further dissolved in HPLC grade or Milli-Q water for analysis.

2.3. High-Performance Liquid Chromatography Quadrupole Time-of-Flight Mass Spectrometry (HPLC-ESI-QTOF-MS)

Identification of individual polyphenols in organic extracts (methanol 80%) was carried on an HPLC-ESI-QTOF-MS instrument according to Torres-Aguirre et al. [12]. Chromatographic separation was performed on an Agilent 1200 series system (Agilent Technologies, Palo Alto, CA, USA). The equipment was equipped with a vacuum degasser, an auto-sampler, and a quaternary pump. Extracts (sugar-free) were firstly separated using a reverse phase C_{18} analytical column (2.1 mm × 50 mm × 1.8 µm particle size; ZORBAX Eclipse Plus), protected with a guard cartridge of the same packing and maintained at 25 °C. The mobile phase [formic acid (0.1%) in Milli-Q water (A) and acetonitrile (B)] was pumped at 0.4 mL/min into the HPLC System. Two microliters were injected, and the sample was eluted following the gradient elution program: 0–4 min (90% A), 4–6 min (70% A), 6–8 min (62% A), 8–8.5min (40% A), 8.5–9.5 min (90% A) and the column was further re-equilibrated for 3 min. The quadrupole time-of-flight mass spectrometer (QTOF-MS; Agilent Technologies, Palo Alto, CA, USA) was coupled to a dual electrospray ionization (ESI) interface. The ESI-QTOF-MS operating conditions were: Source temperature (120 °C), gas desolvation temperature (340 °C), drying, nebulizing and collision gas (nitrogen; 13 L/min), capillary voltage (4.5 kV) and mass scan (100–1000 *m/z*).

Before analysis, all samples (three batches x triplicate, $n = 9$) were filtered and concentrated by solid-phase extraction (Oasis HLB micro Elution plates, 96-well, 30 µm; Waters, Milford, MA, USA). Individual polyphenol identification was done by comparing the exact mass and molecular composition of the pseudo-molecular ion and/or quantification was performed by comparing with retention times (*rt*), UV-Vis spectra and molecular ion mass [*m/z* ± 0.1, $M{-}H^+$ (anthocyanins and rutin) or $M{-}H^-$ (all other polyphenols) mode] of pure phenolic standards (freshly prepared from stock solutions for each measurement), using the Mass Hunter Workstation Data Acquisition Software (ver. B.07.00; Agilent Technologies, Inc.) and an open-access MS-library (MassBank; https://massbank.eu/MassBank). The individual concentration of phenolic compounds was expressed in µg/g extract. Three different batches by triplicate ($n = 9$) from each fruit were evaluated.

2.4. Total Antioxidant Capacity

Trolox equivalent antioxidant capacity (TEAC) of organic extracts (pure methanol; 1:20 w/v) per sample (three batches by quadruplicate, $n = 12$) was evaluated by the DPPH method (515 nm), the ferric ion reducing antioxidant power assay (FRAP, 630 nm) and the oxygen radical absorbance capacity [ORAC; fluorescein: 10 nM, (excitation (485 nm)/emission (520 nm), AAPH (240 mM)], as previously described [13], using a FLUOstar™ OMEGA spectrophotometer (BMG LABTECH; Chicago, IL, USA) in UV/VIS (DPPH, FRAP) and fluorescence (ORAC) modes. For all three assays a trolox standard curve [0.006–0.2 µmol/mL, $R^2 \geq 0.95$] was used. Values were expressed as mg or µmol of trolox equivalents (TE) per g (DPPH, FRAP) or µmole 1×10^{10} (ORAC) per g of freeze-dried sample ± standard deviation ($n = 12$) and as percentage considering blackberry antioxidant titers as 100% (sample with the highest total polyphenol content) [7].

2.5. In Vitro Digestion

The method reported by Campos-Vega et al. [14] with minor modifications was used. For the oral stage, three otherwise healthy subjects were invited to participate in the study, providing written informed consent prior to participation. In fasting conditions and after brushing their teeth without toothpaste, each subject chewed each freeze-dried fruit (1 g × three batches, $n = 3$) 15 times for approximately 15 s. Chewed samples were collected into a beaker containing 5 mL of distilled water and subjects rinsed their mouths with another 5 mL of distilled water for 60 s. The volume of saliva +

water was considered for data correction. For the gastric stage, pooled salivary samples per subject and sample were re-mixed per participant (n = 3) in an aseptic vessel and an aliquot (10 mL) was adjusted to pH 2 using HCl solution (2 N). Pepsin from porcine gastric mucosa (55 mg ≥ 250 units/mg protein, Sigma-Aldrich) dissolved in 0.94 mL of 20 mM hydrochloric acid was added to each sample and incubated for 2 h at 37 °C with constant agitation. For intestinal stage, a simulated intestinal extract was prepared 30 min before use by dissolving gall Ox (3 mg of bovine bile; CAS: 8008-63-7, Sigma-Aldrich) and porcine pancreatin (2.6 mg, 8 × USP, Sigma-Aldrich, St. Louis, MO, USA) in 5 mL Krebs–Ringer buffer (118 mM NaCl, 25 mM $NaHCO_3$, 11 mM glucose, 4.7 mM KCl, 2.5 mM $CaCl_2$, 1.2 mM $MgSO_4$, 1.2 mM KH_2PO_4; pH 6.8]. Five mL of this solution were added to each sample coming from the gastric stage, pH adjusted to 7.2–7.4 with NaOH (2 M) and incubated for 2 h at 37 °C with constant agitation. All digested samples [three (pooled samples from gastric-to-intestinal phases per berry fruit) × triplicate, n = 9], were immediately transferred to the ex vivo bioanalytical system.

2.6. Rat Everted Gut Sacs

Six small intestinal gut sacs were obtained from six young male albino rats (~300 g BW) which were fasted overnight (16–20 h) and anesthetized with an intraperitoneal injection of sodium pentobarbital (70 mg/kg BW, Pisabental, Guadalajara, Jalisco, Mexico), before surgical procedures and euthanasia, as suggested by Campos-Vega et al. [14]. Briefly, the intestine was exposed by a midline abdominal incision and 20–25 cm of the jejunal section was excised and placed in a gasified (CO_2) Krebs-Ringer buffer solution at 37 °C. Each gut sac was gently washed externally with the same buffer and everted over a glass rod, re-excised into 6 cm-segments, filled (basolateral side) with 1 mL of Krebs-Ringer buffer (to avoid tissue denaturation) and fastened with braided silk sutures to a final length of approximately 4 cm [15]. Experiments were performed by triplicate and a blank was prepared using distilled water instead of in vitro digested sample. The experimental protocol was approved by the Animal Experimentation Ethics Committee of the Autonomous University of Ciudad Juarez (Code FO-CIP-01/254063; 30 July 2016) and animals cared according to the corresponding Mexican regulations (NOM-062-ZOO-1999) and the National Institutes of Health (NIH) guide for the care and use of laboratory animals.

2.7. Real-Time Monitoring of Phenolic First-Pass Metabolism

The ex vivo first-pass metabolism biosystem (Figure 1) consisted of a 15 mL of pre-digested (oral-gastric-intestinal) samples (n = 3 per dried berry) and three 4 cm-closed everted duodenal sacs incubated in an oscillating (60–80 cycles/ min) water bath at 37 °C for 2 h, in an anaerobic chamber. Bioaccesible (from Section 2.6) and biotransformed polyphenols, withdrawn from the apical side (Figure 1, "out"), were tracked by three independent analytical methods:

2.7.1. Spectrophotometry (Method 1)

Total polyphenols were quantified spectrophotometrically (765 nm) with the FC method (TP_{FC}) at the end of the experiment (120 min; t_{120}); values were expressed as mean ± SD values [3 independent samples x triplicate, n = 9; mg of gallic acid equivalents (GAE)/ mL] as previously reported [7]. The apparent permeability coefficient (P_{app}; Equation (1), efflux (ER; Equation (2) and uptake (UR; Equation (3)) ratios were calculated as follows, using the concentration of total polyphenols (Method 1) inside (basolateral; B) and outside (apical; A) the everted sacs at 120 min ($_{t120}$):

$$P_{app} = (\Delta Q/\Delta t) \times (1-AC_0)^{-1} \tag{1}$$

$$ER = (B \to A) \times (A \to B)^{-1} \tag{2}$$

$$UR = (A \to B) \times (B \to A)^{-1} \tag{3}$$

where $\Delta Q/\Delta t$ is the steady-state flux ($mg{\cdot}s^{-1}$) of polyphenols transported across the membrane per second, A (cm^2) is the surface area available for permeation and C_0 (mg/mL) represents the initial concentration of total polyphenols in the donor chamber (apical side of everted sacs; Figure 1). P_{app} (mean ± SD) values were calculated and expressed in 10^{-5} cms^{-1}.

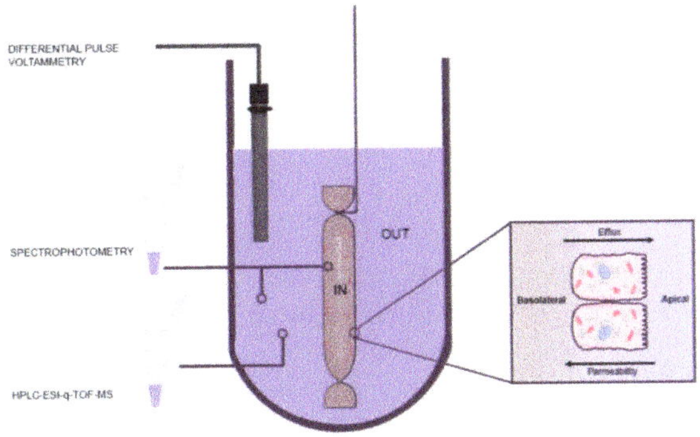

Figure 1. Real-time ex vivo monitoring of first-pass phenolic metabolism using the everted gut sac technique. The apparent permeability (P*app*) and biotransformation of bioaccesible phenolic compounds (PC) and their associated first-pass metabolites were followed by spectrophotometry (Folin–Ciocalteau; 120 min), HPLC-ESI-QTOF-MS (120 min) and, differential pulse voltammetry (DPV; 0 to 120 min). Diffusion from apical (A; out) to basolateral (B; in) and B-A were considered permeability and efflux, respectively.

2.7.2. Differential Pulse Voltammetry (DPV; Method 2)

Differential pulse voltammetry (DPV) real-time (0–120 min) measurements of mixed polyphenols (parent + metabolites) were monitored (by triplicate) using a potentiostat (BASi® EC Epsilon potentiostat/galvanostat; West Lafayette, IN, USA) and voltammetric measurements were carried out with a standard three-electrode electrochemical cell [working (glassy carbon, carefully polished with diamond spray, particle size 1 μM), counter (platinum wire) and reference (Ag|AgCl|KCl; 3M) electrodes]. Experimental conditions were: room temperature, pH, (7.2–7.4), scan range (0–600 mV) and rate (5 mVs^{-1}), pulse width (70 ms) and amplitude (50 mV); these conditions were selected to avoid the interference of electrochemical species other than polyphenols and the current density ($\mu A \times 10^{-5}$) from the first (t_0) oxidation peak (current *maxima*; mean = 203 mV, range 180 to 216 mV) was chosen as the reference value to estimate total polyphenols by DPV (TP_{DPV})

2.7.3. HPLC-ESI-QTOF-MS (Method 3)

Non-targeted mass spectral identification (MassBank; https://massbank.eu/MassBank) and semi-quantification [as ion abundance (IA)] at t_0 and t_{120} of parent polyphenols and their metabolites was performed by HPLC-ESI-QTOF-MS as reported above, following Koistinen et al. [10] recommendations. To avoid phytochemical loss by direct drying, individual samples (1 mL) were cleanup and concentrated by solid-phase extraction in Oasis HLB micro Elution 96-well plates (30 μm; Waters). Considering that real-time oxidation/reduction reactions readily occur within the ex vivo bioanalytical system used in this study, molecular ion identification was performed under the following considerations: $m/z \pm 0.3$, $M{-}H^+$ (anthocyanins and rutin) or $M{-}H^-$ (all other polyphenols) mode].

2.8. Statistical Analysis

Results were expressed as mean ± standard deviation (SD) obtained from at least by triplicate. Inter-group (Red Globe grape, raspberry, blackberry) comparisons were performed by one-way-ANOVA followed by Tukey's post hoc test and the statistical significance was defined at $p < 0.05$. When needed,

Pearson's product-moment correlation (r) was used to establish any possible correlation between response variables. Quadratic/cubic regression curves were constructed to explain electrochemical data (DPV). All statistics were performed using the statistical program NCSS 2007 (NCSS, Statistical Software, Kaysville, UT, USA).

3. Results and Discussion

3.1. Phenolic Profile of Berry Samples

Edible berries are rich in flavan-3-ols and anthocyanins that are barely affected during processing [16], although their content and molecular diversity is cultivar dependent [6]. In a preceding paper [7] we reported the spectrophotometric estimation (per g DW) of polyphenol subgroups in freeze-dried Red Globe grape, raspberry, and blackberry as follows: total polyphenols 9.4, 17.6 and 22.7 mg GAE, flavonoids 7.0, 13.1 and 35.3 mg quercetin equivalents (QE), monomeric anthocyanins (0.01, 0.49 and 0.67 mg cyanidin-3-*O*-glycoside equivalents), proanthocyanidins (0.22, 0.23 and 0.06 mg QE) and hydrolysable phenols (3.7, 7.2, 11.5 mg GAE). In the present study, we confirmed that these polyphenol subgroup titers correlate ($r \geq 0.76$) with the overall content (ratio 0.07-0.14-1.0) and molecular diversity of flavan-3-ols and anthocyanins in the same fruits (blackberry> raspberry > Red Globe grape; Table 1).

Table 1. HPLC-ESI-q-TOF-MS and cheminformatics of polyphenols from three edible berries [1,2].

Compound	*rt*	*m/z*	Grape	Raspberry	Blackberry	TPSA	LogP
Catechin	1.5	289.1	86 ± 10	–	–	110	1.37
Epicatechin	2.7	289.1	10 ± 5 [c]	451 ± 6 [b]	1121 ± 95 [a]	110	1.37
Cyanidin-3-*O*-β-glucoside	4.2	450.1	–	–	2762 ± 31	181	0.34
Cyanidin-3-*O*-arabinoside	5.6	420.2	–	–	21 ± 0	161	−2.37
Pelargonidin	6.6	272.1	67 ± 4 [a]	47 ± 1 [b]	65 ± 2 [a]	82	−0.26
Pelargonidin-3-*O*-glucoside	7.1	433.2	–	20 ± 3 [a]	15 ± 0 [b]	171.2	−2.30
Cyanidin-3,5-*O*-diglucoside	7.9	612.4	134 ± 8 [a]	58 ± 4 [c]	111 ± 3 [b]	270.6	−4.61
Total polyphenols			297 ± 27 [c]	576 ± 14 [b]	4095 ± 131 [a]		

[1] Results are expressed as mean ($n \geq 9$) ± standard deviation (μg /g dry weight basis); different superscript letters between samples for a same compound means statistical differences ($p < 0.05$); retention time (*rt*, min), molecular ion [*m/z* ± 0.3, positive (anthocyanins) or negative (flavan-3-ols) mode], below quantification limit (–). [2] Total polar surface area (TPSA. Å^2) and octanol/water partition coefficient (LogP) values were retrieved from Molinspiration chemoinformatics (https://www.molinspiration.com/), using each compound's canonical SMILE sequence retrieved from PubChem (https://pubchem.ncbi.nlm.nih.gov/.

Our data also indicate that epicatechin, pelargonidin, and cyanidin-3,5-*O*-diglucoside (cyanin) were present in the evaluated samples, but catechin (Red Globe grape), kuromanin (cyanidin-3-*O*-β-glucoside) and cyanidin-3-*O*-arabinoside (blackberry) and callistephin (pelargonidin-3-*O*-glycoside; raspberry and blackberry) were fruit-specific.

Several anthocyanidins (aglycones) and related anthocyanins (3-*O*-glycosides and acyl glycosides) have been identified in grapes and shrubby berries. Colombo et al. [17] identified several flavan-3-ols (e.g., catechin, epicatechin, proanthocyanidin di/trimers), flavanols (quercetin and derivates), anthocyanins (all but pelargonidin glycosides), *cis*-resveratrol and caftaric acid (esterified phenolic acid) in Red Globe grape. However, pelargonidin (3,5,7,4′-tetrahydroxyflavium) and its 3-O-glycoside (callistephin) were not reported by these authors although they have been reported, in trace amounts, in certain grape varieties [18]. Also, blackberry and raspberry were better sources of polyphenols as compared to Red Globe grape, particularly in anthocyanin content. It is well-known that these berries are good sources of flavones (e.g., apigenin, chrysin), flavonols (e.g., kaempferol), phenolic acids (e.g., ellagic acid, caffeic acid), ellagitannins (e.g., sanguiin H-6, lambertianin C), anthocyanidins (all but peonidin) and anthocyanins such as cyanidin, delphinidin and pelargonidin glycosides [3,16,19].

The amount and natural occurrence of anthocyanidins (aglycone) and derived anthocyanins (glycosylated forms) are influenced by many factors including the type of cultivar and pre/postharvest handling of grapes [18], blackberries and raspberries [6]. Besides this, biosynthesis of anthocyanins in berry fruits is tightly controlled during the transcription of several genes involved in the flavan-3-ol proanthocyanidin pathway, in a fruit-specific manner [18,20]. Taking this into consideration, blackberry and raspberry are more valuable than Red Globe grape from a nutraceutical standpoint [1], even if their parent polyphenols are biotransformed into other ones [20,21]; the specific phenolic fingerprint of these two berries may be related to different yet complementary effects on preventing several non-communicable chronic diseases including certain types of cancers, cardiovascular disease, type II diabetes, inflammation and oxidative stress [2].

3.2. *Antioxidant Capacity of Berry Samples*

The antioxidant capacity of a given molecule (or a complex antioxidant mixture) is defined by its ability to reduce free reactive species (pro-oxidants or free radicals). The evaluation of the antioxidant capacity of plant-based foods by simultaneously using more than one method is a recommended practice in food science and technology [13].In this study, we used a single electron-transfer (SET; FRAP), a hydrogen atom transfer (HAT; ORAC) and a combined SET/HAT (DPPH) method to assay the antioxidant capacity in the studied samples (Figure 2).

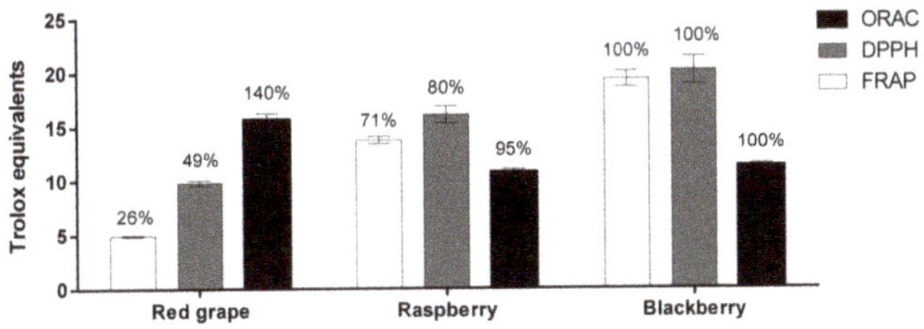

Figure 2. Antioxidant capacity of hydroalcoholic extracts from commercial *Red Globe* grape, raspberry and blackberry. Values were expressed as mean [$n \geq 9$; mg TE/ gDW (DPPH, FRAP) or 1×10^1 μmol TE/g DW (ORAC)]; percentages above bars indicate differences between samples considering blackberry antioxidant titers as 100%.

FRAP and DPPH values ranged between 5 (Red Globe grape)-19 (blackberry) and 10 (Red Globe grape)-20 (blackberry) mg TE/ g DW and from 11 (raspberry) to 16 (Red Globe grape) $\times 10^1$ μmol TE/g DW with the ORAC method; similar trends in antioxidant capacity have been reported by other authors for the same berry fruits [3,16]. Also, the antioxidant capacity trend (blackberry > raspberry > Red Globe grape) was directly proportional to their phenolic content (Table 1), as measured by FRAP (100%-71%-26%; $r = 0.86$) and DPPH (100%-80%-49%; $r = 0.84$) methods but not with ORAC (100%-95%-140%; $r = -0.42$). DPPH and FRAP titers also correlate ($r \geq 0.95$) with all polyphenol subgroups (total polyphenols, flavonoids, and anthocyanins) reported in our previous study [7] and same results have been reported for other polyphenol-rich fruits [6].

Conventionally, the higher the content of polyphenols in berry fruits, the higher their antioxidant capacity. It is important to point out that the radical scavenging capacity of most polyphenols is mediated by HAT rather than SET mechanisms. However, the antioxidant capacity is also related to the number and position of hydroxyl groups, the O–H bond dissociation enthalpy and conjugation/resonance effects [21]. Although the main antioxidant capacity mechanism in complex phytochemical mixtures is difficult to establish, the observed antioxidant capacity pattern (mostly blackberry > raspberry > Red Globe grape) apparently is the result of synergism between flavonoid species, the number of their available hydroxyl groups (particularly *O*-hydroxyls in A & B rings) and their level of glycosylation [21,22]. In support of this, Rice-Evans et al. [23] reported the following trend in antioxidant capacity with the ABTS radical (mixed SET/HAT mechanism): cyaniding > epicatechin/catechin >

oenin > pelargonidin, and so, major drivers of the overall antioxidant capacity in the studied samples seem to be catechin (Red Globe grape), epicatechin (raspberry, blackberry) and cyanidin glycosides (blackberry). Lastly, cyanidin has a higher antioxidant capacity than its derived glycosides in the ORAC assay [24] and the catechin content shows a better lineal relationship with ORAC values than that observed with other flavan-3-ols [25]; whether these arguments justify Red Globe grape's antioxidant activity in the ORAC method (Figure 2 merits future study.

3.3. Apparent Permeability of Berry Polyphenols

The net bioaccesibility (TP_{FC} = oral + gastric + intestinal) of polyphenols (as mg GAE per g DW) from Red Globe grape, raspberry, and blackberry was 2.0, 3.6 and 4.2 (A_{t0}, Table 2) which represents 21.3, 20.4 and 18.6% of their original content [7].

Table 2. Apparent permeability of phenolic compounds from selected berries.[1,2]

Parameter	Red Globe Grape	Raspberry	Blackberry
A_{t0} (TP_{FC})	2 ± 0.0^{c}	3.6 ± 0.1^{b}	4.2 ± 0.1^{a}
A_{t120}	1.3 ± 0.3^{b}	1.5 ± 0.1^{b}	2.4 ± 0.1^{a}
B_{t120}	0.10 ± 0.0^{b}	0.09 ± 0.0^{b}	0.13 ± 0.0^{a}
Absorptive P_{app} ($A_{t120} \rightarrow B_{t120}$)	1.20	0.06	0.07
Secretory P_{app} ($B_{t120} \rightarrow A_{t120}$)	1.55	0.98	1.38
Efflux ratio $(B \rightarrow A)^*(A \rightarrow B)^{-1}$	1.29	16.12	19.12
Uptake ratio $(A \rightarrow B)^*(B \rightarrow A)^{-1}$	0.78	0.06	0.05
p (ER vs. UR)	0.02	0.002	<0.0001

[1] Results are expressed as mean ± standard deviation ($n \geq 9$; mg GAE /g dry weight; Folin-Ciocalteu method), different superscript letters within a same row means statistical differences ($p < 0.05$). [2] Total polyphenol content by the Folin-Ciocalteu method (TP_{FC}), basal (t_0) and final (t_{120}) apical (A) or basolateral (B) concentration. Apparent permeability coefficient (P_{app}; $cm^*s^{-1} \times 10^{-5}$). Statistical difference between efflux (ER) vs. uptake (UR) ratios as determined by t-student test ($p < 0.05$).

In our preceding study we also reported that anthocyanins, but no other flavonoids were pH-unstable under simulated intestinal conditions (pH 7.0); similar results have been reported for strawberry [26], maqui berry [27] and blueberry [28]. After intestinal digestion (Table 2), 1.3, 1.5 and 2.4 mg GAE per g DW remain in the apical side (A_{t120}) suggesting 35%, 58% and, 43% of net polyphenol absorption; however, the basolateral (serosal) concentration of polyphenols at 2h (B_{t120}) was blackberry (0.13) > Red Globe grape/ raspberry (~0.095) and the absorptive ($A_{t120} \rightarrow B_{t120}$) and secretory ($B_{t120} \rightarrow A_{t120}$) P_{app} ($cm \cdot s^{-1} \times 10^{-5}$) were Red Globe grape (1.20)> blackberry (0.06)/ raspberry (0.07) and Red Globe grape (1.55)/blackberry (1.38) > raspberry (0.98), respectively. In consequence, efflux (19.1, 16.1, 1.3) were higher than uptake (0.05, 0.06, 0.78) ratios were fruit-specific ($p \leq 0.02$) and concentration-dependent (blackberry > raspberry > Red Globe grape).

Many transport mechanisms seem to be involved in the uptake/efflux behavior of polyphenols. Dixit et al. [29] used a standardized everted sac-based biosystem to evaluate the permeability behavior of atenolol (Pubchem CID: 2249; XLogP3 = 0.2, simple paracellular transport), metoprolol (Pubchem CID: 4171; XLogP3 = 1.9, transcellular transport) and propranolol (Pubchem CID: 4946; XLogP3 = 3.0; passive diffusion), reporting absorptive (A→B) P_{app} values of 0.054, 0.84 and 1.64 $cms^{-1} \times 10^{-5}$, respectively. Considering miLogP values reported in Table 1, and the fact that anthocyanins and anthocyanidins cannot cross cell membranes passively, passive (Red Globe grape) and paracellular transport (blackberry/raspberry) may be major transport mechanisms in the studied samples.

Molecular bioinformatics provided information on the capability of each berry polyphenol to be absorbed by the intestinal epithelia. Most polyphenols listed in Table 1 had a topological polar surface area [TPSA; 92.1 (pelargonidin) to 270.6 (cyanin)] and octanol/ water partition coefficient [LogP; −4.61 (cyanin) to 1.37 (catechin/epicatechin)]. It is known that phytochemicals with a TPSA > 140 $Å^2$ or ≤ 60 $Å^2$ have low and high permeability, respectively and those neutral or with a LogP > 2.0 easily permeate by passive diffusion [30]. The "pH partition" hypothesis postulates that non-ionized (neutral) molecules are passively transported through the lipid membranes [30] and according to the modified *theoretical transcellular permeability* model proposed by Farrell et al. [31], the P_{app} of monomeric polyphenols is well explained by *logP* and molecular weight. Based on this, active more than passive transport and regular permeability of polyphenols from the assayed berries should be expected [8,9].

However, P-glycoprotein (*P-gp* or MDR-1) and breast cancer resistance protein (BCRP) can simultaneously reduce the first-pass bioavailability of certain polyphenols by acting as efflux modulators [5,32], as it has been proposed for polyphenols coming from spent coffee [14] and mango bagasse [33]. In fact, certain polyphenols may also act as competitive inhibitors of *P-gp* since it has a high affinity toward molecules with a planar ring system ranging from 200 to 1900 Da [32]. However, uptake/efflux behavior is also an asymmetric function that depends on the chemical structure of a particular polyphenol to be absorbed and the rate to which it is bioaccesible during *in vivo* or *in vitro* gastrointestinal conditions [14,33,34]; but such dynamic changes in transport behavior could not be observed with our endpoint assay (Method 1).

3.4. Ex Vivo Biotransformation of Berry Polyphenols

As previously discussed, several in vitro and ex vivo permeability models have been developed to mimic specific aspects of gastrointestinal metabolism, each one with advantages and disadvantages [15,29]. Particularly, the sensitivity and reliability of the everted gut sac technique increase when combined with high-throughput detection methods such as electrochemical and HPLC/MS-based methods [10,34]. On the other hand, experimental data such as retention time (*rt*), mass-spectra ($m/z^{+/-}$) of a particular standard molecule is often used in targeted metabolomics (Level 1) to objectively identify and quantify the same molecule in a given biological sample that presumably contains it. Conversely, untargeted metabolomics (Level 2) focuses on the global detection and relative quantitation of metabolites of unknown chemical nature, and their putative identification and semi-quantification rely upon spectral matching with databases (e.g., MassBank) or as suggested in earlier publications [10]. In this study, a level 2 strategy was used to partially identify (HPLC-MS-QTOF-MS) and semi-quantify (DPV current *maxima* and ion counts) of major [effluxed (B-A)] metabolites coming from the *ex vivo* biotransformation of parent polyphenols (Table 1).

3.4.1. DPV

Voltammetric methods are useful to predict the antioxidant activity of plant foods [11] and biological samples. Particularly, *maxima* DPV anodic current highly correlates with the antioxidant capacity (FRAP, ABTS and, DPPH) and concentration (HPLC and spectrophotometry) of pure polyphenols [35,36] but it is more sensitive (μM vs. mM) to quantify total polyphenols (TP_{DPV}) in grapes, raspberries and red wines [11,19]. DPV has been also used to evaluate the P_{app} of specific drugs in vitro using intestinal cell monolayers or ex vivo using intestinal reperfusion models. However, to our knowledge, reports on the electrodynamics of polyphenols during their first-pass metabolism using an everted gut sac have not been reported before. DPV voltammograms (vs. Ag|AgCl|KCl reference electrode) of pre-digested berries at neutral pH (Figure 3) showed the following trend [potential peak (mV)/ anodic current density ($\mu A \times 10^{-5}$)]: 180/0.24 (*Red Globe* grape) > 216/0.34 (raspberry) > 212/0.44 (blackberry); the estimated TP_{DPV} ratio was 0.6:0.8:1.0 which linearly correlated ($r \geq 0.95$) with TP_{FC} (Table 2) and antioxidant capacity (Figure 2; DPPH and FRAP).

The oxidation potential of monomeric polyphenols depends on the amount and position (*ortho* or *para* > *meta*) of reactive hydroxyl groups in benzene ring(s) and the *ortho*-effect between two hydroxyl groups or hydroxyl/carbonyl groups [35]. Alcalde et al. [36] evaluated the relationship between the molecular structure and electrochemical behavior of fifteen polyphenols by DPV, demonstrating that polyphenols with higher sensitivities (lower DPV potential) are strong antioxidants and that flavonoids are more electroactive (lower potentials) than phenolic acids (higher potentials); they also found that catechin exhibits two oxidation peaks at ~200 (A ring) and ~600 (B ring) mV (vs. Ag|AgCl|KCl reference electrode) at pH 5.0, while at pH 7.5 [37] and 3.6 [11] it exhibits both oxidation peaks at 148/537 mV and ~450/750 mV, respectively.

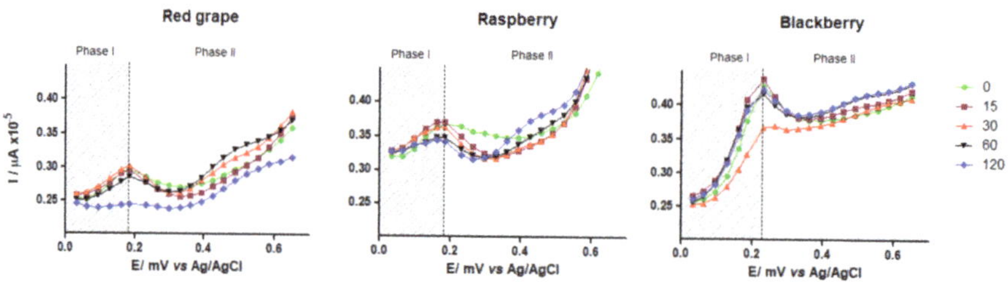

Figure 3. Differential pulse voltammograms of post-digested berry samples, during their *ex vivo* apparent permeability and biotransformation. Polarization rate 5mV*s^{-1} (abscissa values × 10^{-3}); phase I ("absorption"; grey rectangle), phase II ("biotransformation" open white), pH = 7.2–7.4.

Kuromanin and cyanin are major anthocyanins in berry fruits [3,16] both exhibiting the same oxidation potential in their catechol group [P1 (current peak): B-ring; SET mechanism, reversible reduced] but different oxidation behavior at their resorcinol moiety (P2: A-ring, SET mechanism, irreversible) associated with the additional glucosylation (shift right: +95 (pH 3.5–4.5), +130 (pH 7.0) mV] and P1 shifts to lower potentials (vs. Ag|AgCl|KCl reference electrode) when increasing pH [20–22].

The aforementioned inverse relationship between pH and oxidation potential has been also demonstrated for several anthocyanins from *Vitis vinifera* [22]. Together, this evidence supports the idea that the observed DPV current peak *maxima* at t_0 in all three voltammograms (Figure 3) may be partially explained by a berry-specific amount of polyprotic species at neutral pH (7.2–7.4), which is the case of the not-methylated anthocyanidins and anthocyanins reported in Table 1. It is worth mentioning that ellagic acid, the most representative phenolic acid in berries, exhibits a very low oxidation potential nearby neutral pH [38] and so, it seems that it does not contribute to the observed current peak *maxima* at t_0 in all three samples (Figure 3).

Since redox processes and apparent permeability of polyphenols were concurrent events in the *ex vivo* bioanalytical system used here, we used time-trend kinetic data to gain more insights on such events. For this purpose, the following procedure was followed: *A)* The real-time electrodynamic behavior exhibited by each pre-digested sample (Figure 3) from 0 (basal) to 120 min was arbitrarily divided into two segments, before and after current *maxima* [Red Globe grape (180 mV), raspberry (216 mV), blackberry (212 mV)], and labeled as "absorption" (stage 1) and "metabolite production" (stages 2); here we postulate that a time-trend reduction of current density before *maxima* current peak is mainly due to an apparent permeability and/or structural modification of parental polyphenols while increments after peak *maxima* are more likely to be due to new chemicals species with reduced electroactivity. *B)* Goodness-of-fit regression models for stage 1 (quadratic) and stage 2 (cubic) were then obtained and, *C)* theoretical current density values (TCD; µA × 10^{-5}) were calculated using the mean potential peak value (203 mV) at each time point (Table S1 and Figure S1).

During stage 1, a time-trend lineal reduction of TCD for Red Globe grape (R^2 = 0.91) but not for raspberry or blackberry was documented (Table S1 and Figure S1); this seems to be related to the lower amount and less diverse polyphenol profile (high catechin) in this table grape (Table 1), to higher electrochemical stability and apparent permeability and to a lower efflux of polyphenols (Table 2) than that observed for raspberry and blackberry. It is known, that anthocyanins>flavan-3-ols are pH-sensitive under simulated intestinal conditions [7,28], that anthocyanidins are more electroactive than their corresponding anthocyanins [24], that anthocyanin di-glycosides are more stable than mono-glycosides [21,22] and that GLUT-2 and SGLT-1 are efficient transporters for anthocyanins and flavan-3-ols at lower apical concentrations while *P-gp*/MCRP efflux systems are activated upon GLUT-2/SGLT-1 saturation [33,39,40].

During stage 2, less electroactive (~400–600 mV) species were erratically produced (X^3-behavior) in a berry-specific trend (Figure 3, Table S1 and Figure S1): raspberry (X^3 range = 3 to 10)> Red Globe grape (X^3 range= −2 to 0.2)> blackberry (X^3 range= −2 to −9). DPV is also useful to study the real-time redox phenomena in vitro (particularly under acidic conditions) such as the time-course photo-degradation of 4-acetamidophenol (a.k.a. acetaminophen) [41] or the electro-Fenton degradation of 3-methyl phenol (*m*-cresol) [42] in which new molecules with higher (acetaminophen) or lower (*m*-cresol) potentials are produced from these synthetic phenolics.

However, to our knowledge, the real-time degradation of natural polyphenols under neutral pH has not been reported yet, nor the use of DPV for monitoring their time-trend enteral biotransformation. Although the evidence points out to a berry-specific ex vivo biotransformation of parental polyphenols, the chemical nature of phenolic metabolites could not be evidenced by this method. Nonetheless, partially oxidized flavan-3-ols and anthocyanins > phenolic acids seem to be predominant antioxidant species at neutral pH within the narrowed potential range used in this study (0–600 mV), a fact previously reported by other authors [35,37].

3.4.2. HPLC-ESI-QTOF-MS

The European Cooperation in Science and Technology Commission (COST; FA-1403 POSITIVe action) recommends the use of high through-output analytical platforms in untargeted metabolomics to evaluate the inter-individual variability in the physiological response to phytochemical intakes [4,10]. Particularly, HPLC-ESI-MS^n is widely used in untargeted polyphenol metabolomics [19,28]; such a platform was used here to track the ex vivo small gut biotransformation (2 h; end-point assay) of parent polyphenols from three berries with graded levels of phenolic compounds [7]. Table 3 shows the chemical nature and apparent content of bio-accessible polyphenols (released by in vitro digestion) before their ex vivo biotransformation (t_0) that substantially differed from those identified in the assayed fruits who were chemically extracted (Table 1).

Table 3. First-pass metabolism of phenolic compounds from selected berries: HPLC-ESI-q-TOF-MS [1,2].

Sample	Phenolic	*rt*	*m/z*	Ion Abundance (IA)		Δ (%)
				t_0	t_{120}	
Raspberry	Quinic acid	0.6	191.1	104,000 ± 1061	45,600 ± 636	−56 ± 0
	Epicatechin	3.0	289.1	9500 ± 707	3100 ± 141	−67 ± 1
	Cy3G	4.2	450.1	16,500 ± 707	4750 ± 354	−71 ± 1
Blackberry	Quinic acid	0.6	191.0	8950 ± 212	4600 ± 566	−47 ± 9
	Chlorogenic acid	1.6	353.1	71,300 ± 1768	47,300 ± 354	−34 ± 1
	Caffeic acid	2.3	179.0	3750 ± 354	23,500 ± 707	530 ± 78
	Ma3G	4.9	494.1	950 ± 71	8450 ± 354	1006 ± 8

[1] Results are expressed as mean ($n \geq 9$; ion counts) ± standard deviation of selected phenolic compounds detected with a signal-to-noise ratio ≥10:1. [2] Retention time (*rt*, min), mass-to-charge ratio [$m/z \pm 0.1$, positive (anthocyanins) or negative (all other polyphenol) ion mode], cyanidin (Cy3G; kuromanin) or malvidin (Ma3G; oenin)-3-*O*-glucosides; initial (At_0), final (At_{120}) and change (Δ (%) = [1-(t_{120}/ t_0)] × 100), apical ion abundance; reduced (−), increased (+).

Such a difference is even more evident when considering the HPLC-ESI-QTOF-MS profile at a signal-to-noise ratio < 10:1 (Table S2; values at t_0). The same has been reported for black and green currants [43] and strawberries [26] when comparing the polyphenolic profile of these berries before and after *in vitro* digestion, it was observed that not only the quantity of parental polyphenols but also their chemical nature differed (also observed in Figure 4). Many biological and analytical factors are involved in this phenomenon, including the pH-instability (neutral > acidic) of polyphenols, particularly anthocyanins [8,20], the REDOX status of parent/metabolites ($m/z \pm 0.3$), HPLC-ESI-QTOF-MS limit of detection (10:1 signal-to-noise) and their reversible interaction with digestive enzymes and mucin [5,10,44].

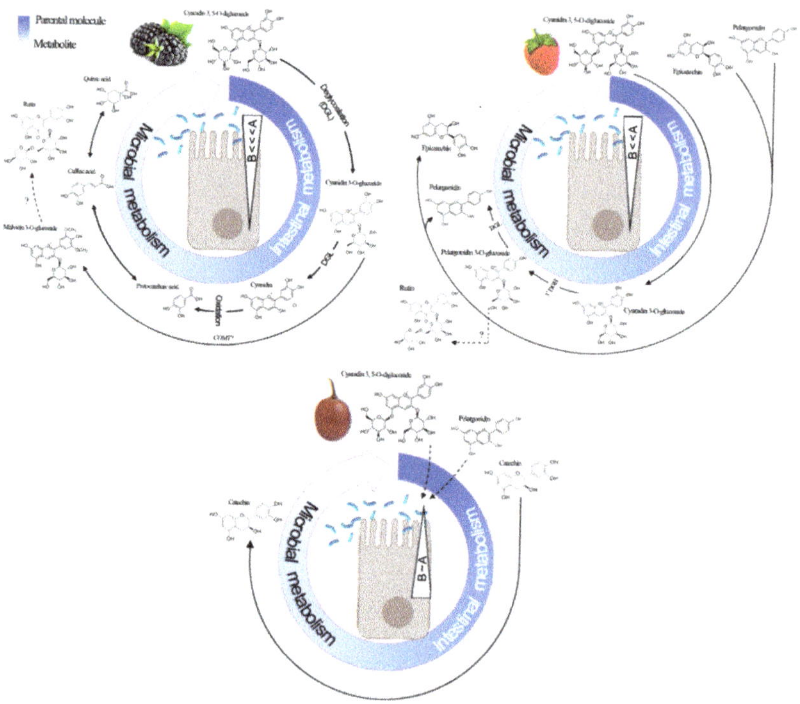

Figure 4. First-pass metabolism of polyphenols from blackberry, raspberry and *Red Globe* grape. Stepwise metabolite production from parental polyphenols (Table 1) detected by HPLC-ESI-QTOF-MS (both, below (Table 3) and over (Table S2) a signal-to-noise 10:1 ratio results from a concerted action of epithelial enzyme machinery and jejunal microflora. Triangle magnitude and direction (A<~>B) is derived from Table 2. Apical (A), basolateral (B), catechol-*O*-methyltransferase (COMT), 3′ hydroxyl removal (3′DOH), deglycosylation (DGL).

Steinert et al. [45] using a $CaCO_2$ monolayer transport system showed that the apical-to-intracellular transport of blackcurrant anthocyanins occur faster than their translocation across the basolateral membrane and that ~11% of all anthocyanins disappeared from the apical chamber within the first twenty minutes; the authors concluded that cell metabolism rather than apparent permeability was involved in the first-pass metabolism of black currant anthocyanin. Kuntz et al. [39] also studied the apparent and apical bioavailability of anthocyanins from grape/blueberry juice and smoothie permeability in transwell chambers with and without Caco-2 cell (ATCC© HTB37™) monolayers, showing that both specific and absolute anthocyanin concentration decreased overtime in apical chambers without cells at neutral (7.4) but not acidic (2.0) pH and that total anthocyanin disappearance were even more evident with cells than without them. Both research groups also documented a structure-specific disappearance rate of anthocyanins due to concurrent absorption and biotransformation processes.

Another plausible explanation comes from microbial biotransformation. As previously mentioned, the absorptive behavior [P_{app} ($A_{t120} \rightarrow B_{t120}$)] and uptake ratio [$(A \rightarrow B) \times (B \rightarrow A)^{-1}$] was inversely related to the fruit-specific polyphenol-richness and luminal biotransformation (Figure 4). This implies that the resident time of parent polyphenols, particularly those from blackberry and raspberry, in the apical side was long enough to be used as substrates for brush border enzymes and possibly by the resident microbiota including but not restricted to *Lactobacillus* sp., *Actinobacterium* sp. and *Clostridium* sp. which together represents ~70% of normal rat duodenal microbiota [46] and whose substrate preference include flavonoids, anthocyanins, and ellagitannins [47]. However, depending on the composition of the microbiota, different metabolites may be produced from the microbial biotransformation (postbiotics) of berry anthocyanins, despite the fact that certain phenolic acids and flavonoids may also act as prebiotics [2]; this double effect of polyphenols enlarges their recognized health benefits [1,3–5].

The 2 h (t_0 vs. t_{120}) ex vivo exposure to the intestinal epithelium reflected both the epithelial in/out interchange discussed above (see dotted triangles in Figure 4) and a great biotransformation phenomenon characterized by a low or no detection of parental anthocyanins and anthocyanidins and higher production of small molecular weight ($\leq$ 354 gmol^{-1}) polyphenols (Table 3, Table S2). Extensive and rapid deglycosylation of anthocyanins occurs in vivo and ex vivo releasing anthocyanidins with a reduced polarity (less TPSA). The resulting anthocyanidins may be either absorbed by passive (paracellular) diffusion or subject to microbial breakdown (particularly on B ring) producing phenolic acids (e.g., protocatechuic, chlorogenic and caffeic acids) and polyols (e.g., quinic acid; Figure 4) [2,9] and C_6-C_3-C_6-derived intermediates [44]. Chen et al. [8] followed the in vitro bioaccessibility and biotransformation of kuromanin under simulated GI conditions showing that this anthocyanin rapidly disappears but a wide range of metabolites (namely protocatechuic acid and derivates, cyanidin, caffeic and ferulic acids) were produced instead, all of them showing different permeability behaviors.

Lastly, since cyanin just have and additional glucose moiety at 5′ when compared to kuromanin, its metabolic fate may be the same after enzymatic deglycosylation [9]. Kuromanin (and possibly cyanin) is partially deglycosylated by β-glucosidase (EC 3.2.1.21) and lactase-phlorizin hydrolase (EC 3.2.1.108) but it undergoes extensive *in vivo* biotransformation to low molecular weight breakdown metabolites and, a wide range of phase II metabolites including anthocyanin methylation [20]; since some of these metabolites are either reported in Table 3 or Table S2, this partially supports an extensive kuromanin biotransformation from blackberry and raspberry (Figure 4). Although a straightforward identification of C_6-C_3-C_6 compounds derived from kuromanin, cyanin or callistephin has not been reported yet, the removal of functional groups *ex vivo* (as hypothesized in this study) may interconvert anthocyanidins (e.g., the loss of a hydroxyl group from the B-ring of cyanidin gives rise to pelargonidin) as it has been shown in vivo [44]. The biotransformation of chlorogenic (Table 3) and protocatechuic (Table S2) acids, two of the most abundant phenolic acids in edible fruits, gives quinic and caffeic acid (Figure 4) whose further methylation gives ferulic and isoferulic acids [48]; however, whether the intestinal or microbial catechol-*O*-methyltransferase (COMT; EC 2.1.1.6) activity is involved in the transformation of kuromanin into callistephin (from blackberry), deserves further study.

4. Conclusions

A moderate permeability (~20%) and a high ex vivo biotransformation of parent polyphenols (molecular breakdown and isomerized anthocyanin products) from the assayed berries were found in this study. This partially suggests fruit-specific health effects, most likely at the intestinal level due to a higher efflux phenomenon. The ex vivo high-through output bioanalytical approached used here [everted gut sac (rat) + three detection methods: spectrophotometry, HPLC-ESI-QTOF-MS, differential pulse voltammetry (DPV)], provided important information on berry polyphenol biotransformation during their ex vivo first-pass metabolism that may help to understand the metabolic fate and effects of the studied berries; however, further studies are needed to understand the biological activities of biotransformed phenolics and not only their parental molecules.

Supplementary Materials: The following are available online at http://www.mdpi.com/2076-3921/9/4/311/s1, Table S1 and Figure S1: Voltamperometric behavior of mixed antioxidants from three berry fruits, Table S2: First-pass metabolism of polyphenols from selected fruits- HPLC-ESI-q-TOF-MS (signal-to-noise ratio < 10:1).

Author Contributions: Conceptualization, A.W.-M.; Data curation, S.M., G.A.G.-A. and M.A.V.-O.; Formal analysis, F.J.O.-A., S.M., G.A.G.-A., M.A.V.-O., J.T.J.Q.-V. and A.W.-M.; Funding acquisition, A.W.-M.; Investigation, F.J.O.-A., E.A.-P. and J.T.J.Q.-V.; Methodology, F.J.O.-A., E.A.-P., M.A.V.-O. and J.T.J.Q.-V.; Project administration, F.J.O.-A. and A.W.-M.; Resources, G.A.G.-A.; Writing – original draft, F.J.O.A. and A.W.-M.; Writing – review & editing, S.M., E.A.-P., G.A.G.-A., M.A.V.-O. and A.W.-M. All authors have read and agreed to the published version of the manuscript.

Funding: National Council of Science and Technology (CONACyT: Basic science project (CB-2015-1/ 254063) and a doctorate scholarship for FJOA

Acknowledgments: All authors acknowledge the financial support from the National Council of Science and Technology (CONACyT) through a granted basic science project (CB-2015-1/ 254063) and a doctorate scholarship for FJOA, is gratefully acknowledged. All authors are indebted to all academic authorities (UACJ, UAQ, and CIAD) and to PROMEP for their institutional support.

Conflicts of Interest: The authors declare no conflict of interest.

References

1. Yang, B.; Kortesniemi, M. Clinical evidence on potential health benefits of berries. *Curr. Opin. Food Sci.* **2015**, *2*, 36–42. [CrossRef]
2. Lavefve, L.; Howard, L.R.; Carbonero, F. Berry polyphenols metabolism and impact on human microbiota and health. *Food Funct.* **2020**, *11*, 45–65. [CrossRef]
3. Skrovankova, S.; Sumczynski, D.; Mlcek, J.; Jurikova, T.; Sochor, J. Bioactive compounds and antioxidant activity in different types of berries. *Int. J. Mol. Sci.* **2015**, *16*, 24673–24706. [CrossRef] [PubMed]
4. Morand, C.; De Roos, B.; Garcia-Conesa, M.T.; Gibney, E.R.; Landberg, R.; Manach, C.; Milenkovic, D.; Rodriguez-Mateos, A.; Van de Wiele, T.; Tomas-Barberan, F. Why interindividual variation in response to consumption of plant food bioactives matters for future personalized nutrition. *Proc. Nutr. Soc.* **2020**, 1–11. [CrossRef]
5. Domínguez-Ávila, J.A.; Wall-Medrano, A.; Velderrain-Rodríguez, G.R.; Chen, O.C.-Y.; Salazar-Lopez, N.J.; Robles-Sánchez, M.; Gonzalez-Aguilar, G. Gastrointestinal interactions, absorption, splanchnic metabolism and pharmacokinetics of orally ingested phenolic compounds. *Food Funct.* **2017**, *8*, 15–38. [CrossRef]
6. Gu, C.; Howell, K.; Dunshea, F.R.; Suleria, H.A. LC-ESI-QTOF/MS characterization of phenolic acids and flavonoids in polyphenol-rich fruits and vegetables and their potential antioxidant activities. *Antioxidants* **2019**, *8*, 405. [CrossRef]
7. Olivas-Aguirre, F.J.; Gaytán-Martínez, M.; Mendoza-Díaz, S.O.; González-Aguilar, G.A.; Rodrigo-García, J.; Martínez-Ruiz, N.D.R.; Wall-Medrano, A. In vitro digestibility of phenolic compounds from edible fruits: Could it be explained by chemometrics? *Int. J. Food Sci. Technol.* **2017**, *52*, 2040–2048. [CrossRef]
8. Chen, Y.; Chen, H.; Zhang, W.; Ding, Y.; Zhao, T.; Zhang, M.; Mao, G.; Feng, W.; Wu, X.; Yang, L. Bioaccessibility and biotransformation of anthocyanin monomers following *in vitro* simulated gastric-intestinal digestion and in vivo metabolism in rats. *Food Funct.* **2019**, *10*, 6052–6061. [CrossRef]
9. Fang, J. Bioavailability of anthocyanins. *Drug Metab. Rev.* **2014**, *46*, 508–520. [CrossRef]
10. Koistinen, V.M.; Bento da Silva, A.; Abrankó, L.; Low, D.; Garcia Villalba, R.; Tomás Barberán, F.; Landberg, R.; Savolainen, O.; Alvarez-Acero, I.; Pascual-Teresa, D. Interlaboratory coverage test on plant food bioactive compounds and their metabolites by mass spectrometry-based untargeted metabolomics. *Metabolites* **2018**, *8*, 46. [CrossRef]
11. Šeruga, M.; Novak, I.; Jakobek, L. Determination of polyphenols content and antioxidant activity of some red wines by differential pulse voltammetry, HPLC and spectrophotometric methods. *Food Chem.* **2011**, *124*, 1208–1216. [CrossRef]
12. Torres-Aguirre, G.A.; Muñoz-Bernal, Ó.A.; Álvarez-Parrilla, E.; Núñez-Gastélum, J.A.; Wall-Medrano, A.; Sáyago-Ayerdi, S.G.; de la Rosa, L.A. Optimización de la extracción e identificación de compuestos polifenólicos en anís (*Pimpinella anisum*), clavo (*Syzygium aromaticum*) y cilantro (*Coriandrum sativum*) mediante HPLC acoplado a espectrometría de masas. *TIP Rev. Esp. Cie. Quim. Biol.* **2018**, *21*, 103–115.
13. Velderrain-Rodríguez, G.; Torres-Moreno, H.; Villegas-Ochoa, M.A.; Ayala-Zavala, J.F.; Robles-Zepeda, R.E.; Wall-Medrano, A.; González-Aguilar, G.A. Gallic acid content and antioxidant mechanism are responsible for the antiproliferative activity of 'Ataulfo'mango peel on LS180 cells. *Molecules* **2018**, *23*, 695. [CrossRef] [PubMed]
14. Campos-Vega, R.; Vázquez-Sánchez, K.; López-Barrera, D.; Loarca-Piña, G.; Mendoza-Díaz, S.; Oomah, B.D. Simulated gastrointestinal digestion and *in vitro* colonic fermentation of spent coffee (*Coffea arabica* L.): Bioaccessibility and intestinal permeability. *Food Res. Int.* **2015**, *77*, 156–161. [CrossRef]
15. Alam, M.A.; Al-Jenoobi, F.I.; Al-mohizea, A.M. Everted gut sac model as a tool in pharmaceutical research: Limitations and applications. *J. Pharm. Pharmacol.* **2012**, *64*, 326–336. [CrossRef]

16. Diaconeasa, Z.; Iuhas, C.I.; Ayvaz, H.; Rugină, D.; Stanilă, A.; Dulf, F.; Bunea, A.; Socaci, S.A.; Socaciu, C.; Pintea, A. Phytochemical characterization of commercial processed blueberry, blackberry, blackcurrant, cranberry, and raspberry and their antioxidant activity. *Antioxidants* **2019**, *8*, 540. [CrossRef]
17. Colombo, F.; Di Lorenzo, C.; Regazzoni, L.; Fumagalli, M.; Sangiovanni, E.; de Sousa, L.P.; Bavaresco, L.; Tomasi, D.; Bosso, A.; Aldini, G.; et al. Phenolic profiles and anti-inflammatory activities of sixteen table grape (*Vitis vinifera* L.) varieties. *Food Funct.* **2019**, *10*, 1797–1807. [CrossRef]
18. He, F.; He, J.-J.; Pan, Q.-H.; Duan, C.-Q. Mass-spectrometry evidence confirming the presence of pelargonidin-3-*O*-glucoside in the berry skins of Cabernet Sauvignon and Pinot Noir (*Vitis vinifera* L.). *Aust. J. Grape Wine Res.* **2010**, *16*, 464–468. [CrossRef]
19. Aguirre, M.J.; Chen, Y.Y.; Isaacs, M.; Matsuhiro, B.; Mendoza, L.; Torres, S. Electrochemical behaviour and antioxidant capacity of anthocyanins from Chilean red wine, grape and raspberry. *Food Chem.* **2010**, *121*, 44–48. [CrossRef]
20. Olivas-Aguirre, F.J.; Rodrigo-García, J.; Martínez-Ruiz, N.D.R.; Cárdenas-Robles, A.I.; Mendoza-Díaz, S.O.; Álvarez-Parrilla, E.; Gonzalez-Aguilar, G.; De la Rosa, L.A.; Ramos-Jimenez, A.; Wall-Medrano, A. Cyanidin-3-O-glucoside: Physical-chemistry, foodomics and health effects. *Molecules* **2016**, *21*, 1264. [CrossRef]
21. Ma, Y.; Feng, Y.; Diao, T.; Zeng, W.; Zuo, Y. Experimental and theoretical study on antioxidant activity of the four anthocyanins. *J. Mol. Struct.* **2020**, *1204*, 127509. [CrossRef]
22. Janeiro, P.; Brett, A.M.O. Redox behavior of anthocyanins present in *Vitis vinifera* L. *Electroanalysis* **2017**, *19*, 1779–1786. [CrossRef]
23. Rice-Evans, C.A.; Miller, N.J.; Paganga, G. Structure-antioxidant activity relationships of flavonoids and phenolic acids. *Free Radic. Biol. Med.* **1996**, *20*, 933–956. [CrossRef]
24. Stintzing, F.C.; Stintzing, A.S.; Carle, R.; Frei, B.; Wrolstad, R.E. Color and antioxidant properties of cyanidin-based anthocyanin pigments. *J. Agric. Food Chem.* **2002**, *50*, 6172–6181. [CrossRef] [PubMed]
25. Roy, M.K.; Koide, M.; Rao, T.P.; Okubo, T.; Ogasawara, Y.; Juneja, L.R. ORAC and DPPH assay comparison to assess antioxidant capacity of tea infusions: Relationship between total polyphenol and individual catechin content. *Int. J. Food Sci. Nutr.* **2010**, *61*, 109–124. [CrossRef] [PubMed]
26. Kosińska-Cagnazzo, A.; Diering, S.; Prim, D.; Andlauer, W. Identification of bioaccessible and uptaken phenolic compounds from strawberry fruits in in vitro digestion/Caco-2 absorption model. *Food Chem.* **2015**, *170*, 288–294. [CrossRef]
27. Lucas-Gonzalez, R.; Navarro-Coves, S.; Pérez-Álvarez, J.A.; Fernández-López, J.; Muñoz, L.A.; Viuda-Martos, M. Assessment of polyphenolic profile stability and changes in the antioxidant potential of maqui berry (*Aristotelia chilensis* (*Molina*) *Stuntz*) during *in vitro* gastrointestinal digestion. *Ind. Crops Prod.* **2016**, *94*, 774–782. [CrossRef]
28. Jiao, X.; Li, B.; Zhang, Q.; Gao, N.; Zhang, X.; Meng, X. Effect of in vitro-simulated gastrointestinal digestion on the stability and antioxidant activity of blueberry polyphenols and their cellular antioxidant activity towards HepG2 cells. *Int. J. Food Sci. Technol.* **2018**, *53*, 61–71. [CrossRef]
29. Dixit, P.; Jain, D.K.; Dumbwani, J. Standardization of an ex vivo method for determination of intestinal permeability of drugs using everted rat intestine apparatus. *J. Pharmacol. Toxicol. Methods* **2012**, *65*, 13–17. [CrossRef]
30. Rastogi, H.; Jana, S. Evaluation of physicochemical properties and intestinal permeability of six dietary polyphenols in human intestinal colon adenocarcinoma Caco-2 cells. *Eur. J. Drug Metab. Pharm.* **2016**, *41*, 33–43. [CrossRef]
31. Farrell, T.L.; Poquet, L.; Dew, T.P.; Barber, S.; Williamson, G. Predicting phenolic acid absorption in Caco-2 cells: A theoretical permeability model and mechanistic study. *Drug Metab. Disp.* **2012**, *40*, 397–406. [CrossRef] [PubMed]
32. Estudante, M.; Morais, J.G.; Soveral, G.; Benet, L.Z. Intestinal drug transporters: An overview. *Adv. Drug Deliv. Rev.* **2013**, *65*, 1340–1356. [CrossRef] [PubMed]
33. Herrera-Cazares, L.A.; Hernández-Navarro, F.; Ramírez-Jiménez, A.K.; Campos-Vega, R.; de la Luz Reyes-Vega, M.; Loarca-Piña, G.; Morales-Sanchez, E.; Wall-Medrano, A.; Gaytan-Martinez, M. Mango-bagasse functional-confectionery: Vehicle for enhancing bioaccessibility and permeability of phenolic compounds. *Food Funct.* **2017**, *8*, 3906–3916. [CrossRef] [PubMed]

34. Luzardo-Ocampo, I.; Campos-Vega, R.; Gaytán-Martínez, M.; Preciado-Ortiz, R.; Mendoza, S.; Loarca-Piña, G. Bioaccessibility and antioxidant activity of free phenolic compounds and oligosaccharides from corn (*Zea mays* L.) and common bean (*Phaseolus vulgaris* L.) chips during *in vitro* gastrointestinal digestion and simulated colonic fermentation. *Food Res. Int.* **2017**, *100*, 304–311. [CrossRef]
35. Gil, E.S.; Couto, R.O. Flavonoid electrochemistry: A review on the electroanalytical applications. *Rev. Bras. Farmacog.* **2013**, *23*, 542–558. [CrossRef]
36. Alcalde, B.; Granados, M.; Saurina, J. Exploring the antioxidant features of polyphenols by spectroscopic and electrochemical methods. *Antioxidants* **2019**, *8*, 523. [CrossRef]
37. Blasco, A.J.; González, M.C.; Escarpa, A. Electrochemical approach for discriminating and measuring predominant flavonoids and phenolic acids using differential pulse voltammetry: Towards an electrochemical index of natural antioxidants. *Anal. Chim. Acta* **2004**, *511*, 71–81. [CrossRef]
38. Guiberteau-Cabanillas, A.; Godoy-Cancho, B.; Bernalte, E.; Tena-Villares, M.; Guiberteau Cabanillas, C.; Martinez-Canas, M.A. Electroanalytical behavior of gallic and ellagic acid using graphene modified screen-printed electrodes. Method for the determination of total low oxidation potential phenolic compounds content in cork boiling waters. *Electroanalysis* **2015**, *27*, 177–184. [CrossRef]
39. Kuntz, S.; Rudloff, S.; Asseburg, H.; Borsch, C.; Fröhling, B.; Unger, F.; Dold, S.; Spengler, B.; Rompp, A.; Kunz, C. Uptake and bioavailability of anthocyanins and phenolic acids from grape/blueberry juice and smoothie in vitro and in vivo. *Brit. J. Nutr.* **2015**, *113*, 1044–1055. [CrossRef]
40. Kamiloglu, S.; Capanoglu, E.; Grootaert, C.; Van Camp, J. Anthocyanin absorption and metabolism by human intestinal Caco-2 cells—A review. *Int. J. Mol. Sci.* **2015**, *16*, 21555–21574. [CrossRef]
41. Berto, S.; Carena, L.; Chiavazza, E.; Marletti, M.; Fin, A.; Giacomino, A.; Malandrino, M.; Barolo, C.; Prenesti, E.; Vione, D. Off-line and real-time monitoring of acetaminophen photodegradation by an electrochemical sensor. *Chemosphere* **2018**, *204*, 556–562. [CrossRef] [PubMed]
42. Bounab, L.; Iglesias, O.; Pazos, M.; Sanromán, M.Á.; González-Romero, E. Effective monitoring of the electro-Fenton degradation of phenolic derivatives by differential pulse voltammetry on multi-walled-carbon nanotubes modified screen-printed carbon electrodes. *Appl. Catal. B Environ.* **2016**, *180*, 544–550. [CrossRef]
43. Barik, S.K.; Russell, W.R.; Moar, K.M.; Cruickshank, M.; Scobbie, L.; Duncan, G.; Hoggard, N. The anthocyanins in black currants regulate postprandial hyperglycaemia primarily by inhibiting α-glucosidase while other phenolics modulate salivary α-amylase, glucose uptake and sugar transporters. *J. Nutr. Biochem.* **2020**, *78*, 108325. [CrossRef] [PubMed]
44. Kalt, W. Anthocyanins and their C6-C3-C6 metabolites in humans and animals. *Molecules* **2019**, *24*, 4024. [CrossRef]
45. Steinert, R.E.; Ditscheid, B.; Netzel, M.; Jahreis, G. Absorption of black currant anthocyanins by monolayers of human intestinal epithelial Caco-2 cells mounted in ussing type chambers. *J. Agric. Food Chem.* **2008**, *56*, 4995–5001. [CrossRef]
46. Wirth, R.; Bódi, N.; Maróti, G.; Bagyánszki, M.; Talapka, P.; Fekete, É.; Bagi, Z.; Kovacs, K. Regionally distinct alterations in the composition of the gut microbiota in rats with streptozotocin-induced diabetes. *PLoS ONE* **2014**, *9*. [CrossRef]
47. Yin, R.; Kuo, H.C.; Hudlikar, R.; Sargsyan, D.; Li, S.; Wang, L.; Wu, R.; Kong, A.N. Gut microbiota, dietary phytochemicals, and benefits to human health. *Curr. Pharmacol. Rep.* **2019**, *5*, 332–344. [CrossRef]
48. Gonthier, M.P.; Verny, M.A.; Besson, C.; Rémésy, C.; Scalbert, A. Chlorogenic acid bioavailability largely depends on its metabolism by the gut microflora in rats. *J. Nutr.* **2003**, *133*, 1853–1859. [CrossRef]

antioxidants

MDPI

Review

Interaction of Polyphenols as Antioxidant and Anti-Inflammatory Compounds in Brain–Liver–Gut Axis

Amritpal Singh, Yu Fung Yau, Kin Sum Leung, Hani El-Nezami and Jetty Chung-Yung Lee *

School of Biological Sciences, The University of Hong Kong, Pokfulam Road, Hong Kong, China; amritpal@connect.hku.hk (A.S.); hyfyau@connect.hku.hk (Y.F.Y.); sam612@connect.hku.hk (K.S.L.); elnezami@hku.hk (H.E.-N.)

* Correspondence: jettylee@hku.hk; Tel.: +852-2299-0318

Received: 2 July 2020; Accepted: 24 July 2020; Published: 26 July 2020

Abstract: Oxidative stress plays an important role in the onset as well as the progression of inflammation. Without proper intervention, acute inflammation could progress to chronic inflammation, resulting in the development of inflammatory diseases. Antioxidants, such as polyphenols, have been known to possess anti-oxidative properties which promote redox homeostasis. This has encouraged research on polyphenols as potential therapeutics for inflammation through anti-oxidative and anti-inflammatory pathways. In this review, the ability of polyphenols to modulate the activation of major pathways of inflammation and oxidative stress, and their potential to regulate the activity of immune cells are examined. In addition, in this review, special emphasis has been placed on the effects of polyphenols on inflammation in the brain–liver–gut axis. The data derived from in vitro cell studies, animal models and human intervention studies are discussed.

Keywords: oxidative stress; inflammation; polyphenols; antioxidant

---

1. Introduction

One of the main innate responses of the immune system is inflammation, which is an important non-specific response to any kind of injury and infection, such as physical wounds, toxins and tissue damage. It is a crucial response to the alteration of tissue integrity, to initiate healing and restore tissue homeostasis [1]. Several types of white blood cells, such as neutrophils and macrophages, and cytokines are involved in the inflammatory process. Cytokines play an enormous part in the inflammatory response and are mainly produced by helper T cells and macrophages [2]. They can be classified into pro-inflammatory cytokines, such as interleukin (IL)-1β and the tumor necrosis factor (TNF)-α, and anti-inflammatory cytokines, such as IL-4 and IL-10 [2]. The regulation and balance between the two types of cytokines is crucial for the immune system. An overproduction of pro-inflammatory cytokines could lead to autoimmune diseases and chronic inflammatory diseases, thus highlighting the need for anti-inflammatory cytokines to prevent chronic inflammatory conditions [2].

The inflammatory response is a multi-stage process which involves a triggering system, a sensor mechanism, signal transmission and the production of inflammatory mediators [1]. The inflammatory response could be triggered by various danger signals, which could be from exogenous, such as invasion by microorganisms, or endogenous sources, such as tissue damage. The exogenous and endogenous signaling molecules are termed pathogen-associated molecular patterns (PAMPs) and damage-associated molecular patterns (DAMPs) respectively [1]. PAMPs and DAMPs are sensed by a variety of pattern recognition receptors (PRRs) which include Toll-like receptors (TLRs), nucleotide-binding oligomerization domain (NOD)-like receptors (NLRs), C-type lectins and receptors for advanced glycation end-products (RAGE) [1]. The activation of PRR triggers intracellular signaling cascades, including kinases, such as mitogen-activated protein kinases (MAPKs), adaptors, the myeloid differentiation primary response protein 88 (MyD88), and transcription factors, such as the nuclear factor kappa B (NF-κB). Furthermore, the activation of NLR could trigger cytokine maturation, which are key to inflammation development, through inflammasomes. For instance, activated NLRP3 are associated to the adaptor protein ASC (apoptosis associated speck-like containing a CARD domain) and caspase-1 to form inflammasome, which promotes the conversion of pro-IL-1β and pro-IL-18 to mature IL-1β and IL-18, respectively [1]. Upon DAMP binding, TLR interacts with MyD88 which activates downstream signaling, resulting in NF-κB and activator protein-1 (AP-1) activation [3]. The signaling pathways mentioned above upregulate the expression of inflammatory mediators like cytokines for inflammation development.

It has been long known that significant oxidative stress could cause cellular damage and modification of genes, which triggers the inflammatory signaling cascade for the onset of inflammation in various inflammatory diseases [4]. As part of the inflammatory response, large amounts of reactive oxygen species (ROS) are generated, which could further promote oxidative stress and chronic inflammation if produced for lengthened periods [1,4]. Besides, several studies have pointed out that oxidants have a significant part in the activation of TLRs [5,6]. The studies have shown that the translocation of TLR4 to the cell membrane was upregulated after exposure to oxidants [5,6]. This enhances the responsiveness of cells to a danger signal for the onset of pro-inflammatory signaling pathways. One of the most discussed pathways in inflammation is the NF-κB pathway. It is a key regulator of inflammation due to its sensitivity to ROS. The activation of NF-κB could result from two pathways, the canonical and alternative pathways [1,7]. In both pathways, NF-κB is freed from its inhibitor, IκB, resulting in NF-κB translocation to the nucleus for the expression of target genes [1].

Under oxidative stress, the expression of antioxidant genes is upregulated, which is modulated by the nuclear factor erythroid 2-related factor 2 (Nrf2) [8]. Nrf2 activation is induced by ROS by the removal of its inhibitor, Kelch-like erythroid CNC homolog-associated protein 1 (Keap1), allowing the translocation of Nrf2 to the nucleus for the expression of genes involved in the antioxidant response [8].

As mentioned earlier, significant oxidative stress results in the propagation of inflammation, which illustrates the importance of redox balance in the resolution, and prevention of inflammation. Redox homeostasis is maintained by antioxidants, which could be from endogenous or exogenous (natural or synthetic) sources. The endogenous sources of antioxidants consist of enzymes such as glutathione peroxidase (GPx), superoxide dismutase (SOD) and catalase (CAT) [4]. Minerals such as zinc, selenium and copper are essential for the activation of the antioxidant enzymes as they are co-factors for these enzymes [9]. The natural exogenous antioxidants include ascorbic acid (vitamin C), α-tocopherol (vitamin E), carotenoids and flavonoids [4,9]. Antioxidants can scavenge free radicals which terminates the chain reaction of oxidation. Besides, they could prevent the initiation of a chain reaction by binding to transition metal ions that catalyze ROS generation [4,9]. As a result, antioxidants are able to reduce oxidative stress.

Polyphenols have been long known to be potent antioxidants. Polyphenols are found in various types of food, including fruits and vegetables, and can be classified into flavonoids and non-flavonoids [10]. Examples of flavonoids include anthocyanins, epigallocatechin gallate (EGCG) and curcumin, while an example of a non-flavonoid is resveratrol (RES) [10–12]. Because of their effects on oxidative stress, researchers have studied the effects of polyphenols in conditions with common underlying factors such as oxidative stress and inflammation. Polyphenols have been studied as potential anti-inflammatory agents in various inflammatory diseases, such as non-alcoholic fatty liver disease (NAFLD), inflammatory bowel disease and neurodegenerative diseases [10–12]. As illustrated in Figure 1, it is hypothesized that polyphenols would modulate the inflammatory signaling pathway via an antioxidant-based mechanism. It is expected that polyphenols would reduce oxidative stress, which would inhibit signal transduction for the production of pro-inflammatory mediators. The aim of this review will be to discuss the effects of polyphenol intervention in experimental and clinical settings on inflammation in specific organs, namely the brain, liver and gut, with reference to their antioxidant and anti-inflammatory properties.

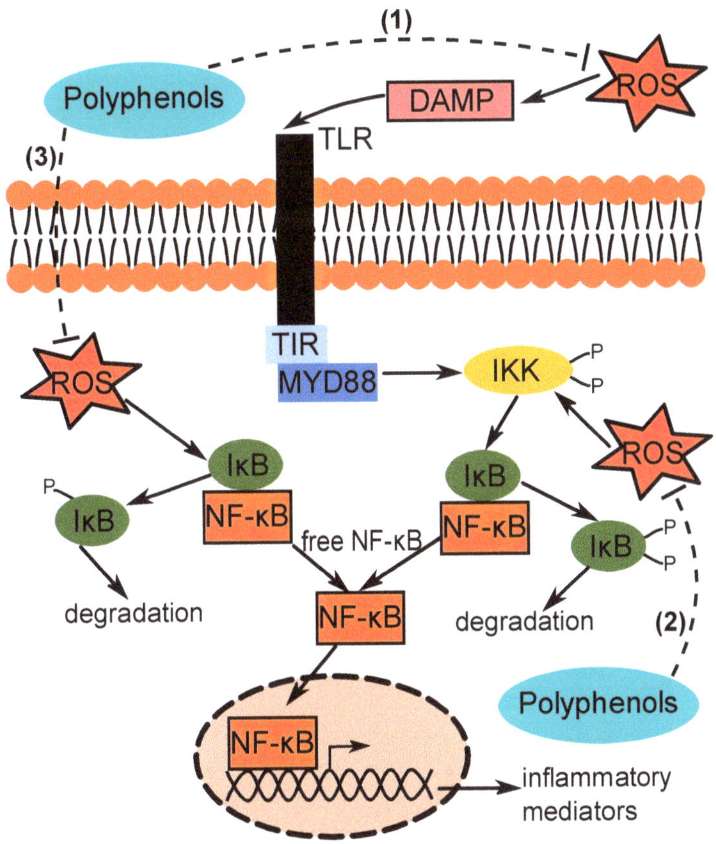

Figure 1. Potential mechanism of action of polyphenols in inflammation inhibition. Polyphenols may target the reactive oxygen species (ROS) to reduce oxidative stress. (1) ROS reduction could reduce the amount of damage-associated molecular patterns (DAMPs); (2) ROS reduction could also arrest phosphorylation of IκB kinase (IKK), which would block the dissociation of IκB from NF-κB; (3) ROS could directly phosphorylate IκB, which could be prevented by polyphenols. These pathways would inhibit nuclear translocation of NF-κB.

2. Polyphenol Intervention in Brain Inflammation

Some of the most common brain disorders include neurodegenerative diseases such as Alzheimer's disease and Parkinson's disease [10–13]. Despite the exact causes of these diseases being unclear, these disorders have been linked to common underlying factors which are high levels of oxidative stress and inflammation [10–13]. Due to these reasons, the use of antioxidants as potential therapeutic agents has emerged. For instance, the effects of polyphenols have been studied on factors that are involved in disease progression. The findings of these studies in the last 15 years (2004–2019) are summarized in Table 1.

Table 1. Summary of the effects of polyphenols on brain inflammation in the last 15 years (2004–2019).

Model of Study	Agent	Effects	Reference
In vitro			
Human neuroblastoma SH-SY5Y cells (oxysterol induced)	Quercetin	↓ TLR4 signaling	[14]
Human PBMC (oxLDL-induced)	Quercetin	↓ TLR2 and TLR4 expressions, NF-κB activation, inflammatory enzymes activity	[15]
Human astrocytes (LPS-induced)	Anthocyanins	↓ IL-6 secretion (low LPS and anthocyanin dose); ↑ IL-6 secretion (high anthocyanin dose in LPS absence)	[16]
Mouse BV2 microglial cells (LPS-induced)	Blueberry extract	↓ NO and TNF-α release, iNOS and COX-2 expressions, NF-κB nuclear translocation	[17–19]
Mouse BV2 microglial cells (LPS-induced)	Anthocyanins	↓ NO, PGE2, TNF-α and IL-1β release, iNOS and COX-2 expressions, NF-κB nuclear translocation	[20,21]
Mouse microglial cells (LPS/IFN-γ-induced)	Anthocyanins	↓ NO and TNF-α release, iNOS expression	[22]
Rat HAPI microglial cells (LPS-induced)	Tart cherry extract	↓ NO and TNF-α release, COX-2 expression; ↔ iNOS expression	[23]
Rat astrocytes (LPS-induced)	Lingonberry extract	↓ ROS production	[24]
Animal			
Mouse model (PD)	GSSE	↓ ROS production, inflammatory markers	[25]
Mouse model (LPS and Aβ-induced microglia neuroinflammation)	RES	↓ TLR4, NF-κB and cytokine secretion	[26]
Mouse model (LPS-impaired adult hippocampal neurogenesis)	EGCG	↓ TLR4 signaling	[27]
Mouse model (LPS-treated)	Anthocyanins	↓ NF-kB, TNF-α, and IL-1β levels	[28]
Mouse model (LPS-treated)	PSPC	↓ TNF-α, IL-6 and IL-1β overproduction, NF-kB activation	[29]
Mouse model (LPS-treated)	Anthocyanins	↓ ROS production, NF-kB activation, TNF-α, and IL-1β levels	[30]
Mouse model (LPS-treated)	Anthocyanins	↓ TNF-α, and IL-1β increase; ↑ IL-10 expression	[31]
Mouse model (high-fat diet)	PSPC	↓ iNOS, COX-2, TNF-α, IL-1β and IL-6 expressions, p38 MAPK and NF-kB activation; ↑ IL-10 levels	[32]
Rat model (MCAO/R)	Anthocyanins	↓ TNF-α, IL-6 and IL-1β levels, NF-kB and NLRP3 expressions	[33]
Human			
Subjects with AD	RES	↓ plasma pro-inflammatory markers	[34]

↑: increase; ↓: decrease; ↔: insignificant change; Aβ: beta-amyloid; AD: Alzheimer's disease; COX: cyclooxygenase; EGCG: epigallocatechin gallate; GSSE: grape seed and skin extract; IFN-γ: interferon gamma; IL: interleukin; iNOS: inducible nitric oxide synthase; LDL: low-density lipoprotein; LPS: lipopolysaccharide; MAPK: mitogen-activated protein kinase; MCAO/R: middle cerebral artery occlusion/reperfusion; NF-κB: nuclear factor kappa B; NLRP: NOD-like receptor protein; NO: nitric oxide; ox: oxidized; PBMC: peripheral blood mononuclear cell; PD: Parkinson's disease; PGE2: prostaglandin E2; PSPC: purple sweet potato color; RES: resveratrol; ROS: reactive oxygen species; TLR: toll-like receptor; TNF-α: tumor necrosis factor alpha.

2.1. *In Vitro Models of Polyphenol Treatment in Brain Inflammation*

Among several in vitro studies, one group reported the inhibition of TLR4 signaling in cells treated with quercetin [14]. Besides, the expressions of TLR2 and TLR4 being reduced by quercetin, it further decreased the production of pro-inflammatory cytokines [15]. Furthermore, quercetin has been shown to attenuate the activity of inflammatory enzymes, including inducible nitric oxide synthase (iNOS) and cyclooxygenase (COX) [15]. In studies that tested anthocyanin-rich extracts in response to lipopolysaccharide (LPS) stimulation, a significant reduction in the production of TNF-α and IL-1β, and expressions of iNOS and COX-2 was reported [17–21]. A study on rat highly aggressively proliferating immortalized (HAPI) cells demonstrated similar results, but with no significant effect on iNOS expression [23]. In mouse microglial cells, anthocyanins reduced NO and TNF-α release and iNOS expression [22]. The reduced levels of pro-inflammatory mediators have been attributed by the modulation of inflammatory signaling pathways, since decreased levels of active p38 MAPK and NF-κB have been observed [17–21]. Besides microglial cells, anthocyanins testing on astrocytes has also revealed similar findings. For instance, rat astrocytes treated with anthocyanin-rich lingonberry extract lowered ROS production, suggesting an anti-oxidative mechanism [24]. In a human astrocyte study, the secretion of IL-6 was reduced at low doses of LPS and anthocyanins [16]. However, in the absence of LPS, IL-6 secretion increased when treated with high concentration of anthocyanin, suggesting antioxidant toxicity with a single compound supplementation [16]. Some studies have highlighted that the synergistic effects of mixed polyphenols provided better outcomes than single compound supplementation. Moreover, it was reported that the anti-inflammatory effects were more significant when the cells were treated with high concentrations of anthocyanins.

2.2. *In Vivo Models of Polyphenol Treatment in Brain Inflammation*

In a mouse model of Parkinson's disease, the neuroprotective effects of grape seed and skin extract (GSSE), which is a mixture of polyphenolic compounds that is mostly comprised of flavonoids namely catechins, gallic acid, vanillin and 2,5-dihydroxybenzoic acid, were studied [25]. A reduction in ROS production, nuclear translocation of NF-κB p65 subunit and loss of SOD was observed [25]. RES inhibited TLR4, NF-κB and cytokine secretion in LPS and β-amyloid (Aβ)-mediated microglia neuroinflammation [26]. EGCG could attenuate TLR4 signaling in LPS-impaired adult hippocampal neurogenesis [27]. In addition, findings from anthocyanin treatment in LPS-induced mouse models have reported a decrease in several markers of oxidative stress and pro-inflammation. For example, levels of pro-inflammatory cytokines, such as TNF-α, IL-6 and IL-1β, were attenuated [28–31]. Furthermore, NF-κB activation was inhibited [28–30]. In one study, the expression of anti-inflammatory cytokines, such as IL-10, increased when the mice were pre-treated with anthocyanins [31]. Purple sweet potato extract showed anti-inflammatory effects in mice fed a high-fat diet (HFD), in which the expressions of iNOS, COX-2, TNF-α, IL-6 and IL-1β attenuated, and the levels of IL-10 increased [32]. Furthermore, the activation of p38 MAPK and NF-κB was inhibited [32]. Similarly, anthocyanin treatment in a rat middle cerebral artery occlusion/reperfusion (MCAO/R) model demonstrated decreased expressions of NF-κB, NLRP3 and pro-inflammatory cytokines [33].

For human studies in relation to polyphenol intervention, a few have been conducted. One study on the administration of RES was shown to decrease neuroinflammation in patients with Alzheimer's disease [34]. The findings from in vitro and in vivo animal studies have demonstrated the benefits of polyphenols on brain inflammation. However, more human studies are required to validate the beneficial effects of polyphenols in brain-related inflammation.

3. Polyphenol Intervention in Liver Inflammation

One of most common liver diseases is NAFLD [35,36]. This disease could further progress to non-alcoholic steatohepatitis (NASH), and can eventually develop into hepatocellular carcinoma [35]. Similar to brain inflammation, the effects of polyphenol interventions in liver inflammation have been extensively researched. The findings of the effects of polyphenols on liver health in the last 15 years (2004–2019) are summarized in Table 2.

Table 2. Summary of the effects of polyphenols on liver inflammation in the last 15 years (2004–2019).

Model of Study	Agent	Effects	Reference
In vitro			
Hepatic stellate cells (glucose-induced)	Curcumin	↓ ROS production; ↑ GCL activity, GSH level	[37]
Human HepG2 cells (fatty acid-induced)	Theaflavins	↓ ROS production	[38]
Human HepG2 cells (glucose-induced)	C3G	↓ ROS production; ↑ GCL activity, GSH level	[39]
Mouse macrophage cells (palmitic acid-induced)	Rutin	↓ ROS production, *MCP-1*, *TNF-α*, *IL-6*, *IFN-γ*, *IL-1β* genes expressions	[40]
Animal			
Mouse model (Western diet)	Quercetin	↓ TBARS, TG and TNF-α levels; ↑ GPx and CAT levels	[41]
Mouse model (MCD)	Quercetin	↓ TLR4 protein concentration, TNF-α, IL-6 and COX-2 mRNA expressions	[42]
Mouse model (HFD)	Rutin	↓ *TNF-α and Mcp1* gene expressions	[40]
Mouse model (HFD)	Troxerutin	↓ ROS levels; ↑ GPx, GSH and SOD levels	[43]
Mouse model (MCDHFD)	Theaflavins	↓ TBARS level, ROS production, TNF-α expressions	[44]
Mouse model (HFD)	Baicalein	↓ MCP-1 and TNF-α levels	[45]
Mouse model (HFD)	RES	↓ macrophage infiltration	[46]
Mouse model (MCD)	Curcumin	↓ ICAM-1, COX-2 and MCP-1 expressions, NF-κB signalling	[47]
Mouse model (MCD)	Curcumin	↓ ROS production	[48]
Mouse model (MCD)	Silibinin	↓ ROS production, iNOS expression, NF-κB activation	[49]
Mouse model (MCD)	Silibinin	↓ Il-6 and TNF-α expressions; ↑ GSH level	[50]
Rat model (HFD)	Quercetin	↓ NF-κB expression; ↑ Nrf2 and HO-1 expressions	[51]
Rat model (HFD)	Rutin	↓ plasma MDA; ↑ GPx expression	[52]
Rat model (HFD)	EGCG	↓ plasma and liver MDA; ↑ GSH level	[53]
Rat model (HFD)	EGCG	↓ iNOS, COX-2 and TNF-α expressions; ↑ GPx and CAT activity	[54]
Rat model (HFD)	Genistein	↓ TNF-α and plasma and liver MDA levels	[55]
Rat model (HCD)	Naringenin	↓ ROS production, TNF-α, IL-6, IL-1β and iNOS expressions	[56]
Rat model (HFD)	Coffee polyphenols	↓ TNF-α, IFN-γ, IL-4 and IL-10 expressions; ↑ GSH/GSSG ratio,	[57]
Rat model (HFD)	Coffee polyphenols	↑ GST expression	[58]
Rat model (fructose-fed)	RES	↓ TBARS level; ↑ SOD activity, Nrf2 and GSH levels	[59]

Table 2. *Cont.*

Model of Study	Agent	Effects	Reference
Human			
Subjects with NAFLD	Catechins	↓ urinary F_{2t}-isoprostane excretion (high dose)	[60]
Subjects with NAFLD	RES	↓ inflammatory markers (TNF-α, CK-18, FGF-21)	[61]
Subjects with NAFLD	RES	↓ inflammatory markers (IL-6, hs-CRP, NF-κB)	[62]
Subjects with NAFLD	Silymarin	↓ NASH score and serum oxidative stress	[63]
Subjects with NAFLD	RES	↑ ALT and AST levels (high dose)	[64]

↑: increase; ↓: decrease; ALT: alanine aminotransferase; AST: aspartate transaminase; C3G: cyanidin-3-glucoside; CAT; catalase; CK: cytokeratin; COX: cyclooxygenase; EGCG: epigallocatechin gallate; FGF: fibroblast growth factor; GCL: glutamate-cysteine ligase; GPx: glutathione peroxidase; GSH: reduced glutathione; GSSG: oxidized glutathione; GST: glutathione s-transferase; HFD: high fat diet; HO: heme oxygenase; hs-CRP: high sensitivity C-reactive protein; ICAM: intercellular adhesion molecule; IFN-γ: interferon gamma; IL: interleukin; iNOS: inducible nitric oxide synthase; MCD: methionine-choline deficient; MCDHFD: methionine-choline deficient high fat diet; MCP: monocyte chemoattractant protein; MDA: malondialdehyde; NAFLD: non-alcoholic fatty liver disease; NASH: non-alcoholic steatohepatitis; NF-κB: nuclear factor kappa B; Nrf: nuclear factor erythroid 2-related factor; RES: resveratrol; ROS: reactive oxygen species; SOD: superoxide dismutase; TBARS: thiobarbituric acid reactive substances; TG: triglyceride; TLR: toll-like receptor; TNF-α: tumour necrosis factor alpha.

3.1. *In Vitro Models of Polyphenol Treatment in Liver Inflammation*

A few in vitro studies have illustrated the beneficial outcomes of polyphenols on oxidative stress and inflammation. Mouse macrophages pre-treated with rutin attenuated oxidative stress and *MCP-1*, *TNF-α*, *IL-6*, *IFN-γ*, *IL-1β* genes expressions [40]. In a study on human hepatocyte HepG2 cells, the glutamate cysteine ligase (GCL) activity and GSH level were improved with cyanidin-3-*O*-β-glucoside (C3G) treatment [39]. Moreover, theaflavins could reduce ROS generation in the same type of cell line [38]. Curcumin treatment also increased GCL activity and GSH level, and attenuated ROS production in glucose-induced hepatic stellate cells [37].

3.2. *In Vivo Models of Polyphenol Treatment in Liver Inflammation*

Much like the in vitro studies, several animal in vivo studies have demonstrated encouraging effects of polyphenol interventions. Quercetin has shown to improve thiobarbituric acid reactive substances (TBARS), GPx and CAT levels, and down-regulate TNF-α, IL-6 and COX-2 mRNA expressions in two mouse models [41,42]. Moreover, it was reported that the TLR4 protein concentration also decreased [42].

In mice fed with HFD, the expressions of TNF-α and MCP-1 genes were attenuated by rutin [40]. In another study, troxerutin increased GPx, SOD and GSH levels, which subsequently reduced ROS production [43]. Theaflavins supplementation lowered TBARS, ROS and TNF-α production in mice fed with a methionine and choline deficient high fat diet (HFD) [44]. Furthermore, the levels of MCP-1 and TNF-α were decreased with baicalein supplementation in HFD mice [45]. Oxidative stress could be reduced through the inhibition of macrophage infiltration when HFD mice are treated with RES [46]. In a few studies in which mice were given a methionine and choline-deficient diet, oxidative stress and inflammation were reduced with curcumin [47,48]. One of the studies reported the attenuation of ICAM-1, COX-2 and MCP-1 expressions and NF-κB signaling [47]. Another phenolic compound, silibinin, was shown to reduce iNOS expression, ROS production and NF-κB activation in mice fed with methionine and choline-deficient diet [49]. In addition, a similar model reported that silibinin supplementation improved GSH levels, and down-regulated IL-6 and TNF-α expressions [50].

Besides mouse models, polyphenol intervention has been studied in rat models. In rats fed with HFD, quercetin supplementation upregulated the expressions of Nrf2 and heme oxygenase 1 (HO-1), and down-regulated the expression of NF-κB [51]. In a similar model, mice supplemented with rutin had higher GPx expression and lower plasma MDA levels [52]. One study showed that, with EGCG supplementation, the GSH level was improved, whereas the plasma and liver MDA levels were reduced in rats fed with HFD [53]. Furthermore, EGCG could increase GPx and CAT activity, and attenuate iNOS, COX-2 and TNF-α expressions [54]. Similarly, genistein treatment decreased TNF-α, and plasma and liver MDA levels in HFD-fed rats [55]. In rats fed with a high cholesterol diet and naringenin, the attenuation of the production of pro-inflammatory cytokines, namely TNF-α, IL-6 and IL-1β, through the inhibition of the NF-κB pathway was also reported [56]. In fructose fed rats, RES increased SOD activity and Nrf2 and GSH levels [59]. Furthermore, RES reduced lipid peroxidation in the same model. The effects of coffee polyphenols have also been studied in a few rat models. The supplementation of coffee polyphenols ameliorated GSH/glutathione disulphide (GSSG) ratio, and attenuated TNF-α and IFN-γ expressions [57]. Besides, coffee polyphenols could reduce hepatic oxidative stress and steatosis in rats fed with HFD [58]. It was also reported that coffee polyphenols suppressed not only the expressions of pro-inflammatory cytokines, but also suppressed the expressions of anti-inflammatory cytokines, such as IL-4 and IL-10 [57].

A few of the polyphenols have also been subject to clinical testing, with the target group being the patients with NAFLD. For instance, supplementation of catechins decreased urinary F_{2t}-isoprostane excretion in the treatment group that was given a higher dose [60]. Moreover, silymarin supplementation ameliorated the NASH score and serum oxidative stress [63]. A few studies have tested the effects of RES in NAFLD patients and have produced mixed results. Several studies observed an improvement in the inflammatory markers, such as TNF-α, cytokeratin 18 (CK-18), IL-6 and NF-κB [61,62]. On the other hand, one study reported the harmful effects of a high concentration of RES, in which the levels of enzymes, like alanine aminotransferase (ALT) and aspartate transaminase (AST) increased [64]. These findings suggest that polyphenols could prevent the progression of NAFLD to NASH. However, supplementation of a high dose could have adverse effects on liver health.

4. Polyphenol Intervention in Gut Inflammation

Compared to the other organs that have been discussed in this review, polyphenol intervention in gut inflammation had not been a subject of profound research until recently. Most of the research has been on berry polyphenols and RES [65,66]. The interventions have targeted the models of inflammatory bowel disease, which is a term used to describe inflammatory disorders in the gut [66]. Some common examples of inflammatory bowel disease include Crohn's disease and ulcerative colitis (UC) [66]. The findings of polyphenol interventions in vitro and in vivo studies in the last 15 years (2004–2019) are summarized in Table 3.

Table 3. Summary of the effects of polyphenols on gut inflammation in the last 15 years (2004-2019).

Model of Study	Agent	Effects	Reference
In vitro			
Human colon epithelial cells (cytokine-treated)	Anthocyanins	↓ IP-10 and TNF-α expression	[67]
Human monocytic THP-1 cells (IFN-γ-treated)	Anthocyanins	↓ IFN-γ receptor 2 expression	[68]
Human intestinal Caco-2 cells (LPS-treated)	RES	↓ COX-2 expression, PGE2 release, NF-κB activation	[69]
Human intestinal Caco-2 cells (LPS-treated)	RES	↓ iNOS and TLR4 expression, NF-κB activation, NO release (high dose)	[70]
Human intestinal Caco-2 cells (IL-1β-treated)	RES	↑ NF-κB activation, p-IκB/IκB ratio, IL-8 production	[71]
Human intestinal Caco-2 cells (TNF-α-treated)	RES	↑ NF-κB activation	[71]
Human colon epithelial HT-29 cells (cytokine-treated)	RES	↓ ROS production, iNOS and COX-2 expression, NO and PGE2 release; ↔ NF-κB activation	[72]
Human colon epithelial HT-29 cells (cytokine-treated)	RES	↑ HO-1 and GCL expression, Nrf2 activation, GSH:GSSG ratio	[73]
Human colon SW480 cells (LPS-treated)	RES	↓ iNOS and TLR4 expression, NF-κB activation, and NO release (high dose)	[70]
Animal			
Mouse model (TNBS-induced colitis)	Anthocyanins	↓ MPO activity, IL-12, TNF-α and IFN-γ increase, NO production; ↑ IL-10 expression	[74]
Mouse model (DSS-induced colitis)	Blueberry extract	↓ COX-2, iNOS, IFN-γ and IL-1β expression, NF-κB activation, neutrophil infiltration, MDA and serum PGE2 levels; ↑ CAT and SOD activity	[75]
Mouse model (DSS-induced colitis)	Black raspberry powder	↓ TNF-α and IL-1β expression, NF-κB and COX-2 activity; ↔ RNS and MDA levels, inflammatory cells infiltration	[76]
Mouse model (DSS-induced colitis)	Anthocyanins	↓ TNF-α and IFN-γ secretion	[77]
Mouse model (DSS-induced colitis)	Black raspberry powder	↓ macrophages and neutrophils infiltration, NF-κB translocation	[78]
Mouse model (DSS-induced colitis)	Cranberry extract or dried cranberries	↓ MPO activity, TNF-α and IL-1β expression	[79]

Table 3. *Cont.*

Model of Study	Agent	Effects	Reference
Mouse model (high fat diet)	Cranberry extract	↓ COX-2 and TNF-α expression, LPS level; ↔ MDA and SOD levels	[80]
Mouse model (DSS-induced colitis)	RES	↓ iNOS, COX-2 and TNF-α levels	[81]
Mouse model (DSS-induced colitis)	RES	↓ iNOS and COX-2 expression, TNF-α and IL-1β levels; ↑ IL-10 level	[82]
Mouse model (DSS-induced colitis)	RES	↓ IL-6, TNF-α, IFN-γ and IL-1β levels, COX-1 and COX-2 expression	[83]
Mouse model (DSS-induced colitis)	RES	↔ MPO activity and TNF-α level	[84]
Mouse model (DSS-induced colitis)	RES	↓ MPO, SOD and GPx activity, MDA level, TNF-α, IFN-γ and IL-8 expression	[85]
Mouse model (DSS-induced colitis)	RES	↔ MPO activity, PGE2, IL-6 and IL-10 levels	[86]
Mouse model (DSS-induced colitis)	RES	↓ iNOS level, NF-κB and IκB activation	[87]
Mouse model (DSS-induced colitis)	RES	↓ TNF-α level, COX-2 and IL-6 expression	[88]
Mouse model (Spontaneous chronic colitis)	RES	↓ IL-6, IL-12, TNF-α, IFN-γ and IL-1β levels	[89]
Rat model (DSS-induced colitis)	Blueberry powder	↓ MPO activity, MDA concentration; ↔ MCP-1 and GRO/CINC-1 levels	[90]
Rat model (TNBS-induced colitis)	RES	↓ MPO activity, VCAM-1, ICAM-1, MDA, NO and GSH levels	[91]
Rat model (TNBS-induced colitis)	RES	↓ MPO activity, GSH level, ICAM-1, MCP-1, CINC-1, TNF-α, IL-1β, IL-6 and IL-12 expression	[92]
Rat model (TNBS-induced colitis)	RES	↓ MPO activity, IL-1β, PGE2 and PGD2 levels	[93]
Rat model (TNBS-induced colitis)	RES	↓ MPO activity, TNF-α level, COX-1, COX-2 and NF-κB p65 expression; ↑ PGE2 level; ↔ PGD2 level	[94]
Rat model (TNBS-induced colitis)	RES	↓ MDA level; ↑ GPx activity; ↔ MPO, SOD, CAT activities	[95]
Rat model (Methotrexate-induced colitis)	RES	↓ MDA and GSH levels, MPO expression	[96]
Rat model (Oxazolone-induced colitis)	RES	↓ MPO activity	[97]

Table 3. *Cont.*

Model of Study	Agent	Effects	Reference
Rat model (DSS-induced colitis)	RES	↓ COX-2, PGE2 and NO levels; ↔ TBARS level	[98]
Rat model (PG-PS-induced colitis)	RES	↓ IL-6, TNF-α and IL-1β expression	[99]
Human			
Subjects with UC	Anthocyanins	↓ TNF-α, IFN-γ and MCP-1 levels, NF-κB activation; ↑ IL-22, IL-10 and IL-17A levels	[68]
Subjects with mild to moderate UC	Anthocyanins	↓ faecal calprotectin level and Riley Index	[100]
Subjects with UC	RES	↓ hs-CRP, TNF-α levels, PBMC NF-κB activation; ↑ IBDQ-9 score; ↔ SCCAI score	[101]
Subjects with mild to moderate UC	RES	↓ MDA level; ↑ SOD activity, IBDQ-9 score; ↔ SCCAI score	[102]

↑: increase; ↓: decrease; ↔: insignificant change; CAT; catalase; CINC: cytokine-induced neutrophil chemoattractant; COX: cyclooxygenase; DSS: dextran sodium sulphate; GCL: glutamate-cysteine ligase; GPx: glutathione peroxidase; GSH: reduced glutathione; GSSG: oxidized glutathione; HO: heme oxygenase; hs-CRP: high sensitivity C-reactive protein; ICAM: intercellular adhesion molecule; IFN-γ: interferon gamma; IL: interleukin; IP: IFN-γ-induced protein; iNOS: inducible nitric oxide synthase; LPS: lipopolysaccharide; MCP: monocyte chemoattractant protein; MDA: malondialdehyde; MPO: myeloperoxidase; NF-κB: nuclear factor kappa B; NO: nitric oxide; Nrf: nuclear factor erythroid 2-related factor; PBMC: peripheral blood mononuclear cell; PG-PS: peptidoglycan-polysaccharide; PGD2: prostaglandin D2; PGE2: prostaglandin E2; RES: resveratrol; RNS: reactive nitrogen species; ROS: reactive oxygen species; SOD: superoxide dismutase; TBARS: thiobarbituric acid reactive substances; TLR: toll-like receptor; TNBS: 2, 4, 6-Trinitrobenzenesulfonic acid; TNF-α: tumour necrosis factor alpha; UC: ulcerative colitis; VCAM: vascular cell adhesion molecule.

4.1. In Vitro Models of Polyphenol Treatment in Gut Inflammation

Much like the brain and liver, quite a few studies have been conducted to evaluate the effects of polyphenols on gut health. The amelioration of oxidative stress and inflammatory markers has been observed in several in vitro studies on human intestinal cell lines. For example, the expressions of TNF-α, IFN-γ-induced protein 10 (IP-10) and IFN-γ receptor 2 were inhibited by anthocyanins [67,68]. One study demonstrated the beneficial effects of RES in human intestinal Caco-2 cells treated with LPS. It was reported that RES pre-treatment suppressed COX-2 expression, prostaglandin E2 (PGE2) release and NF-κB activation [69]. In a similar study, iNOS and TLR4 expression, NF-κB activation and NO release were reduced but only in high RES concentration treatment groups [70]. The same results were obtained from RES treatment in human colon SW480 cells [70]. Moreover, RES upregulated HO-1 and GCL expression through the activation of the Nrf2 pathway and improved the GSH/GSSG ratio in human colon epithelial HT-29 cells [73]. Besides, ROS production, iNOS and COX-2 expression, and NO and PGE2 release could be attenuated with RES. However, it had no significant effect on the activation of NF-κB [72]. In Caco-2 cells stimulated by IL-1β or TNF-α, the phosphorylation of IκB could not be inhibited by RES, which allowed the activation of NF-κB [71]. This shows that stimulation by pro-inflammatory cytokines could trigger the inflammatory cascade even in the presence of an anti-inflammatory agent like RES.

4.2. In Vivo Models of Polyphenol Treatment in Gut Inflammation

In vivo animal models have also been used to better understand the impact of polyphenols. In a rat model of colitis induced by dextran sodium sulphate (DSS), blueberry powder reduced myeloperoxidase (MPO) activity and MDA concentration, but had an insignificant effect on MCP-1 level [90]. MPO activity is used to determine neutrophil infiltration, where a high activity would represent higher oxidative stress [65]. In another rat model of DSS-induced colitis, RES treatment decreased COX-2, PGE2 and NO levels. However, no significant change occurred in the TBARS level [98]. Another substance used to induce colitis in animal models is 2, 4, 6-trinitrobenzenesulfonic acid (TNBS). One study observed the attenuation of MPO activity, and VCAM-1, ICAM-1, MDA, NO and GSH levels with RES administration [91]. In addition to these effects, RES administration suppressed the expressions of MCP-1, cytokine-induced neutrophil chemoattractant 1 (CINC-1), TNF-α, IL-1β, IL-6 and IL-12 in rats with TNBS-induced colitis [92]. Pre-treatment of rats with RES lowered MPO activity, and IL-1β, PGE2 and prostaglandin D2 (PGD2) levels from TNBS-induced colitis [93]. In one study, even though the expression of certain inflammatory mediators was downregulated with RES treatment, the level PGE2 was increased [94]. Moreover, RES reduced colon MDA level and promoted GPx activity. Despite that, the MPO, SOD and CAT activities were unaffected [95]. In rats with methotrexate-induced or oxazolone-induced colitis, MPO activity was suppressed by RES administration [96,97]. Furthermore, the expression of pro-inflammatory cytokines, like IL-6, TNF-α and IL-1β, was decreased in a peptidoglycan-polysaccharide-induced colitis rat model [99].

Similar effects were demonstrated in mice models. Anthocyanins inhibited MPO activity, and the increase of pro-inflammatory cytokines in mice with TNBS-induced colitis. At the same time, the expression of the anti-inflammatory cytokine, IL-10, was upregulated [74]. Same effects were observed in both acute and chronic inflammatory conditions in a different study [77]. One study reported the suppression of inflammatory mediators and neutrophil infiltration, and the increase of CAT and SOD activity with blueberry extract. However, these effects were more prominent in the pre-treated group than the post-treated group [75]. Likewise, black raspberry extract attenuated TNF-α and IL-1β expressions, and NF-κB and COX-2 activity in mice with DSS-induced colitis. Yet, it had no significant effect on MDA and inflammatory cells infiltration [76]. Another study showed the inhibition of macrophages and neutrophils infiltration, and NF-κB nuclear translocation [78]. In a study on cranberry extract and dried cranberries treatment, dried cranberries attenuated MPO activity and pro-inflammatory cytokines production [79]. In mice that were given an HFD, cranberry extract downregulated the expression of inflammatory mediators but had an insignificant effect on MDA and

SOD levels [80]. RES intervention has also been studied in various DSS-induced colitis mouse models. One study recorded a decrease in protein levels of iNOS, COX-2 and TNF-α [81]. Similar effects were observed in another study alongside an increase in IL-10 level [82]. In addition, a few studies showed an attenuation in the levels of pro-inflammatory cytokines and inflammatory enzymes with RES administration [83,88]. Moreover, RES supplementation reduced iNOS protein levels and NF-κB activation in colons of mice with DSS-induced colitis [87]. In mice with spontaneous chronic colitis, RES administration for 28 weeks decreased the levels of pro-inflammatory cytokines in the colon and serum [89]. On the other hand, RES had no significant impact on MPO activity and levels of TNF-α, PGE2, IL-6 and IL-10 in mouse models of DSS-induced colitis [84,86]. Although the MPO activity and the expressions of pro-inflammatory cytokines were downregulated, Yao et al. found the SOD and GPx activities to be suppressed as well [85].

Several studies have been conducted on polyphenol intervention in patients with UC. For instance, anthocyanins were reported to reduce TNF-α, IFN-γ and MCP-1 levels and NF-κB activation. In the same study, the levels of IL-22, IL-10 and IL-17A were elevated [68]. One study on subjects with mild to moderate UC showed a reduction in fecal calprotectin level with anthocyanins treatment, which suggests that neutrophil migration was lowered. However, an increase in disease activity was observed after the termination of the treatment [100]. Besides anthocyanins, RES has also been tested in subjects with UC. One study reported a decrease in high-sensitivity CRP and TNF-α levels, and the suppression of NF-κB activation [101]. Additionally, RES administration in patients with mild to moderate UC ameliorated plasma SOD activity and lessened plasma MDA level [102]. Even though there have been encouraging results from some clinical interventions, more clinical trials are required to consider polyphenols as potential therapeutics for gut inflammatory diseases.

5. Research Gap

Although polyphenols have demonstrated anti-inflammatory properties in vitro and in vivo animal studies, there is inconclusive evidence of their effects in humans. Currently, there is insufficient evidence to support the use of polyphenols as therapeutics in subjects with inflammatory diseases. There is a need for more human trials on polyphenol intervention to gain more conclusive evidence. Moreover, it is worth noting that the available human studies have only demonstrated symptom amelioration in subjects with inflammatory diseases. For instance, RES administration reduced inflammation in Alzheimer's disease patients [34]. Other studies also reported the amelioration of symptoms with polyphenol supplementation in subjects with liver inflammatory diseases [60–63]. Furthermore, remission was observed from clinical interventions related to polyphenol administration in gut inflammatory disease patients [68,100–102]. Albeit these trials demonstrated beneficial effects, they could not demonstrate complete resolution of inflammation. As a result, it is rather premature to use polyphenols to treat inflammatory diseases.

In addition, a few studies have shown that there could be harmful effects associated with polyphenol treatment. For example, studies have reported an increase in pro-inflammatory cytokine secretion and NF-κB activation [16,71,94]. A study also reported the downregulation of antioxidant enzymes [85]. In another report, it showed RES supplementation resulted in the deterioration of liver health in patients with NAFLD [64]. Most of these adverse effects have been related to the administered dose of polyphenols. Therefore, further research on the appropriate dosage of polyphenols to produce beneficial effects and prevent adverse effects is required to reach a consensus.

6. Conclusions

Several in vitro and in vivo animal studies have demonstrated the antioxidant and anti-inflammatory effects of polyphenols in the brain–liver–gut axis. Polyphenols have been shown to target different stages of the inflammatory cascade to reduce the severity of inflammation. In general, the natural antioxidants seem to be more useful in the prevention of inflammation rather than in resolution. Although some antioxidants have had promising effects in vitro and animal studies, those results could not be extrapolated to human studies. As a result, further research is needed on polyphenol intervention in human trials, and on ways to improve the bioavailability and efficacy of polyphenols in subjects with inflammatory diseases.

Author Contributions: A.S. wrote the manuscript. Y.F.Y., K.S.L., H.E.-N. and J.C.-Y.L. conceptualized the manuscript. J.C.-Y.L. revised the manuscript. All authors have read and agreed to the published version of the manuscript.

Funding: This research received no external funding.

Conflicts of Interest: The authors declare no conflict of interest.

References

1. Lugrin, J.; Rosenblatt-Velin, N.; Parapanov, R.; Liaudet, L. The role of oxidative stress during inflammatory processes. *Biol. Chem.* **2014**, *395*, 203–230. [CrossRef] [PubMed]
2. Zhang, J.M.; An, J. Cytokines, inflammation and pain. *Int. Anesthesiol. Clin.* **2007**, *45*, 27–37. [CrossRef] [PubMed]
3. Song, D.H.; Lee, J.O. Sensing of microbial molecular patterns by Toll-like receptors. *Immunol. Rev.* **2012**, *250*, 216–229. [CrossRef] [PubMed]
4. Chatterjee, S. Oxidative stress, inflammation, and disease. In *Oxidative Stress and Biomaterials*; Dziubla, T., Butterfield, D.A., Eds.; Academic Press: Cambridge, MA, USA, 2016; pp. 35–58.
5. Powers, K.A.; Szászi, K.; Khadaroo, R.G.; Tawadros, P.S.; Marshall, J.C.; Kapus, A.; Rotstein, O.D. Oxidative stress generated by hemorrhagic shock recruits Toll-like receptor 4 to the plasma membrane in macrophages. *J. Exp. Med.* **2006**, *203*, 1951–1961. [CrossRef] [PubMed]
6. Nakahira, K.; Kim, H.P.; Geng, X.H.; Nakao, A.; Wang, X.; Murase, N.; Drain, P.F.; Wang, X.; Sasidhar, M.; Nabel, E.G.; et al. Carbon monoxide differentially inhibits TLR signaling pathways by regulating ROS-induced trafficking of TLRs to lipid rafts. *J. Exp. Med.* **2006**, *203*, 2377–2389. [CrossRef]
7. Lawrence, T. The nuclear factor NF-κB pathway in inflammation. *Cold Spring Harb. Perspect. Biol.* **2009**, *1*, a001651. [CrossRef]
8. Ramos-Tovar, E.; Muriel, P. Free radicals, antioxidants, nuclear factor-E2-related factor-2 and liver damage. *J. Appl. Toxicol.* **2020**, *40*, 151–168. [CrossRef]
9. Pham-Huy, L.A.; He, H.; Pham-Huy, C. Free radicals, antioxidants in disease and health. *Int. J. Biomed. Sci.* **2008**, *4*, 89–96.
10. Magrone, T.; Magrone, M.; Russo, M.A.; Jirillo, E. Recent Advances on the Anti-Inflammatory and Antioxidant Properties of Red Grape Polyphenols: In Vitro and In Vivo Studies. *Antioxidants* **2020**, *9*, 35. [CrossRef]
11. Winter, A.N.; Bickford, P.C. Anthocyanins and Their Metabolites as Therapeutic Agents for Neurodegenerative Disease. *Antioxidants* **2019**, *8*, 333. [CrossRef]
12. Silva, R.F.; Pogačnik, L. Polyphenols from Food and Natural Products: Neuroprotection and Safety. *Antioxidants* **2020**, *9*, 61. [CrossRef] [PubMed]
13. Devassy, J.G.; Leng, S.; Gabbs, M.; Monirujjaman, M.; Aukema, H.M. Omega-3 polyunsaturated fatty acids and oxylipins in neuroinflammation and management of Alzheimer disease. *Adv. Nutr.* **2016**, *7*, 905–916. [CrossRef] [PubMed]
14. Testa, G.; Gamba, P.; Badilli, U.; Gargiulo, S.; Maina, M.; Guina, T.; Calfapietra, S.; Biasi, F.; Cavalli, R.; Poli, G.; et al. Loading into nanoparticles improves quercetin's efficacy in preventing neuroinflammation induced by oxysterols. *PLoS ONE* **2014**, *9*, e96795. [CrossRef] [PubMed]
15. Bhaskar, S.; Shalini, V.; Helen, A. Quercetin regulates oxidized LDL induced inflammatory changes in human PBMCs by modulating the TLR-NF-κB signaling pathway. *Immunobiology* **2011**, *216*, 367–373. [CrossRef]
16. Casedas, G.; Bennett, A.C.; Gonzalez-Burgos, E.; Gomez-Serranillos, M.P.; Lopez, V.; Smith, C. Polyphenol-associated oxidative stress and inflammation in a model of LPS-induced inflammation in glial cells: Do we know enough for responsible compounding? *Inflammopharmacology* **2019**, *27*, 189–197. [CrossRef]
17. Carey, A.N.; Fisher, D.R.; Rimando, A.M.; Gomes, S.M.; Bielinski, D.F.; Shukitt-Hale, B. Stilbenes and anthocyanins reduce stress signaling in BV-2 mouse microglia. *J. Agric. Food Chem.* **2013**, *61*, 5979–5986. [CrossRef]

18. Lau, F.C.; Bielinski, D.F.; Joseph, J.A. Inhibitory effects of blueberry extract on the production of inflammatory mediators in lipopolysaccharide-activated BV2 microglia. *J. Neurosci. Res.* **2007**, *85*, 1010–1017. [CrossRef]
19. Lau, F.C.; Joseph, J.A.; McDonald, J.E.; Kalt, W. Attenuation of iNOS and COX2 by blueberry polyphenols is mediated through the suppression of NF-κB activation. *J. Funct. Foods* **2009**, *1*, 274–283. [CrossRef]
20. Poulose, S.M.; Fisher, D.R.; Larson, J.; Bielinski, D.F.; Rimando, A.M.; Carey, A.N.; Schauss, A.G.; Shukitt-Hale, B. Anthocyanin-rich acai (*Euterpe oleracea* Mart.) fruit pulp fractions attenuate inflammatory stress signaling in mouse brain BV-2 microglial cells. *J. Agric. Food Chem.* **2012**, *60*, 1084–1093. [CrossRef]
21. Jeong, J.W.; Lee, W.S.; Shin, S.C.; Kim, G.Y.; Choi, B.T.; Choi, Y.H. Anthocyanins downregulate lipopolysaccharide-induced inflammatory responses in BV2 microglial cells by suppressing the NF-kappaB and Akt/MAPKs signaling pathways. *Int. J. Mol. Sci.* **2013**, *14*, 1502–1515. [CrossRef]
22. Zhao, L.; Chen, S.; Liu, T.; Wang, X.; Huang, H.; Liu, W. Callistephin enhances the protective effects of isoflurane on microglial injury through downregulation of inflammation and apoptosis. *Mol. Med. Rep.* **2019**, *20*, 802–812. [CrossRef] [PubMed]
23. Shukitt-Hale, B.; Kelly, M.E.; Bielinski, D.F.; Fisher, D.R. Tart Cherry Extracts Reduce Inflammatory and Oxidative Stress Signaling in Microglial Cells. *Antioxidants* **2016**, *5*, 33. [CrossRef] [PubMed]
24. Pacheco, S.M.; Azambuja, J.H.; De Carvalho, T.R.; Soares, M.S.P.; Oliveira, P.S.; Da Silveira, E.F.; Stefanello, F.M.; Braganhol, E.; Gutierres, J.M.; Spanevello, R.M. Glioprotective Effects of Lingonberry Extract Against Altered Cellular Viability, Acetylcholinesterase Activity, and Oxidative Stress in Lipopolysaccharide-Treated Astrocytes. *Cell. Mol. Neurobiol.* **2018**, *38*, 1107–1121. [CrossRef] [PubMed]
25. Ben Youssef, S.; Brisson, G.; Doucet-Beaupré, H.; Castonguay, A.M.; Gora, C.; Amri, M.; Lévesque, M. Neuroprotective benefits of grape seed and skin extract in a mouse model of Parkinson's disease. *Nutr. Neurosci.* **2019**, *25*, 1–15. [CrossRef] [PubMed]
26. Capiralla, H.; Vingtdeux, V.; Zhao, H.; Sankowski, R.; Al-Abed, Y.; Davies, P.; Marambaud, P. Resveratrol mitigates lipopolysaccharide- and Aβ-mediated microglial inflammation by inhibiting the TLR4/NF-κB/STAT signaling cascade. *J. Neurochem.* **2012**, *120*, 461–472. [CrossRef] [PubMed]
27. Seong, K.J.; Lee, H.G.; Kook, M.S.; Ko, H.M.; Jung, J.Y.; Kim, W.J. Epigallocatechin-3-gallate rescues LPS-impaired adult hippocampal neurogenesis through suppressing the TLR4-NF-κB signaling pathway in mice. *Korean J. Physiol. Pharmacol.* **2016**, *20*, 41–51. [CrossRef] [PubMed]
28. Khan, M.S.; Ali, T.; Kim, M.W.; Jo, M.H.; Chung, J.I.; Kim, M.O. Anthocyanins Improve Hippocampus-Dependent Memory Function and Prevent Neurodegeneration via JNK/Akt/GSK3beta Signaling in LPS-Treated Adult Mice. *Mol. Neurobiol.* **2019**, *56*, 671–687. [CrossRef]
29. Wang, Y.; Zheng, Y.; Lu, J.; Chen, G.; Wang, X.; Feng, J.; Ruan, J.; Sun, X.; Li, C.; Sun, Q. Purple sweet potato color suppresses lipopolysaccharide-induced acute inflammatory response in mouse brain. *Neurochem. Int.* **2010**, *56*, 424–430. [CrossRef]
30. Khan, M.S.; Ali, T.; Kim, M.W.; Jo, M.H.; Jo, M.G.; Badshah, H.; Kim, M.O. Anthocyanins protect against LPS-induced oxidative stress-mediated neuroinflammation and neurodegeneration in the adult mouse cortex. *Neurochem. Int.* **2016**, *100*, 1–10. [CrossRef]
31. Carvalho, F.B.; Gutierres, J.M.; Bueno, A.; Agostinho, P.; Zago, A.M.; Vieira, J.; Fruhauf, P.; Cechella, J.L.; Nogueira, C.W.; Oliveira, S.M.; et al. Anthocyanins control neuroinflammation and consequent memory dysfunction in mice exposed to lipopolysaccharide. *Mol. Neurobiol.* **2017**, *54*, 3350–3367. [CrossRef]
32. Li, J.; Shi, Z.; Mi, Y. Purple sweet potato color attenuates high fat-induced neuroinflammation in mouse brain by inhibiting MAPK and NF-kappaB activation. *Mol. Med. Rep.* **2018**, *17*, 4823–4831. [PubMed]
33. Pan, Z.; Cui, M.; Dai, G.; Yuan, T.; Li, Y.; Ji, T.; Pan, Y. Protective Effect of Anthocyanin on Neurovascular Unit in Cerebral Ischemia/Reperfusion Injury in Rats. *Front. Neurosci.* **2018**, *12*, 947. [CrossRef] [PubMed]
34. Moussa, C.; Hebron, M.; Huang, X.; Ahn, J.; Rissman, R.A.; Aisen, P.S.; Turner, R.S. Resveratrol regulates neuro-inflammation and induces adaptive immunity in Alzheimer's disease. *J. Neuroinflamm.* **2017**, *14*, 1. [CrossRef] [PubMed]
35. Elvira-Torales, L.I.; García-Alonso, J.; Periago-Castón, M.J. Nutritional importance of carotenoids and their effect on liver health: A review. *Antioxidants* **2019**, *8*, 229. [CrossRef]
36. Salomone, F.; Godos, J.; Zelber-Sagi, S. Natural antioxidants for non-alcoholic fatty liver disease: Molecular targets and clinical perspectives. *Liver Int.* **2016**, *36*, 5–20. [CrossRef] [PubMed]

37. Lin, J.; Tang, Y.; Kang, Q.; Feng, Y.; Chen, A. Curcumin inhibits gene expression of receptor for advanced glycation end-products (RAGE) in hepatic stellate cells in vitro by elevating PPARgamma activity and attenuating oxidative stress. *Br. J. Pharmacol.* **2012**, *166*, 2212–2227. [CrossRef]
38. Lin, C.L.; Huang, H.C.; Lin, J.K. Theaflavins attenuate hepatic lipid accumulation through activating AMPK in human HepG2 cells. *J. Lipid Res.* **2007**, *48*, 2334–2343. [CrossRef]
39. Zhu, W.; Jia, Q.; Wang, Y.; Zhang, Y.; Xia, M. The anthocyanin cyanidin-3-O-beta-glucoside, a flavonoid, increases hepatic glutathione synthesis and protects hepatocytes against reactive oxygen species during hyperglycemia: Involvement of a cAMP-PKA-dependent signaling pathway. *Free. Radic. Biol. Med.* **2012**, *52*, 314–327. [CrossRef]
40. Gao, M.; Ma, Y.; Liu, D. Rutin suppresses palmitic acids- triggered inflammation in macrophages and blocks high fat diet-induced obesity and fatty liver in mice. *Pharm. Res.* **2013**, *30*, 2940–2950. [CrossRef]
41. Kobori, M.; Masumoto, S.; Akimoto, Y.; Oike, H. Chronic dietary intake of quercetin alleviates hepatic fat accumulation associated with consumption of a Western-style diet in C57/BL6J mice. *Mol. Nutr. Food Res.* **2011**, *55*, 530–540. [CrossRef]
42. Marcolin, E.; San-Miguel, B.; Vallejo, D.; Tieppo, J.; Marroni, N.; González-Gallego, J.; Tuñón, M.J. Quercetin treatment ameliorates inflammation and fibrosis in mice with nonalcoholic steatohepatitis. *J. Nutr.* **2012**, *142*, 1821–1828. [CrossRef] [PubMed]
43. Zhang, Z.-F.; Fan, S.-H.; Zheng, Y.-L.; Lu, J.; Wu, D.-M.; Shan, Q.; Hu, B. Troxerutin improves hepatic lipid homeostasis by restoring NAD(+)-depletion-mediated dysfunction of lipin 1 signaling in high-fat diet-treated mice. *Biochem. Pharmacol.* **2014**, *91*, 74–86. [CrossRef] [PubMed]
44. Luo, X.Y.; Takahara, T.; Hou, J.; Kawai, K.; Sugiyama, T.; Tsukada, K.; Takemoto, M.; Takeuchi, M.; Zhong, L.; Li, X.-K. Theaflavin attenuates ischemia-reperfusion injury in a mouse fatty liver model. *Biochem. Biophys. Res. Commun.* **2012**, *417*, 287–293. [CrossRef] [PubMed]
45. Pu, P.; Wang, X.A.; Salim, M.; Zhu, L.-H.; Wang, L.; Chen, K.-J.; Xiao, J.-F.; Deng, W.; Shi, H.-W.; Jiang, H.; et al. Baicalein, a natural pro- duct, selectively activating AMPKalpha(2) and ameliorates metabolic disorder in diet-induced mice. *Mol. Cell Endocrinol.* **2012**, *362*, 128–138. [CrossRef]
46. Jeon, B.T.; Jeong, E.A.; Shin, H.J.; Lee, Y.; Lee, N.H.; Kim, H.J.; Kang, S.S.; Cho, G.J.; Choi, W.S.; Roh, G.S. Resveratrol attenuates obesity-associated peripheral and central inflammation and improves memory deficit in mice fed a high-fat diet. *Diabetes* **2012**, *61*, 1444–1454. [CrossRef]
47. Leclercq, I.A.; Farrell, G.C.; Sempoux, C.; Dela Pena, A.; Hors-mans, Y. Curcumin inhibits NF-kappaB activation and reduces the severity of experimental steatohepatitis in mice. *J. Hepatol.* **2004**, *41*, 926–934. [CrossRef]
48. Vizzutti, F.; Provenzano, A.; Galastri, S.; Milani, S.; Delogu, W.; Novo, E.; Caligiuri, A.; Zamara, E.; Arena, U.; Laffi, G.; et al. Curcumin limits the fibrogenic evolution of experimental steatohepatitis. *Lab. Investig.* **2010**, *90*, 104–115. [CrossRef]
49. Salamone, F.; Galvano, F.; Cappello, F.; Mangiameli, A.; Barbagallo, I.; Volti, G.L. Silibinin modulates lipid homeostasis and inhibits nuclear factor kappa B activation in experimental nonalcoholic steatohepatitis. *Transl. Res.* **2012**, *159*, 477–486. [CrossRef]
50. Salamone, F.; Galvano, F.; Marino Gammazza, A.; Paternostro, C.; Tibullo, D.; Bucchieri, F.; Mangiameli, A.; Parola, M.; Bugianesi, E.; Volti, G.L. Silibinin improves hepatic and myocardial injury in mice with nonalcoholic steatohepatitis. *Dig. Liver Dis.* **2012**, *44*, 334–342. [CrossRef]
51. Panchal, S.K.; Poudyal, H.; Brown, L. Quercetin ameliorates cardiovascular, hepatic, and metabolic changes in diet- induced metabolic syndrome in rats. *J. Nutr.* **2012**, *142*, 1026–1032. [CrossRef] [PubMed]
52. Panchal, S.K.; Poudyal, H.; Arumugam, T.V.; Brown, L. Rutin attenuates metabolic changes, nonalcoholic steatohepatitis, and cardiovascular remodeling in high-carbohydrate, high-fat diet-fed rats. *J. Nutr.* **2011**, *141*, 1062–1069. [CrossRef] [PubMed]
53. Kuzu, N.; Bahcecioglu, I.H.; Dagli, A.F.; Ozercan, I.H.; Ustundag, B.; Sahin, K. Epigallocatechin gallate attenuates experimental non-alcoholic steatohepatitis induced by high fat diet. *J. Gastroenterol. Hepatol.* **2008**, *23*, e465–e470. [CrossRef]
54. Xiao, J.; Ho, C.T.; Liong, E.C.; Nanji, A.A.; Leung, T.M.; Lau, T.Y.H.; Fung, M.L.; Tipoe, G.L. Epigallocatechin gallate attenuates fibrosis, oxidative stress, and inflammation in non-alcoholic fatty liver disease rat model through TGF/ SMAD, PI3 K/Akt/FoxO1, and NF-kappa B pathways. *Eur. J. Nutr.* **2014**, *53*, 187–199. [CrossRef] [PubMed]

55. Yalniz, M.; Bahcecioglu, I.H.; Kuzu, N.; Poyrazoğlu, O.K.; Bulmus, Ö.; Celebi, S.; Ustundag, B.; Ozercan, I.H.; Sahin, K. Preventive role of genistein in an experimental non-alcoholic steatohepatitis model. *J. Gastroenterol. Hepatol.* **2007**, *22*, 2009–2014. [CrossRef] [PubMed]
56. Chtourou, Y.; Fetoui, H.; Jemai, R.; Ben Slima, A.; Makni, M.; Gdoura, R. Naringenin reduces cholesterol-induced hepatic inflammation in rats by modulating matrix metalloproteinases-2, 9 via inhibition of nuclear factor kappaB pathway. *Eur. J. Pharmacol.* **2015**, *746*, 96–105. [CrossRef]
57. Vitaglione, P.; Morisco, F.; Mazzone, G.; Amoruso, D.C.; Ribecco, M.T.S.; Romano, A.; Fogliano, V.; Caporaso, N.; D'Argenio, G. Coffee reduces liver damage in a rat model of steatohepatitis: The underlying mechanisms and the role of polyphenols and melanoidins. *Hepatology* **2010**, *52*, 1652–1661. [CrossRef] [PubMed]
58. Salomone, F.; Volti, G.L.; Vitaglione, P.; Morisco, F.; Fogliano, V.; Zappalà, A.; Palmigiano, A.; Garozzo, D.; Caporaso, N.; D'Argenio, G.; et al. Coffee enhances the expression of chaperones and antioxidant proteins in rats with nonalcoholic fatty liver disease. *Transl. Res.* **2014**, *163*, 593–602. [CrossRef] [PubMed]
59. Bagul, P.K.; Middela, H.; Matapally, S.; Padiya, R.; Bastia, T.; Madhusudana, K.; Reddy, B.R.; Chakravarty, S.; Banerjee, S.K. Attenuation of insulin resistance, metabolic syndrome and hepatic oxidative stress by resveratrol in fructose-fed rats. *Pharmacol. Res.* **2012**, *66*, 260–268. [CrossRef]
60. Sakata, R.; Nakamura, T.; Torimura, T.; Ueno, T.; Sata, M. Green tea with high-density catechins improves liver function and fat infiltration in non-alcoholic fatty liver disease (NAFLD) patients: A double-blind placebo-controlled study. *Int. J. Mol. Med.* **2013**, *32*, 989–994. [CrossRef]
61. Chen, S.; Zhao, X.; Ran, L.; Wan, J.; Wang, X.; Qin, Y.; Shu, F.; Gao, Y.; Yuan, L.; Zhang, Q.; et al. Resveratrol improves insulin resistance, glucose and lipid metabolism in patients with non-alcoholic fatty liver disease: A randomized controlled trial. *Dig. Liver Dis.* **2015**, *47*, 226–232. [CrossRef]
62. Faghihzadeh, F.; Adibi, P.; Rafiei, R.; Hekmatdoost, A. Resveratrol supplementation improves inflammatory biomarkers in patients with nonalcoholic fatty liver disease. *Nutr. Res.* **2014**, *34*, 837–843. [CrossRef]
63. Stiuso, P.; Scognamiglio, I.; Murolo, M.; Ferranti, P.; De Simone, C.; Rizzo, M.R.; Tuccillo, C.; Caraglia, M.; Loguercio, C.; Federico, A. Serum oxidative stress markers and lipidomic profile to detect NASH patients responsive to an antioxidant treatment: A pilot study. *Oxid. Med. Cell Longev.* **2014**, *2014*, 169216. [CrossRef] [PubMed]
64. Chachay, V.S.; Macdonald, G.A.; Martin, J.H.; Whitehead, J.; O'Moore–Sullivan, T.M.; Lee, P.; Franklin, M.; Klein, K.; Taylor, P.J.; Ferguson, M.; et al. Resveratrol does not benefit patients with nonalcoholic fatty liver disease. *Clin. Gastroenterol. Hepatol.* **2014**, *12*, 2092–2103. [CrossRef] [PubMed]
65. Lavefve, L.; Howard, L.R.; Carbonero, F. Berry polyphenols metabolism and impact on human gut microbiota and health. *Food Funct.* **2020**, *11*, 45–65. [CrossRef]
66. Nunes, S.; Danesi, F.; Del Rio, D.; Silva, P. Resveratrol and inflammatory bowel disease: The evidence so far. *Nutr. Res. Rev.* **2018**, *31*, 85–97. [CrossRef] [PubMed]
67. Triebel, S.; Trieu, H.L.; Richling, E. Modulation of inflammatory gene expression by a bilberry (*Vaccinium myrtillus* L.) extract and single anthocyanins considering their limited stability under cell culture conditions. *J. Agric. Food Chem.* **2012**, *60*, 8902–8910. [CrossRef]
68. Roth, S.; Spalinger, M.R.; Gottier, C.; Biedermann, L.; Zeitz, J.; Lang, S.; Weber, A.; Rogler, G.; Scharl, M. Bilberry-derived anthocyanins modulate cytokine expression in the intestine of patients with ulcerative colitis. *PLoS ONE* **2016**, *11*, e0154817. [CrossRef]
69. Cianciulli, A.; Calvello, R.; Cavallo, P.; Dragone, T.; Carofiglio, V.; Panaro, M.A. Modulation of NF-κB activation by resveratrol in LPS treated human intestinal cells results in downregulation of PGE2 production and COX-2 expression. *Toxicol. In Vitro* **2012**, *26*, 1122–1128. [CrossRef]
70. Panaro, M.A.; Carofiglio, V.; Acquafredda, A.; Cavallo, P.; Cianciulli, A. Anti- inflammatory effects of resveratrol occur via inhibition of lipopolysaccharide-induced NF-κB activation in Caco-2 and SW480 human colon cancer cells. *Br. J. Nutr.* **2012**, *108*, 1623–1632. [CrossRef]
71. Romier, B.; Van De Walle, J.; During, A.; Larondelle, Y.; Schneider, Y.-J. Modulation of signalling nuclear factor-κB activation pathway by polyphenols in human intestinal Caco-2 cells. *Br. J. Nutr.* **2008**, *100*, 542–551. [CrossRef]
72. Serra, D.; Rufino, A.T.; Mendes, A.F.; Almeida, L.M.; Dinis, T.C.P. Resveratrol modulates cytokine-induced Jak/STAT activation more efficiently than 5-aminosalicylic acid: An in vitro approach. *PLoS ONE* **2014**, *9*, e109048. [CrossRef] [PubMed]

73. Serra, D.; Almeida, L.M.; Dinis, T.C. Anti-inflammatory protection afforded by cyanidin-3-glucoside and resveratrol in human intestinal cells via Nrf2 and PPAR-γ: Comparison with 5-aminosalicylic acid. *Chem. Biol. Interact.* **2016**, *260*, 102–109. [CrossRef] [PubMed]
74. Wu, L.H.; Xu, Z.L.; Dong, D.; He, S.A.; Yu, H. Protective effect of anthocyanins extract from blueberry on TNBS-induced IBD model of mice. *Evid. Based Complementary Altern. Med.* **2011**, *2011*, 525462. [CrossRef] [PubMed]
75. Pervin, M.; Hasnat, M.A.; Lim, J.H.; Lee, Y.M.; Kim, E.O.; Um, B.H.; Lim, B.O. Preventive and therapeutic effects of blueberry (*Vaccinium corymbosum*) extract against DSS-induced ulcerative colitis by regulation of antioxidant and inflammatory mediators. *J. Nutr. Biochem.* **2016**, *28*, 103–113. [CrossRef]
76. Montrose, D.C.; Horelik, N.A.; Madigan, J.P.; Stoner, G.D.; Wang, L.S.; Bruno, R.S.; Park, H.J.; Giardina, C.; Rosenberg, D.W. Anti-inflammatory effects of freeze-dried black raspberry powder in ulcerative colitis. *Carcinogenesis* **2011**, *32*, 343–350. [CrossRef]
77. Piberger, H.; Oehme, A.; Hofmann, C.; Dreiseitel, A.; Sand, P.G.; Obermeier, F.; Schoelmerich, J.; Schreier, P.; Krammer, G.; Rogler, G. Bilberries and their anthocyanins ameliorate experimental colitis. *Mol. Nutr. Food Res.* **2011**, *55*, 1724–1729. [CrossRef]
78. Wang, L.S.; Kuo, C.T.; Stoner, K.; Yearsley, M.; Oshima, K.; Yu, J.; Huang, T.H.; Rosenberg, D.; Peiffer, D.; Stoner, G.; et al. Dietary black raspberries modulate DNA methylation in dextran sodium sulfate (DSS)-induced ulcerative colitis. *Carcinogenesis* **2013**, *34*, 2842–2850. [CrossRef]
79. Xiao, X.; Kim, J.; Sun, Q.; Kim, D.; Park, C.S.; Lu, T.S.; Park, Y. Preventive effects of cranberry products on experimental colitis induced by dextran sulphate sodium in mice. *Food Chem.* **2015**, *167*, 438–446. [CrossRef]
80. Anhê, F.F.; Roy, D.; Pilon, G.; Dudonné, S.; Matamoros, S.; Varin, T.V.; Garofalo, C.; Moine, Q.; Desjardins, Y.; Levy, E.; et al. A polyphenol-rich cranberry extract protects from diet-induced obesity, insulin resistance and intestinal inflammation in association with increased Akkermansia spp. population in the gut microbiota of mice. *Gut* **2015**, *64*, 872–883. [CrossRef]
81. Cui, X.; Jin, Y.; Hofseth, A.B.; Pena, E.; Habiger, J.; Chumanevich, A.; Poudyal, D.; Nagarkatti, M.; Nagarkatti, P.S.; Singh, U.P.; et al. Resveratrol suppresses colitis and colon cancer associated with colitis. *Cancer Prev. Res.* **2010**, *3*, 549–559. [CrossRef]
82. Sánchez-Fidalgo, S.; Cárdeno, A.; Villegas, I.; Talero, E.; Alarcón, C. Dietary supplementation of resveratrol attenuates chronic colonic inflammation in mice. *Eur. J. Pharmacol.* **2010**, *633*, 78–84. [CrossRef] [PubMed]
83. Singh, U.P.; Singh, N.P.; Singh, B.; Hofseth, L.J.; Price, R.L.; Nagarkatti, M.; Nagarkatti, P.S. Resveratrol (trans-3,5,4′-trihydroxystilbene) induces silent mating type information regulation-1 and down-regulates nuclear transcription factor-κB activation to abrogate dextran sulfate sodium-induced colitis. *J. Pharmacol. Exp. Ther.* **2010**, *332*, 829–839. [CrossRef] [PubMed]
84. Wagnerova, A.; Babickova, J.; Liptak, R.; Vlkova, B.; Celec, P.; Gardlik, R. Sex differences in the effect of resveratrol on DSS-induced colitis in mice. *Gastroenterol. Res. Pract.* **2017**, *2017*, 8051870. [CrossRef]
85. Yao, J.; Wang, J.Y.; Liu, L.; Li, Y.-X.; Xun, A.-Y.; Zeng, W.-S.; Jia, C.-H.; Wei, X.-X.; Feng, J.-L.; Zhao, L.; et al. Antioxidant effects of resveratrol on mice with DSS-induced ulcerative colitis. *Arch. Med. Res.* **2010**, *41*, 288–294. [CrossRef] [PubMed]
86. Larrosa, M.; Tomé-Carneiro, J.; Yáñez-Gascón, M.J.; Alcántara, D.; Selma, M.V.; Beltrán, D.; Garcia-Conesa, M.T.; Urbán, C.; Lucas, R.; Tomás-Barberán, F.A.; et al. Preventive oral treatment with resveratrol pro-prodrugs drastically reduce colon inflammation in rodents. *J. Med. Chem.* **2010**, *53*, 7365–7376. [CrossRef] [PubMed]
87. Youn, J.; Lee, J.-S.; Na, H.-K.; Kundu, J.K.; Dong, Z. Resveratrol and piceatannol inhibit iNOS expression and NF-κB activation in dextran sulfate sodium-induced mouse colitis. *Nutr. Cancer* **2009**, *61*, 847–854. [CrossRef] [PubMed]
88. Altamemi, I.; Murphy, E.A.; Catroppo, J.F.; Zumbrun, E.E.; Zhang, J.; McClellan, J.L.; Singh, U.P.; Nagarkatti, P.S.; Nagarkatti, M. Role of microRNAs in resveratrol-mediated mitigation of colitis-associated tumorigenesis in ApcMin/+ mice. *J. Pharmacol. Exp. Ther.* **2014**, *350*, 99–109. [CrossRef] [PubMed]
89. Singh, U.P.; Singh, N.P.; Singh, B.; Hofseth, L.J.; Taub, D.D.; Price, R.L.; Nagarkatti, P.S.; Nagarkatti, M. Role of resveratrol-induced $CD_{11}b^+$ $Gr\text{-}1^+$ myeloid derived sup-pressor cells (MDSCs) in the reduction of $CXCR3^+$ T cells and amelioration of chronic colitis in IL-10(−/−) mice. *Brain Behav. Immun.* **2012**, *26*, 72–82. [CrossRef]
90. Osman, N.; Adawi, D.; Ahrné, S.; Jeppsson, B.; Molin, G. Probiotics and blueberry attenuate the severity of dextran sulfate sodium (DSS)-induced colitis. *Digest. Dis. Sci.* **2008**, *53*, 2464–2473. [CrossRef]

91. Abdallah, D.M.; Ismael, N.R. Resveratrol abrogates adhesion molecules and protects against TNBS-induced ulcerative colitis in rats. *Can. J. Physiol. Pharmacol.* **2011**, *89*, 811–818.
92. Lozano-Pérez, A.A.; Rodriguez-Nogales, A.; Ortiz-Cullera, V.; Algieri, F.; Zorrilla, P.; Rodríguez-Cabezas, M.E.; Mesa, N.G.; Utrilla, M.P.; De Matteis, L.; Mesa, J.G.; et al. Silk fibroin nanoparticles constitute a vector for controlled release of resveratrol in an experimental model of inflammatory bowel disease in rats. *Int. J. Nanomed.* **2014**, *9*, 4507–4520.
93. Martín, A.R.; Villegas, I.; La Casa, C.; De La Lastra, A. Resveratrol, a polyphenol found in grapes, suppresses oxidative damage and stimulates apoptosis during early colonic inflammation in rats. *Biochem. Pharmacol.* **2004**, *67*, 1399–1410. [PubMed]
94. Martín, A.R.; Villegas, I.; Sánchez-Hidalgo, M.; Alarcón, C. The effects of resveratrol, a phytoalexin derived from red wines, on chronic inflammation induced in an experimentally induced colitis model. *Br. J. Pharmacol.* **2006**, *147*, 873–885. [CrossRef] [PubMed]
95. Yildiz, G.; Yildiz, Y.; Ulutas, P.A.; Yaylali, A.; Ural, M. Resveratrol pretreatment ameliorates TNBS colitis in rats. *Recent Pat. Endocr. Metab. Immune Drug Discov.* **2015**, *9*, 134–140. [CrossRef]
96. Arslan, A.; Ozcicek, F.; Cimen, F.K.; Altuner, D.; Yarali, O.; Kurt, N.; Tumkaya, L.; Ozturk, C.; Suleyman, H. Protective effect of resveratrol against methotrexate- induced oxidative stress in the small intestinal tissues of rats. *Int. J. Clin. Exp. Med.* **2015**, *8*, 10491–10500. [PubMed]
97. Abdin, A.A. Targeting sphingosine kinase 1 (SphK1) and apoptosis by colon-specific delivery formula of resveratrol in treatment of experimental ulcerative colitis in rats. *Eur. J. Pharmacol.* **2013**, *718*, 145–153. [CrossRef] [PubMed]
98. Larrosa, M.; Yañéz-Gascón, M.J.; Selma, M.V.; González-Sarrías, A.; Toti, S.; Cerón, J.; Tomás-Barberán, F.A.; Dolara, P.; Espín, J.C. Effect of a low dose of dietary resveratrol on colon microbiota, inflammation and tissue damage in a DSS-induced colitis rat model. *J. Agric. Food Chem.* **2009**, *57*, 2211–2220. [CrossRef]
99. Rahal, K.; Schmiedlin-Ren, P.; Adler, J.; Dhanani, M.; Sultani, V.; Rittershaus, A.C.; Reingold, L.; Zhu, J.; McKenna, B.J.; Christman, G.M.; et al. Resveratrol has antiinflammatory and antifibrotic effects in the peptidoglycan-polysaccharide rat model of Crohn's disease. *Inflamm. Bowel Dis.* **2012**, *18*, 613–623. [CrossRef]
100. Biedermann, L.; Mwinyi, J.; Scharl, M.; Frei, P.; Zeitz, J.; Kullak-Ublick, G.A.; Vavricka, S.R.; Fried, M.; Weber, A.; Humpf, H.U.; et al. Bilberry ingestion improves disease activity in mild to moderate ulcerative colitis—An open pilot study. *J. Crohns Colitis* **2013**, *7*, 271–279. [CrossRef]
101. Samsami-Kor, M.; Daryani, N.E.; Asl, P.R.; Hekmatdoost, A. Anti-inflammatory effects of resveratrol in patients with ulcerative colitis: A randomized, double-blind, placebo-controlled pilot study. *Arch. Med. Res.* **2015**, *46*, 280–285. [CrossRef]
102. Samsami-Kor, M.; Daryani, N.E.; Asl, P.R.; Hekmatdoost, A. Resveratrol supplementation and oxidative/anti-oxidative status in patients with ulcerative colitis: A randomized, double-blind, placebo-controlled pilot study. *Arch. Med. Res.* **2016**, *47*, 304–309. [CrossRef] [PubMed]

antioxidants

MDPI

Article

Dietary Lipids Influence Bioaccessibility of Polyphenols from Black Carrots and Affect Microbial Diversity under Simulated Gastrointestinal Digestion

Chunhe Gu [1], Hafiz A. R. Suleria [1], Frank R. Dunshea [1,2] and Kate Howell [1,*]

1 School of Agriculture and Food, Faculty of Veterinary and Agricultural Sciences, The University of Melbourne, Parkville 3010, VIC, Australia; chunheg@student.unimelb.edu.au (C.G.); hafiz.suleria@unimelb.edu.au (H.A.R.S.); fdunshea@unimelb.edu.au (F.R.D.)

2 Faculty of Biological Sciences, The University of Leeds, Leeds LS2 9JT, UK

* Correspondence: khowell@unimelb.edu.au; Tel.: +61-470-439-67

Received: 22 July 2020; Accepted: 14 August 2020; Published: 17 August 2020

Abstract: The bioaccessibility and activity of polyphenols is dependent on their structure and entrapment in the food matrix. While dietary lipids are known to transit into the colon, the impact of different lipids on the microbiome, and their interactions with dietary polyphenols are largely unknown. Here, we investigated the effect of dietary lipids on the bioaccessibility of polyphenols from purple/black carrots and adaptation of the gut microbiome in a simulated in vitro digestion-fermentation. Coconut oil, sunflower oil, and beef tallow were selected to represent common dietary sources of medium-chain fatty acids (MCFAs), long-chain polyunsaturated fatty acids (PUFAs), and long-chain polysaturated fatty acids (SFAs), respectively. All lipids promoted the bioaccessibility of both anthocyanins and phenolic acids during intestinal digestion with coconut oil exhibiting the greatest protection of anthocyanins. Similar trends were shown in antioxidant assays (2,2-Diphenyl-1-pricrylhydrazyl (DPPH), ferric reducing ability (FRAP), and total phenolic content (TPC)) with higher phytochemical bioactivities observed with the addition of dietary lipids. Most bioactive polyphenols were decomposed during colonic fermentation. Black carrot modulated diversity and composition of a simulated gut microbiome. Dramatic shifts in gut microbiome were caused by coconut oil. Inclusion of sunflower oil improved the production of butyrate, potentially due to the presence of PUFAs. The results show that the impact of polyphenols in the digestive tract should be considered in the context of other components of the diet, particularly lipids.

Keywords: anthocyanin; phenolic acid; antioxidant capacity; medium chain fatty acids; short chain fatty acids; gut microbiome

1. Introduction

Polyphenols are naturally occurring compounds found in fruits and vegetables, and have been widely studied as beneficial compounds for health based on their antioxidant activity. In the human body, absorbed polyphenols prevent the generation of free radicals and reactive oxygen species (ROS) and, therefore, increase the plasma antioxidant defences [1,2]. Cardiovascular diseases are related with excessive absorption of triglycerides and triglyceride-rich lipoproteins [3]. Absorbed polyphenols can prevent accumulation of oxidised plasma triglycerides and lipoproteins [4]. Therefore, bioavailable polyphenols can potentially mitigate the adverse effects of high levels of plasma triglycerides, and are considered beneficial health compounds which can lower the risks of cardiovascular disease in humans. After consumption, polyphenols are released from the food matrix in the stomach and small intestine and are absorbed by intestinal enterocytes [5]. However, most polyphenols bound to the food matrix escape small intestinal digestion to be transported to the colon for fermentation and further metabolism by the gut microbiota [6]. Therefore, bioaccessibility of polyphenols released at different digestive stages as well as the interactions of polyphenols with macronutrients of the food matrix and gut microbiota are crucial factors for understanding polyphenol bioavailability.

High plasma triglycerides concentrations may be a result of excessive dietary lipid consumption [7]. Although lipids are absorbed in the small intestine, some dietary fats escape absorption and reach the colon, leading to structure shifts in the gut microbiome [8]. Fats in the food matrix increase the bioaccessibility of polyphenols during digestion. Ortega, et al. [9] reported an increased recovery of phenolic compounds from cocoa during an in vitro duodenal digestion and additionally that poly-unsaturated fatty acids (PUFAs) released from the food matrix during digestion enhanced the stability of phenols from a carob flour [9]. Shishikura, et al. [10] verified that tea polyphenols could interact with olive oil and shifted the extent of oil emulsification during digestion. However, the effects of the type of lipids on bioaccessibility of polyphenols during digestion as well as the reciprocal interactions between a polyphenol-lipid matrix and gut microbiota remain largely unclear.

In this study, a standardised static in vitro digestive system was used. Black carrots (*Daucus carota ssp. sativus var. atrorubens Alef.*), which are rich in anthocyanins and phenolic acids [11], were used as a source of polyphenols. Coconut oil, sunflower oil, and beef tallow were used to represent different dietary sources of medium-chain fatty acids (MCFAs), longer chain PUFAs, and long-chain polysaturated fatty acids (SFAs), respectively (Table 1). Antioxidant activity and bioaccessibility of individual black carrot polyphenols with the presence of fats in gastric, small intestinal and colonic digestive stages were measured. The effects of different polyphenol-fat combinations on a fermentation of a gut microbiome was investigated. The present study unravels the role of lipid in the polyphenol digestion-fermentation process and the corresponding shift in microbial diversity and metabolites production in a gut fermentation.

Table 1. Fatty acid composition of dietary lipid preparations [12].

Fatty Acid (%)	Coconut Oil	Sunflower Oil	Beef Tallow
C8	7.0	-	-
C10	8.0	-	-
C12	48.0	-	-
C14	16.0	-	3.0
C16	9.5	5.0	26.0
C16:1	-	-	3.0
C18	-	6.0	14.0
C18:1	6.5	30.0	47.0
C18:2	-	59.0	3.0
C18:3	-	-	1.0

2. Materials and Methods

2.1. Chemicals and Reagents

All chemicals were analytical grade or analytical standards and were purchased from Sigma-Aldrich (Castle Hill, NSW, Australia), unless stated. Potassium chloride, potassium dihydrogen phosphate, sodium hydrogen carbonate, sodium chloride, magnesium dichloride, ammonium carbonate, sodium hydroxide, hydrogen chloride and calcium chloride were used for the digestive electrolyte stock solution preparation. Pepsin from porcine gastric mucosa, pancreatin from porcine pancreas, and porcine bile extract were the enzymes used for simulating gastric and small intestinal digestions. Pancreatin from porcine pancreas was purchased from US Biological (Assay Matrix Pty Ltd., Australia). Starch, peptone, tryptone, yeast extract, pectin, mucin, casein, L-Cysteine hydrochloride anhydrous, magnesium sulphate, guar, dipotassium phosphate, bile salts, and Tween-80 were used for the colonic basal media preparation. Analytical grade methanol, ethanol and formic acid were used for polyphenols extraction. Folin-Ciocalteu reagent, sodium carbonate, gallic acid,

2,2-Diphenyl-1-pricrylhydrazyl (DPPH), ascorbic acid, sodium acetate, 2,4,6-tripytidyl-s-triazine (TPTZ), and ferric chloride were used for the antioxidant assays. The 96-well plates (flat bottoms with 300 μL total volume) (Corning Inc., Corning, NY, USA) were purchased from Thermo Fisher Scientific (Scoresby, Australia). HPLC analytical grade chlorogenic acid, caffeic acid, *p*-coumaric acid, ferulic acid, cyanidin-3-*O*-glucoside, acetic acid and acetonitrile were for chromatographic analysis. The 1.5 mL HPLC vials were purchased from Agilent Technologies (Mulgrave, Australia).

2.2. Sample Preparation

Black carrots (*Daucus carota sativus var. atrorubens*) and commercial oils, including sunflower oil, coconut oil, and beef tallow were purchased from a local market in Melbourne, Victoria. Black carrots were grown in Melbourne, Victoria. About 300 g carrots were peeled and weighed, followed by being chopped into 1 cm^3 cubes, which were then put into a mortar and instantly frozen in liquid nitrogen. The frozen carrot cubes were then ground into fine powder using a coffee grinder (Sunbeam® Multi Grinder—EM0405) and lyophilised to constant weight. The powder was then weighed and a dry/wet weight ratio of 0.1 was gathered. Afterwards, aliquots of 0.2 g carrot powder were weighed in a 50 mL centrifuge tube. When the lipid preparations were included, tallow and coconut oil were melted by heating at 40 °C, and then 1 g fat of each type was weighed and mixed with the carrot powder to make combinations of black carrot with coconut oil (BC), sunflower oil (BS), and tallow (BT) samples. The setting of carrot/fat ratio was according to the dietary guidelines recommended by the National Health and Medical Research Council of Australian Government (2015), where the ratio of vegetables to other meal components containing fat was about 0.2. The percentage of the fat content of the basal diet was estimated as 10% by weight basis. Therefore, the ratio of dry carrot powder/fat was approximate 0.2. Controls with black carrot only (0.2 g) (B) and fat only (1 g) of coconut oil (C), sunflower oil (S) and tallow (T) were also prepared. The samples were then homogenised at 5000 rpm for 2 min using an IKA Ultra-Turrax® T25 homogenizer (Rawang, Selangor, Malaysia) ready for the in vitro digestive treatments. For extraction of polyphenols from carrots, the frozen carrot powder was extracted using a solvent of methanol/formic acid/water (80:1:19, *v/v/v*) at a ratio of 50 mg/mL. The extraction was taken in a shaking incubator (ZWYR-240, Labwit, Ashwood, Australia) at 120 rpm, 4 °C for 10 h in darkness followed by centrifugation at 10,000× *g* for 10 min. The supernatant was collected for instrumental analysis.

2.3. In Vitro Gastrointestinal Digestion

Samples were subjected to simulated in vitro gastro-intestinal digestive systems described by Pérez-Burillo et al. [13]. Briefly, the gastro-intestinal digestion was composed of gastric phase (2 h at 37 °C with pepsin 2000 U/mL, pH 3.0) and small intestinal phase (2 h at 37 °C with pancreatin 13.37 mg/mL, pH 7.0). All procedures were carried out in a shaking incubator at 100 rpm in darkness. The reactions were terminated by instantly freezing the samples in liquid nitrogen which were then stored in −80 °C for further analysis.

2.4. In Vitro Colonic Fermentation

The in vitro colonic fermentation procedure was carried out according to the protocol of Fu, et al. [14] with modifications. Pig faeces were used as sources of gut microbiome as a substitute of human faeces, as pigs and humans are primarily colonic fermenters sharing a comparable gut microbiome [15]. Five female large White × landrace grower pigs with liveweight of about 50 kg and raised in animal house of University of Melbourne. The selected pigs were fed with standard grower diet for two weeks. The faeces were taken immediately after the pigs defecated, and were put into an anaerobic chamber. The faeces were mixed together, then 20 g of the pooled faeces was weighed, added to 80 g sterilised pre-nitrogen flushed 0.1 M phosphate buffer (pH 7.0) and homogenised for 5 min in a stomacher mixer (MiniMix® Lab Blender, Thomas Scientific, Swedenborg, NJ, USA) to make 20% faecal slurry (*w:w*). The homogenised mixture was filtered through sterile muslin cloth to remove particulate matter. Sediments from the small intestinal digestion were prepared after centrifugation at 10,000× *g* for 10 min. To test the effects of the partial undigested lipids on gut microbiome, 10% of supernatant after small intestine treatments was added to the sediments, as physiologically, 10% of the soluble fraction during digestion enters the large intestine [13]. Afterwards, aliquots of 5 mL faecal slurry were added into the tubes with the sediments followed by adding 5 mL of the basal media. All tubes were flushed with nitrogen and then incubated with shaking at 100 rpm for 20 h in darkness. The tubes were centrifuged at 10,000× *g*, 5 °C for 10 min and the supernatant was taken from the sediment. Supernatants were stored in −80 °C for HPLC and phytochemical bioactivity analyses, and faecal sediments were ready for DNA extraction.

2.5. Phytochemical and Antioxidant Assays

The antioxidant potential of the supernatant from each digestive compartment was measured according to a standardised high throughput 96-well plate method [16]. The supernatants were centrifuged at 10,000× *g* for 20 min and 50 µL from the aqueous layer was taken and diluted 10-fold with methanol. Total phenolic content (TPC), 2,2-Diphenyl-1-pricrylhydrazyl (DPPH), and ferric reducing ability (FRAP) were measured to describe the antioxidant potential of the bioaccessible portions of the digesta in stomach, small intestine, and colon. For the fat-oil combined samples, the results were calculated with eliminations of the fat controls.

2.6. Separation and Analysis of Polyphenols

Aliquots of 4 mL of the supernatant from the aqueous layer of each digestive compartment after centrifugation (at 10,000× *g* for 20 min) were taken and freeze dried for 24 h, and the solids were added with 1 mL extraction solution (methanol/formic acid/water, 80:1:19, *v/v/v*). Shaking incubated 120 rpm, 4 °C for 10 h in darkness. The extracts were then centrifuged at 10,000× *g* for 5 min. Afterwards, 0.5 mL hexane was added to remove the residual micellised lipids to guard against column clogging. The mixture was left to stand for 2 min until the solvent was separated into two layers. Here, 0.5 mL was taken from the methanol-phenol layer and stored before ready LC separation and analysis.

Separation and quantification of individual polyphenols were performed on Agilent 1200 series HPLC (Agilent Technologies, Santa Clara, CA, USA) equipped with a diode array detector (DAD). A Synergi Hydro-RP (250 × 4.6 mm i.d.) reverse phase column with a particle size of 4 µm (Phenomenex, Lane Cove, NSW, Australia) was protected by a Phenomenex 4.0 × 2.0 mm i.d. C18 ODS guard column. Instrument control, data acquisition and processing were performed using MassHunter workstation software (Qualitative Analysis, version B.03.01, Agilent).

Anthocyanin quantification followed the method of Padayachee, et al. [17] with some modifications using a wavelength of 520 nm. The mobile phase consisted of water/formic acid/acetonitrile (87:10:3, *v/v/v*; eluent A) and acetonitrile/formic acid/water (50:10:40, *v/v/v*; eluent B). The gradient profile was isocratic 10% B (0–5 min), 10-16% B (5–8 min), 16–23% B (8–19 min), 23–100% B (19–22 min), isocratic 100% B (22–27 min), 100–10% B (27–30 min). The flow rate was 0.5 mL/min, and 20 µL was injected for each sample.

Phenolic acid quantification followed the protocol of Gu, et al. [16] using a wavelength of 320 nm. The mobile phase consisted of water/acetic acid (98:2, v/v; eluent A) and acetonitrile/acetic acid/water (100:1:99, *v/v/v*; eluent B). The gradient profile was 0% B (0–5 min), 10–25% B (5–25 min), 25–35% B (25–35 min), 35–40% B (35–45 min), 40–55% B (45–75 min), 55–80% B (75–80 min), 80–90% B (80–82 min), 90–100% B (82–85 min), 100–0% B (85–88 min), isocratic 0% B (88–90 min). The flow rate was 0.8 mL/min and a volume of 20 µL was injected for each sample.

Peak identification was performed in both negative and positive modes with conditions reported by Gu, et al. [16]. All the standard curves were established with five concentrations with $R^2 > 0.9$. The standards were tentatively characterized using an Agilent 6520 Accurate-Mass Q-TOF LC-MS (Agilent Technologies, Santa Clara, CA, USA) under the same conditions as the samples. Cyanidin-3-*O*-glucoside was used as the standard for anthocyanins. The calculations of the anthocyanins in samples were according to Chandra, et al. [18] and expressed as an equivalence to the applied external standard.

2.7. The 16S rRNA Sequencing and Analysis

DNA was extracted from 0.25 g of the fermented samples using a DNeasy® PowerSoil® kit (QIAGEN GmbH, Hilden, Germany). The V3-V4 regions of 16S rRNA gene was amplified using Bakt 341F primers (forward primer CCTAYGGGRBGCASCAG and reverse primer GGACTACNNGGGTATCTAAT). The procedure of 16S rRNA amplicons preparation followed a standard protocol. High throughput sequencing was performed on the Illumina MiSeq platform by Australian Genome Research Facilities (AGRF, Australia). Paired-ends reads were imported and demultiplexed in QIIME 2 (2019.2) [19]. Sequence quality control was performed using DADA2 pipeline (1.12) [20] in R (software version 3.6.2), where the paired reads were filtered, trimmed and merged, and sequence table was constructed with chimeras removed. Phylogenetic tree generation, alpha and beta diversity analysis were performed using the align-to-tree-mafft-fasttree and core-metrics-phylogenetic pipelines of the plugins in QIIME 2 (2019.1). Taxonomic analysis was performed in QIIME 2 (2019.1) using a trained classifier based on SILVA SSU 138 Ref NR 99 sequences and taxonomy [21].

2.8. Short Chain Fatty Acids Analysis

The short chain fatty acids (SCFAs) production of the colonic digesta was assessed according to the protocol described by Gu, et al. [11]. Briefly, the acidified post-fermented faeces samples were extracted with water/formic acid (99:1, *v/v*) and analysed using gas chromatography (7890B Agilent, Santa Clara, USA) equipped with a flame ionisation detector (FID). The column was capillary (SGE BP21, 12 × 0.53 mm internal diameter (ID) with 0.5 μm film thickness, SGE International, Ringwood, VIC, Australia, P/N 054473). Helium was used as carrier gas, and the makeup gas consisted with nitrogen, hydrogen and air. The injection volume was 1 μL. 4-methyl-valeric acid was used as the internal standard. Acetic, propionic and butyric acids were analysed for each sample. All results were expressed as mmol/L.

2.9. Statistical Analysis

Different carrot-fat treatments were performed with biological quadruplicate digestion-fermentations. One-way analysis of variance (ANOVA) was performed to test the statistical significance among samples at $p < 0.05$ significance level using Minitab® 18 Statistical software (Minitab Inc., State College, PA, USA). GraphPad Prism version 8.0 and XLSTAT (2019) were used for data visualisation. Statistical analysis of gut microbiome data followed the method of Tian, et al. [22], the p value for significance was validated by Bonferroni adjustment and reported as $p < 0.002$ with 28 comparisons among 8 different treatments (the $p < 0.05$ value was divided by 28). Multivariate statistical treatments including principal component analysis (PCA) utilising DPPH, FRAP, TPC, and individual polyphenols from different treatment groups and principal coordinate analysis (PCoA) utilising phylogenetic weighted UniFrac distance of microbial beta diversity, which was gathered in QIIME 2, as a measure of sample dissimilarity were performed and visualised using XLSTAT (2019).

3. Results and Discussion

This study investigated the role of interactions between lipids and polyphenols in a simulated digestion. Bioaccessible polyphenols and antioxidant activities (DPPH, FRAP, and TPC) were measured to show the digestion of black carrot at each digestive compartment and the influence from dietary lipids. The 16S rRNA sequencing and SCFAs measurement were performed to illustrate the metabolism status of different carrot-lipid matrix.

3.1. Evaluation of Bioaccessibility of Individual Phenolic Compounds and Effects of Lipids in Different Digestive Compartments

Five anthocyanins were separated in raw black carrots and the digestive fluids at different digestive compartments (Table 2 and Figure S1). Individual anthocyanins were identified referencing Padayachee, et al. [17]. All were cyanidins and derivatives as reported by Kammerer et al. [23]. Bioaccessibility was calculated as the ratio of the anthocyanins released to the digestive fluids to the corresponding polyphenols extracted from raw carrots. Table 2 shows that at the gastric stage, bioaccessibility of total anthocyanins was 32.7%. The majority of anthocyanins would not decompose during stomach digestion, as anthocyanins are very stable in the acidic condition of the gastric compartment at pH 3 [24]. The remaining portion (67.3%) are non-bioaccessible bound anthocyanins remaining in the food matrix. This result was comparable with other studies where the available anthocyanins were around 30% during the gastric digestion in a plant matrix [25,26]. In comparison, higher bioaccessibility (54.7%) of total phenolic acids was measured (Table 2). Phenolic acids similarly have higher stability at low pH (pH 3) [27], and may be released from their glycosidically bound forms under the actions of hydrolytic enzymes to result in higher measured bioaccessibility [28].

Limited effects of lipids on polyphenols bioaccessibility were shown at the gastric stage except where tallow (BT) had a significantly ($p < 0.05$) lower availability for both anthocyanins and phenolic acids compared to other treatments (Table 2). This could be a result of the high melting point of tallow (~45 °C), which is higher than the temperature during digestion (37 °C), leading to the mixture solidifying and preventing the carrot powder from contacting the digestive fluids. Therefore, most polyphenols would not be released, as lipids are not digested in stomach. No significant differences were seen in coconut (BC) or sunflower (BS) groups compared to the control (B), indicating the undigested lipids would not affect the release of polyphenols from food matrix during gastric digestion.

A decline of anthocyanin bioaccessibility after small intestine treatments was observed with only 0.98% in black carrot control (B) group released (Table 2). It has been verified that there are major reductions of anthocyanins during the small intestinal digestion [29,30], due to low stability of anthocyanins at pH 7. The flavylium cation of anthocyanin molecules is likely to transform to a colourless chalcone, which is an anthocyanin pseudobase and more stable when pH > 5 [26]. Additionally, partial oxidation and degradation of anthocyanins may also occur and form phenolic acids of the B ring during the small intestinal digestion [31]. For phenolic acids, decrease in neochlorogenic acid and chlorogenic acid contents were observed. Similar results were obtained by Gumienna, et al. [27] and Kim, et al. [29], and due to the partial hydrolysis of chlorogenic acids in the small intestine under the influence of intestinal esterase. On the other hand, increased caffeic acid and ferulic acid contents were generated together with newly detected *p*-coumaric acid. Part of these phenolic acids might be the acyl moieties hydrolysed from the corresponding acylated anthocyanins and chlorogenic acid under the actions of intestinal esterase [32]. However, given the relatively high resistance of acylated anthocyanins to gastrointestinal digestion, the newly generated phenolic acids were likely released from bound phenolic acids associated with the plant cell wall of the black carrots under the digestive conditions [32,33].

Lipids significantly affected bioaccessibility during small intestinal digestion as higher ($p < 0.05$) anthocyanin and phenolic acid contents were found preparations containing lipids (BC, BS and BT) compared to the control (B) (Table 2). This result indicated a protective effect of lipids on bioaccessible polyphenols and was in line with previous studies, where the stability and recovery of polyphenols increased with the presence of fat in the food matrix [9,33]. Lipids are emulsified by bile salts and biliary phosphatidylcholine (PC) and break into micelles under the actions of lipase before they are absorbed [34]. The protective effect on polyphenols could be attributed to molecular phenol-lipid interactions. Hydrophobic interactions can exist between anthocyanins and lipids, resulting in incorporation of anthocyanins into the lipid phase of the micelles which prevents the degradation of anthocyanins [10]. For both anthocyanins and phenolic acids, hydrogen bonds could be formed between the hydroxyl groups of polyphenols and the hydrophilic head of PC/digestive enzymes, which could enhance the stability of polyphenols [35,36].

Among the lipid treatments, coconut oil (BC) showed the highest protective effect of anthocyanins with significantly higher bioaccessibility ($p < 0.05$) than sunflower oil (BS) and tallow (BT) at intestinal stage (Table 2). Sunflower and tallow consist primarily of long-chain fatty acids, contrasting with medium-chain fatty acids (MCFAs) forming the majority of fatty acids in coconut oil. The greater protective effects on polyphenols of medium-chain triglycerides could be possibly due to these triglycerides being digested to a greater degree than longer chain triglycerides, as MCFAs possess a higher dispersibility in the aqueous phase [37]. The free MCFAs digested from triglycerides could migrate into the surrounding aqueous phase rapidly and not inhibit the interfacial lipase reaction, instead, long-chain free fatty acids tend to accumulate at the oil-water interface and inhibit lipase activity [38]. As a result, the higher micellarisation efficiency of MCFAs could favour the formation of hydrogen bonds between polyphenols and the hydrophilic head of emulsifiers surrounding the lipid droplets and thus increased the stability of phenols and their bioaccessibility. No differences in anthocyanin bioaccessibility were observed between BS (rich in PUFA) and BT (rich in SFA). Huo, et al. [39] pointed out that the degree of unsaturation of lipids might have more influence on the in vivo absorption of lipophilic bioactive compounds rather than the lipid micellarisation during digestion. Similar results were observed by Colle, et al. [40] who found no significant effects of the degree of unsaturation of lipids on in vitro transfer of lycopene from the food matrix to mixed micelles. For phenolic acids, no statistical differences between types of lipids could be described to explain the bioaccessibility in the small intestinal (Table 2), which might be related to the higher stability of phenolic acids in slight alkaline environment compared to anthocyanins, making the impacts from polyphenol-lipids interactions relatively less. Similar results were gathered by Ortega, et al. [33], where fat content did not affect the digestibility of and stability of the phenolic acids during in vitro duodenal digestion.

Most of the target polyphenols were not detected after colonic fermentation except ferulic acids in the BC treatment with 66.7% residual bioaccessibility (Table 2), suggesting the polyphenols were metabolised into other compounds by the colonic microbiota. Indeed, most polyphenols can be metabolised by microbes. For example, anthocyanins can be degraded by *Lactobacillus casei* to phenolic acids as seen by degradation to protocatechuic acid in the colon [41]. Cyanidin dioxaloylhexoside, *p*-coumaric acid, ferulic acid, and syringic acid were also reported as metabolites of anthocyanins in other in vitro colonic digestion studies [29,42,43]. In our study, we did not detect phenolic acids, which were most likely to further decomposed into smaller molecules during the extended 20 h colonic fermentation time. *Escherichia*, *Lactobacillus*, and *Bifidobacterium* can transform both chlorogenic acid and neochlorogenic acid to caffeic acid after 5 h fermentation [44,45], while further degradation of caffeic acid would occur after 5 h with reduction of the double bond to generate dihydrocaffeic acid and then 3-hydroxylpheylpropionia acid [46]. The absence of ferulic acid and *p*-coumaric acid after 20 h of fermentation in the descending colon has been reported when phenolic acid-rich fruits and vegetables are ingested [29,47]. The measurement of ferulic acid in BC group (Table 2) could be due to the protective effect of coconut oil on anthocyanins during intestinal digestion. As a result, ferulic acid might be produced during colonic fermentation as an important metabolite of anthocyanins [48]. Other than that, ferulic acid in conjugated forms protected by coconut oil during small intestinal stage might also be catabolised and produce ferulic acid during the colonic fermentation [49]. Generally, during the in vitro digestion, inclusion of lipids promoted polyphenols bioaccessibility with coconut oil showed the highest protective effects against polyphenols degradation.

Table 2. Quantity of individual polyphenols in the black carrot and oil combinations at different digestive compartments (ug/g black carrot DW).

Identified Compound	R value	Gastric Fraction			
		B (BA%)	BC (BA%)	BS (BA%)	BT (BA%)
Cy 3-xylglcgal	110.9	18.1^{b} (16.3%)	42.2^{a} (38.1%)	45.0^{a} (40.6%)	10.1^{b} (9.10%)
Cy 3-xylgal	850.3	64.1^{bc} (7.50%)	209.5^{a} (24.6%)	194.7^{ab} (22.9%)	48.8^{c} (5.70%)
Cy 3-xyl (sin) glcgal	126.4	39.6^{a} (31.3%)	51.0^{a} (40.3%)	45.7^{a} (36.1%)	8.50^{b} (6.7%)
Cy 3-xyl (fer) glcgal	1483.8	711.6^{a} (48.0%)	823.9^{a} (55.5%)	732.6^{a} (49.4%)	140.5^{b} (9.5%)
Cy 3-xyl (cmr) glcgal	55.6	26.6^{ab} (47.9%)	39.9^{a} (71.7%)	32.5^{a} (58.4%)	6.90^{b} (12.3%)
Total anthocyanins	2627	859.9^{a} (32.7%)	1166^{a} (44.4%)	1050^{a} (40.0%)	214.7^{b} (8.20%)
Neochlorogenic acid	996.2	13.0 (1.30%)	17.1 (1.70%)	17.5 (1.80%)	7.60 (0.80%)
Chlorogenic acid	2470.8	1899^{ab} (76.9%)	2303^{a} (93.2%)	2002^{a} (81.0%)	861.7^{b} (34.9%)
Caffeic acid	109.4	63.5^{a} (58.0%)	51^{b} (46.6%)	47.1^{b} (43.0%)	14.3^{c} (13.0%)
Ferulic acid	135.4	55.8^{a} (41.2%)	40.3^{b} (29.8%)	44.3^{b} (32.8%)	13.2^{c} (9.80%)
p-coumaric acid	-	-	-	-	-
Total phenolic acids	3712	2031^{a} (54.7%)	2411^{a} (65.0%)	2111^{a} (56.9%)	896.9^{b} (24.2%)
Identified compound	**R value**	**Small Intestinal Fraction**			
		B (BA%)	**BC (BA%)**	**BS (BA%)**	**BT (BA%)**
Cy 3-xylglcgal	110.9	0.25^{b} (0.23%)	1.76^{a} (1.59%)	1.35^{a} (1.22%)	1.70^{a} (1.54%)
Cy 3-xylgal	850.3	0.95^{b} (0.11%)	9.78^{a} (1.15%)	3.68^{b} (0.43%)	3.02^{b} (0.36%)
Cy 3-xyl (sin) glcgal	126.4	0.80^{c} (0.63%)	3.54^{a} (2.8%)	2.84^{ab} (2.24%)	2.09^{b} (1.66%)
Cy 3-xyl (fer) glcgal	1483.8	23.0^{c} (1.55%)	117^{a} (7.9%)	73.3^{b} (4.94%)	69.1^{b} (4.66%)
Cy 3-xyl (cmr) glcgal	55.6	0.84^{c} (1.51%)	3.00^{a} (5.4%)	1.47^{bc} (2.65%)	1.77^{b} (3.18%)
Total anthocyanins	2627	25.8^{c} (0.98%)	135.3^{a} (5.15%)	82.7^{b} (3.15%)	77.7^{b} (2.96%)
Neochlorogenic acid	996.2	7.73^{c} (0.78%)	5.65^{c} (0.57%)	23.7^{b} (2.37%)	39.3^{a} (3.95%)
Chlorogenic acid	2470.8	101.6^{b} (4.1%)	805.6^{a} (32.6%)	599.5^{a} (24.3%)	862.5^{a} (34.9%)
Caffeic acid	109.4	172.7^{b} (157.9%)	202.3^{a} (184.9%)	210.1^{a} (192.1%)	206.8^{a} (189.1%)
Ferulic acid	135.4	247.5^{b} (182.8%)	236.1^{b} (174.4%)	308.2^{a} (227.7%)	246.0^{b} (181.7%)
p-coumaric acid	-	144.5^{c} (n.a.)	168.5^{b} (n.a.)	204.0^{a} (n.a.)	178.5^{b} (n.a.)
Total phenolic acids	3712	673.9^{b} (18.2%)	1418^{a} (38.2%)	1346^{a} (36.3%)	1533^{a} (41.3%)
Identified compound	**R value**	**Colonic Fraction**			
		B (BA%)	**BC (BA%)**	**BS (BA%)**	**BT (BA%)**
Cy 3-xylglcgal	110.9	-	-	-	-
Cy 3-xylgal	850.3	-	-	-	-
Cy 3-xyl (sin) glcgal	126.4	-	-	-	-
Cy 3-xyl (fer) glcgal	1483.8	-	-	-	-
Cy 3-xyl (cmr) glcgal	55.6	-	-	-	-
Total anthocyanins	2627	-	-	-	-
Neochlorogenic acid	996.2	-	-	-	-
Chlorogenic acid	2470.8	-	-	-	-
Caffeic acid	109.4	-	-	-	-
Ferulic acid	135.4	-	90.3^{a} (66.7%)	-	-
p-coumaric acid	-	-	-	-	-
Total phenolic acids	3712	-	90.3^{a} (2.43%)	-	-

Different letters (a–c) in the same row indicate significant differences at $p < 0.05$ amongst the 4 treatments (B, BC, BS, and BT). Abbreviations: R (raw carrot powder extracts), B (black carrot only), BC (black carrot and coconut oil mixture), BS (black carrot and sunflower oil mixture), BT (black carrot and tallow mixture), and BA% (bioaccessibility%). n.a. (not applicable). Cy-3-xylglcgal (cyanidin 3-xylosyl(glucosyl)galactoside), cy-3-xylgal (cyanidin 3-xylosylgalactoside), cy-3-xyl(sin)glcgal (cyanidin 3-xylosyl(sinapoylglucosyl)galactoside), cy-3-xyl(fer)glcgal (cyanidin 3-xylosyl(feruloylglucosyl)galactoside), cy-3-xyl(cmr)glcgal (cyanidin 3-xylosyl(coumaroylgluco-syl)galactoside).

3.2. Antioxidant and Phytochemical Assays

Overall increases in bioaccessible antioxidant capacity (DPPH and FRAP) and total phenolic contents were observed in lipid treatments and in line with the results of polyphenols measurements. Gastric fractions showed no impact from presence of lipids except for the tallow group due to the solid state of mixture (Figure 1). During the intestinal digestion, oil groups showed significantly higher DPPH values ($p < 0.05$) compared to the carrot control (Figure 1a) but were not dependent on oil composition as no significantly different DPPH values of the overall antioxidant compounds were shown among different lipids treatments. However, the DPPH assay also takes other antioxidant compounds such as the phenolic acids in glycolysed forms, flavonoids, and ascorbic acids contained in black carrots into account which may explain this result [16,50]. Decreased DPPH activities were obtained in the colonic fraction, indicating the decomposition of most antioxidants related compounds. Strong positive correlations exist between DPPH and FRAP assays confirming these results [16]. It was noticeable that BC in colonic fraction exhibited statistically lower FRAP activity than the carrot control group. This could be related to the shift of gut microbiome during fermentation with coconut treatment (Figure 3), which would result in decreased bacterial activity and lower level of metabolites with reducing capability. In terms of TPC, a range of 18.1 to 31.9 mg gallic acid equivalents (GAE)/g carrot on dry weight (DW) basis were observed at gastric stage with significant lower value in BT group (Figure 1c). No statistical differences were seen at intestinal stages, and this was probably due to the reducing compounds other than antioxidants such as reducing sugars, since Folin–Ciocalteu assay does not specify to antioxidants only but any reducing substance [51]. Thus, the antioxidant assays largely reflected the LC results above and confirmed that inclusion of lipids help increase bioaccessible bioactivity during in vitro digestion.

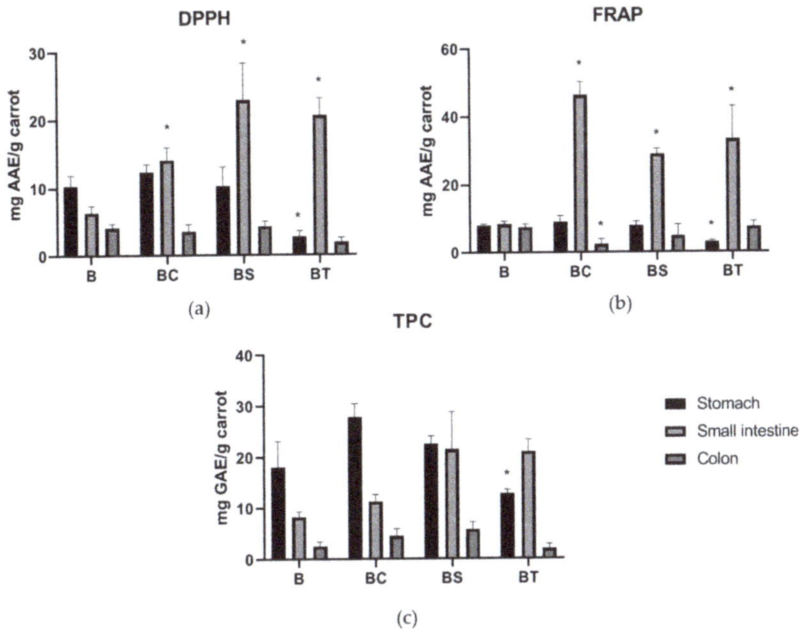

Figure 1. Bioaccessible phytochemicals were increased when associated with lipids. DPPH (**a**), FRAP (**b**), and total phenolic content (**c**) of carrot control and different carrot-oil mixtures from gastric, small intestinal and colonic fractions. Abbreviations: B (black carrot only), BS (black carrot and sunflower oil mixture), BT (black carrot and tallow mixture), and BC (black carrot and coconut oil mixture). * Significant differences at $p < 0.05$ compared to control (B) were highlighted. No significant differences were shown among different types of oil treatments. Results of DPPH and FRAP were expressed as mg ascorbic acid equivalent (AAE) per g of carrot on dry weight basis, and results of TPC were expressed as mg gallic acid equivalent (GAE) per g of carrot on dry weight basis.

The effect of the type of lipids on black carrot polyphenol digestion at different digestive stages was considered by a principal component analysis (PCA) of the dataset of the HPLC-DAD quantified individual polyphenols and phytochemical bioactivities (DPPH, FRAP and TPC; Figure 2a). Gastric, small intestinal and colonic fractions could be well separated with 89.8% variability explained with the two principal components and suggested significant roles of intestinal lipid protection and colonic microbiome metabolism of the antioxidant compounds. Figure 2b shows the impact of the different lipid interferences with 65.7% variability explained. To eliminate the impact of tallow solidification in the gastric phase, only small intestinal and colonic fractions datasets were used. The three lipid treatments were separated from the carrot control, indicating that lipids impact on the digestion of bioactive compounds. BS and BT groups were very similar, while BC samples were separated in the PCA space, suggesting a significant role of coconut oil in protecting polyphenols from degradation.

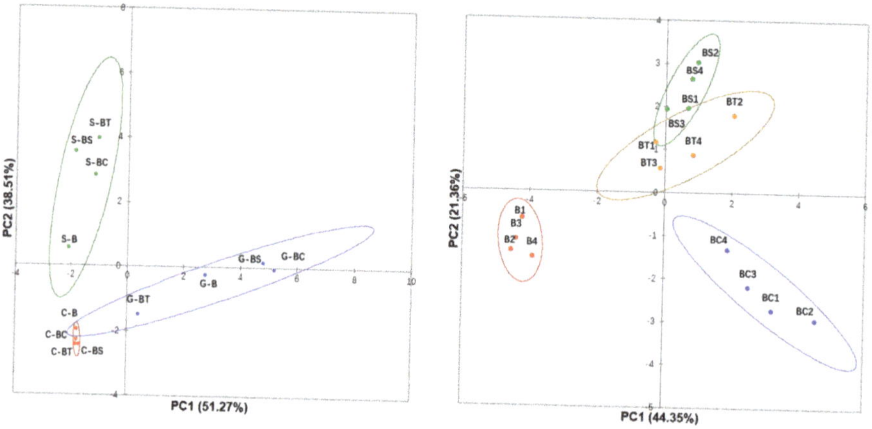

(a) (b)

Figure 2. Bioaccessibility of polyphenols and antioxidant measures during digestion of black carrots are explained by the inclusion of lipids. Principal component analysis (PCA) of the dataset of individual polyphenols and phytochemical bioactivities (DPPH, FRAP, and TPC) dataset. Panel (**a**) shows the separation of gastric, small intestinal, and colonic digestive fluids. Panel (**b**) shows the separation of different carrot and oil combinations. Different groups are denoted by colours as shown in the legend. Abbreviations: G (gastric fraction), S (small intestinal fraction), and C (colonic fraction); B (black carrot only), BS (black carrot and sunflower oil mixture), BT (black carrot and tallow mixture), and BC (black carrot and coconut oil mixture).

3.3. The Carrot-Oil Matrix Modulates the Gut Microbiota and Affects Short Chain Fatty Acid Production

In general, dietary fats are degraded and absorbed in small intestine, however, it has been reported that a portion of ingested fats would enter the human colon and are metabolised by colonic bacteria [8]. To test the interference of lipids on adaptation of the gut microbiota when treated with a digested black carrot matrix, 16S rRNA sequencing was carried out on the faeces samples after 20 h in vitro fermentation. *Megasphaera, Escherichia, Muribaculaceae, Prevotella,* and *Streptococcus* were the most abundant genera (Figure 3a). Significant differences were found between these major genera between the different treatment groups ($p < 0.0001$) (Table 3). It should be noted that this test aims to give a view of the role of gut microbiome on phenol-lipids matrix digestion. In vivo experiments with larger sample size would be required to test more accurate gut microbiome composition changes.

Presence of black carrots changed the abundance and composition of the microbial population and was different to the control (Ctrl) (Figure 3b). The shift in microbial diversity was most likely promoted by dietary fibre and any bound polyphenols that escaped the small intestinal digestion. Beneficial bacteria including *Prevotella*, *Prevotellaceae* NK3B31 *group*, and *Lactobacillus* had higher relative abundance in black carrot diets, while genera such as *Streptococcus* were suppressed (Figure 3a and Table 3). Both *Prevotella* and *Prevotellaceae* NK3B31 *group* were associated with pectin metabolism [52,53]. The metabolism of anthocyanins might promote the growth of *Lactobacillus* [54]. Generally, our results verified that intake of polyphenol-rich dietary polysaccharides shifted microbial diversity to include genera with beneficial consequences.

Lipid preparations played a significant role in modulating the gut microbiota and were largely unaffected by presence of carrots as the carrot-lipid mixed samples closely clustered with the corresponding lipid controls (Figure 3b). The type of lipid was important as seen by the separation of coconut positive control (C) and BC groups from other treatments, while much more clustered and similar microbiome composition were shown between sunflower (S and BS) and tallow (T and BT) groups (Figure 3a,b). *Escherichia* significantly increased ($p < 0.002$) in coconut-containing treatments (Figure 3a and Table 3). Increased *Escherichia* is an indicator of ecosystem disturbance of intestinal microbiota, which is linked with intestinal diseases like constipation and diarrhoea [55]. Zentek, et al. [56] showed that supplementing with the MCFAs capric acid (C10:0), lauric acid (C12:0), and caprylic acid (C8:0) enriched colonic pathogenic *Escherichia* in an in vivo study in piglets. On the other hand, sunflower oil and beef tallow resulted in reduced relative abundance in *Streptococcus* (Table 3). Both long-chain PUFA (especially linoleic acid) and SFA (stearic acid and palmitic acid) have been reported with antibacterial activity against *Streptococcus* [57,58].

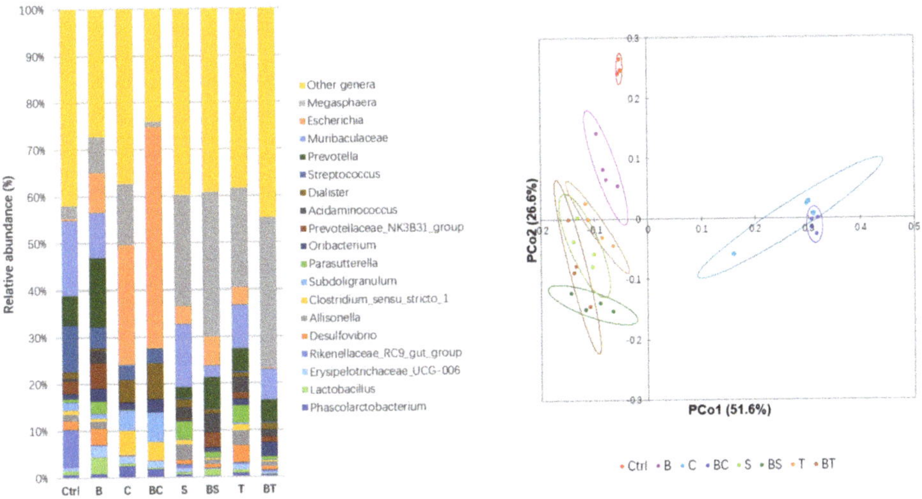

(a) (b)

Figure 3. Microbial diversity and abundance were altered by inclusion of lipids in the model diet. Relative abundance of most abundant (>1%) microbial genera (**a**) and weighted UniFrac-principal coordinate analysis (PCoA) analysis of beta diversity among all profiled samples (**b**). Different groups are denoted by colours as shown in the legend. Abbreviations: Ctrl (negative control with faeces only), B (black carrot only), C (coconut oil), BC (black carrot and coconut oil mixture), S (sunflower oil), BS (black carrot and sunflower oil mixture), T (tallow), and BT (black carrot with tallow mixture).

Inclusion of coconut oil decreased SCFA production (Table 3; C and BC). Coconut oil contains lauric acid with antibacterial activity [59], and so the low SCFA production of coconut treatment could be a result of the inhibition of the colonic bacteria. Addition of black carrots promoted total SCFAs production compared to the Ctrl as a result of microbial metabolism of the dietary fibre, majorly cellulose and pectin, in black carrot [17]. SCFAs production was not significantly increased by the inclusion of black carrots and consistent with small changes in black carrot inclusion to microbial diversity and composition (Figure 3a,b). Similar results were obtained in our previous in vivo study using a pig model [11]. Inclusion of fat in the diet can decrease colonic microbial cellulolytic and pectinolytic activities [60] and ameliorate any microbial metabolism of dietary fibre. The destruction of the plant cell wall of black carrots during sample preparation before submission to the simulated digestion could contribute to a decreased metabolic rate, as colonic bacteria favour the local microenvironment provided by the cell junctions within large cell clusters compared to exposed cell surfaces for SCFA production [61]. Among the SCFAs, butyrate is an important beneficial SCFA as an energy source for non-transformed colon epithelial cells and can inhibit proliferation of colon cancer cells [62] and is linked to the positive effects of long-term intake of plant-based food rich in dietary fibres [63]. Butyrate production was the highest with sunflower oil (S and BS) with 34.2 mmol/L (Table 3). A study on in vitro colonic fermentation of nuts showed that nuts rich in PUFAs significantly increased the butyrate ratio to the total SCFAs production [64]. Here, the increased butyrate production verified the beneficial effects from PUFAs contained in the sunflower oil.

Table 3. Relative abundance (%) of microbiota (>1%) and short chain fatty acid (SCFA) productions (mmol/L) of controls and samples after colonic fermentation.

Genera	Controls				Samples				p Value
	Ctrl	C	S	T	B	BC	BS	BT	
Megasphaera	2.87^{b}	15.2^{ab}	31.1^{ab}	26.8^{ab}	8.39^{b}	1.14^{b}	44.7^{a}	47.0^{a}	<0.0001
Escherichia	0.39^{c}	29.4^{b}	5.01^{c}	4.81^{c}	9.15^{c}	47.8^{a}	8.91^{c}	0.42^{c}	<0.0001
Muribaculaceae	16.4^{a}	0.06^{c}	17.9^{a}	11.8^{ab}	10.6^{ab}	0.01^{c}	3.83^{bc}	9.84^{ab}	<0.0001
Prevotella	6.61^{bc}	0.06^{cd}	3.02^{bcd}	6.07^{bcd}	15.85^{a}	0.01^{d}	9.61^{ab}	6.93^{b}	<0.0001
Streptococcus	10.3^{a}	3.66^{bc}	0.38^{c}	0.55^{c}	4.82^{b}	3.18^{bc}	0.26^{c}	0.29^{c}	<0.0001
Dialister	1.32^{c}	5.50^{a}	2.22^{bc}	1.24^{c}	0.48^{c}	7.77^{ab}	1.06^{c}	2.20^{bc}	<0.0001
Acidaminococcus	0.63^{c}	0.01^{c}	3.13^{abc}	4.21^{ab}	2.95^{abc}	-	6.00^{a}	2.18^{bc}	<0.0001
Prevotellaceae NK3B31 *group*	2.71^{b}	-	0.39^{c}	1.60^{bc}	5.93^{a}	-	4.64^{a}	1.78^{bc}	<0.0001
Oribacterium	1.13^{b}	1.79^{ab}	0.33^{b}	1.85^{ab}	2.99^{ab}	2.86^{ab}	1.25^{b}	4.37^{a}	<0.0001
Parasutterella	0.82^{b}	0.04^{b}	4.99^{a}	4.68^{a}	2.78^{ab}	0.01^{b}	1.93^{ab}	1.07^{b}	<0.0001
Subdoligranulum	1.79^{bc}	5.09^{ab}	0.26^{c}	0.42^{c}	1.20^{bc}	6.42^{a}	0.21^{c}	0.18^{c}	<0.0001
Clostridium sensu stricto 1	0.84^{b}	5.87^{a}	1.13^{b}	1.61^{b}	0.57^{b}	3.84^{ab}	0.37^{b}	0.45^{b}	<0.0001
Allisonella	1.41^{c}	0.11^{c}	4.35^{a}	4.09^{ab}	1.77^{bc}	0.07^{c}	1.42^{c}	1.39^{c}	<0.0001
Desulfovibrio	1.97^{bc}	0.20^{c}	1.39^{c}	4.69^{a}	3.74^{ab}	0.06^{c}	1.17^{c}	1.27^{c}	<0.0001
Rikenellaceae RC9 gut group	8.27^{a}	-	1.07^{b}	0.47^{b}	0.18^{b}	-	0.08^{b}	0.50^{b}	<0.0001
Erysipelotrichaceae UCG-006	0.79	1.93	1.08	1.38	2.62	1.52	0.16	0.58	0.003
Lactobacillus	0.78^{b}	0.60^{b}	0.63^{b}	0.76^{b}	4.00^{a}	0.33^{b}	2.09^{ab}	0.16^{b}	<0.0001
Phascolarctobacterium	0.74^{b}	2.81^{a}	0.74^{b}	1.14^{ab}	0.83^{b}	1.81^{ab}	0.39^{b}	0.71^{b}	<0.0001
Short chain fatty acids									
Acetate	64.3^{a}	31.2^{d}	32.3^{cd}	42.6^{bc}	67.5^{a}	37.2^{bcd}	40.0^{bcd}	46.8^{b}	<0.0001
Propionate	15.0^{c}	4.00^{d}	19.7^{bc}	16.3^{c}	24.1^{b}	3.17^{d}	23.3^{b}	31.4^{a}	<0.0001
Butyrate	22.1^{c}	3.57^{d}	34.5^{a}	27.2^{bc}	24.5^{bc}	2.99^{d}	34.1^{a}	30.1^{ab}	<0.0001
Total SCFAs	101.4^{abc}	37.9^{d}	86.4^{c}	86.1^{c}	116.0^{a}	43.4^{d}	97.3^{bc}	108.3^{ab}	<0.0001

Different letters (^{a}–^{d}) in the same row indicate significant differences at $p < 0.002$ compared among 8 different treatments (Ctrl, C, S, T, B, BC, BS and BT). Abbreviations: Ctrl (negative control with faeces only), C (coconut oil), S (sunflower oil), T (tallow), B (black carrot only), BC (black carrot and coconut oil mixture), BS (black carrot and sunflower oil mixture), BT (black carrot with tallow mixture). SCFA (short chain fatty acid).

Overall, the results from this study suggested inclusion of all the three types of lipids increased bioaccessibility of polyphenols during simulated digestion, with coconut oil showing the highest protective effect. At the same time, however, colonic metabolism of coconut oil resulted in increased health detrimental bacteria *Escherichia* and suppressed the production of beneficial colonic SCFAs which are associated with positive cardiovascular effects [65]. Although coconut oil showed the potential to promote absorption of antioxidative polyphenols, it also has been related with increased serum triglycerides and low density lipoproteins (LDL), which are associated with higher risks of cardiovascular diseases [66]. The exact health impact of coconut oil on cardiovascular health remains unclear, and larger clinical intervention studies are required [67]. Our study suggests that the effects of coconut oil on health, especially on cardiovascular health should be assessed with caution. Sunflower oil, on the other hand, also showed protective effects on polyphenols, and it reduced the relative abundance of *Streptococcus*, which is associated with multiple diseases including diarrhoea and obesity [68]. Meanwhile, inclusion of sunflower oil increased butyrate as health beneficial SCFA. PUFA, as the major composition of sunflower oil, has been related with cardioprotective and anti-inflammatory effects [69]. Therefore, without considering effects from compounds other than lipids and polyphenols, from the perspective of promoting polyphenols bioaccessibility and positive gut microbiome adaptation, consuming polyphenol-rich dietary foods coupled with PUFA could be a healthy approach to potentially improve polyphenols absorption with lowest negative effects. Further in vivo experiments with larger sample size and human as subjects are expected to verify this conclusion.

4. Conclusions

This study reports the effects of dietary lipids on the bioaccessibility of black carrot polyphenols and microbial diversity in a simulated in vitro digestion and fermentation. Lipids promoted the bioaccessibility of both anthocyanins and phenolic acids during small intestinal digestion, since polyphenol-lipid interactions enhanced the stability of the polyphenols. Coconut oil showed the highest protection of anthocyanins due in part to the medium-chain fatty acids in the coconut oil. Anthocyanins and phenolic acids were largely degraded during colonic fermentation. Inclusion of dietary carrots modulated the gut microbiota and increased the relative abundances of beneficial bacteria, including *Prevotella*, *Prevotellaceae* NK3B31 *group* and *Lactobacillus*, and reduced the detrimental genera *Streptococcus*. Coconut oil appeared to alter microbial growth in the simulated digestion and fermentation with an increase in the abundance of *Escherichia* and suppressed the production of SCFAs. Inclusion of sunflower oil and tallow reduced relative abundance in health detrimental *Streptococcus*. Sunflower oil improved the production of butyrate, potentially due to the presence of PUFAs. The results show that the impact of polyphenols in the digestive tract should be considered in the context of other components of the diet, particularly the lipid component. Coupling polyphenol-rich foods with appropriate amount of PUFA-rich oil could be a beneficial approach to potentially increase polyphenols absorption.

Supplementary Materials: The following are available online at http://www.mdpi.com/2076-3921/9/8/762/s1, Figure S1: HPLC-DAD profile of anthocyanins and phenolic acids.

Author Contributions: Conceptualization, C.G., K.H. and F.R.D.; methodology, C.G., K.H., and H.A.R.S.; software, C.G. and K.H.; formal analysis, C.G.; investigation, C.G., K.H., and H.A.R.S.; resources, K.H., H.A.R.S., and F.R.D.; data curation, C.G. and K.H.; writing—original draft preparation, C.G.; writing—review and editing, K.H., H.A.R.S., and F.R.D.; visualization, C.G. and K.H.; supervision, K.H., H.A.R.S., and F.R.D.; project administration, K.H. and F.R.D.; funding acquisition, K.H. and F.R.D. All authors have read and agreed to the published version of the manuscript.

Funding: This research received no external funding.

Acknowledgments: We would like to thank Nicholas Williamson and Shuai Nie from the Mass Spectrometry and Proteomics Facility, Bio21 Molecular Science and Biotechnology Institute, the University of Melbourne, VIC, Australia for providing access and support for the use of HPLC and -LC-ESI-QTOF/MS and data analysis.

Conflicts of Interest: The authors declare no conflict of interest.

References

1. Selby-Pham, S.N.; Cottrell, J.J.; Dunshea, F.R.; Ng, K.; Bennett, L.E.; Howell, K.S. Dietary phytochemicals promote health by enhancing antioxidant defence in a pig model. *Nutrients* **2017**, *9*, 758. [CrossRef] [PubMed]
2. Cheng, Y.-C.; Sheen, J.-M.; Hu, W.L.; Hung, Y.-C. Polyphenols and oxidative stress in atherosclerosis-related ischemic heart disease and stroke. *Oxid. Med. Cell. Longev.* **2017**, *2017*. [CrossRef] [PubMed]
3. Nordestgaard, B.G.; Varbo, A. Triglycerides and cardiovascular disease. *Lancet* **2014**, *384*, 626–635. [CrossRef]
4. Kwon, S.-H.; Ahn, I.-S.; Kim, S.-O.; Kong, C.-S.; Chung, H.-Y.; Do, M.-S.; Park, K.-Y. Anti-obesity and hypolipidemic effects of black soybean anthocyanins. *J. Med. Food* **2007**, *10*, 552–556. [CrossRef]
5. Kemperman, R.A.; Bolca, S.; Roger, L.C.; Vaughan, E.E. Novel approaches for analysing gut microbes and dietary polyphenols: Challenges and opportunities. *Microbiology* **2010**, *156*, 3224–3231. [CrossRef]
6. Padayachee, A.; Day, L.; Howell, K.; Gidley, M. Complexity and health functionality of plant cell wall fibers from fruits and vegetables. *Crit. Rev. Food Sci. Nutr.* **2017**, *57*, 59–81. [CrossRef]
7. Lambert, J.E.; Parks, E.J. Postprandial metabolism of meal triglyceride in humans. *Biochim. Biophys. Acta (BBA) Mol. Cell Biol. Lipids* **2012**, *1821*, 721–726. [CrossRef]

8. Agans, R.; Gordon, A.; Kramer, D.L.; Perez-Burillo, S.; Rufián-Henares, J.A.; Paliy, O. Dietary fatty acids sustain the growth of the human gut microbiota. *Appl. Environ. Microbiol.* **2018**, *84*, e01525-18. [CrossRef]
9. Ortega, N.; Macià, A.; Romero, M.-P.; Reguant, J.; Motilva, M.-J. Matrix composition effect on the digestibility of carob flour phenols by an in-vitro digestion model. *Food Chem.* **2011**, *124*, 65–71. [CrossRef]
10. Shishikura, Y.; Khokhar, S.; Murray, B.S. Effects of tea polyphenols on emulsification of olive oil in a small intestine model system. *J. Agric. Food Chem.* **2006**, *54*, 1906–1913. [CrossRef]
11. Gu, C.; Howell, K.; Padayachee, A.; Comino, T.; Chhan, R.; Zhang, P.; Ng, K.; Cottrell, J.J.; Dunshea, F.R. Effect of a polyphenol-rich plant matrix on colonic digestion and plasma antioxidant capacity in a porcine model. *J. Funct. Foods* **2019**, *57*, 211–221. [CrossRef]
12. US Department of Agriculture, Agricultural Research Service. USDA National Nutrient Database for Standard Reference, Release 28; Nutrient Data Laboratory. Available online: https://www.ars.usda.gov/northeast-area/beltsville-md-bhnrc/beltsville-human-nutrition-research-center/methods-and-application-of-food-composition-laboratory/mafcl-site-pages/sr11-sr28 (accessed on 16 August 2020).
13. Pérez-Burillo, S.; Rufián-Henares, J.; Pastoriza, S. Towards an improved global antioxidant response method (GAR+): Physiological-resembling in vitro digestion-fermentation method. *Food Chem.* **2018**, *239*, 1253–1262. [CrossRef] [PubMed]
14. Fu, S.; Augustin, M.A.; Sanguansri, L.; Shen, Z.; Ng, K.; Ajlouni, S. Enhanced bioaccessibility of curcuminoids in buttermilk yogurt in comparison to curcuminoids in aqueous dispersions. *J. Food Sci.* **2016**, *81*, H769–H776. [CrossRef] [PubMed]
15. Guilloteau, P.; Zabielski, R.; Hammon, H.M.; Metges, C.C. Nutritional programming of gastrointestinal tract development. Is the pig a good model for man? *Nutr. Res. Rev.* **2010**, *23*, 4–22. [CrossRef] [PubMed]
16. Gu, C.; Howell, K.; Dunshea, F.R.; Suleria, H.A. Lc-esi-qtof/ms characterisation of phenolic acids and flavonoids in polyphenol-rich fruits and vegetables and their potential antioxidant activities. *Antioxidants* **2019**, *8*, 405. [CrossRef]
17. Padayachee, A.; Netzel, G.; Netzel, M.; Day, L.; Zabaras, D.; Mikkelsen, D.; Gidley, M. Binding of polyphenols to plant cell wall analogues–Part 1: Anthocyanins. *Food Chem.* **2012**, *134*, 155–161. [CrossRef]
18. Chandra, A.; Rana, J.; Li, Y. Separation, identification, quantification, and method validation of anthocyanins in botanical supplement raw materials by HPLC and HPLC-MS. *J. Agric. Food Chem.* **2001**, *49*, 3515–3521. [CrossRef]
19. Bolyen, E.; Rideout, J.R.; Dillon, M.R.; Bokulich, N.A.; Abnet, C.C.; Al-Ghalith, G.A.; Alexander, H.; Alm, E.J.; Arumugam, M.; Asnicar, F. Reproducible, interactive, scalable and extensible microbiome data science using QIIME 2. *Nat. Biotechnol.* **2019**, *37*, 852–857. [CrossRef]
20. Callahan, B.J.; McMurdie, P.J.; Rosen, M.J.; Han, A.W.; Johnson, A.J.A.; Holmes, S.P. DADA2: High-resolution sample inference from Illumina amplicon data. *Nat. Methods* **2016**, *13*, 581. [CrossRef]
21. Quast, C.; Pruesse, E.; Yilmaz, P.; Gerken, J.; Schweer, T.; Yarza, P.; Peplies, J.; Glöckner, F.O. The SILVA ribosomal RNA gene database project: Improved data processing and web-based tools. *Nucleic Acids Res.* **2012**, *41*, D590–D596. [CrossRef]
22. Tian, G.; Wu, X.; Chen, D.; Yu, B.; He, J. Adaptation of gut microbiome to different dietary nonstarch polysaccharide fractions in a porcine model. *Mol. Nutr. Food Res.* **2017**, *61*, 1700012. [CrossRef] [PubMed]
23. Kammerer, D.; Carle, R.; Schieber, A. Quantification of anthocyanins in black carrot extracts (Daucus carota ssp. sativus var. atrorubens Alef.) and evaluation of their color properties. *Eur. Food Res. Technol.* **2004**, *219*, 479–486. [CrossRef]
24. McDougall, G.J.; Fyffe, S.; Dobson, P.; Stewart, D. Anthocyanins from red cabbage–stability to simulated gastrointestinal digestion. *Phytochemistry* **2007**, *68*, 1285–1294. [CrossRef] [PubMed]
25. Padayachee, A.; Netzel, G.; Netzel, M.; Day, L.; Mikkelsen, D.; Gidley, M.J. Lack of release of bound anthocyanins and phenolic acids from carrot plant cell walls and model composites during simulated gastric and small intestinal digestion. *Food Funct.* **2013**, *4*, 906–916. [CrossRef] [PubMed]
26. Gumienna, M.; Lasik, M.; Czarnecki, Z. Bioconversion of grape and chokeberry wine polyphenols during simulated gastrointestinal in vitro digestion. *Int. J. Food Sci. Nutr.* **2011**, *62*, 226–233. [CrossRef] [PubMed]
27. Sakakibara, H.; Honda, Y.; Nakagawa, S.; Ashida, H.; Kanazawa, K. Simultaneous determination of all polyphenols in vegetables, fruits, and teas. *J. Agric. Food Chem.* **2003**, *51*, 571–581. [CrossRef]

28. Kim, I.; Moon, J.K.; Hur, S.J.; Lee, J. Structural changes in mulberry (Morus Microphylla. Buckl) and chokeberry (Aronia melanocarpa) anthocyanins during simulated in vitro human digestion. *Food Chem.* **2020**, *318*. [CrossRef]
29. Bouayed, J.; Hoffmann, L.; Bohn, T. Total phenolics, flavonoids, anthocyanins and antioxidant activity following simulated gastro-intestinal digestion and dialysis of apple varieties: Bioaccessibility and potential uptake. *Food Chem.* **2011**, *128*, 14–21. [CrossRef]
30. Pérez-Vicente, A.; Gil-Izquierdo, A.; García-Viguera, C. In vitro gastrointestinal digestion study of pomegranate juice phenolic compounds, anthocyanins, and vitamin C. *J. Agric. Food Chem.* **2002**, *50*, 2308–2312. [CrossRef]
31. Woodward, G.; Kroon, P.; Cassidy, A.; Kay, C. Anthocyanin stability and recovery: Implications for the analysis of clinical and experimental samples. *J. Agric. Food Chem.* **2009**, *57*, 5271–5278. [CrossRef]
32. Oliveira, H.; Perez-Gregório, R.; de Freitas, V.; Mateus, N.; Fernandes, I. Comparison of the in vitro gastrointestinal bioavailability of acylated and non-acylated anthocyanins: Purple-fleshed sweet potato vs red wine. *Food Chem.* **2019**, *276*, 410–418. [CrossRef]
33. Ortega, N.; Reguant, J.; Romero, M.-P.; Macia, A.; Motilva, M.-J. Effect of fat content on the digestibility and bioaccessibility of cocoa polyphenol by an in vitro digestion model. *J. Agric. Food Chem.* **2009**, *57*, 5743–5749. [CrossRef]
34. Jakobek, L. Interactions of polyphenols with carbohydrates, lipids and proteins. *Food Chem.* **2015**, *175*, 556–567. [CrossRef] [PubMed]
35. Erlejman, A.; Verstraeten, S.; Fraga, C.; Oteiza, P. The interaction of flavonoids with membranes: Potential determinant of flavonoid antioxidant effects. *Free Radic. Res.* **2004**, *38*, 1311–1320. [CrossRef] [PubMed]
36. Charlton, A.J.; Baxter, N.J.; Khan, M.L.; Moir, A.J.; Haslam, E.; Davies, A.P.; Williamson, M.P. Polyphenol/peptide binding and precipitation. *J. Agric. Food Chem.* **2002**, *50*, 1593–1601. [CrossRef] [PubMed]
37. Porter, C.J.; Trevaskis, N.L.; Charman, W.N. Lipids and lipid-based formulations: Optimizing the oral delivery of lipophilic drugs. *Nat. Rev. Drug Discov.* **2007**, *6*, 231–248. [CrossRef] [PubMed]
38. Li, Y.; Hu, M.; McClements, D.J. Factors affecting lipase digestibility of emulsified lipids using an in vitro digestion model: Proposal for a standardised pH-stat method. *Food Chem.* **2011**, *126*, 498–505. [CrossRef]
39. Huo, T.; Ferruzzi, M.G.; Schwartz, S.J.; Failla, M.L. Impact of fatty acyl composition and quantity of triglycerides on bioaccessibility of dietary carotenoids. *J. Agric. Food Chem.* **2007**, *55*, 8950–8957. [CrossRef] [PubMed]
40. Colle, I.J.; Van Buggenhout, S.; Lemmens, L.; Van Loey, A.M.; Hendrickx, M.E. The type and quantity of lipids present during digestion influence the in vitro bioaccessibility of lycopene from raw tomato pulp. *Food Res. Int.* **2012**, *45*, 250–255. [CrossRef]
41. Marín, L.; Miguélez, E.M.; Villar, C.J.; Lombó, F. Bioavailability of dietary polyphenols and gut microbiota metabolism: Antimicrobial properties. *BioMed Res. Int.* **2015**, *2015*. [CrossRef]
42. Fernandes, I.; Faria, A.; de Freitas, V.; Calhau, C.; Mateus, N. Multiple-approach studies to assess anthocyanin bioavailability. *Phytochem. Rev.* **2015**, *14*, 899–919. [CrossRef]
43. Lecumberri, E.; Mateos, R.; Izquierdo-Pulido, M.; Rupérez, P.; Goya, L.; Bravo, L. Dietary fibre composition, antioxidant capacity and physico-chemical properties of a fibre-rich product from cocoa (Theobroma cacao L.). *Food Chem.* **2007**, *104*, 948–954. [CrossRef]
44. Budryn, G.; Nebesny, E. Phenolic acids-their properties, occurrence in plant materials, absorption and metabolism. *Bromatol. Chem. Toksykol.* **2006**, *39*, 103.
45. Olthof, M.R.; Hollman, P.C.; Buijsman, M.N.; Van Amelsvoort, J.M.; Katan, M.B. Chlorogenic acid, quercetin-3-rutinoside and black tea phenols are extensively metabolized in humans. *J. Nutr.* **2003**, *133*, 1806–1814. [CrossRef]
46. Rechner, A.R.; Smith, M.A.; Kuhnle, G.; Gibson, G.R.; Debnam, E.S.; Srai, S.K.S.; Moore, K.P.; Rice-Evans, C.A. Colonic metabolism of dietary polyphenols: Influence of structure on microbial fermentation products. *Free Radic. Biol. Med.* **2004**, *36*, 212–225. [CrossRef]
47. Khairallah, J.; Sadeghi Ekbatan, S.; Sabally, K.; Iskandar, M.M.; Hussain, R.; Nassar, A.; Sleno, L.; Rodes, L.; Prakash, S.; Donnelly, D.J. Microbial biotransformation of a polyphenol-rich potato extract affects antioxidant capacity in a simulated gastrointestinal model. *Antioxidants* **2018**, *7*, 43. [CrossRef]

48. De Ferrars, R.; Czank, C.; Zhang, Q.; Botting, N.; Kroon, P.; Cassidy, A.; Kay, C. The pharmacokinetics of anthocyanins and their metabolites in humans. *Br. J. Pharmacol.* **2014**, *171*, 3268–3282. [CrossRef]
49. Gao, Y.; Ma, S.; Wang, M.; Feng, X.-Y. Characterization of free, conjugated, and bound phenolic acids in seven commonly consumed vegetables. *Molecules* **2017**, *22*, 1878. [CrossRef]
50. Howard, L.; Wong, A.; Perry, A.; Klein, B. β-Carotene and ascorbic acid retention in fresh and processed vegetables. *J. Food Sci.* **1999**, *64*, 929–936. [CrossRef]
51. Escarpa, A.; González, M. Approach to the content of total extractable phenolic compounds from different food samples by comparison of chromatographic and spectrophotometric methods. *Anal. Chim. Acta* **2001**, *427*, 119–127. [CrossRef]
52. Ferrario, C.; Statello, R.; Carnevali, L.; Mancabelli, L.; Milani, C.; Mangifesta, M.; Duranti, S.; Lugli, G.A.; Jimenez, B.; Lodge, S. How to feed the mammalian gut microbiota: Bacterial and metabolic modulation by dietary fibers. *Front. Microbiol.* **2017**, *8*, 1749. [CrossRef] [PubMed]
53. Nograšek, B.; Accetto, T.; Fanedl, L.; Avguštin, G. Description of a novel pectin-degrading bacterial species Prevotella pectinovora sp. nov., based on its phenotypic and genomic traits. *J. Microbiol.* **2015**, *53*, 503–510. [CrossRef] [PubMed]
54. Hidalgo, M.; Oruna-Concha, M.J.; Kolida, S.; Walton, G.E.; Kallithraka, S.; Spencer, J.P.; de Pascual-Teresa, S. Metabolism of anthocyanins by human gut microflora and their influence on gut bacterial growth. *J. Agric. Food Chem.* **2012**, *60*, 3882–3890. [CrossRef] [PubMed]
55. Verbeke, K.A.; Boesmans, L.; Boets, E. Modulating the microbiota in inflammatory bowel diseases: Prebiotics, probiotics or faecal transplantation? *Proc. Nutr. Soc.* **2014**, *73*, 490–497. [CrossRef]
56. Zentek, J.; Buchheit-Renko, S.; Männer, K.; Pieper, R.; Vahjen, W. Intestinal concentrations of free and encapsulated dietary medium-chain fatty acids and effects on gastric microbial ecology and bacterial metabolic products in the digestive tract of piglets. *Arch. Anim. Nutr.* **2012**, *66*, 14–26. [CrossRef]
57. Zheng, C.J.; Yoo, J.-S.; Lee, T.-G.; Cho, H.-Y.; Kim, Y.-H.; Kim, W.-G. Fatty acid synthesis is a target for antibacterial activity of unsaturated fatty acids. *FEBS Lett.* **2005**, *579*, 5157–5162. [CrossRef]
58. Nalina, T.; Rahim, Z. The crude aqueous extract of Piper betle L. and its antibacterial effect towards Streptococcus mutans. *Am. J. Biochem. Biotechnol.* **2007**, *3*, 10–15. [CrossRef]
59. Božik, M.; Hovorková, P.; Klouček, P. Antibacterial effect of carvacrol and coconut oil on selected pathogenic bacteria. *Sci. Agric. Bohem.* **2018**, *49*, 46–52. [CrossRef]
60. Falcão-e-Cunha, L.; Peres, H.; Freire, J.P.; Castro-Solla, L. Effects of alfalfa, wheat bran or beet pulp, with or without sunflower oil, on caecal fermentation and on digestibility in the rabbit. *Anim. Feed. Sci. Technol.* **2004**, *117*, 131–149. [CrossRef]
61. Day, L.; Xu, M.; Øiseth, S.K.; Hemar, Y.; Lundin, L. Control of morphological and rheological properties of carrot cell wall particle dispersions through processing. *Food Bioprocess Technol.* **2010**, *3*, 928–934. [CrossRef]
62. Hinnebusch, B.F.; Meng, S.; Wu, J.T.; Archer, S.Y.; Hodin, R.A. The effects of short-chain fatty acids on human colon cancer cell phenotype are associated with histone hyperacetylation. *J. Nutr.* **2002**, *132*, 1012–1017. [CrossRef] [PubMed]
63. Yang, J.; Rose, D.J. Long-term dietary pattern of fecal donor correlates with butyrate production and markers of protein fermentation during in vitro fecal fermentation. *Nutr. Res.* **2014**, *34*, 749–759. [CrossRef] [PubMed]
64. Schlörmann, W.; Birringer, M.; Lochner, A.; Lorkowski, S.; Richter, I.; Rohrer, C.; Glei, M. In Vitro fermentation of nuts results in the formation of butyrate and c9, t11 conjugated linoleic acid as chemopreventive metabolites. *Eur. J. Nutr.* **2016**, *55*, 2063–2073. [CrossRef] [PubMed]
65. Den Besten, G.; van Eunen, K.; Groen, A.K.; Venema, K.; Reijngoud, D.-J.; Bakker, B.M. The role of short-chain fatty acids in the interplay between diet, gut microbiota, and host energy metabolism. *J. Lipid Res.* **2013**, *54*, 2325–2340. [CrossRef] [PubMed]
66. Parry, J. Pacific islanders pay heavy price for abandoning traditional diet. *World Health Organ. Bull. World Health Organ.* **2010**, *88*, 484.
67. Wallace, T.C. Health effects of coconut oil—A narrative review of current evidence. *J. Am. Coll. Nutr.* **2019**, *38*, 97–107. [CrossRef]

68. Jones, R.B.; Alderete, T.L.; Kim, J.S.; Millstein, J.; Gilliland, F.D.; Goran, M.I. High intake of dietary fructose in overweight/obese teenagers associated with depletion of Eubacterium and Streptococcus in gut microbiome. *Gut Microbes* **2019**, *10*, 712–719. [CrossRef]
69. Endo, J.; Arita, M. Cardioprotective mechanism of omega-3 polyunsaturated fatty acids. *J. Cardiol.* **2016**, *67*, 22–27. [CrossRef]

antioxidants

MDPI

Article

Beneficial Regulatory Effects of Polymethoxyflavone—Rich Fraction from Ougan (*Citrus reticulata* cv. *Suavissima*) Fruit on Gut Microbiota and Identification of Its Intestinal Metabolites in Mice

Jiebiao Chen [1,†], Yue Wang [1,†], Tailin Zhu [1], Sijia Yang [1], Jinping Cao [1], Xian Li [1], Li-Shu Wang [2] and Chongde Sun [1,*]

1 Laboratory of Fruit Quality Biology/Zhejiang Provincial Key Laboratory of Horticultural Plant Integrative Biology/The State Agriculture Ministry Laboratory of Horticultural Plant Growth, Development and Quality Improvement, Zhejiang University, Zijingang Campus, Hangzhou 310058, China; jiebiaochen@zju.edu.cn (J.C.); fruit@zju.edu.cn (Y.W.); flannery@zju.edu.cn (T.Z.); 3170100347@zju.edu.cn (S.Y.); caojinpingabc@126.com (J.C.); xianli@zju.edu.cn (X.L.)

2 Division of Hematology and Oncology, Department of Medicine, Medical College of Wisconsin, Milwaukee, WI 53226, USA; liswang@mcw.edu

* Correspondence: adesun2006@zju.edu.cn; Tel.: +86-0571-88982229

† These author contributed equally to this work.

Received: 30 July 2020; Accepted: 4 September 2020; Published: 6 September 2020

Abstract: Polymethoxyflavones (PMFs) are special flavonoids in citrus fruits that have been suggested to be beneficial to human health. However, whether PMFs in citrus fruit alter human gut microbiota is not well understood. The aim of the present study was to investigate the effects of PMF-rich fraction from Ougan (*Citrus reticulata* cv. *Suavissima*) on gut microbiota and evaluate the intestinal metabolic profile of PMFs in Institute of Cancer Research mice. The main components of the PMF-rich fraction were nobiletin, tangeretin, and 5-demethylnobiletin. The composition of the gut microbiota was analyzed using 16S ribosomal DNA sequencing. The results showed that after oral administration, the composition of mice gut microbiota was significantly altered. The relative abundance of two probiotics, *Lactobacillus* and *Bifidobacterium*, were found to increase significantly. A total of 21 metabolites of PMFs were detected in mice intestinal content by high performance liquid chromatography electrospray ionization tandem mass spectrometry, and they were generated through demethylation, demethoxylation, hydroxylation, and glucuronidation. Our results provided evidence that PMFs have potential beneficial regulatory effects on gut microbiota that in turn metabolize PMFs, which warrants further investigation in human clinical trials.

Keywords: Ougan (*Citrus reticulata* cv. *Suavissima*); polymethoxyflavones; gut microbiota; metabolism in vivo; metabolite identification; beneficial regulatory effect

1. Introduction

The gut microbiota is a complex ecosystem inhabiting in the gastrointestinal tract and consisting of a diverse microbiotic community living in symbiosis with the host [1]. Mounting studies have suggested that the gut microbiota possesses many vital functions, functioning as an indispensable part of the human body to maintain our health, including regulating the energy metabolism, immune function, and hormonal balance [2–4]. The gut microbiota can directly shape our health states. A critical part of gut immune system is the significant role it plays in stabilizing the host defense by protecting against pathogens, and there will be a risk of immune system-mediated diseases if it is out of balance [4]. The microbial compositions and abundance are directly and indirectly changed by the dietary patterns, antibiotics, probiotics, and lifestyle of the individual, among which the diet is considered the primary modulator of the gut microbiota [2]. This is not surprising because distinct food components selectively enrich microorganisms that are able to utilize these nutrients and support microbial metabolic cross-feeding, leading to the maintenance of a diverse and balanced community [4].

Fruit is an important component of the human daily diet that is abundant in phenolic compounds. As an extensively distributed kind of phytochemical, it is well accepted that phenolic compounds possess many bioactivities [5]. Recent studies have reported that some fruits such as cranberry [6], pomegranate [7], blackcurrant [8], apple, grape [9], and blueberry [10] are able to alter the gut microbiota, either by increasing the abundance of beneficial bacteria or reducing the abundance of harmful bacteria. Moreover, the gut microbiota regulation abilities are believed to be attributed to the abundant phenolic compounds in these fruits [11]. Flavonoids are naturally occurring phenolic compounds that can be divided into six classes according to their molecular structure: flavones, flavanones, flavonols, isoflavones, anthocyanidins, and flavanols [12]. Most plant-derived flavonoids are present in a glycoside form and are conjugated with sugars. Due to their hydrophilicity, they are not easily absorbed and metabolized by the upper gastrointestinal tract. Evidence has shown that after oral administration, a large fraction of flavanone reaches the colon intact, where it is metabolized into various small absorbable phenolics by the gut microbiota [13]. The gut microbiota also play an important role in metabolizing the flavonoids that cannot be metabolized by human enzymatic hydrolysis, such as *C*-glycosyl flavonoids [14] and rhamno-glucosides [13].

Citrus fruits are rich sources of flavonoids, and those specifically derived from citrus fruits are called citrus flavonoids. Among the flavonoids, flavanones are the highest in content [15], while the content of polymethoxyflavones (PMFs) varies among different types and tissues of citrus fruit [16]. Studies have been carried out to investigate the interaction between the intestinal metabolism of citrus flavanones and gut microbiota. For example, the representative citrus flavanone glycosides hesperidin and naringin are mainly metabolized by intestinal bacteria, resulting in the generation of their aglycone forms and other smaller phenolics. As a result, these citrus flavanones and their metabolites influence the composition and activity of the gut microbiota [17]. However, only a few studies have investigated the bacterial metabolism of PMFs and the influence of PMFs on the composition of gut microbiota.

PMFs are flavones that contain four or more methoxy groups (OCH_3) on their basic benzo-γ-pyrone (15-carbon, C6-C3-C6) skeleton with a carbonyl group at the C4 position. In general, PMFs exclusively exist in citrus fruits, and they are principally identified in the flavedo of certain categories. The major PMFs distributed in citrus fruit are nobiletin and tangeretin [18]. The research community has been interested in PMFs for many years because of its broad spectrum of bioactivities. Plenty of studies have indicated that PMFs show potent protective effects on humans' physical health, such as anti-inflammation [19], anticancer effects [20], neuroprotection [21], and metabolic disorder regulating functions [22]. Many properties are partly related by their antioxidant bioactivity, and there is a growing interest in the usage of citrus PMFs as a strategy to prevent oxidative damage in various health disorders [23]. Bioavailability is a crucial factor that determines the biological activity of a certain substance. Experiments in vivo and in vitro have been carried out to study the absorption and metabolism of PMFs. The metabolites of PMFs closely associate with their bioactivities, and a growing body of evidence has suggested that their metabolites may have greater activity compared to parental compounds [24]. PMFs undergo a series of complex biotransformations in vivo, and there is evidence that the gut microbiota plays a significant role in this process [25].

In this study, PMF-rich fraction was extracted from Ougan (*Citrus reticulata* cv. *Suavissima*), a characteristic citrus variety in Zhejiang province, which is abundant in PMFs according to our previous findings [16]. The main objective of the present study was to study the effects of PMFs from Ougan on the composition of the gut microbiota, as well as how PMFs are converted in the gastrointestinal tract of mice.

2. Materials and Methods

2.1. Materials and Animals

Ougan fruits at commercial maturity were collected in December 2017, from Lishui city, Zhejiang province, China. Fruits without mechanical damage, diseases, or pests were selected and stored at 4 °C for further experiments.

Four-week-old male Institute of Cancer Research (ICR) mice (n = 70) were bred with four animals per cage in the Laboratory Animal Center of Zhejiang University. ICR mice are a general-purpose model that are used in particular in toxicology, neurobiology, oncology, epidemiology, infection, and pharmacology testing, as well as in product safety testing. The mice were housed in a controlled environment (temperature controlled at 23 ± 3 °C, 12 h daylight cycle) with food and water ad libitum. All the protocols in this study were approved by the Committee on the Ethics of Animal Experiments of Zhejiang University (permission number: ZJU20200045). The mouse feed was purchased from Fbsh Bio-Pharmaceutical Co., Ltd. (Shanghai, China), and was flavonoid-free.

2.2. Extraction of PMF-Rich Fraction from Ougan Fruit

The extraction of PMF-rich fraction was performed according to our previous study [26] with modifications. After washing, the peel of Ougan fruit was separated and collected. A total of 100 g of peel was accurately weighed and grinded in 500 mL of 80% ethanol using a juice extractor. Grinded peel was then ultrasonically extracted with a frequency of 53 kHz for 1 h. Four-layer gauze and filter paper were subsequently used for filtering, and the filtrate was dried in a vacuum rotatory evaporator at 37 °C. After that, the residue was dissolved in double-distilled H_2O (ddH_2O). Sep-pak C18 cartridge columns (20 cc, 5 g sorbent, Waters Corp., Milford, MA, USA) were used for enrichment of the PMF fraction. After the samples were loaded, ddH_2O of 20 bed volume (BV) was eluted to remove the organic acids and sugars. Then, the flavonoid components with high polarities were removed by 8 BV 30% and 1 BV 35% aqueous methanol solution successively. Next, the PMF-rich fraction was eluted by 1 BV 100% methanol. The eluent was vacuum-dried for further experiments.

2.3. Collection of the Intestinal Contents

After 1 week of acclimation, 60 mice were administrated with PMF-rich fraction at a dose of 200 $mg{\cdot}kg^{-1}$ $BW{\cdot}d^{-1}$ by gavage. The contents in the small intestine, cecum, and colon were collected and combined 1 h, 2 h, 3 h, 6 h, 12 h, and 24 h after the PMF administration, with 10 mice at each time point. The controls were administrated with equal volume of pure water (n = 10), and the intestinal contents were collected 1 h after gavage. All the intestinal contents were snap frozen in liquid nitrogen and then kept at −80 °C for subsequent experiments.

2.4. Extraction of Flavonoid Components from the Intestinal Contents

The intestinal contents from five mice of each group were randomly selected for the extraction of flavonoid components. For each sample, 0.2 g of intestinal contents was accurately weighed and ultrasonically extracted with 2 mL of 80% ethanol, with a frequency of 53 kHz for 30 min. The mixture was then centrifuged at 5000 rpm for 10 min at room temperature, and the supernatant was collected. The extraction steps were repeated three times, and the supernatant was combined and dried in a vacuum rotatory evaporator at a temperature of 37 °C. The residue was dissolved in 3 mL of ddH_2O. Sep-pak C18 cartridge columns (1 cc, 0.1 g sorbent, Waters Corp., Milford, MA, USA) were used for enrichment of the flavonoid components. After the samples were loaded, we eluted ddH_2O of 20 BV to remove the non-flavonoid fraction. Then, the flavonoid components were eluted by 2 BV 100% methanol. Next, 200 µL eluent of each sample was pipetted and combined. The mixed solution was vacuum dried at 37 °C, and the residue was dissolved in 200 µL of HPLC-grade methanol for further analysis. Likewise, 200 µL eluent of each control group sample was mixed and vacuum-dried, and then dissolved in 200 µL of HPLC-grade methanol as the background control.

2.5. Identification of Flavonoids

High performance liquid chromatography electrospray ionization tandem mass spectrometry (HPLC–ESI–MS/MS) was performed to identify the flavonoid components in PMF-rich fraction and intestinal content. Two mixed samples were used for the identification of intestinal metabolites. Detailed conditions were described in our previous publication with some modifications [16].

HPLC was performed on a SunFire C18 (5 μm, 4.6 × 250 mm) column. The linear gradient elution was performed using a mobile phase composed of mixture of solution A (water containing 0.1% formic acid) and B (HPLC grade acetonitrile). The optimized condition is listed as follows: solution B: 0–5 min, 20%; 5–10 min, 20–27%; 10–15 min, 27%; 15–25 min, 27–40%; 25–35 min, 40–60%; 35–40 min, 60–80%; 40–42 min, 80–100%; 42–45 min, 100–20%; 45–50 min, 20%. Detection temperature was 25 °C, injection volume was 10 μL, the flow rate was 1 mL/min, and the compounds were detected between 200 and 500 nm.

Mass spectrometric analysis was operated on a AB Triple TOF 5600^{plus} system. Briefly, the nebulizer pressure was set to 45 psi, and the drying gas flow rate was 5 L/min. The flow rate and the temperature of the sheath gas were 11 L/min and 350 °C, respectively. All the produced ions were introduced into the TOF–MS instrument for accurate mass determination.

2.6. DNA Extraction and 16S ribosomal DNA Sequencing of Intestinal Content from Mice

The intestinal contents from the remaining five mice of each group were used for 16S ribosomal DNA sequencing to study the composition of the gut microbiota. The DNA of the intestinal content was extracted using MagPure Stool DNA KF kit B (Magen, Hong Kong, China) following the manufacturer's instructions. The extracted DNA was quantified with a Qubit Fluorometer, and the V3–V4 hypervariable regions of the bacteria 16S ribosomal DNA gene were amplified by PCR with barcode-indexed primers 515F (5′-GTGCCAGCMGCCGCGGTAA-3′) and 806R (5′-GGACTACHVGGGTWTCTAAT-3′). The validated libraries were used for sequencing on the Illumina HiSeq 2500 platform (BGI, Shenzhen, China). Detailed PCR conditions are listed in Table S1.

2.7. Sequencing Data Processing

After the raw reads were filtered, we added paired-end reads to tags by FLASH (v1.2.11) to obtain the tags. The tags were clustered into operational taxonomic units (OTUs) with a cutoff value of 97%, then out-representative sequences were taxonomically classified using RDP Classifier v.2.2 with a minimum confidence threshold of 0.6, and trained on the Greengenes database v201305 by QIIME v1.8.0. The USEARCH_global was used to compare all tags back to OTU to obtain the OTU abundance statistic table of each sample.

2.8. Statistical Analysis

The HPLC–ESI–MS/MS data was analyzed by PeakView software (version 1.2, AB SCIEX, Toronto, ON, Canada).

Significant species were determined by R (v3.4.1) on the basis of the Wilcoxon test or Kruskal–Wallis test. All the results were considered statistically significant at $p < 0.05$. Alpha and beta diversity were estimated by MOTHUR (v1.31.2) and QIIME (v1.8.0) at the OTU level. Principal component analysis (PCA) in OTUs was plotted with R package "ade4". Sample cluster was conducted by QIIME (v1.8.0) on the basis of UPGMA. The heat map of different classification levels was plotted with R package "gplots".

3. Results

3.1. Identification of PMF-Rich Fraction

A total of 11 flavonoid compounds were identified in PMF-rich fraction (Table 1 and Figure 1a), including two flavanone-O-glycosides (neohesperidin and poncirin) and nine PMFs. The top three components in this fraction were nobiletin, tangeretin, and 5-demethylnobiletin, and they were in proportions of 48.85%, 31.16%, and 5.05% respectively; the three combined was 85.06%. We previously reported that after C18 Solid Phase Extraction (SPE) enrichment, the proportion of combined nobiletin, tangeretin, and 5-demethylnobiletin increased from 58.7% to 85.3% [26]. Thus the result from this current study was consistent with our previous finding.

Table 1. Flavonoids identification in Ougan polymethoxyflavone (PMF)-rich fraction by HPLC–ESI–MS/MS.

Peak No.	Retention Time (min)	$[M + H]^+$ or $[M - H]^-$ (m/z)	Error (ppm)	Formula	Fragment Ions (m/z)	Tentative Compounds
1	13.4438	609.1827 (-)	0.5	$C_{28}H_{34}O_{15}$	489, 343, 301, 286, 242	Neohesperidin
2	21.2998	593.1876 (-)	0.0	$C_{28}H_{34}O_{14}$	285	Poncirin
3	29.1432	373.1291	2.2	$C_{20}H_{20}O_7$	358, 343, 327, 315, 299, 181, 163, 153	Isosinensetin
4	29.3402	389.1231	2.6	$C_{20}H_{20}O_8$	374, 359, 344, 331, 313, 298, 287, 211, 183	Monohydroxy-pentamethoxyflavone
5	31.3811	373.1282	0.9	$C_{20}H_{20}O_7$	357, 343, 329, 312, 297, 153	Sinensetin
6	31.8459	343.1176	0.8	$C_{19}H_{18}O_6$	328, 313, 285, 257, 181, 153	Tetramethyl-O-isoscutellarein
7	33.4704	403.1387	0.6	$C_{21}H_{22}O_8$	388, 373, 358, 355, 327, 211, 183	Nobiletin
8	33.853	375.1074	1.7	$C_{19}H_{18}O_8$	360, 345, 330, 327, 317, 302, 197, 169, 149	5,4′-Dihydroxyl-3,7,8,3′-tetramethoxyflavone
9	34.0789	343.1176	2.0	$C_{19}H_{18}O_6$	328, 313, 285, 257, 181, 153	Tetramethyl-O-scutellarein
10	35.9401	373.1282	0.3	$C_{20}H_{20}O_7$	358, 343, 328, 297, 211, 183	Tangeretin
11	37.8647	389.1231	0.5	$C_{20}H_{20}O_8$	374, 359, 341, 331, 197	5-demethylnobiletin

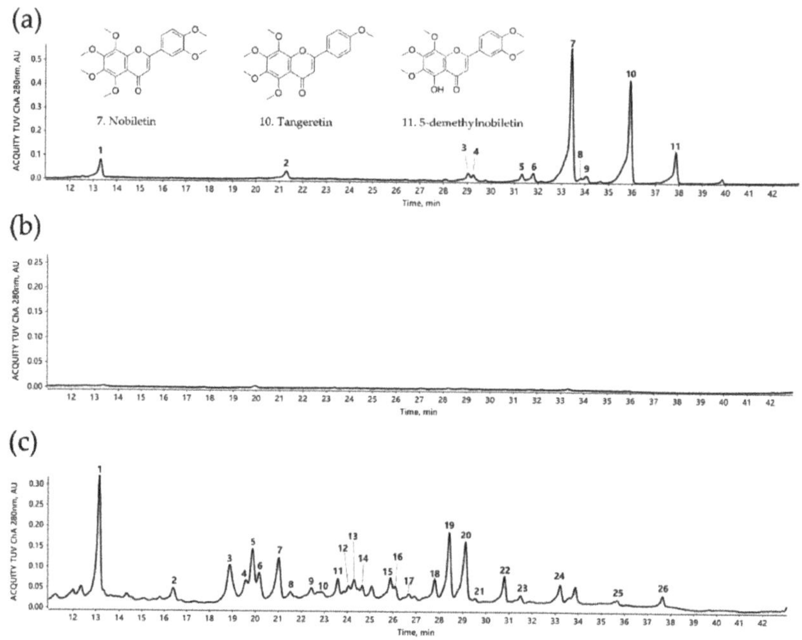

Figure 1. HPLC chromatogram of Ougan PMF-rich fraction and mice intestinal content samples (λ = 280nm). (**a**) HPLC chromatogram of Ougan PMF-rich fraction. 1: neohesperidin; 2: poncirin; 3: isosinensetin; 4: monohydroxy-pentamethoxyflavone; 5: sinensetin; 6: tetramethyl-*O*-isoscutellarein; 7: nobiletin; 8: 5,4′-dihydroxyl-3,7,8,3′-tetramethoxyflavone; 9: tetramethyl-*O*-scutellarein; 10: tangeretin; 11: 5-demethylnobiletin. Nobiletin (7), tangeretin (10), and 5-demethylnobiletin (11) are the three major components in Ougan PMF-rich fraction. (**b**) HPLC chromatogram of intestinal content of the mixture of control group. (**c**) HPLC chromatogram of intestinal content of the mixture of all groups. 1: neohesperidin; 2: nobiletin-*O*-glucuronide (1); 3: tangeretin-*O*-glucuronide (1); 4: nobiletin-*O*-glucuronide (2); 5: nobiletin-*O*-glucuronide (3); 6: dihydroxy-tetramethoxyflavone (1); 7: dihydroxy-monomethoxyflavone; 8: dihydroxy-trimethoxyflavone; 9: dihydroxy-tetramethoxyflavone (2); 10: monohydroxy-trimethoxyflavone; 11: tangeretin-*O*-glucuronide (2); 12: dihydroxy-tetramethoxyflavone (3); 13: dihydroxy-tetramethoxyflavone (4); 14: nobiletin-*O*-glucuronide (4); 15: monohydroxy-pentamethoxyflavone (1); 16: trihydroxy-flavone; 17: monohydroxy-tetramethoxyflavone (1); 18: monohydroxy-pentamethoxyflavone (2); 19: monohydroxy-tetramethoxyflavone (2); 20: monohydroxy-pentamethoxyflavone (3); 21: dihydroxy-dimethoxyflavone; 22: monohydroxy-tetramethoxyflavone (3); 23: tetramethoxyflavone; 24: nobiletin; 25: tangeretin; 26: 5-demethylnobiletin.

Interestingly, isomerization was frequently found among these substances, for example, sinensetin, isosinensetin and tangeretin, tetramethyl-*O*-scutellarein and tetramethyl-*O*-isoscutellarein, monohydroxy-pentamethoxyflavone, and 5-demethylnobiletin.

3.2. Identification of Metabolites of PMFs in Intestinal Content

The HPLC profile of the mixture of samples from the control group is shown in Figure 1b, which served as a background control, with the HPLC profile of the mixture of all samples that contained control and PMF-treated groups being displayed in Figure 1c. Compared with Figure 1c, the chromatogram in Figure 1b is much smoother and steadier, which indicates that the flavonoids in the intestinal content of control group were in an undetectable level. Moreover, the difference between these two figures suggested that the administration of PMF-rich fraction caused a dramatic change in intestinal content by producing many flavonoid metabolites.

We determined 26 compounds using HPLC–ESI–MS/MS (Figure 1c and Table 2). Among the 26 detected compounds, 5 of them were identified to be parental compounds of PMF-rich fraction (neohesperidin, monohydroxy-pentamethoxyflavone, nobiletin, tangeretin, and 5-demethylnobiletin), which suggested that PMFs had not been completely metabolized 1 h after they were consumed. Further, 21 compounds were tentatively predicted to be the metabolites of the PMF-rich fraction. These metabolites are mainly generated through demethylation, demethoxylation, hydroxylation, and glucuronidation. Metabolites varied with the number and position of chemical groups participating in demethylation, demethoxylation, and hydroxylation, including monohydroxy-trimethoxyflavone, monohydroxy-tetramethoxyflavone, monohydroxy-pentamethoxyflavone, dihydroxy-monomethoxyflavone, dihydroxy-dimethoxyflavone, dihydroxy-trimethoxyflavone, dihydroxy-tetramethoxyflavone, trihydroxy-flavone, and tetramethoxyflavone. Further, glucuronidation was found to occur after the demethylation of nobiletin and tangeretin, generating corresponding conjugates.

Isomerization was widely found among the metabolites. Specifically, two monohydroxy-tetramethoxyflavones, two monohydroxy-pentamethoxyflavones, four dihydroxy-tetramethoxyflavones, four nobiletin-*O*-glucuronides, and two tangeretin-*O*-glucuronides were detected in the intestinal content. Therefore, this finding suggested that the reaction position was of great diversity. The co-existence of multiple isomers created hurdles for accurately determining the structure of some metabolites. Moreover, it should be noted that the structure of the three main parental compounds (nobiletin, tangeretin, and 5-demethylnobiletin) was of high similarity, which caused barriers to match the metabolites with their corresponding parental compounds. In addition, there were also possibilities that some metabolites were generated through a series of reactions successively. Thus, further study was required to dissect the metabolic pathway of PMFs.

Table 2. Identification of intestinal metabolites of Ougan PMF-rich fraction by HPLC–ESI–MS/MS.

Peak No.	Retention Time (min)	$[M + H]^+$ or $[M - H]^-$ (m/z)	Formula	Fragment Ions (m/z)	Tentative Compounds
1	13.1284	609.1825 (-)	$C_{28}H_{34}O_{15}$	489, 343, 325, 301, 286, 257, 242, 164	Neohesperidin
2	16.3972	563.1460 (-)	$C_{26}H_{28}O_{14}$	387, 372, 357, 342, 314, 299, 175, 113	Nobelitin-*O*-glucuronide (1)
3	18.8731	535.1449	$C_{25}H_{26}O_{13}$	359, 344, 329, 314, 311, 301, 298, 286, 283, 257	Tangeretin-*O*-glucuronide (1)
4	19.5825	565.1555	$C_{26}H_{28}O_{14}$	389, 374, 359, 345, 344, 343, 341, 331, 328, 327, 316, 315, 313	Nobelitin-*O*-glucuronide (2)
5	19.8712	565.1551	$C_{26}H_{28}O_{14}$	389, 374, 359, 345, 344, 343, 341, 339, 331, 329, 328, 327, 316, 315, 313, 310, 301, 298	Nobelitin-*O*-glucuronide (3)
6	20.1621	375.1078	$C_{19}H_{18}O_8$	360, 345, 330, 327, 317, 302, 299, 197, 169, 149	Dihydroxy-tetramethoxyflavone (1)
7	21.0102	287.0917	$C_{16}H_{14}O_5$	161, 153, 135, 133, 125, 121, 118, 111, 103, 97, 69, 67	Dihydroxy-monomethoxyflavone
8	21.5526	345.0974	$C_{18}H_{16}O_7$	330, 315, 297, 287, 272, 197, 169	Dihydroxy-trimethoxyflavone
9	22.4321	375.1081	$C_{19}H_{18}O_8$	360, 345, 330, 327, 317, 302, 197, 169	Dihydroxy-tetramethoxyflavone (2)
10	22.6898	327.0874 (-)	$C_{18}H_{16}O_6$	327, 312, 297, 282, 269, 254, 226, 182, 177, 117	Monohydroxy-trimethoxyflavone
11	23.5766	535.1449	$C_{25}H_{26}O_{13}$	359, 344, 329, 311	Tangeretin-*O*-glucuronide(2)
12	24.0618	375.1081	$C_{19}H_{18}O_8$	360, 345, 331, 330, 327, 317, 314, 302, 301, 299, 287, 285, 274, 273, 271, 211, 183, 168, 165, 147, 139, 137, 135, 134, 127	Dihydroxy-tetramethoxyflavone (3)
13	24.2942	375.1084	$C_{19}H_{18}O_8$	360, 345, 331, 330, 327, 325, 317, 314, 313, 302, 299, 287, 285, 274, 273, 271, 230, 211, 183, 168, 137, 135, 127	Dihydroxy-tetramethoxyflavone (4)
14	24.6514	565.1555	$C_{26}H_{28}O_{14}$	389, 374, 359, 356, 341	Nobelitin-*O*-glucuronide (4)
15	25.8876	389.1237	$C_{20}H_{20}O_8$	374, 359, 356, 341, 331, 316, 285, 244, 197, 169, 163, 148, 113	Monohydroxy-pentamethoxyflavone (1)
16	26.0809	271.0607	$C_{15}H_{10}O_5$	253, 243, 215, 197, 169, 153, 149, 115, 91	Trihydroxy-flavone
17	26.6764	359.1135	$C_{19}H_{18}O_7$	344, 343, 341, 329, 327, 325, 315, 301, 300, 298, 297, 283, 272, 255, 227, 181, 153	Monohydroxy-tetramethoxyflavone (1)
18	27.7937	389.1238	$C_{20}H_{20}O_8$	374, 359, 344, 343, 341, 331, 316, 197, 169, 165, 163	Monohydroxy-pentamethoxyflavone (2)
19	28.4142	359.1128	$C_{19}H_{18}O_7$	344, 329, 315, 314, 311, 309, 301, 298, 297, 286, 285, 283, 271, 268, 258, 257, 255, 240, 230, 228, 215, 214, 212, 211, 200, 193, 187, 183, 168, 165, 139, 131, 127, 121	Monohydroxy-tetramethoxyflavone (2)
20	29.1096	389.1234	$C_{20}H_{20}O_8$	374, 359, 345, 344, 343, 341, 339, 331, 328, 326, 316, 315, 313, 310, 301, 299, 288, 287, 285, 270, 260, 245, 230, 217, 211, 193, 183, 168, 151, 127	Monohydroxy-pentamethoxyflavone (3)

Table 2. *Cont.*

Peak No.	Retention Time (min)	$[M + H]^+$ or $[M - H]^-$ (m/z)	Formula	Fragment Ions (m/z)	Tentative Compounds
21	29.5447	359.0772	$C_{17}H_{14}O_6$	344, 329, 314, 311, 301, 286, 283, 258, 242, 230, 214, 202, 193, 177, 174, 133	Dihydroxy-dimethoxyflavone
22	30.7912	359.1135	$C_{19}H_{18}O_7$	344, 329, 314, 311, 301, 286, 283, 197, 169, 133	Monohydroxy-tetramethoxyflavone (3)
23	31.4917	343.1182	$C_{19}H_{18}O_6$	328, 313, 285, 181, 153, 133	Tetramethoxyflavone
24	33.2004	403.1395	$C_{21}H_{22}O_8$	388, 387, 373, 359, 358, 357, 355, 353, 345, 343, 341, 339, 330, 327, 325, 315, 313, 311, 301, 299, 259, 257, 244, 231, 211, 193, 183, 175, 168, 163, 162	Nobiletin
25	35.6699	373.1291	$C_{20}H_{20}O_7$	358, 343, 325, 315, 300, 297, 283, 271, 269, 229, 211, 193, 183, 168, 135	Tangeretin
26	37.6439	389.1241	$C_{20}H_{20}O_8$	374, 359, 356, 341, 343, 328, 316, 313, 197, 169, 163	5-demethylnobiletin

3.3. Effects of PMF-Rich Fraction on Gut Microbiota

We obtained a total of 2,137,084 high-quality 16S ribosomal DNA gene sequence tags from intestinal contents from 35 mice with an average read length of 252 bp (SD = 1 bp). Filtered tags were clustered into OTU by using the Greengene database at 97% similarity. Alpha diversity of each sample was analyzed and the results are displayed in Figure 2a and Figure S1. The diversity of species of the gut microbial community reflected by Shannon and Simpson indexes showed no significant difference among seven groups ($p > 0.05$, p (Shannon) = 0.20873, p (Simpson) = 0.51176), whereas the species richness indicated by Ace and Chao indexes was significantly increased in the PMF groups compared with the control group ($p < 0.05$, p (Ace) = 0.00671, p (Chao) = 0.01441), except the PMF 12h group.

Principle component analysis (PCA) was performed to show the differences of microbial composition among different groups at the OTU level. As depicted in Figure 2b, samples from groups PMFs.12h and PMFs.24h formed a cluster that was different from metagenomes derived from clusters formed by the remaining five groups. Therefore, administration of PMF-rich fraction led to the alterations of the gut microbial community 12–24 h after it was consumed.

Similarity in composition of species among samples was evaluated. The clustering results were shown in Figure 2c. As shown in this figure, groups PMFs.12h and PMFs.24h were grouped into one cluster (with a green background), and the other five groups formed the other cluster. This suggested that the gut microbial compositions between groups PMFs.12h and PMFs.24h were similar, and they were distinguished from the control group and other early time points.

Similarly, we carried out heat map analysis on the basis of the relative abundance of each species at the genus and species levels. The distance algorithm was "Euclidean", and the clustering method was "complete". As displayed in Figure 2e and Figure S2, at the genus and species levels, most of the samples of group PMFs.12h and group PMFs.24h were clustered into the same cluster, which indicated that the microbial compositions of these two groups have high similarity.

We further performed analysis of beta diversity to evaluate the difference of samples in the complexity of species. The distance algorithm was "weighted unifrac". The heat map of beta diversity distribution is shown in Figure 2d. Groups PMFs.12h and PMFs. 24h were clustered into the same cluster, suggesting a high similarity in their microbial compositions, which was consistent with the results of PCA, clustering analysis, and heat map analysis.

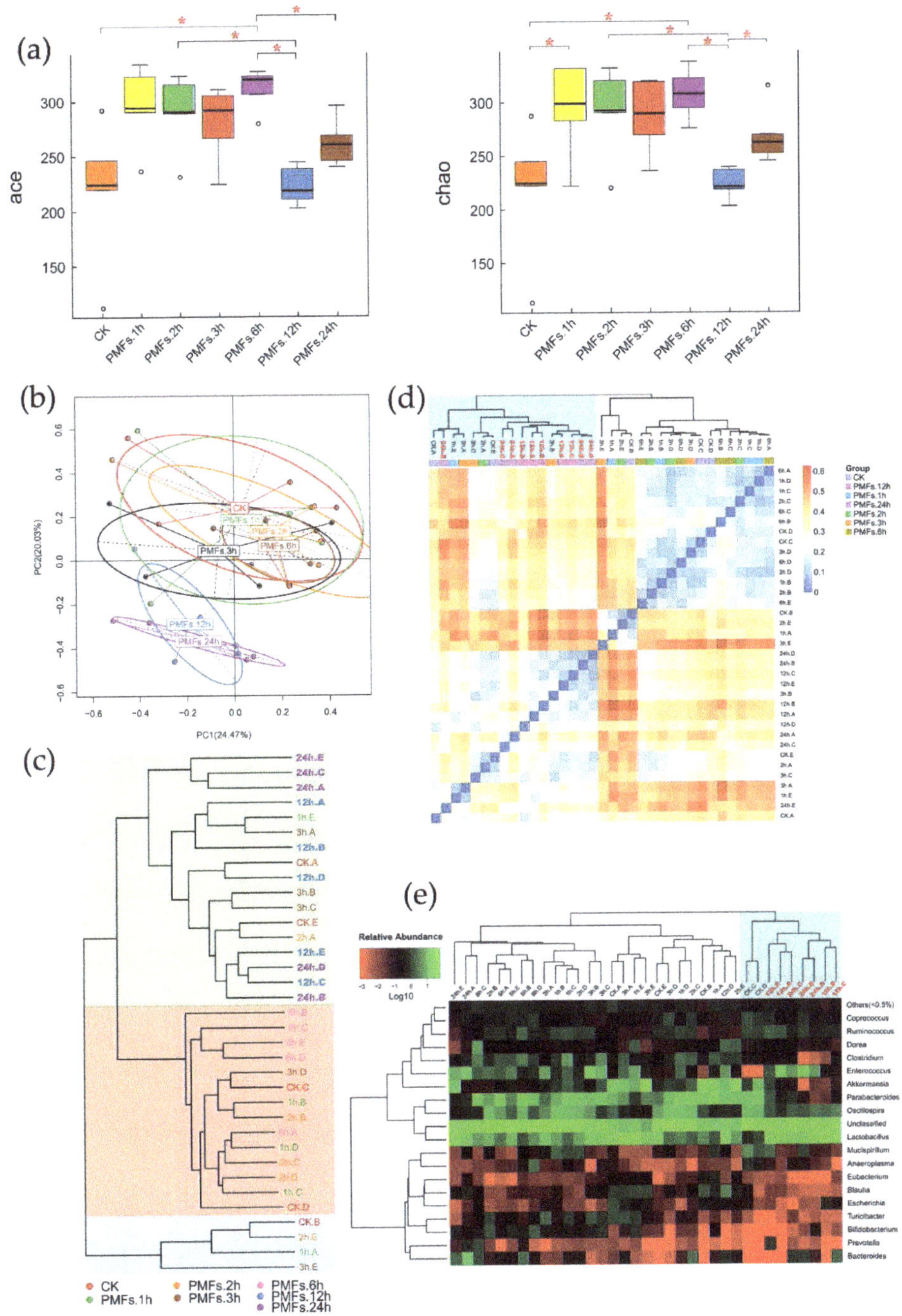

Figure 2. Administration of PMF-rich fraction significantly altered the composition of mice gut microbiota. (**a**) Alpha diversity indices boxplot among groups. Five lines from bottom to top is the minimum value, the first quartile, median, the third quartile, and the maximum value, and the abnormal values are outliers shown as "o"; * $p < 0.05$. (**b**) Principal component analysis (PCA) based on operational taxonomic unit (OTU) abundance. *x*-axis, first principal component, and *y*-axis, second

principal component. Numbers in brackets represent contributions of principal components to differences among samples. A dot represents each sample, and different colors represent different groups. (**c**) Samples clustering (weighted unifrac). The word with the same color represents the samples in the same group, the shorter distance between samples represents high similarity. Samples in the same background color represent a high similarity. (**d**) Beta diversity heat map (weighted unifrac). Weighted unifrac value was used to measure beta diversity. Groups PMFs.12h and PMFs.24 h were clustered into the same group highlighted in blue in the top-left corner. (**e**) Log-scaled percentage heat map based on the relative abundance of each species in each sample (genus level). Longitudinal clustering indicates the similarity of all species among different samples, and the horizontal clustering indicates the similarity of certain species among different samples. The closer the distance, the shorter the branch length, and the more similar the species composition between the samples. Relative abundance values were all log-transformed. Most of the samples of groups PMFs.12h and PMFs.24h were clustered into the same group, highlighted in blue in the top-right corner.

In order to assess the specific effect of PMF-rich fraction on the composition and abundance of microbial communities, we performed species annotation on the basis of OTU at the phylum, genus, and species levels. The distribution of taxonomic composition histograms of each sample at different levels are shown in Figure 3a and Figure S3. We further chose species with high relative abundance of each taxonomic rank to perform difference analysis, and the results are displayed in Figure 3 and Figure S4.

At the phylum level (Figure 3b and Figure S4a), no significant difference was observed in the composition of Firmicutes ($p = 0.5237$) and Bacteroidetes ($p = 0.2403$) among groups, whereas the relative abundance of Verrucomicrobia ($p = 0.04$, Figure 3b) was significantly altered. Compared with the control group, the relative abundance of Verrucomicrobia was significantly decreased in PMFs.6h ($p = 0.0122$).

At the genus level (Figure 3c and Figure S4b), no significant shift was observed in the compositions of *Oscillospira* ($p = 0.4076$) and *Blautia* ($p = 0.3281$), while the compositions of *Akkermansia* ($p = 0.0382$), *Lactobacillus* ($p = 0.0166$), *Parabacteroides* ($p = 0.0063$), *Bifidobacterium* ($p = 0.0034$), and *Enterococcus* ($p = 0.0097$) were significantly changed. The relative abundance of *Akkermansia* in group PMFs.6h was significantly decreased compared to the control group ($p = 0.0122$), PMFs.1h ($p = 0.01219$), and PMFs.3h ($p = 0.0367$). The relative abundance of *Lactobacillus* in group PMFs.12h was significantly increased compared to the control group ($p = 0.0367$), PMFs.2h ($p = 0.0122$), and PMFs.6h ($p = 0.0122$). As for *Parabacteroides*, the relative abundance of this species in PMFs.24h was significantly decreased compared to the control group, PMFs.1h, PMFs.2h, PMFs.3h, and PMFs.6h ($p = 0.0122$). Further, the relative abundance of *Enterococcus* was significantly increased after the administration of PMF-rich fraction at all time points when it was compared to the control group ($p < 0.05$). Moreover, compared to the control group, the relative abundance of *Bifidobacterium* in PMFs.1h ($p = 0.0112$), PMFs.2h ($p = 0.0345$), and PMFs.6h ($p = 0.0112$) was significantly increased. However, the relative abundance of *Bifidobacterium* was significantly decreased in PMFs.12h ($p = 0.0367$) and PMFs.24h ($p = 0.0345$) compared to PMFs.6h.

Further analysis at the species level (Figure 3d) revealed that the species that significantly changed in the genus *Lactobacillus* were *Lactobacillus reuteri* ($p = 0.0129$) and *Lactobacillus salivarius* ($p = 0.0067$), and that significantly changed in the genus *Bifidobacterium* was *Bifidobacterium pseudolongum* ($p = 0.0034$).

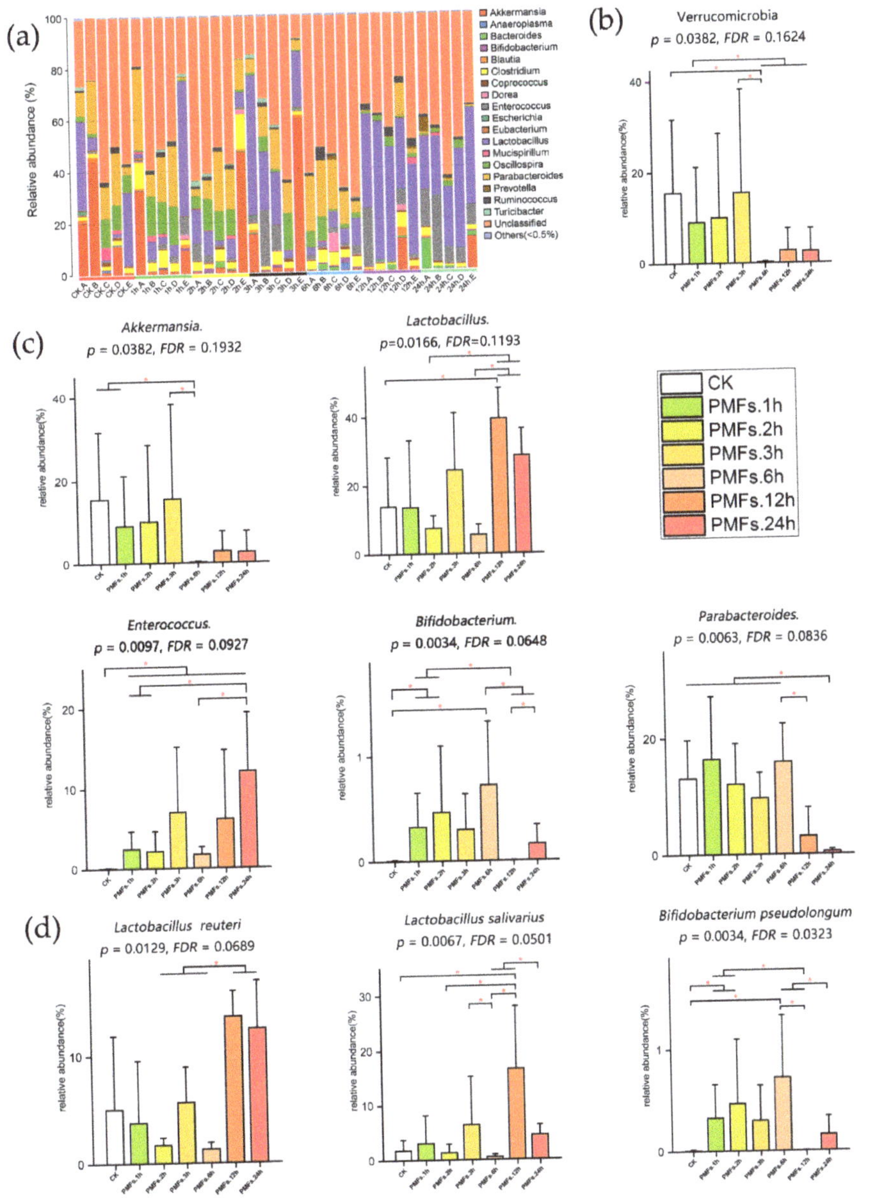

Figure 3. Specific effect of PMF-rich fraction on the composition and abundance of microbial communities. (**a**) The taxonomic composition distribution (genus level). "Others (<0.5%)" includes all the taxonomic groups with relative abundance less than 0.5%. (**b**) Effects of PMF-rich fraction on the composition of gut microbiota at the phylum level. (**c**) Effects of PMF-rich fraction on the composition of gut microbiota at the genus level. (**d**) Effects of PMF-rich fraction on the composition of gut microbiota at the species level. Data were expressed as mean ± SD. Kruskal–Wallis test was used for multi-group comparations, wherein the significance level was 0.05 and the *p*-value was adjusted in the false discovery rate (FDR) method. Wilcoxon rank-sum test was used for two group comparations. * $p < 0.05$, $n = 5$ individuals per group.

4. Discussion

Several studies have been carried out to investigate the metabolism of PMFs in vivo. Nevertheless, urine, feces, and blood plasma of animals were used in the majority of these studies, with only few studies measuring the metabolites in the intestinal contents. Goncalves et al. [27] identified seven metabolites in rat urine after the administration of PMFs, including tangeretin-4′-glucuronide, 3′,4′-dihydroxy-5,6,7-trimethoxy-4′-glucuronide, 4′-demethyltangeretin, 3′-demethylsinensetin, nobiletin-*O*-glucuronide, 4′-demethylnobiletin, and 3′,4′-dihydroxy-3,5,6,7,8-pentamethoxyflavone. Zheng et al. [28] found seven metabolites in mice urine after oral administration of nobiletin, which were 3′-demethylnobiletin, 4′-demethylnobiletin, 5-demethylnobiletin, 5,3′-demethylnobiletin, 5,4′-demethylnobiletin, 3′,4′-demethylnobiltin, and 5,3′,4′,-demethylnobiletin. Similar results were obtained in the study by Li et al. [29]. They found that 4′-demethylnobiletin was the main metabolite of nobiletin in mice urine, while there were also small amounts of 3′-demethylnobiletin and other dimethyl products. They also reported that the metabolites further underwent glucuronidation, sulfation, and demethylation. Zeng et al. [30] proposed that the metabolic pathway of nobiletin and tangeretin was systematically highly similar; the parental compounds underwent demethlylation, hydroxylation, and demethoxylation as the first step, and the metabolites generated in the first step further underwent glucuronidation.

In our study, intestinal content was used to investigate the metabolism of PMFs. A total of 21 metabolites of PMFs were identified. It is likely that these metabolites were generated mostly through demethylation, hydroxylation, demethoxylation, and glucuronidation. Our result agreed with the previous literature to some extent [27–30]. However, due to the co-existence of multiple isomers and the trace amount of some metabolites, the chemical structures of these metabolites are required to be further confirmed. Moreover, the high structural similarity of the parental compounds added another layer of barriers to identify the metabolic pathway of PMFs. It is possible that the same metabolites are derived from different parental compounds. Further research is required to match the metabolites with corresponding parental compounds by administrating individual parental compounds. It also needs to be confirmed as to whether a certain metabolite is generated through one-step-reactions or chain reactions. We speculated that the gut microbiota played a crucial role in metabolism of PMFs. However, the enzymes present in intestinal enterocytes may also be involved in the metabolic process. In addition, the metabolism of PMFs is likely to be influenced by the enterohepatic circulation. Germ-free or pseudo-germ-free animal models are needed to study the specific role of the gut microbiota on the metabolism of PMFs.

Our results showed that the administration of Ougan PMFs altered the gut microbiota composition. The administration of PMFs significantly increased the richness of the microbial community. However, no significant change was found in the diversity of the microbial community. Different analyses between different time points were performed to predict the trend of the microbial changes at different taxonomic levels. At the phylum level, the relative abundance of Verrucomicrobia was significantly decreased after 6 h oral administration of PMFs. At the genus level, a significant decrease in the composition of *Parabacteroides* was observed 24 h after PMF administration, while that of *Akkermansia* was significantly decreased 6 h after consuming PMFs. A significant increase was found in the composition of *Enterococcus*. The increase began 1 h after oral PMFs and its effect lasted at least until 24 h. Moreover, PMFs also caused a significant increase in the relative abundance of *Lactobacillus* 12 h later, and that of *Bifidobacterium* was found to significantly increase 6 h later. Further analysis on the species level revealed that *Lactobacillus reuteri* and *Lactobacillus salivarius* in *Lactobacillus*, and *Bifidobacterium pseudolongum* in *Bifidobacterium*, were significantly increased. Accordingly, it was as early as 1 h after oral PMFs that the gut bacteria were altered, and a more comprehensive spectrum of the changes were observed in 24 h.

There are only few studies investigating the influence of PMFs on gut microbiota. A recent study carried out by Tung et al. [31] examined the effects of citrus peel extracts on the gut microbiota of high-fat diet-induced obesity in C57BL/6 mice. Their results indicated that PMFs in citrus peel extracts were able to reverse the unbalanced gut microbiota caused by a high fat diet, mainly by increasing the abundance of *Prevotella* and decreasing the abundance of *rc4-4*. The differences between the results of this study and our current study can be caused by the mouse species, pathological states of mice, composition of PMFs, and treatment duration. Nevertheless, both studies indicated the potential health-promoting property of citrus PMFs through beneficially altering the composition of gut microbiota. A more recent study carried out by Man Zhang et al. [32] investigated the effects of aged citrus peel (*chenpi*) extract, which was rich in PMFs on colonic microbiota in high-fat diet induced obese mice. The results showed that after 11 weeks' treatment, *chenpi* extract demonstrated a prebiotic effect, evidenced by the promoted *Lactobacillus* and *Bifidobacterium*. Their results were consistent with ours, which indicated that citrus PMFs could exert a prebiotic effect, both in a short-term and long-term manner. Moreover, their study also demonstrated the dynamics of *Akkermansia* in a dose- and time-dependent manner. However, in our study, the relative abundance of *Akkermansia* was found to be significantly decreased 6h after consuming PMFs, and it was not time-dependent. The difference of the changes in *Akkermansia* may be because of that 24 h was too short for *Akkermansia* to present a time-dependent manner in the present study.

The Food and Agriculture Organization of the United Nations has defined probiotics as live microorganisms that are able to promote human health when administrated in adequate amounts [33]. Bacteria classified in the genera *Lactobacillus* and *Bifidobacterium* are universally acknowledged as important probiotics [34]. They are butyrate-producing bacteria, and mounting studies have shown that they play significant roles in regulating and preventing diseases [35,36]. An et al. [37] indicated that some species of *Bifidobacterium* are related to fat loss, which could be utilized as a potential method for weight loss in obese individuals. Tuohy et al. [38] drew a conclusion that some species of *Lactobacillus* and *Bifidobacterium* reduced intestinal absorption of cholesterol, thus lowering the risks of cardiovascular disease. Yunes et al. [39] indicated that various species of *Lactobacillus* and *Bifidobacterium* produce neurotransmitter γ-gamma-amino butyric acid (GABA), which is beneficial in regulating different physiological activities after passing through the gut–brain axis. Hsu et al. [40] found that the supplement of *Lactobacillus reuteri* mitigated hepatic inflammation and apoptosis. In our current study, the supplement of Ougan PMFs significantly increased the relative abundances of *Lactobacillus* and *Bifidobacterium* in gut microbiota. *Lactobacillus* and *Bifidobacterium* were positively correlated with body health, and thus it could be implied that the citrus PMFs were able to exert health-promoting and disease-preventing ability, partly through beneficially regulating the gut microbiota.

On the basis of the obtained results, we believe that there were possible interactions between Ougan PMFs and gut microbiota. In the intestinal contents, the main components of Ougan PMFs, nobiletin, tangeretin, and 5-demethylnobiletin, were metabolized and various metabolites were generated. These metabolites were mainly generated through demethylation, demethoxylation, hydroxylation, and glucuronidation. Although the metabolites may be generated through many metabolic pathways, such as enterocyte cell metabolism, enterohepatic circulation, and microbial metabolism, we speculated that the gut microbiota played the most important role. In turn, the administration of PMFs significantly altered the composition of gut microbiota, especially *Akkermansia*, *Lactobacillus*, *Parabacteroides*, *Bifidobacterium*, and *Enterococcus* at the genus level. The proposed interactions between Ougan PMFs and gut microbiota were shown in Figure 4. Further investigations using antibiotics or germ-free animals that feed Ougan PMFs are needed to confirm if gut microbiota are required to metabolize Ougan PMFs.

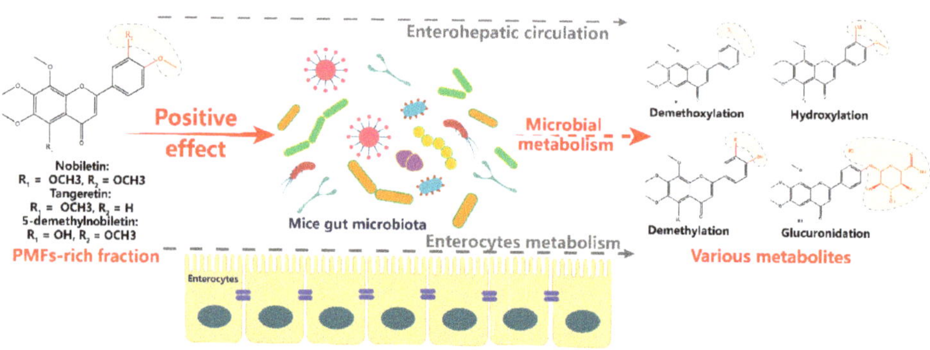

Figure 4. Hypothetical scheme of interactions between Ougan PMFs and the gut microbiota.

5. Conclusions

In conclusion, PMF-rich fraction extracted from Ougan fruit, with nobiletin, tangeretin, and 5-demethylnobiletin as main components, greatly altered the gut microbiota composition of mice. Two important probiotics, *Lactobacillus* and *Bifidobacterium*, were significantly increased after PMF administration, indicating a beneficial health-promoting effect of citrus PMFs through beneficially regulating the gut microbiota. Various metabolites of PMFs were detected and were generated through demethylation, hydroxylation, demethoxylation, and glucuronidation in the mouse gastrointestinal tract. However, there are still some limitations of the present study. Due to the lack of a single administration of each compound of the PMF-rich fraction, we found it difficult to match the metabolites with corresponding parental compounds. Moreover, it also caused hurdles for us to understand if all the single parental compounds had the same regulatory effects on the gut microbiota. Our study provided strong rationale for using PMFs as a food supplement to enhance human health, which warrants testing in the settings of clinical trials.

Supplementary Materials: The following are available online at http://www.mdpi.com/2076-3921/9/9/831/s1, Table S1: PCR conditions. Figure S1: Alpha diversity indices boxplot among groups. Figure S2: Log-scaled percentage heat map based on the relative abundance of each species in each sample (species level). Figure S3: The taxonomic composition distribution. Figure S4: Effects of PMF-rich fraction on the composition of gut microbiota.

Author Contributions: J.C. (Jiebiao Chen) carried out experiments and wrote the original manuscript. T.Z., S.Y., and Y.W. participated in the separation experiment. Y.W., J.C. (Jinping Cao), X.L., and L.-S.W. edited and revised the manuscript. C.S. conceived and designed the study. All authors have read and agreed to the published version of the manuscript.

Funding: This research was funded by the Natural Science Foundation of Zhejiang Province (LR17C200001), the 111 project (B17039), the Fundamental Research Funds for the Central Universities, and the Agricultural Outstanding Talents and Innovation Team of the State Agricultural Ministry on Health and Nutrition of Fruit.

Acknowledgments: The work was supported by the Natural Science Foundation of Zhejiang Province (LR17C200001), National Key R&D Program of China (2017YFD0400200), the 111 project (B17039), the Fundamental Research Funds for the Central Universities, and the Agricultural Outstanding Talents and Innovation Team of the State Agricultural Ministry on Health and Nutrition of Fruit.

Conflicts of Interest: The authors declare that there is no conflict of interest.

References

1. Hansen, N.W.; Sams, A. The Microbiotic Highway to Health-New Perspective on Food Structure, Gut Microbiota, and Host Inflammation. *Nutrients* **2018**, *10*, 1590. [CrossRef]
2. Ganesan, K.; Chung, S.K.; Vanamala, J.; Xu, B. Causal Relationship between Diet-Induced Gut Microbiota Changes and Diabetes: A Novel Strategy to Transplant Faecalibacterium prausnitzii in Preventing Diabetes. *Int. J. Mol. Sci.* **2018**, *19*, 3720. [CrossRef]
3. Kristek, A.; Schar, M.Y.; Soycan, G.; Alsharif, S.; Kuhnle, G.G.C.; Walton, G.; Spencer, J.P.E. The gut microbiota and cardiovascular health benefits: A focus on wholegrain oats. *Nutr. Bull.* **2018**, *43*, 358–373. [CrossRef]

4. Zopf, Y.; Reljic, D.; Dieterich, W. Dietary Effects on Microbiota-New Trends with Gluten-Free or Paleo Diet. *Med. Sci.* **2018**, *6*, 92. [CrossRef] [PubMed]
5. Sun, C.; Zhao, C.; Guven, E.C.; Paoli, P.; Simal–Gandara, J.; Ramkumar, K.M.; Wang, S.; Buleu, F.; Pah, A.; Turi, V. Dietary polyphenols as antidiabetic agents: Advances and opportunities. *Food Front.* **2020**, *1*, 18–44. [CrossRef]
6. Anhe, F.F.; Roy, D.; Pilon, G.; Dudonne, S.; Matamoros, S.; Varin, T.V.; Garofalo, C.; Moine, Q.; Desjardins, Y.; Levy, E.; et al. A polyphenol-rich cranberry extract protects from diet-induced obesity, insulin resistance and intestinal inflammation in association with increased *Akkermansia* spp. population in the gut microbiota of mice. *Gut* **2015**, *64*, 872–883. [CrossRef] [PubMed]
7. Neyrinck, A.M.; Van Hee, V.F.; Bindels, L.B.; De Backer, F.; Cani, P.D.; Delzenne, N.M. Polyphenol-rich extract of pomegranate peel alleviates tissue inflammation and hypercholesterolaemia in high-fat diet-induced obese mice: Potential implication of the gut microbiota. *Br. J. Nutr.* **2013**, *109*, 802–809. [CrossRef] [PubMed]
8. Molan, A.-L.; Liu, Z.; Kruger, M. The ability of blackcurrant extracts to positively modulate key markers of gastrointestinal function in rats. *World J. Microbiol. Biotechnol.* **2010**, *26*, 1735–1743. [CrossRef]
9. Sembries, S.; Dongowski, G.; Mehrlaender, K.; Will, F.; Dietrich, H. Physiological effects of extraction juices from apple, grape, and red beet pomaces in rats. *J. Agric. Food. Chem.* **2006**, *54*, 10269–10280. [CrossRef]
10. Molan, A.L.; Lila, M.A.; Mawson, J.; De, S. In vitro and in vivo evaluation of the prebiotic activity of water-soluble blueberry extracts. *World J. Microbiol. Biotechnol.* **2009**, *25*, 1243–1249. [CrossRef]
11. Yanhui, H.; Hang, X. Whole food-based approaches to modulating gut microbiota and associated diseases. *Annu. Rev. Food Sci. Technol.* **2020**, *11*, 119–143.
12. Peterson, J.; Dwyer, J. Flavonoids: Dietary occurrence and biochemical activity. *Nutr. Res.* **1998**, *18*, 1995–2018. [CrossRef]
13. Najmanová, I.; Vopršalová, M.; Saso, L.; Mladěnka, P. The pharmacokinetics of flavanones. *Crit. Rev. Food Sci. Nutr.* **2019**, 1–17. [CrossRef] [PubMed]
14. Courts, F.L.; Williamson, G. The Occurrence, Fate and Biological Activities of C-glycosyl Flavonoids in the Human Diet. *Crit. Rev. Food Sci. Nutr.* **2015**, *55*, 1352–1367. [CrossRef] [PubMed]
15. Benavente-García, O.; Castillo, J.; Marin, F.R.; Ortuño, A.; Del Río, J.A. Uses and Properties of Citrus Flavonoids. *J. Agric. Food. Chem.* **1997**, *45*, 4505–4515. [CrossRef]
16. Wang, Y.; Qian, J.; Cao, J.; Wang, D.; Liu, C.; Yang, R.; Li, X.; Sun, C. Antioxidant Capacity, Anticancer Ability and Flavonoids Composition of 35 Citrus (*Citrus reticulata* Blanco) Varieties. *Molecules* **2017**, *22*, 1114. [CrossRef] [PubMed]
17. Stevens, Y.; Van Rymenant, E.; Grootaert, C.; Van Camp, J.; Possemiers, S.; Masclee, A.; Jonkers, D. The Intestinal Fate of Citrus Flavanones and Their Effects on Gastrointestinal Health. *Nutrients* **2019**, *11*, 1464. [CrossRef]
18. Li, S.; Wang, H.; Guo, L.; Zhao, H.; Ho, C.-T. Chemistry and bioactivity of nobiletin and its metabolites. *J. Funct. Foods* **2014**, *6*, 2–10. [CrossRef]
19. Choi, S.-Y.; Ko, H.-C.; Ko, S.-Y.; Hwang, J.-H.; Park, J.-G.; Kang, S.-H.; Han, S.-H.; Yun, S.-H.; Kim, S.-J. Correlation between flavonoid content and the NO production inhibitory activity of peel extracts from various citrus fruits. *Biol. Pharm. Bull.* **2007**, *30*, 772–778. [CrossRef]
20. Dong, Y.; Cao, A.; Shi, J.; Yin, P.; Wang, L.; Ji, G.; Xie, J.; Wu, D. Tangeretin, a citrus polymethoxyflavonoid, induces apoptosis of human gastric cancer AGS cells through extrinsic and intrinsic signaling pathways. *Oncol. Rep.* **2014**, *31*, 1788–1794. [CrossRef]
21. Shu, Z.; Yang, B.; Zhao, H.; Xu, B.; Jiao, W.; Wang, Q.; Wang, Z.; Kuang, H. Tangeretin exerts anti-neuroinflammatory effects via NF-kappa B modulation in lipopolysaccharide-stimulated microglial cells. *Int. Immunopharmacol.* **2014**, *19*, 275–282. [CrossRef] [PubMed]
22. Mulvihill, E.E.; Assini, J.M.; Lee, J.K.; Allister, E.M.; Sutherland, B.G.; Koppes, J.B.; Sawyez, C.G.; Edwards, J.Y.; Telford, D.E.; Charbonneau, A.; et al. Nobiletin Attenuates VLDL Overproduction, Dyslipidemia, and Atherosclerosis in Mice With Diet-Induced Insulin Resistance. *Diabetes* **2011**, *60*, 1446–1457. [CrossRef] [PubMed]
23. Gao, Z.; Gao, W.; Zeng, S.-L.; Li, P.; Liu, E.H. Chemical structures, bioactivities and molecular mechanisms of citrus polymethoxyflavones. *J. Funct. Foods* **2018**, *40*, 498–509. [CrossRef]
24. Li, S.; Sang, S.; Pan, M.-H.; Lai, C.-S.; Lo, C.-Y.; Yang, C.S.; Ho, C.-T. Anti-inflammatory property of the urinary metabolites of nobiletin in mouse. *Bioorg. Med. Chem. Lett.* **2007**, *17*, 5177–5181. [CrossRef]

25. Kim, M.; Kim, N.; Han, J. Metabolism of Kaempferia parviflora Polymethoxyflavones by Human Intestinal Bacterium Bautia sp MRG-PMF1. *J. Agric. Food. Chem.* **2014**, *62*, 12377–12383. [CrossRef]
26. Wang, Y.; Zang, W.; Ji, S.; Cao, J.; Sun, C. Three Polymethoxyflavones Purified from Ougan (*Citrus reticulata* cv. *Suavissima*) Inhibited LPS-Induced NO Elevation in the Neuroglia BV-2 Cell Line via the JAK2/STAT3 Pathway. *Nutrients* **2019**, *11*, 791. [CrossRef]
27. Goncalves, D.R.; Manthey, J.A.; da Costa, P.I.; Rodrigues, M.C.M.; Cesar, T.B. Analysis of Fluorescence Spectra of Citrus Polymethoxylated Flavones and Their Incorporation into Mammalian Cells. *J. Agric. Food. Chem.* **2018**, *66*, 7531–7541. [CrossRef]
28. Zheng, J.; Bi, J.; Johnson, D.; Sun, Y.; Song, M.; Qiu, P.; Dong, P.; Decker, E.; Xiao, H. Analysis of 10 Metabolites of Polymethoxyflavones with High Sensitivity by Electrochemical Detection in High-Performance Liquid Chromatography. *J. Agric. Food. Chem.* **2015**, *63*, 509–516. [CrossRef]
29. Li, S.M.; Wang, Z.Y.; Sang, S.M.; Huang, M.T.; Ho, C.T. Identification of nobiletin metabolites in mouse urine. *Mol. Nutr. Food Res.* **2006**, *50*, 291–299. [CrossRef]
30. Zeng, S.-L.; Duan, L.; Chen, B.-Z.; Li, P.; Liu, E.H. Chemicalome and metabolome profiling of polymethoxylated flavonoids in Citri Reticulatae Pericarpium based on an integrated strategy combining background subtraction and modified mass defect filter in a Microsoft Excel Platform. *J. Chromatogr. A* **2017**, *1508*, 106–120. [CrossRef]
31. Tung, Y.-C.; Chang, W.-T.; Li, S.; Wu, J.-C.; Badmeav, V.; Ho, C.-T.; Pan, M.-H. Citrus peel extracts attenuated obesity and modulated gut microbiota in mice with high-fat diet-induced obesity. *Food Funct.* **2018**, *9*, 3363–3373. [CrossRef] [PubMed]
32. Zhang, M.; Zhu, J.; Zhang, X.; Zhao, D.G.; Ma, Y.Y.; Li, D.; Ho, C.T.; Huang, Q. Aged citrus peel (chenpi) extract causes dynamic alteration of colonic microbiota in high-fat diet induced obese mice. *Food Funct.* **2020**, *11*, 2667–2678. [CrossRef] [PubMed]
33. Lee, E.-S.; Song, E.-J.; Nam, Y.-D.; Lee, S.-Y. Probiotics in human health and disease: From nutribiotics to pharmabiotics. *J. Microbiol.* **2018**, *56*, 773–782. [CrossRef]
34. Robles Alonso, V.; Guarner, F. Linking the gut microbiota to human health. *Br. J. Nutr.* **2013**, *109*, S21–S26. [CrossRef] [PubMed]
35. Fukuda, S.; Toh, H.; Hase, K.; Oshima, K.; Nakanishi, Y.; Yoshimura, K.; Tobe, T.; Clarke, J.M.; Topping, D.L.; Suzuki, T.; et al. Bifidobacteria can protect from enteropathogenic infection through production of acetate. *Nature* **2011**, *469*, U543–U791. [CrossRef] [PubMed]
36. Levy, M.; Thaiss, C.A.; Elinav, E. Metabolites: Messengers between the microbiota and the immune system. *Genes Dev.* **2016**, *30*, 1589–1597. [CrossRef]
37. An, H.M.; Park, S.Y.; Lee, D.K.; Kim, J.R.; Cha, M.K.; Lee, S.W.; Lim, H.T.; Kim, K.J.; Ha, N.J. Antiobesity and lipid-lowering effects of *Bifidobacterium* spp. in high fat diet-induced obese rats. *Lipids Health Dis.* **2011**, *10*, 116. [CrossRef]
38. Tuohy, K.M.; Fava, F.; Viola, R. The way to a man's heart is through his gut microbiota-dietary pro- and prebiotics for the management of cardiovascular risk. *Proc. Nutr. Soc.* **2014**, *73*, 172–185. [CrossRef]
39. Yunes, R.A.; Poluektova, E.U.; Dyachkova, M.S.; Klimina, K.M.; Kovtun, A.S.; Averina, O.V.; Orlova, V.S.; Danilenko, V.N. GABA production and structure of gadB/gadC genes in Lactobacillus and Bifidobacterium strains from human microbiota. *Anaerobe* **2016**, *42*, 197–204. [CrossRef]
40. Hsu, T.-C.; Huang, C.-Y.; Liu, C.-H.; Hsu, K.-C.; Chen, Y.-H.; Tzang, B.-S. Lactobacillus paracasei GMNL-32, Lactobacillus reuteri GMNL-89 and L-reuteri GMNL-263 ameliorate hepatic injuries in lupus-prone mice. *Br. J. Nutr.* **2017**, *117*, 1066–1074. [CrossRef]

antioxidants

MDPI

Article

Dendropanax morbifera Leaf Extracts Improved Alcohol Liver Injury in Association with Changes in the Gut Microbiota of Rats

Taekil Eom [1,†], Gwangpyo Ko [2,†], Kyeoung Cheol Kim [3], Ju-Sung Kim [3] and Tatsuya Unno [1,2,*]

1 Subtropical/Tropical Organism Gene Bank, SARI, Jeju National University, Jeju 63243, Korea; taekil7@hanmail.net

2 Faculty of Biotechnology, College of Agriculture & Life Sciences, SARI, Jeju National University, Jeju 63243, Korea; rhkdvy1004@gmail.com

3 Majors in Plant Resource and Environment, College of Agriculture & Life Sciences, SARI, Jeju National University, Jeju 63243, Korea; cheolst@jejunu.ac.kr (K.C.K.); aha2011@jejunu.ac.kr (J.-S.K.)

* Correspondence: tatsu@jejunu.ac.kr; Tel.: +82-64-754-3354; Fax: +82-64-756-3351

† These authors contributed equally to this work.

Received: 21 August 2020; Accepted: 22 September 2020; Published: 24 September 2020

Abstract: This study evaluated the protective effects of *Dendropanax morbifera* leaf (DML) extracts in the liver due to excessive ethanol consumption. Our results showed that the ethanol extract had better antioxidant activity than the water extract, likely due to the higher levels of total flavonoid and phenolic compounds in the former. We found that the main phenolic acid was chlorogenic acid and the major flavonoid was rutin. Results from the animal model experiment showed concentration-dependent liver protection with the distilled water extract showing better liver protection than the ethanol extract. Gut microbiota dysbiosis induced by alcohol consumption was significantly shifted by DML extracts through increasing mainly *Bacteroides* and *Allobaculum*. Moreover, predicted metabolic activities of biosynthesis of beneficial monounsaturated fatty acids such as oleate and palmitoleate were enhanced. Our results suggest that these hepatoprotective effects are likely due to the increased activities of antioxidant enzymes and partially promoted by intestinal microbiota shifts.

Keywords: *Dendropanax morbifera*; hepatoprotective; leaf extract; liver damage; alcoholism; oxidative damage

1. Introduction

The worldwide increase in alcohol consumption has led to alcoholic liver disease (ALD) accounting for more than 5% of all diseases, with more than three million deaths being estimated to be from ALD. ALDs include alcoholic fatty liver disease, alcoholic hepatitis, and alcoholic cirrhosis, and these can later develop into liver cancer [1,2].

The liver is an organ that absorbs, metabolizes, and stores various substances that enter the body, with 80% of the alcohol consumed by humans being detoxified by the liver. When alcohol is consumed, it is absorbed from the small intestine and metabolized by enzymes such as alcohol dehydrogenase (ADH) and aldehyde dehydrogenase (ALDH) in the liver, but when excessive alcohol consumption exceeds the metabolic ability of ADH and ALDH, the expression of Cytochrome P450 2E1 (CYP2E1) enzymes is induced, leading to alcoholic metabolism [3]. This can lead to production of acetaldehyde or reactive oxygen species (ROS) and toxic substances, causing oxidative damage and inflammation to liver cells, resulting in ALD [4]. ROS produced by excessive alcohol consumption are known to be removed by antioxidant enzymes to prevent oxidative damage caused by alcohol. Antioxidant enzymes such as catalase (CAT), superoxide dismutase (SOD), glutathione *S*-transferas (GST), and glutathione reductase (GR) play a direct role in removing ROS [5].

The intestines are normally home to various types of bacteria, with the term "gut microbiota" referring to the various microorganisms that exist in the human gut. Although the functions of the gut microbiota are not completely known, their effects on human health have recently been actively studied [6,7]. In particular, the gut microbiome is involved in inhibiting the growth of pathogenic bacteria, stimulating the production of total and pathogen-specific mucosal IgA, nutrient production in mucosal cells, the development and regulation of the immune system, and immunological resistance. Thus, the gut microbiome appears to play an essential role in maintaining human homeostasis by engaging in bidirectional interactions with the host [8]. Gut microbiomes vary depending on race, age, dietary habits, and drug use, and the impact of these factors on the health of the host due to changes in the gut microbiome is referred to as dysbiosis [9]. Several recent studies have shown that liver damage caused by alcohol consumption and gut microbiota interact very closely with each other, which is referred to as the gut-liver axis [10–12].

Recently, many studies using various medicinal plant resources have been conducted on foods and medicines that prevent liver damage due to excessive alcohol intake [13,14]. However, studies on changes in the gut microbiome caused by medicinal plants and their effects on the prevention of liver damage are insufficient. *Dendropanax morbifera*, which belongs to the Araliaceae family, is a medicinal plant that grows on the southern coast of Korea and is known to have various physiological functions such as anti-inflammatory, anti-cancer, anti-diabetic, and immune regulatory effects [15–18]. In addition, a previous study conducted by our research team showed that *D. morbifera* extract, with its maximal antioxidant activity, inhibited liver damage caused by oxidative stress [19]. Therefore, this study aims to verify the prevention of alcoholic liver damage by *D. morbifera* extract and to analyze its relationship with the gut microbiome, thereby examining its mechanism of action.

2. Materials and Methods

2.1. Extraction

The *D. morbifera* leaves and stems used in the experiment were purchased from Jeju Hwangchil (Jeju, Korea), and were dried in the shade, crushed, and stored in a refrigerator. To extract the *D. morbifera* leaves and stems, 1 L of 70% ethanol or distilled water was added per 100 g of the sample, followed by three rounds of extraction for 3 h at 80 °C. After the extraction, the sample was filtered under reduced pressure, concentrated using a rotary evaporator (Hei-VAP Precision, Heidolph, Schwabach, Germany), and dried using a freeze dryer (SCANVAC, Stockholm, Sweden), then used in the experiment.

2.2. Total Polyphenol and Flavonoid Contents

The polyphenol and flavonoid compounds contained in the extracts were quantified using the Folin-Ciocalteu method and the aluminum chloride method as previously described [20].

2.3. Antioxidant Activity

The antioxidant activity of the extracts was compared and evaluated using the 1,1-diphenyl-2-picrylhydrazyl (DPPH) radical scavenging assay, as well as the total equivalent antioxidant capacities (TEAC) and ferric reducing antioxidant power (FRAP) methods as previously described [19].

2.4. HPLC Analysis of Phenolic Acid and Flavonoid Components of *D. morbifera* Extracts

Flavonoid and phenolic acid contents were quantified using the prominence HPLC system with an SPD-M20A PDA detector (Agilent infinity 1260 series, Munich, Germany). The analysis was performed with a Triart C-18 column (250 mm × 4.6 mm, 5 μm) from YMC Co., Ltd. (Tokyo, Japan). The flavonoid analysis conditions are column temperature 35 °C, flow rate 0.8 mL/min, injection volume 10 μL and detection wavelength 280 nm. The gradient program was designed as follows: Distilled water containing 0.1% trifluoroacetic acid (Sigma-Aldrich, Steinheim, Germany) was mixed with acetonitrile (Sigma-Aldrich) containing 0.1% trifluoroacetic acid with the concentrations increased from 10% to 20% for the first 5 min, held for 20 min, 20% to 25% for 10 min, held for 15 min, 25% to 30% for 5 min, held for 10 min, 30% to 60% for 5 min, held for 5 min, 60% to 80% for 5 min, and held for 5 min (total 80 min).

The phenolic acid analysis conditions are column temperature 45 °C, flow rate 1.0 mL/min, injection volume 10 µl and detection wavelength 245 nm. The gradient program was designed as follows: Distilled water containing 0.1% formic acid (Sigma-Aldrich) for 4 min, and mixed with methanol (Sigma-Aldrich) containing 0.1% formic acid with the concentrations increasing from 0% to 15% for 10 min, held for 3 min, increased from 15% to 16.5% for 7 min, 16.5% to 18% for 4 min, 19% to 25% for 2 min, 25% to 28% for 6 min, 28% to 30% for 2 min, held for 3 min, increased from 30% to 40% for 5 min, 40% to 48% for 2 min, 48% to 53% for 5 min, 53% to 60% for 10 min, 60% to 70% for 2 min, and held for 5 min (total 70 min).

2.5. Animal Experiments

Five-week old Sprague Dawley (SD) rats were purchased from Daehan Biolink (Eumsung, Korea) and used after a one-week acclimation period. The animals were kept in a room with a temperature of 20–22 °C, a humidity of 50%, and a 12 h light-dark cycle. All experiments were approved by the Jeju University Institutional Animal Care and Use Committee (IACUC) (Approval number: 2016-0056).

In order to see whether the *Dendropanax morbifera* leaf (DML) extracts protected against alcoholic acute liver damage, the experiment was conducted in 8 groups as follows. Dose of ethanol and DML extracts were determined as previously described [19,21]. An overview of the experimental design is outlined in Figure S1. Before the feeding trial, all mice were acclimated for 7 days by feeding normal diet ad libitum. Feeding trial was conducted for 10 days. At the first day, animals were divided into 3 groups and fecal materials were collected. (Control (CTL), fed normal diet during the trial; Ethanol, fed normal diet for 10 days and ethanol was supplied from day 8 to day 10; and DML extracts group, fed normal diet with various concentrations of DML extracts for 10 days and ethanol was supplied from day 8 to day 10). Ethanol (5 g/kg) and DML extracts (100, 300, 500 mg/kg) were administered orally. On the 10th day, 8 h after the last ethanol intake, fecal materials were collected, and rats were sacrificed for dissection.

2.6. Serum Biochemical Analysis

The blood samples obtained were left on ice for 20 min and then centrifuged at 570× *g* for 10 min, after which the upper layer of the serum was collected. The separated serum was stored at −80 °C prior to use in the experiments. Serum ethanol and acetaldehyde concentrations were measured and quantified using an ethanol assay kit (Megazyme, Bray, Ireland) and an acetaldehyde assay kit (Megazyme, Bray, Ireland), respectively. Serum aspartate aminotransferase (AST) and alanine aminotransferase (ALT) concentrations were measured using an automated hematology analyzer.

2.7. Antioxidant Enzyme Activity

For the analysis of antioxidant enzyme activity in liver tissue, a sample of liver tissue was homogenized using a homogenizer in 50 mM phosphate buffer (pH 7.4). The homogenized suspension was centrifuged at 3000× *g* for 20 min at 4 °C, and the supernatant was used to measure the enzyme activity. CAT, SOD, GR, and GST were considered as oxidative stress markers [5] and measured as follows: SOD activity was analyzed using the McCord & Fridovich method [22]; and CAT activity was determined using the protocol by Aebi [23]; GST and GR activity was measured using a protocol by Koneru et al. [24]. The amount of protein was quantified using bicinchoninic acid (BCA) to measure the enzyme activity.

2.8. Hepatic Histopathological Observation

For histopathological observations, a portion of the middle lobe of the liver was dissected and fixed in 10% buffered formalin solution (Sigma-Aldrich, Steinheim, Germany) for 24 h. After the fixation, the tissue was embedded in paraffin, cut to a thickness of 5 μm, and stained with hematoxylin (Sigma-Aldrich, Steinheim, Germany) and eosin (Sigma-Aldrich, Steinheim, Germany) (H&E) for observation using an optical microscope with 100-fold magnification.

2.9. Analysis of Gut Microbiota

For microbial community analysis, the V4 hypervariable region of the 16S rRNA gene was amplified, and a library for Illumina Miseq (250 bp × 2) was constructed with two-step PCR. Briefly, first PCR was performed using a KAPA HiFi HotStart ReadyMix PCR kit (Roche, South San Francisco, CA, USA) as follows: 95 °C for 3 min, 25 cycles of 95 °C for 30 s, 55 °C for 30 s, and 72 °C for 30 s, and 72 °C for 5 min. The obtained PCR products were further purified using a HiAccuBead (AccuGene, Seoul, Korea). The purified PCR products were again subjected for PCR to attach barcode-sequences. After PCR products were purified in the same manner, an equimolar of the final PCR amplicons were pooled and sent for sequencing with MiSeq according to the manufacturer's instructions at Macrogen Inc. (Seoul, Korea).

Sequence data was analyzed using MOTHUR [25]. In brief, raw reads paired-end assembly was done with make.contigs, and then aligned to the SILVA Database [26]. After eliminating the singleton, the pre.cluster MOTHUR subroutine was performed to correct the error for rare sequences. Chimeric sequences were detected using VSEARCH [27]. Taxonomic classification was done using Ribosome database project (RDP version 16) database [28]. Sequences classified to undesired taxa (i.e., Chloroplast and Mitochondria) were removed using remove.lineages MOTHUR subroutine. Clustering was performed with 97% similarity using Opti.clust [29] and designated as operational taxonomic units (OTUs). Number of reads per sample was normalized to 20,000 for downstream analyses. Species richness and evenness were evaluated using Chao [30] and Shannon indices [31], respectively. The non-metric multidimensional scaling (NMDS) analysis was conducted based on the Bray-Curtis distance [32]. The linear discriminant analysis effect size (LEfSe) [33] was used to identify significantly increased or decreased taxa after the DML extracts treatments. Analysis of molecular variance (AMOVA) was applied to test significant difference between microbiota. Metabolic activities were predicted using Phylogenetic Investigation of Communities by Reconstruction of Unobserved States (PICRUSt) 2 [34] and abundance comparison was performed using the ALDEx2 package in R software (https://www.r-project.org/) [35]. Spearman correlation analysis was performed to estimate associations between the LEfSe-selected OTUs and ALDEx2-selected predicted metabolic activities.

2.10. Statistical Analysis

Analysis of all data is expressed as ± standard deviation, and significance was determined using ANOVA and verified using the Duncan test. Minitab ver. 17 (Minitab Inc., IL, State College, PA, USA) was used for all statistical analysis.

3. Results

3.1. Total Polyphenol and Flavonoid Contents

The total polyphenol and flavonoid contents were higher in *D. morbifera* leaf (DML) extracts than stem extracts, and higher in the 70% ethanol (EtOH) extracts than the water extracts (Table S1). Flavonoid content specifically was 1.5 times higher in 70% EtOH than in water for leaf extracts, and seven times higher in 70% EtOH than in water for stem extracts.

3.2. Antioxidant Activity

Three widely used antioxidation experiments were used to measure the antioxidant activity of DML and stem extracts, and the results are shown in Table S2. The DPPH radical scavenging assay is the simplest method of measuring antioxidant activity, wherein electrons or protons are provided to unstable DPPH radicals, and the ability to scavenge the radicals is measured. DPPH radical scavenging assays showed similar radical scavenging activities for water and EtOH extracts, but leaf extracts showed more than twice the radical scavenging activity of stem extracts. FRAP activity measures antioxidant activity based on the reducing power of the sample, and samples with a large reducing power are considered to have excellent antioxidant activity, as they can effectively reduce ROS with high oxidation states. The results from the FRAP activity assays were consistent with those obtained from the DPPH radical scavenging assay, with the leaf extract demonstrating more than twice the reducing power of the stem extract, and the water and ethanol extracts having similar reducing powers. TEAC activity compares and evaluates the antioxidant activities of samples with an equal amount of Trolox, which is used as an antioxidant. The TEAC value showed a similar trend, but the difference between the leaf and the stem extracts was not as significant as observed with the DPPH and FRAP assays. The antioxidant activity evaluated was in the order of *D. morbifera* leaf ethanol extract (DMLEE), *D. morbifera* leaf distilled water extract (DMLDE), *D. morbifera* stem ethanol extract (DMSEE), and *D. morbifera* stem distilled water extract (DMSDE), which was consistent with the total phenolic and flavonoid contents.

3.3. Flavonoid and Phenolic Acid Analyses of D. morbifera Leaf Extract

Content analysis was performed for 18 flavonoids and 12 phenolic acids using HPLC (Figure S2). A smaller number of flavonoids were identified in the water extracts compared to the ethanol extracts, and the flavonoid with the highest content was rutin, a quercetin glycoside (Table 1). The content of rutin was the highest in DMLEE (44.88 ± 0.436 mg/g), followed by DMLDE (15.723 ± 0.005 mg/g), DMSEE (1.085 ± 0.059 mg/g) and DMSDE (0.122 ± 0.095 mg/g). After rutin, taxifolin was the flavonoid compound with the highest content, which was the highest in DMLEE (11.705 ± 0.029 mg/g), followed by DMLDE (4.369 ± 0.014 mg/g), DMSEE (1.593 ± 0.023 mg/g) and DMSDE (1.253 ± 0.218 mg/g).

Table 1. Individual flavonoid contents in the *Dendropanax morbifer* leaf and stem extracts (unit: ug/g of extracts).

Flavonoids	DMLDE	DMLEE	DMSDE	DMSEE
	Flavonoid glycoside (mg/g of extracts)			
Naringin	0.848 ± 1.075	0.017 ± 0.014	0.116 ± 0.010	0.104 ± 0.184
Narirutin	0.176 ± 0.012	2.808 ± 0.015	0.329 ± 0.117	0.227 ± 0.173
Neohesperidin	0.185 ± 0.003	3.308 ± 0.009	0.351 ± 0.149	0.053 ± 0.037
Hesperidin	0.021 ± 0.002	1.256 ± 0.179	0.484 ± 0.003	0.374 ± 0.004
Rutin	15.723 ± 0.005	44.88 ± 0.436	0.122 ± 0.095	1.085 ± 0.059
	Flavonoid aglycone (mg/g of extracts)			
Apigenin	-	-	-	-
Hesperetin	-	-	-	-
Isorhaemnetin	0.428 ± 0.001	-	0.385 ± 0.076	0.282 ± 0.096
Kaempferol	-	-	-	-
Luteorin	-	-	-	-
Myricetin	0.245 ± 0.001	0.306 ± 0.001	0.236 ± 0.004	0.203 ± 0.001
Naringenin	0.546 ± 0.007	-	0.406 ± 0.013	0.499 ± 0.034
Quercetin	0.149 ± 0.002	0.148 ± 0.001	0.020 ± 0.001	0.139 ± 0.001
Rhaemnetin	-	-	-	-
Taxifolin	4.369 ± 0.014	11.705 ± 0.029	1.253 ± 0.218	1.593 ± 0.023

Table 1. *Cont.*

Flavonoids	DMLDE	DMLEE	DMSDE	DMSEE
	Polymethoxyflavone (mg/g of extracts)			
Nobiletin	0.615 ± 0.012	0.574 ± 0.001	-	-
Sinesetin	0.542 ± 0.001	0.461 ± 0.073	-	-
Tangeretin	0.265 ± 0.002	0.579 ± 0.095	-	-

Means ± SD of determinations were made in triplicate experiments.

All 12 Phenolic acids were identified in the extracts, and the major phenolic acids were caffeic acid and chlorogenic acid, a caffeic acid glycoside. However, there was more caffeic acid in the ethanol extracts than the distilled water extracts, and more chlorogenic acid in the distilled water extracts than the ethanol extracts (Table 2). DMLDE (5.165 ± 0.004 mg/g) contained the most chlorogenic acid, followed by DMLEE (2.945 ± 0.119 mg/g), DMSDE (1.532 ± 0.637 mg/g), and DMSEE (0.811 ± 0.024 mg/g). Caffeic acid content was the highest in DMLEE (21.824 ± 1.356 mg/g), followed by DMLDE (13.850 ± 0.024 mg/g), DMSEE (17.446 ± 0.286 mg/g) and DMSDE (11.072 ± 0.178 mg/g).

Table 2. Individual phenolic acid contents in the *Dendropanax morbifer* leaf and stem extracts (unit: ug/g of extracts).

Derivatives	DMLDE	DMLEE	DMSDE	DMSEE
	Benzoic acid derivative (mg/g of extracts)			
Benzoic acid	0.460 ± 0.002	1.081 ± 0.022	0.702 ± 0.073	0.638 ± 0.033
p-Hydroxybenzoic acid	0.294 ± 0.146	0.196 ± 0.040	0.167 ± 0.016	0.153 ± 0.003
Protocatechuic acid	0.663 ± 0.001	0.553 ± 0.001	0.466 ± 0.187	0.248 ± 0.001
Vanillic acid	0.611 ± 0.322	0.333 ± 0.053	0.843 ± 0.086	0.966 ± 0.045
Syringic acid	0.297 ± 0.002	0.371 ± 0.115	1.054 ± 0.534	0.454 ± 0.045
Gallic acid	0.653 ± 0.007	0.732 ± 0.003	0.597 ± 0.062	0.364 ± 0.315
	Cinnamic acid derivative (mg/g of extracts)			
Cinnamic acid	0.164 ± 0.001	0.105 ± 0.001	0.161 ± 0.043	0.111 ± 0.001
p-Coumaric acid	0.679 ± 0.018	2.092 ± 0.247	0.654 ± 0.178	0.892 ± 0.078
Caffeic acid	13.850 ± 0.024	21.824 ± 1.356	11.072 ± 0.178	17.446 ± 0.286
Ferullic acid	0.440 ± 0.001	0.916 ± 0.003	0.452 ± 0.020	0.475 ± 0.006
Sinapinic acid	0.417 ± 0.001	0.359 ± 0.001	0.428 ± 0.016	0.563 ± 0.001
Chlorogenic acid	5.165 ± 0.004	2.945 ± 0.119	1.532 ± 0.637	0.811 ± 0.024

Means ± SD of determinations were made in triplicate experiments.

3.4. Effects of DML Extracts on Suppression of Liver Damage

The effect of DML extracts on liver damage due to alcohol consumption was examined by orally administering DML extract for seven days and then excessive alcohol in a binge rat model [36]. After sacrificing the rats, the activity of serum ALT and AST enzymes were compared as indicators of liver damage, as was liver weight. During the experiment, rats in all groups showed weight gain, and recent studies reported that continuous consumption of excessive alcohol causes fat metabolism in the liver and formation of fatty liver tissue, which results in liver hypertrophy [37]. In this study, the alcohol-administered groups also showed decrease in weight compared to the group that was not administered alcohol. Body and liver weight gain for the group given DML extracts did not show significant difference from that of Ethanol groups (Table S3).

Serum AST and ALT activities were investigated and used as an index to judge the extent of direct liver damage, as AST and ALT are released from the liver to blood in response to liver damage, increasing their enzymatic activity. Results from Figure 1 show that repeated intake of high concentrations of alcohol was shown to increase serum AST and ALT activities ($p < 0.05$ compared to the Control group). However, the groups with DML extract intakes had inhibited AST and ALT activities after alcohol consumption. In particular, the group that consumed the 500 mg/kg ethanol extract had a 40% decrease in AST activity and a 58% decrease in ALT activity compared to the alcohol intake group. The group that consumed the 500 mg/kg distilled water extract showed AST activity that was reduced by 53% and ALT activity that was reduced by 65% compared to the alcohol intake group.

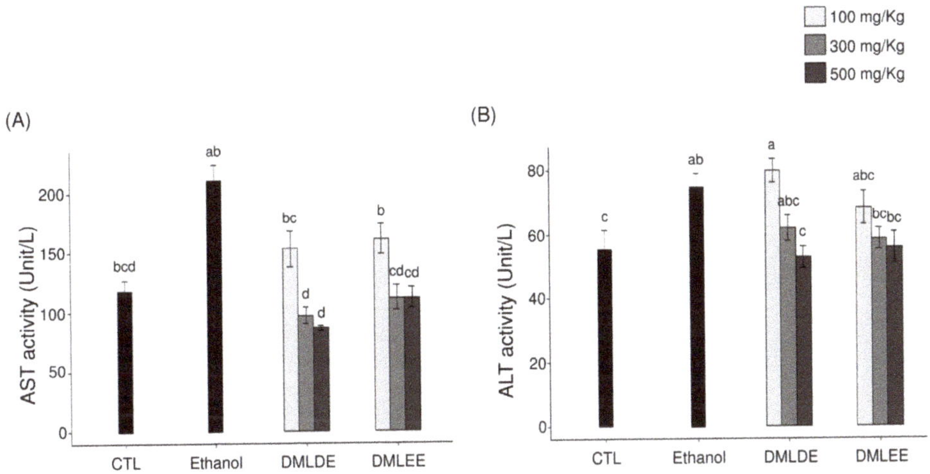

Figure 1. Effect of the *Dendropanax morbifer* leaf ethanol and D.W extracts on serum ALT and AST activities in alcohol-fed rat. (**A**) AST activity and (**B**) ALT activity. The data are expressed as the mean ± SD ($n = 6$), and different letters indicate (a > b > c > d) a significant difference at $p < 0.05$, as determined by a Duncan's multiple range test.

The damage to the liver tissue due to excessive ethanol consumption represented by histochemical staining of the liver is shown in Figure 2. The control group with no ethanol consumption showed a normal hepatocyte structure, whereas the group with ethanol consumption showed reduced boundaries between the hepatocytes, and enlargement of hepatocytes due to fat accumulation and infiltration of neutrophils were observed. However, the symptoms and the histopathological changes were improved in the groups that were administered with DML extract.

3.5. Reduction of Ethanol and Acetaldehyde in Blood with DML Extract Intake

Oral administration of excessive alcohol increases absorption of ethanol in the small intestines, and it is metabolized to acetaldehyde through ADH and ALDH in the liver. After liver damage, the decomposition of alcohol by ADH and ALDH decreases, so the concentrations of ethanol and acetaldehyde in blood can be used as an indirect indicator for examining liver damage caused by ethanol. As shown in Figure 3, the group with excessive ethanol intake had 1.8 times higher serum ethanol level ($p < 0.05$) and 2.1 times higher acetaldehyde level than the ethanol group. However, the group being given DML extracts showed a concentration-dependent decrease in blood ethanol and acetaldehyde concentration, and for a given concentration, the administration of the distilled water extracts showed a significant reduction compared to the use of the ethanol extracts.

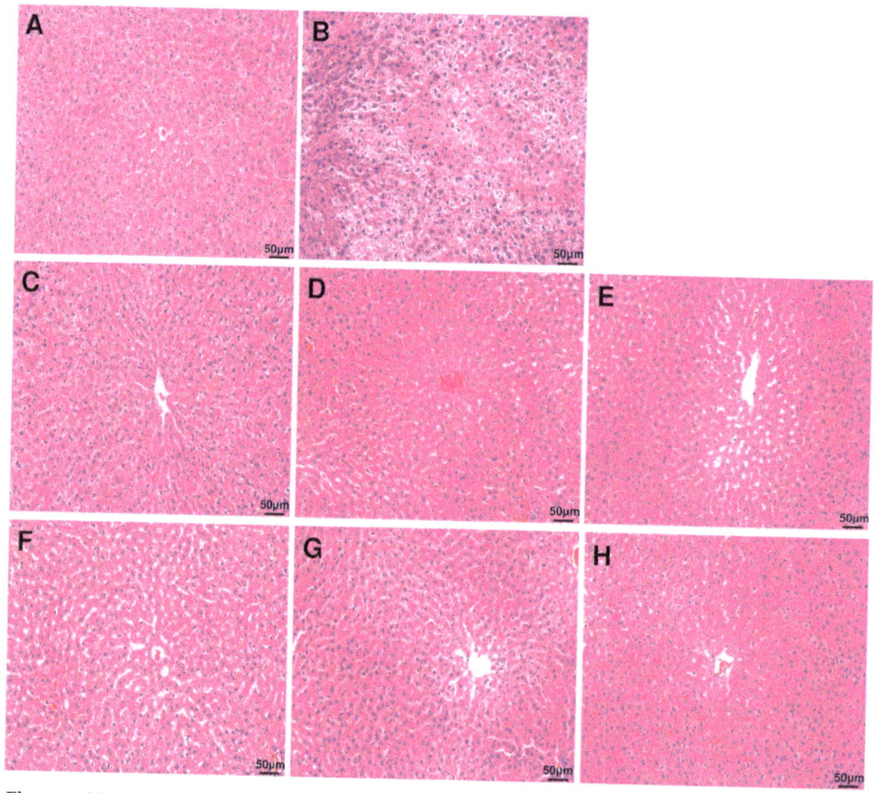

Figure 2. Histopathological evaluation (H&E staining) of rat liver. Liver section images of (**A**) control group, (**B**) ethanol group, (**C**) EtOH + DMLDE 100 mg/kg, (**D**) EtOH + DMLDE 300 mg/kg and (**E**) EtOH + DMLDE 500 mg/kg (**F**) EtOH + DMLEE 100 mg/kg, (**G**) EtOH + DMLEE 300 mg/kg and (**H**) EtOH + DMLEE 500 mg/kg display histopathological changes in liver.

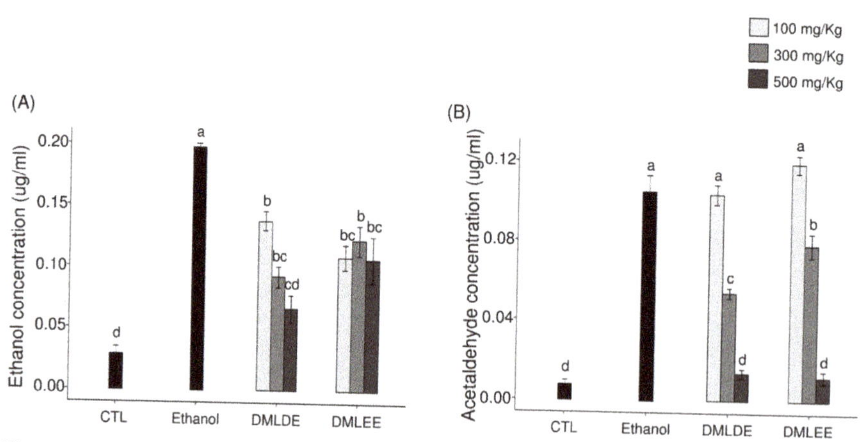

Figure 3. Effect of the *Dendropanax morbifer* leaf ethanol and D.W extracts on serum concentration of ethanol and acetaldehyde. (**A**) Ethanol concentration and (**B**) acetaldehyde concentration. The data are expressed as the mean ± SD ($n = 6$), and different letters indicate (a > b > c > d) a significant difference at $p < 0.05$, as determined by a Duncan's multiple range test.

3.6. Effects of DML Extracts on Antioxidant Enzymes in Liver Tissues

ROS produced by continuous alcohol intake weakens the defense mechanisms associated with oxidative stress in the body and induces oxidative damage. Therefore, the impact of DML extract intake on the activities of antioxidant enzymes related to other antioxidant defenses were evaluated. Figure 4 shows the activities of the antioxidant enzymes, CAT, SOD, GST, and GR. Among the ROS produced in the body, superoxide is converted to hydroxyl radicals by SOD, and the produced hydroxyl radicals are decomposed into water by CAT. CAT activity decreased in the control group compared to the control group, but with no significant difference. In addition, groups with 100 or 300 mg/kg intakes of DML extract did not show differences in CAT activity compared to the ethanol group, but the group with an intake of 500 mg/kg had significantly increased CAT activity. SOD activity was significantly decreased in the ethanol group compared to the control group, and the enzyme activity significantly increased with the intake of DML extracts.

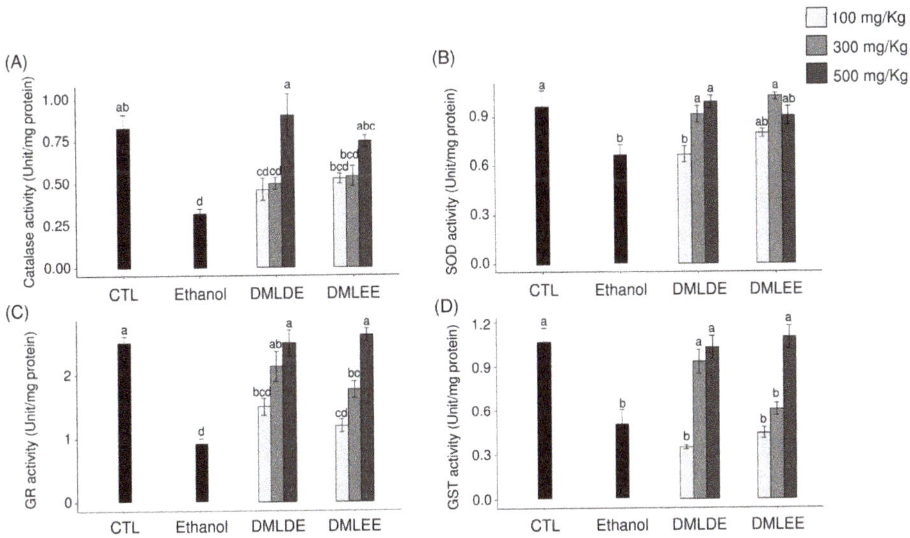

Figure 4. Effects of the *Dendropanax morbifer* leaf D.W and ethanol extracts on antioxidant enzyme activity in alcohol-fed rat. (**A**) Catalase activity, (**B**) SOD activity, (**C**) GR activity and (**D**) GST activity. The data are expressed as the mean ± SD ($n = 6$), and different letters indicate (a > b > c > d) a significant difference at $p < 0.05$, as determined by a Duncan's multiple range test.

3.7. Microbiota Shifted by DML Extracts

Various concentrations of DML extracts affected microbiota similarly except for DMLEE at 300 mg/kg and DMLDE at 100 mg/kg (Figure S3). With a few exceptions, most of the DML treatments did not show significant differences between the different types of extracts nor concentrations of DML extracts (Table S4). Among the treatments, we selected rats treated with highest concentrations of DMLDE and DMLEE for the further gut microbiota analyses, because they showed most distinctive improvements from the liver damages. Results from Figure 5 show that species evenness between the CTL and DMLEE groups was significantly different, while there was no significant difference in species richness. AMOVA and NMDS indicated that no significant microbial community differences between each other across groups before the treatment (Figure S4), but significant differences has detected after the treatment between CTL and Ethanol as well as Ethanol and DML treated samples (Table S4, Figure 5C). These results suggest that alcohol ingestion and DML treatments can significantly shift the microbiota of rats, while extraction methods did not affect the total microbiota shifts.

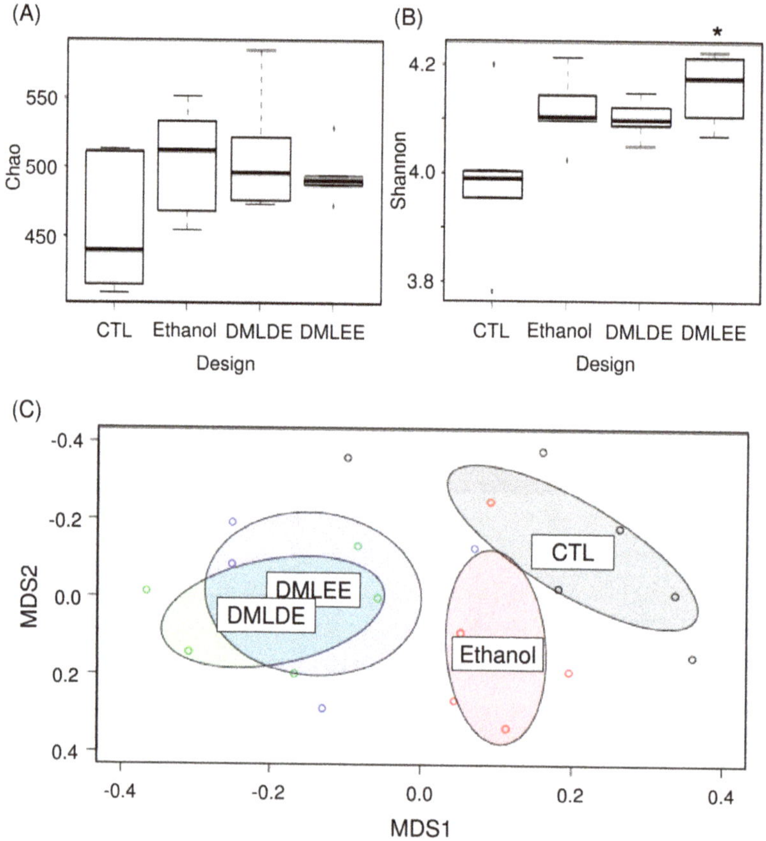

Figure 5. The comparative analysis of gut microbiota using ecological indices and non-metric multidimensional scaling (NMDS). (**A**) Species richness, (**B**) species evenness, and (**C**) NMDS. * indicates significant difference between CTL and DMLEE.

We investigated taxonomic compositions of rats' gut microbiota at the phylum, family and genus levels (Figure S5). Two major phyla, Bacteroidetes and Firmicutes were detected at the phylum level. At the lower taxonomic levels, the family Prevotellaceae and the genus *Prevotella* were the most abundant in the rats' gut microbiota. We observed the family Sutterellaceae and the genera *Prevotella* and *Romboutsia* were significantly decrease by alcohol consumption. However, taxonomic composition did not cluster samples according to the treatment groups.

Then differential abundance test was performed to investigate significantly increased or decreased OTUs. Results from Figure 6 show that alcohol consumption significantly increased the abundance of members in the family Porphyromonadaceae, the genus *Alloprevotella, Clostridium, Turicibacter* and *Romboutsia*, while decreased the abundance of other OTUs belonging to the family Porphyromonadaceae, the genus *Prevotella*, and *Parasutterella*. DMLDE increased *Allobaculum* and one unclassified OTU belonging to the phylum Candidatus Saccharibacteria, while decreased 12 OTUs including various unclassified OTUs mostly belonging to the family Porphyromonadaceae. On the other hand, there are seven and five OTUs increased and decreased by DMLEE treatments, respectively. Most of the OTUs were the unclassified members of the family Porphyromonadaceae. Two OTUs (Otu0022 and Otu0037) were decreased and one OTU (Otu0012) was increased by both DML treatments, but other OTUs were specific to each treatment. There were no OTUs whose abundance were changed due to alcohol consumption but recovered by DML treatments.

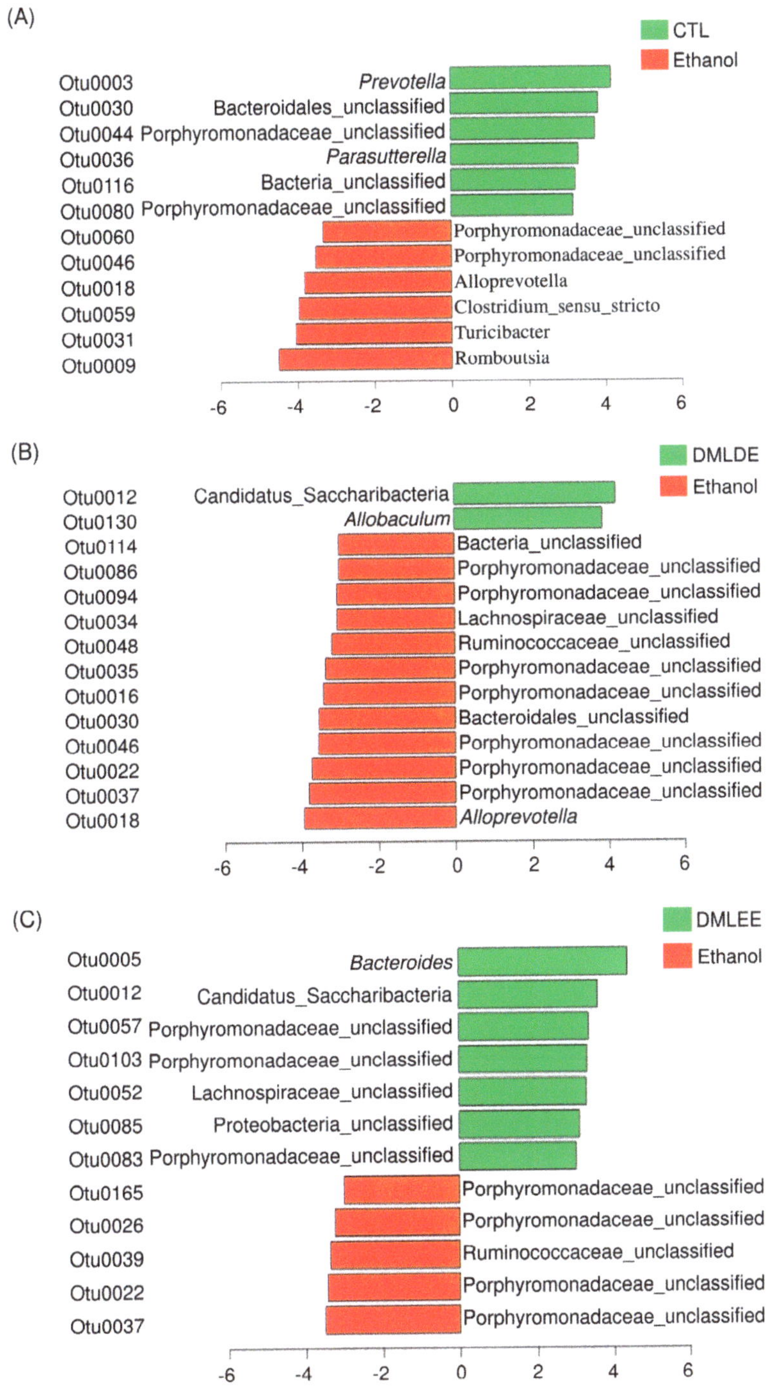

Figure 6. The relative abundance increased or decreased bacteria in OTU level in the comparison between ethanol and CTL or DML extracts. Abundance analysis was performed using linear discriminant analysis effect size (LEfSe). (**A**) Ethanol v/s CTL, (**B**) ethanol v/s DMLDE, (**C**) ethanol v/s DMLEE.

PICRUSt2 predicted several metabolic activity changes by alcohol consumption and DML extracts treatments. Results in Figure 7 show that seven and six metabolic activities were predicted to be enriched and depleted due to alcohol consumption, respectively. On the other hand, there are five metabolic activities (biosynthesis of oleic acid, (5Z)-dodec-5-enoate, palmitate, cytochrome c aerobic respiration, and CMP-legionaminate) predicted to be enriched by DML treatments. Beside these, there are four and 12 metabolic activities predicted to be enriched by DMLEE and DMLDE, respectively. Our results showed there were more metabolic activities predicted to be enriched by DMLDE than by DMLEE.

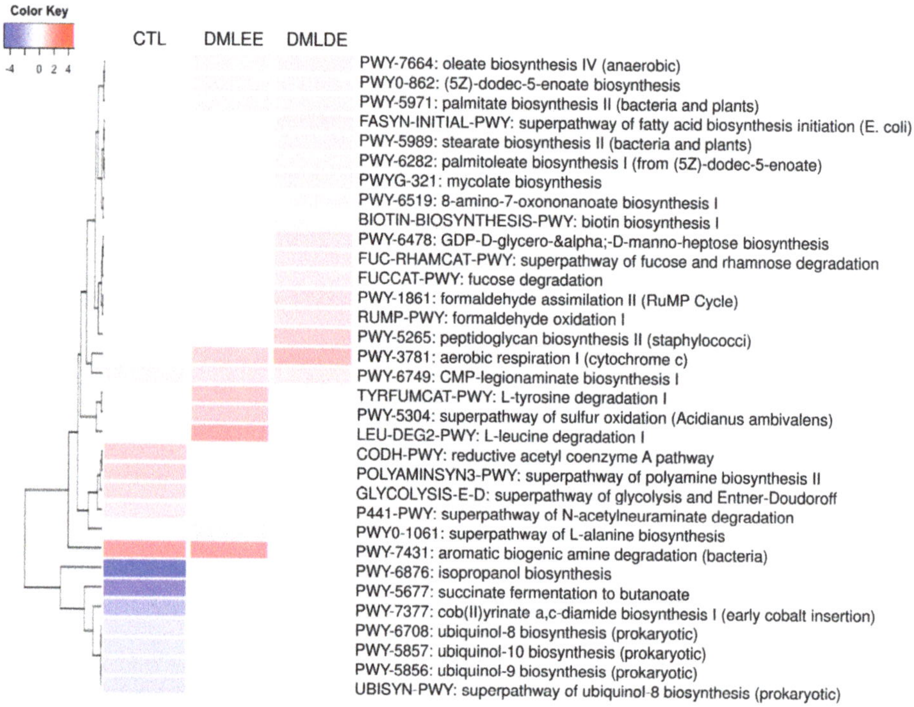

Figure 7. Heatmap analysis of predicted metabolic pathways that were significantly increased or decreased in the comparison between ethanol and CTL or DML extracts. Abundance analysis was performed using ALDEx2 ($p < 0.05$).

A total of 23 OTUs that were affected by either of both DML treatments in this study. We investigated associations of these OTUs with 21 predicted metabolic activities that were also affected by DML treatments. Results from Figure 8 show that seven OTUs were significantly correlated with one or more of the 17 predicted metabolic pathways. Three OTUs (Otu0005, Otu0012, and Otu0130) were associated with the most of the predicted metabolic pathways, while the other OTUs were associated with only one or two. The three OTUs are classified to be the genera *Bacteroides* and *Allobaculum* and one unclassified belonging to the phylum Candidatus Saccharibacteria. Predicted metabolic pathways associated with these OTUs were mostly related with monounsaturated fatty acids (oleate and palmitoleate) synthesis pathways.

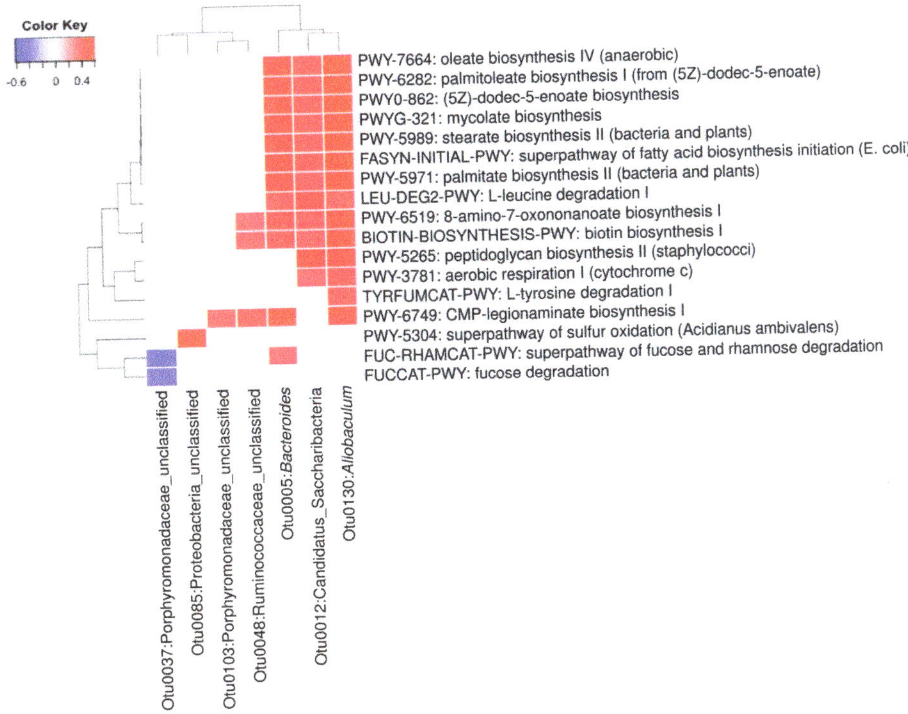

Figure 8. Heatmap analysis of operational taxonomic units (OTUs) associated with the predicted metabolic pathways shifted by DML extracts ($p < 0.01$).

4. Discussion

The liver is an organ that plays an important role in the metabolism of various toxic substances, including ethanol. In the body, 80–90% of alcohol is first decomposed into acetaldehyde by ADH present in liver cells, then metabolized by ALDH to form acetic acid, which finally undergoes complete decomposition via hydrolysis into carbon dioxide and water. Nicotinamide adenine dinucleotide phosphate (NADP), acetaldehyde, and various ROS produced during alcohol metabolism in the liver react with DNA and proteins to act as major mediators of liver damage [38]. Therefore, various antioxidants such as plant extracts have been recently suggested for use as medicines and functional foods to prevent liver damage caused by oxidative stress by inhibiting ROS production due to ethanol [39]. It was also found in several previous studies that DML and branch extracts have strong antioxidant activities, and so their function for reducing liver damage was studied. First, it was confirmed that the antioxidant activity of DML extract was higher than that of the stem extract. Antioxidant activity showed a close relationship to the total phenol and flavonoid content in the extract [17,19]. Total phenolic compounds have a hydroxyl group attached to an aromatic ring, while flavonoids have a basic C6-C3-C6 skeleton [40]. These compounds have a common aromatic hydroxyl group with excellent ability to donate electrons, with donation being stabilized by a resonance effect, leading to excellent antioxidant activity [41]. The antioxidant activity of DML extracts in this study may also be due to the effects of these total phenolics and flavonoids. Previous studies on the effect of harvesting time on the antioxidant activities of DML extracts showed that antioxidant activity was correlated with total phenolic and flavonoid content, and that DML extract had greater antioxidant activity than the branch or bark extracts, which was consistent with the findings of this study [17]. We measured the prevention of liver damage due to ethanol using DML extracts with excellent antioxidant activity, and found that both ethanol and distilled water extracts prevented liver damage, with the latter showing a much greater effect as opposed to in vitro experiments.

Chronic ethanol intake promotes liver damage by reducing the activity of antioxidants such as glutathione (GSH) and various antioxidant enzymes in the body [5]. The effects of GST and GR related to the metabolism of GSH on the enzymatic activity were confirmed. GSH, which acts as an antioxidant in the body, reacts with radicals via GST to form inactive glutathione disulfide (GSSG), with the cycle then closing via conversion of GSSG back into active GSH by GR [42]. Both GST and GR activities were significantly decreased in the Ethanol group compared to the control group, but the intake of DML extract led to the recovery of their activities up to that of the control group. In addition, DMLDE showed a greater effect than DMLEE groups of the same doses.

Activities of antioxidant enzyme such as SOD, GST, and GR were reduced in the Ethanol group compared to the control group, and the groups that were administered DML extracts showed a concentration-dependent increase in the activities of all antioxidant enzymes. In a study by Bae et al. [21], subjects were given 100 or 300 mg/kg of DMLDE for four weeks and ethanol for the next three weeks, and it was found that the activities of antioxidant enzymes related to glutathione metabolism were reduced in the EtOH-treated group. For the treatment group that received 300 mg/kg of DMLDE, a similar recovery of antioxidant activity was observed to that seen for a positive control group treated with silymarin.

Plant-derived flavonoids and phenolic acids have strong antioxidant properties and are known to prevent inflammatory diseases, cancer, cardiovascular diseases, and various degenerative diseases associated with oxidative stress [13,43]. Previous studies have found that rutin and chlorogenic acid are the main components of *D. morbifera* leaves, which is consistent with the findings of this study [19]. Rutin in particular has been reported to exhibit anti-inflammatory activity against various inflammatory diseases and to protect the liver from various liver damage models [44]. Intake of chlorogenic acid, a type of caffeic acid glycoside that is present in coffee beans, is known to prevent liver damage caused by ethanol consumption by increasing inflammation and antioxidant enzymatic activities in liver [45]. In this study, the rutin, chlorogenic acid, and caffeic acid that are present in the extract are known to prevent liver damage.

In this study, we observed that alcohol consumption increased *Prevotella* and *Parasutterella* and decreased *Clostridium*, *Turcibacter* and *Romboutsia*, which is in line with the previous reports [46–49]. Moreover, alcohol consumption also increased the abundance of some of the gram-negative bacteria (i.e, Porphyromonadaceae and *Alloprevotella*). It has been reported that alcohol consumption may compromise intestinal permeability, allowing the intestinal lipopolysaccharide (LPS) to flow into the liver through the portal vein [50]. In contrast, treatment of DML extracts decreased relative abundance of these gram-negative bacteria. Moreover, it has been reported that DML extracts suppressed the production of LPS-induced pro-inflammatory mediators and cytokines [51] We observed that DMLDE increased the genus *Allobaculum*, known as a butyrate producer [52]. Butyrate was reported to improve alcohol-derived liver damage through enhancing intestinal epithelial barrier function in mice [53,54]. On the other hand, DMLEE increased the abundance of OTUs belonging to the family Porphyromonadaceae and *Bacteroides* and decreased the abundance of other members of the family Porphyromonadaceae. Previous study has reported that gut microbiota of mice ingested with pectin restored *Bacteroides* and completely prevented alcohol-derived liver damage, suggesting that abundance of *Bacteroides* may be associated with the recovery from alcohol-derived liver damage [55]. Although many members of the family Porphyromonadaceae include short chain fatty acids producers such as *Butyricimonas*, *Coprobacter* and *Macellibacteroides* [56–58], this family has been reported to be associated with alcohol consumption and resulting complications in chronic liver diseases [59,60]. To date, due to the lack of genetic information, further studies are required at the genus or species level to investigate roles of these OTUs belonging to the family Porphyromonadaceae.

Our data suggested that isopropanol biosynthesis and succinate fermentation to butanoate may have been largely enriched by alcohol consumption due to secondary alcohol dehydrogenase and succinate semialdehyde produced by *clostridium* strains. Ubiquitous bacteria produce ubiquinone which can be used to catalyze ethanol to a ubiquinol and acetaldehyde. Therefore, there is a possibility that the alcohol consumption may have enriched the metabolic activities of bacterial ubiquinol biosynthesis. Cmp-legionaminate biosynthesis was also depleted by alcohol consumption, but our data also suggested DML treatments may have recovered both of them. The legionaminic acid has been reported to be a virulence-associated cell-surface glycoconjugate in *Campylobacter*, an intestinal pathogenic bacteria [61], however another study has reported that this pathway was also predicted to be enriched in the cecum of probiotics treated broilers [62]. In our data, DMLEE showed enriched aromatic biogenic amine degradation which was depleted by alcohol consumption. Many of aromatic biogenic amines are involved in gastrointestinal pathology such as intestinal bowel syndrome and inflammatory bowel disease [63]. In this study, our results suggested that this biogenic amine degradation could be associated with the enrichment of L-tyrosine degradation. Besides Cmp-legionaminate biosynthesis, there were four more enriched metabolic pathways by both DML treatments, including aerobic respiration I (cytochrome c), oleate biosynthesis IV, (5z)-dodec-5-enoate biosynthesis, and palmitate biosynthesis II. Alcohol consumption is known to be negatively associated with activity of mitochondria [64], thus enriched metabolic activities of cytochrome c suggest that DML treatments may have reduced the damage caused by alcohol consumption. The rest of the three enriched metabolic activities by DML treatments were similar to each other, suggesting these might be predicted from the abundance of similar bacterial species. Oleate and palmitoleate are the major monounsaturated fatty acids in olive oils and its consumption has been reported to improve liver secretory activity, reduce inflammation and oxidative stress [65]. Our data indicated that DMLDE may have enhanced the metabolic activities of biosynthesis of stearate, palmitoleate, and biotin. Stearate is converted to oleate by stearoyl-CoA desaturase 1 (SCD1) and palmitoleate is produced from palmitate, therefore enhancement of these metabolic pathways may bring similar benefits as olive oils do. It has been reported that the amount of SCD1 is substantially low in germ free mice, suggesting that SCD1 is regulated by gut microbiota [66]. Biotin biosynthesis was also predicted to be enhanced by DMLDE. Biotin is a B-complex vitamin that acts as an essential coenzyme. Low biotin levels can occur in elderly individuals, excessive alcohol consumers and smokers. Glucokinase decreased by alcohol consumption can be increased by supplementation of biotin [67,68]. DMLDE also enriched fucose degradation, suggesting increased intestinal bacteria that consume fucose. Fucose is a monosaccharide abundant in mammalian gut, and a degradation of fucose plays an important role in maintaining gut homeostasis resulting in reduced inflammation and controlled hepatic bile acid synthesis [69]. Collectively, our data suggest that DML extracts, especially DMLDE may have enriched metabolic activities predicted to be beneficial to the gut and liver.

Furthermore, we investigated the association of the differentially abundant bacteria with the predicted metabolic activities. We observed that many of the predicted metabolic changes were positively associated with *Bacteroides*, *Allobaculum* and one OTU of Candidatus Saccharibacteria. Previously, it has been reported alcoholic patients have significantly low abundance of *Bacteroides* in human gut [70]. Xiao et al. [71] reported that chronic alcohol consumption shifts gut microbiota which may be a partial cause of alcohol withdrawal-induced anxiety. The decreased abundance of *Allobaculum* was the most obvious shift caused by excessive alcohol consumption. Our results suggest that DML extracts significantly increased the abundance of these genera, suggesting that DML extracts may partially recover the gut microbiota shifted by alcohol consumption. Millman et al. [72] reported that the ingestion of extra virgin olive oils significantly increased these two genera. In addition, they reported that the abundance of *Bacteroides* was found to be associated with loss of plasma triglyceride concentration. Houghton et al. [73] suggested that *Allobaculum* may improve age-related mitochondrial dysfunction through producing butyrate. With the increasing evidence of beneficial effects of these bacteria, our results suggest that DML extracts may improve intestinal functions through shifting gut microbiota impaired by alcohol consumption.

In this study, we investigated antioxidative effects of DML extracts on rats fed ethanol. Our results showed that DML extracts significantly improve the damage cause d by alcohol consumption. However, there are limitations in the study. Metabolomic shifts were only estimated by gut microbiota shifts using PICRUSt2, which could be more accurate if mass-spectrometry metabolomics was applied. In addition, further study should proceed for the identification of bioactive compounds in the DML extracts. Nevertheless, the study presents the effects of DML extracts not only on rats' physiology but also on gut microbiota, providing a fundamental information of how DML extracts may act as functional food.

5. Conclusions

Summarizing these results, we believe that the large amounts of polyphenol compounds contained in DML extracts promote ethanol metabolism in the liver, preventing liver damage. Our results showed hot water extracts showed better improvement in vivo, suggesting that water extracts may provide more bioavailable materials, resulting in enriching beneficial intestinal metabolic activities. These results suggest that DML extracts, especially ones extracted with hot water, may be used as a material for new functional foods or medicines that can prevent and treat liver toxicity caused by ethanol.

Supplementary Materials: The following are available online at http://www.mdpi.com/2076-3921/9/10/911/s1, Figure S1: Animal experiment design, Table S1: Total polyphenol and total flavonoid contents of *Dendropanax morbifer* leaf and stem extracts, Table S2: Antioxidant activity of the *Dendropanax morbifer* leaf and stem extracts, Figure S2: HPLC chromatogram of *Dendropanax morbifer* leaf and stem extracts, Table S3: Body weight change and liver-body weight ratio of *Dendropanax morbifer* leaf ethanol and D.W extracts in alcohol-fed rats, Figure S3: Comparison of microbiota according to the concentrations of DML extracts. Numbers indicate the concentrations of DML treatments, Figure S4: Comparison of microbiota before DML extract treatment, Table S4: Analysis of molecular variances (AMOVA), Figure S5: Comparison of the taxonomic composition.

Author Contributions: T.E., G.K. and T.U. conceptualization; T.E., G.K. and T.U. data curation; T.E., G.K. and T.U. formal analysis; J.-S.K. and T.U. funding acquisition; T.E., G.K. and K.C.K. investigation; T.E., G.K., K.C.K., J.-S.K. and T.U. methodology; T.U. project administration; T.E., G.K. and T.U. resources; G.K. and T.U. software; T.U. supervision; T.E., G.K. and K.C.K. validation; T.E., G.K. and T.U. visualization; T.E., G.K. and T.U. writing—original draft; T.E., G.K. and T.U. writing—review & editing. All authors have read and agreed to the published version of the manuscript.

Funding: This research was supported by the Basic Science Research Program through the National Research Foundation of Korea (NRF) funded by the Ministry of Education (2016R1A6A1A03012862). This research was also partially supported by the Cooperative R&D between Industry, Academy, and Research Institute funded by Korea Ministry of SMEs and Startups in 2015 (grant number C0297060).

Acknowledgments: We are grateful to Sustainable Agriculture Research Institute (SARI) in Jeju National University for providing the experimental facilities.

Conflicts of Interest: The authors declare no conflict of interest.

References

1. Sookoian, S.; Pirola, C.J. Systems biology elucidates common pathogenic mechanisms between nonalcoholic and alcoholic-fatty liver disease. *PLoS ONE* **2013**, *8*, e58895. [CrossRef] [PubMed]
2. Ratna, A.; Mandrekar, P. Alcohol and cancer: Mechanisms and therapies. *Biomolecules* **2017**, *7*, 61. [CrossRef] [PubMed]
3. Druesne-Pecollo, N.; Tehard, B.; Mallet, Y.; Gerber, M.; Norat, T.; Hercberg, S.; Latino-Martel, P. Alcohol and genetic polymorphisms: Effect on risk of alcohol-related cancer. *Lancet Oncol.* **2009**, *10*, 173–180. [CrossRef]
4. Seitz, H.K.; Stickel, F. Molecular mechanisms of alcohol-mediated carcinogenesis. *Nat. Rev. Cancer* **2007**, *7*, 599–612. [CrossRef] [PubMed]
5. Han, K.H.; Hashimoto, N.; Fukushima, M. Relationships among alcoholic liver disease, antioxidants, and antioxidant enzymes. *World J. Gastroenterol.* **2016**, *22*, 37–49. [CrossRef]
6. Alonso, V.R.; Guarner, F. Linking the gut microbiota to human health. *Brit. J. Nutr.* **2013**, *109*, S21–S26. [CrossRef]
7. Clemente, J.C.; Ursell, L.K.; Parfrey, L.W.; Knight, R. The impact of the gut microbiota on human health: An integrative view. *Cell* **2012**, *148*, 1258–1270. [CrossRef]

8. Arrieta, M.C.; Finlay, B.B. The commensal microbiota drives immune homeostasis. *Front. Immunol.* **2012**, *3*, 33. [CrossRef]
9. Wieland, A.; Frank, D.N.; Harnke, B.; Bambha, K. Systematic review: Microbial dysbiosis and nonalcoholic fatty liver disease. *Aliment. Pharmacol. Ther.* **2015**, *42*, 1051–1063. [CrossRef]
10. Bajaj, J.S. Alcohol, liver disease and the gut microbiota. *Nat. Rev. Gastroenterol. Hepatol.* **2019**, *16*, 235–246. [CrossRef]
11. Vassallo, G.; Mirijello, A.; Ferrulli, A.; Antonelli, M.; Landolfi, R.; Gasbarrini, A.; Addolorato, G. Alcohol and gut microbiota-the possible role of gut microbiota modulation in the treatment of alcoholic liver disease. *Aliment. Pharmacol. Ther.* **2015**, *41*, 917–927. [CrossRef] [PubMed]
12. Tripathi, A.; Debelius, J.; Brenner, D.A.; Karin, M.; Loomba, R.; Schnabl, B.; Knight, R. The gut-liver axis and the intersection with the microbiome. *Nat. Rev. Gastroenterol. Hepatol.* **2018**, *15*, 397–411. [CrossRef] [PubMed]
13. Ding, R.B.; Tian, K.; Huang, L.L.; He, C.W.; Jiang, Y.; Wang, Y.T.; Wan, J.B. Herbal medicines for the prevention of alcoholic liver disease: A review. *J. Ethnopharmacol.* **2012**, *144*, 457–465. [CrossRef]
14. Seeff, L.B.; Bonkovsky, H.L.; Navarro, V.J.; Wang, G.Q. Herbal products and the liver: A review of adverse effects and mechanisms. *Gastroenterology* **2015**, *148*, 517–532. [CrossRef] [PubMed]
15. Choo, G.S.; Lim, D.P.; Kim, S.M.; Yoo, E.S.; Kim, S.H.; Kim, C.H.; Woo, J.S.; Kim, H.J.; Jung, J.Y. Anti-inflammatory effects of *Dendropanax morbifera* in lipopolysaccharide-stimulated raw264.7 macrophages and in an animal model of atopic dermatitis. *Mol. Med. Rep.* **2019**, *19*, 2087–2096. [CrossRef]
16. Jung, H.Y.; Chung, T.H.; Hwang, I.K. *Dendropanax morbifera* leveille extract ameliorates memory impairments and inflammatory responses in the hippocampus of streptozotocin-induced type 1 diabetic rats. *Mol. Cell. Toxicol.* **2016**, *12*, 429–436. [CrossRef]
17. Hyun, T.K.; Kim, M.O.; Lee, H.; Kim, Y.; Kim, E.; Kim, J.S. Evaluation of anti-oxidant and anti-cancer properties of *Dendropanax morbifera* leveille. *Food Chem.* **2013**, *141*, 1947–1955. [CrossRef]
18. Park, J.U.; Kang, B.Y.; Kim, Y.R. Ethyl acetate fraction from *Dendropanax morbifera* leaves increases t cell growth by upregulating nf-at-mediated il-2 secretion. *Am. J. Chin. Med.* **2018**, *46*, 453–467. [CrossRef]
19. Eom, T.; Kim, K.C.; Kim, J.S. *Dendropanax morbifera* leaf polyphenolic compounds: Optimal extraction using the response surface method and their protective effects against alcohol-induced liver damage. *Antioxidants* **2020**, *9*, 120. [CrossRef]
20. Hyun, T.K.; Kim, H.C.; Ko, Y.J.; Kim, J.S. Antioxidant, alpha-glucosidase inhibitory and anti-inflammatory effects of aerial parts extract from korean crowberry (*Empetrum nigrum* var. *Japonicum*). *Saudi J. Biol. Sci.* **2016**, *23*, 181–188. [CrossRef]
21. Bae, D.; Kim, J.; Lee, S.Y.; Choi, E.J.; Jung, M.A.; Jeong, C.S.; Na, J.R.; Kim, J.J.; Kim, S. Hepatoprotective effects of aqueous extracts from leaves of *Dendropanax morbifera* leveille against alcohol-induced hepatotoxicity in rats and in vitro anti-oxidant effects. *Food Sci. Biotechnol.* **2015**, *24*, 1495–1503. [CrossRef]
22. McCord, J.M.; Fridovich, I. Superoxide dismutase. An enzymic function for erythrocuprein (hemocuprein). *J. Biol. Chem.* **1969**, *244*, 6049–6055. [PubMed]
23. Aebi, H. Catalase invitro. *Methods Enzymol.* **1984**, *105*, 121–126. [CrossRef] [PubMed]
24. Koneru, M.; Sahu, B.D.; Gudem, S.; Kuncha, M.; Ravuri, H.G.; Kumar, J.M.; Kilari, E.K.; Sistla, R. Polydatin alleviates alcohol-induced acute liver injury in mice: Relevance of matrix metalloproteinases (mmps) and hepatic antioxidants. *Phytomedicine* **2017**, *27*, 23–32. [CrossRef]
25. Schloss, P.D.; Westcott, S.L.; Ryabin, T.; Hall, J.R.; Hartmann, M.; Hollister, E.B.; Lesniewski, R.A.; Oakley, B.B.; Parks, D.H.; Robinson, C.J.; et al. Introducing mothur: Open-source, platform-independent, community-supported software for describing and comparing microbial communities. *Appl. Environ. Microbiol.* **2009**, *75*, 7537–7541. [CrossRef]
26. Quast, C.; Pruesse, E.; Yilmaz, P.; Gerken, J.; Schweer, T.; Yarza, P.; Peplies, J.; Glöckner, F.O. The silva ribosomal RNA gene database project: Improved data processing and web-based tools. *Nucleic Acids Res.* **2013**, *41*, D590–D596. [CrossRef] [PubMed]
27. Rognes, T.; Flouri, T.; Nichols, B.; Quince, C.; Mahé, F. Vsearch: A versatile open source tool for metagenomics. *PeerJ* **2016**, *4*, e2584. [CrossRef]
28. Cole, J.R.; Chai, B.; Farris, R.J.; Wang, Q.; Kulam-Syed-Mohideen, A.; McGarrell, D.M.; Bandela, A.; Cardenas, E.; Garrity, G.M.; Tiedje, J.M. The ribosomal database project (RDP-II): Introducing myrdp space and quality controlled public data. *Nucleic Acids Res.* **2007**, *35*, D169–D172. [CrossRef]

29. Westcott, S.L.; Schloss, P.D. Opticlust, an improved method for assigning amplicon-based sequence data to operational taxonomic units. *mSphere* **2017**, *2*, e00073-17. [CrossRef]
30. Chao, A.; Chazdon, R.L.; Colwell, R.K.; Shen, T.J. A new statistical approach for assessing similarity of species composition with incidence and abundance data. *Ecol. Lett.* **2005**, *8*, 148–159. [CrossRef]
31. Shannon, C.E. A mathematical theory of communication. *SIGMOBILE Mob. Comput. Commun. Rev.* **2001**, *5*, 3–55. [CrossRef]
32. Beals, E.W. Bray-curtis ordination: An effective strategy for analysis of multivariate ecological data. In *Advances in Ecological Research*; Academic Press: Cambridge, MA, USA, 1984; pp. 1–55. [CrossRef]
33. Segata, N.; Izard, J.; Waldron, L.; Gevers, D.; Miropolsky, L.; Garrett, W.S.; Huttenhower, C. Metagenomic biomarker discovery and explanation. *Genome Biol.* **2011**, *12*, 1–18. [CrossRef] [PubMed]
34. Douglas, G.M.; Maffei, V.J.; Zaneveld, J.R.; Yurgel, S.N.; Brown, J.R.; Taylor, C.M.; Huttenhower, C.; Langille, M.G.I. PICRUSt2 for prediction of metagenome functions. *Nat. Biotechnol.* **2020**, 1–5. [CrossRef]
35. Fernandes, A.D.; Reid, J.N.; Macklaim, J.M.; McMurrough, T.A.; Edgell, D.R.; Gloor, G.B. Unifying the analysis of high-throughput sequencing datasets: Characterizing rna-seq, 16S rRNA gene sequencing and selective growth experiments by compositional data analysis. *Microbiome* **2014**, *2*, 15. [CrossRef] [PubMed]
36. Obernier, J.A.; Bouldin, T.W.; Crews, F.T. Binge ethanol exposure in adult rats causes necrotic cell death. *Alcohol. Clin. Exp. Res.* **2002**, *26*, 547–557. [CrossRef] [PubMed]
37. Bailey, S.M. Emerging role of circadian clock disruption in alcohol-induced liver disease. *Am. J. Physiol. Gastrointest. Liver Physiol.* **2018**, *315*, G364–G373. [CrossRef]
38. Dunn, W.; Shah, V.H. Pathogenesis of alcoholic liver disease. *Clin. Liver Dis.* **2016**, *20*, 445–456. [CrossRef]
39. Wang, F.; Li, Y.; Zhang, Y.J.; Zhou, Y.; Li, S.; Li, H.B. Natural products for the prevention and treatment of hangover and alcohol use disorder. *Molecules* **2016**, *21*, 64. [CrossRef]
40. Halbwirth, H. The creation and physiological relevance of divergent hydroxylation patterns in the flavonoid pathway. *Int. J. Mol. Sci.* **2010**, *11*, 595–621. [CrossRef]
41. Eom, T.K.; Senevirathne, M.; Kim, S.K. Synthesis of phenolic acid conjugated chitooligosaccharides and evaluation of their antioxidant activity. *Environ. Toxicol. Pharmacol.* **2012**, *34*, 519–527. [CrossRef]
42. Desideri, E.; Ciccarone, F.; Ciriolo, M.R. Targeting glutathione metabolism: Partner in crime in anticancer therapy. *Nutrients* **2019**, *11*, 1926. [CrossRef] [PubMed]
43. Ma, X.; Jiang, Y.X.; Zhang, W.W.; Wang, J.B.; Wang, R.L.; Wang, L.F.; Wei, S.Z.; Wen, J.X.; Li, H.T.; Zhao, Y.L. Natural products for the prevention and treatment of cholestasis: A review. *Phytother. Res.* **2020**, *34*, 1291–1309. [CrossRef]
44. Patel, K.; Patel, D.K. The beneficial role of rutin, a naturally occurring flavonoid in health promotion and disease prevention: A systematic review and update. In *Bioactive Food as Dietary Interventions for Arthritis and Related Inflammatory Diseases*; Academic Press: Cambridge, MA, USA, 2019; pp. 457–479. [CrossRef]
45. Li, L.; Su, C.P.; Chen, X.Y.; Wang, Q.; Jiao, W.C.; Luo, H.; Tang, J.Y.; Wang, W.; Li, S.; Guo, S.Z. Chlorogenic acids in cardiovascular disease: A review of dietary consumption, pharmacology, and pharmacokinetics. *J. Agric. Food Chem.* **2020**, *68*, 6464–6484. [CrossRef] [PubMed]
46. Caslin, B.; Maguire, C.; Karmakar, A.; Mohler, K.; Wylie, D.; Melamed, E. Alcohol shifts gut microbial networks and ameliorates a murine model of neuroinflammation in a sex-specific pattern. *Proc. Natl. Acad. Sci. USA* **2019**, *116*, 25808–25815. [CrossRef] [PubMed]
47. Zhang, X.; Wang, H.; Yin, P.; Fan, H.; Sun, L.; Liu, Y. Flaxseed oil ameliorates alcoholic liver disease via anti-inflammation and modulating gut microbiota in mice. *Lipids Health Dis.* **2017**, *16*, 1–10. [CrossRef]
48. Yu, L.; Wang, L.; Yi, H.; Wu, X. Beneficial effects of lrp6-crispr on prevention of alcohol-related liver injury surpassed fecal microbiota transplant in a rat model. *Gut Microbes* **2020**, 1–15. [CrossRef]
49. Gu, Z.; Wu, Y.; Wang, Y.; Sun, H.; You, Y.; Piao, C.; Liu, J.; Wang, Y. *Lactobacillus rhamnosus* granules dose-dependently balance intestinal microbiome disorders and ameliorate chronic alcohol-induced liver injury. *J. Med. Food* **2020**, *23*, 114–124. [CrossRef]
50. Brandl, K.; Kumar, V.; Eckmann, L. Gut-liver axis at the frontier of host-microbial interactions. *Am. J. Physiol. Gastrointest. Liver Physiol.* **2017**, *312*, G413–G419. [CrossRef]
51. Hyun, T.K.; Ko, Y.-J.; Kim, E.-H.; Chung, I.-M.; Kim, J.-S. Anti-inflammatory activity and phenolic composition of *Dendropanax morbifera* leaf extracts. *Ind. Crop. Prod.* **2015**, *74*, 263–270. [CrossRef]
52. Greetham, H.L.; Gibson, G.R.; Giffard, C.; Hippe, H.; Merkhoffer, B.; Steiner, U.; Falsen, E.; Collins, M.D. *Allobaculum stercoricanis* gen. Nov., sp. Nov., isolated from canine feces. *Anaerobe* **2004**, *10*, 301–307. [CrossRef]

53. Cresci, G.A.; Bush, K.; Nagy, L.E. Tributyrin supplementation protects mice from acute ethanol-induced gut injury. *Alcohol. Clin. Exp. Res.* **2014**, *38*, 1489–1501. [CrossRef] [PubMed]
54. Cresci, G.A.; Glueck, B.; McMullen, M.R.; Xin, W.; Allende, D.; Nagy, L.E. Prophylactic tributyrin treatment mitigates chronic-binge ethanol-induced intestinal barrier and liver injury. *J. Gastroenterol. Hepatol.* **2017**, *32*, 1587–1597. [CrossRef]
55. Ferrere, G.; Wrzosek, L.; Cailleux, F.; Turpin, W.; Puchois, V.; Spatz, M.; Ciocan, D.; Rainteau, D.; Humbert, L.; Hugot, C.; et al. Fecal microbiota manipulation prevents dysbiosis and alcohol-induced liver injury in mice. *J. Hepatol.* **2017**, *66*, 806–815. [CrossRef] [PubMed]
56. Sakamoto, M.; Takagaki, A.; Matsumoto, K.; Kato, Y.; Goto, K.; Benno, Y. *Butyricimonas synergistica* gen. Nov., sp. Nov. and *Butyricimonas virosa* sp. Nov., butyric acid-producing bacteria in the family '*Porphyromonadaceae*' isolated from rat faeces. *Int. J. Syst. Evol. Microbiol.* **2009**, *59*, 1748–1753. [CrossRef] [PubMed]
57. Shkoporov, A.N.; Khokhlova, E.V.; Chaplin, A.V.; Kafarskaia, L.I.; Nikolin, A.A.; Polyakov, V.Y.; Shcherbakova, V.A.; Chernaia, Z.A.; Efimov, B.A. *Coprobacter fastidiosus* gen. nov., sp. nov., a novel member of the family *Porphyromonadaceae* isolated from infant faeces. *Int. J. Syst. Evol. Microbiol.* **2013**, *63*, 4181–4188. [CrossRef] [PubMed]
58. Jabari, L.; Gannoun, H.; Cayol, J.-L.; Hedi, A.; Sakamoto, M.; Falsen, E.; Ohkuma, M.; Hamdi, M.; Fauque, G.; Ollivier, B.; et al. *Macellibacteroides fermentans* gen. nov., sp. nov., a member of the family *Porphyromonadaceae* isolated from an upflow anaerobic filter treating abattoir wastewaters. *Int. J. Syst. Evol. Microbiol.* **2012**, *62*, 2522–2527. [CrossRef]
59. Huang, H.; Lin, Z.; Zeng, Y.; Lin, X.; Zhang, Y. Probiotic and glutamine treatments attenuate alcoholic liver disease in a rat model. *Exp. Ther. Med.* **2019**, *18*, 4733–4739. [CrossRef] [PubMed]
60. Bajaj, J.S.; Ridlon, J.M.; Hylemon, P.B.; Thacker, L.R.; Heuman, D.M.; Smith, S.; Sikaroodi, M.; Gillevet, P.M.J. Linkage of gut microbiome with cognition in hepatic encephalopathy. *Am. J. Physiol. Gastrointest. Liver Physiol.* **2012**, *302*, G168–G175. [CrossRef]
61. Schoenhofen, I.C.; Vinogradov, E.; Whitfield, D.M.; Brisson, J.R.; Logan, S.M. The cmp-legionaminic acid pathway in campylobacter: Biosynthesis involving novel gdp-linked precursors. *Glycobiology* **2009**, *19*, 715–725. [CrossRef]
62. Rodrigues, D.R.; Briggs, W.; Duff, A.; Chasser, K.; Murugesan, R.; Pender, C.; Ramirez, S.; Valenzuela, L.; Bielke, L. Cecal microbiome composition and metabolic function in probiotic treated broilers. *PLoS ONE* **2020**, *15*, e0225921. [CrossRef]
63. Fernández-Reina, A.; Urdiales, J.L.; Sánchez-Jiménez, F. What we know and what we need to know about aromatic and cationic biogenic amines in the gastrointestinal tract. *Foods* **2018**, *7*, 145. [CrossRef] [PubMed]
64. Hoek, J.B.; Cahill, A.; Pastorino, J.G. Alcohol and mitochondria: A dysfunctional relationship. *Gastroenterology* **2002**, *122*, 2049–2063. [CrossRef] [PubMed]
65. Bermudez, B.; Lopez, S.; Ortega, A.; Varela, L.M.; Pacheco, Y.M.; Abia, R.; Muriana, F.J. Oleic acid in olive oil: From a metabolic framework toward a clinical perspective. *Curr. Pharm. Des.* **2011**, *17*, 831–843. [CrossRef] [PubMed]
66. Singh, V.; Chassaing, B.; Zhang, L.; San Yeoh, B.; Xiao, X.; Kumar, M.; Baker, M.T.; Cai, J.; Walker, R.; Borkowski, K.; et al. Microbiota-dependent hepatic lipogenesis mediated by stearoyl coa desaturase 1 (scd1) promotes metabolic syndrome in tlr5-deficient mice. *Cell Metab.* **2015**, *22*, 983–996. [CrossRef]
67. Chauhan, J.; Dakshinamurti, K. Transcriptional regulation of the glucokinase gene by biotin in starved rats. *J. Biol. Chem.* **1991**, *266*, 10035–10038.
68. Kim, J.Y.; Hwang, J.-Y.; Lee, D.Y.; Song, E.H.; Park, K.J.; Kim, G.H.; Jeong, E.A.; Lee, Y.J.; Go, M.J.; Kim, D.J.; et al. Chronic ethanol consumption inhibits glucokinase transcriptional activity by Atf3 and triggers metabolic syndrome in vivo. *J. Biol.* **2014**, *289*, 27065–27079. [CrossRef]
69. Ke, J.; Li, Y.; Han, C.; He, R.; Lin, R.; Qian, W.; Hou, X. Fucose ameliorate intestinal inflammation through modulating the crosstalk between bile acids and gut microbiota in a chronic colitis murine model. *Inflamm. Bowel Dis.* **2020**, *26*, 863–873. [CrossRef]
70. Tsuruya, A.; Kuwahara, A.; Saito, Y.; Yamaguchi, H.; Tsubo, T.; Suga, S.; Inai, M.; Aoki, Y.; Takahashi, S.; Tsutsumi, E.; et al. Ecophysiological consequences of alcoholism on human gut microbiota: Implications for ethanol-related pathogenesis of colon cancer. *Sci. Rep.* **2016**, *6*, 27923. [CrossRef]
71. Xiao, H.W.; Ge, C.; Feng, G.X.; Li, Y.; Luo, D.; Dong, J.L.; Li, H.; Wang, H.; Cui, M.; Fan, S.J. Gut microbiota modulates alcohol withdrawal-induced anxiety in mice. *Toxicol. Lett.* **2018**, *287*, 23–30. [CrossRef]

72. Millman, J.; Okamoto, S.; Kimura, A.; Uema, T.; Higa, M.; Yonamine, M.; Namba, T.; Ogata, E.; Yamazaki, S.; Shimabukuro, M.; et al. Metabolically and immunologically beneficial impact of extra virgin olive and flaxseed oils on composition of gut microbiota in mice. *Eur. J. Nutr.* **2019**. [CrossRef]
73. Houghton, D.; Stewart, C.J.; Stamp, C.; Nelson, A.; Aj Ami, N.J.; Petrosino, J.F.; Wipat, A.; Trenell, M.I.; Turnbull, D.M.; Greaves, L.C. Impact of age-related mitochondrial dysfunction and exercise on intestinal microbiota composition. *J. Gerontol. A Biol. Sci. Med. Sci.* **2018**, *73*, 571–578. [CrossRef] [PubMed]

antioxidants

MDPI

Review

Insight into Polyphenol and Gut Microbiota Crosstalk: Are Their Metabolites the Key to Understand Protective Effects against Metabolic Disorders?

Mireille Koudoufio [1,2,3], Yves Desjardins [3], Francis Feldman [1,2,3], Schohraya Spahis [1,2,3], Edgard Delvin [1,4] and Emile Levy [1,2,3,5,*]

1 Research Centre, Sainte-Justine University Health Center, Montreal, QC H3T 1C5, Canada; mireille.koudoufio@umontreal.ca (M.K.); francis.feldman@umontreal.ca (F.F.); schohraya.spahis@gmail.com (S.S.); delvine@sympatico.ca (E.D.)
2 Department of Nutrition, Université de Montréal, Montreal, QC H3T 1J4, Canada
3 Institute of Nutrition and Functional Foods, Laval University, Quebec City, QC G1V 0A6, Canada; yves.desjardins@fsaa.ulaval.ca
4 Department of Biochemistry, Université de Montréal, Montreal, QC H3T 1J4, Canada
5 Department of Pediatrics, Université de Montréal, Montreal, QC H3T 1J4, Canada
* Correspondence: emile.levy@recherche-ste-justine.qc.ca; Tel.: +1-514-345-7783

Received: 16 September 2020; Accepted: 30 September 2020; Published: 13 October 2020

Abstract: Lifestyle factors, especially diet and nutrition, are currently regarded as essential avenues to decrease modern-day cardiometabolic disorders (CMD), including obesity, metabolic syndrome, type 2 diabetes, and atherosclerosis. Many groups around the world attribute these trends, at least partially, to bioactive plant polyphenols given their anti-oxidant and anti-inflammatory actions. In fact, polyphenols can prevent or reverse the progression of disease processes through many distinct mechanisms. In particular, the crosstalk between polyphenols and gut microbiota, recently unveiled thanks to DNA-based tools and next generation sequencing, unravelled the central regulatory role of dietary polyphenols and their intestinal micro-ecology metabolites on the host energy metabolism and related illnesses. The objectives of this review are to: (1) provide an understanding of classification, structure, and bioavailability of dietary polyphenols; (2) underline their metabolism by gut microbiota; (3) highlight their prebiotic effects on microflora; (4) discuss the multifaceted roles of their metabolites in CMD while shedding light on the mechanisms of action; and (5) underscore their ability to initiate host epigenetic regulation. In sum, the review clearly documents whether dietary polyphenols and micro-ecology favorably interact to promote multiple physiological functions on human organism.

Keywords: dietary polyphenols; metabolites; oxidative stress; inflammation; epigenetics; microflora; cardiometabolic complications

1. Introduction

Polyphenols, synthesized in a wide variety of fruits, legumes, and herbs, interestingly serve to protect against biotic and abiotic stresses caused by insects and parasites [1–3]. However, much attention is presently given to dietary polyphenols in view of their evident health benefits. These phenolic compounds are intrinsically strong free radical scavengers [4] and exhibit obvious potential in alleviating oxidative stress (OxS)-related illnesses. In this context, epidemiological and animal food intervention studies have established a solid association between the consumption of polyphenol-rich foods or beverages and their preventive influence on complex diseases, including insulin resistance (IR) [5–8], type 2 diabetes (T2D) [9–12], obesity [5,9,13–16], cardiovascular diseases (CVD) [17–22], neurodegenerative disorders [23–26], and cancer [22,27–29]. Recently, polyphenols have also been suggested as plausible adjunctive therapeutic agents for the COVID-19-induced inflammatory storm [30–32].

The growing evidence as to the prevention and management of different disorders by dietary polyphenols has led to multiple studies focusing on their dietary sources [33], efficacy [20], and bioavailability [34–36]. However, much remains to be learned about their potential mechanisms of action. In particular, their dynamic interaction with intestinal microbiota is of utmost importance as ~90% of the amount ingested reach the colon to directly influence microbial ecology and, at the same time, undergo microflora-induced metabolic modifications [37,38]. These interactions represent the specific topic of the present critical review. First, we will discuss the prebiotic action of polyphenols. Second, we will focus on their transformation and physiological impact on the gastrointestinal (GI) tract. Third, we will emphasize the role of their metabolites resulting from their catabolism by colonic bacteria. Fourth, we will analyze their cardiometabolic effects while highlighting their mechanisms of action and underscoring epigenetic regulation. Clearly, the ultimate goal of this review is to show that the health benefits of dietary polyphenols are through the action of their bioactive metabolites. Nonetheless, for the reader's understanding, it is essential to provide a brief introduction, which describes the structure, classification, and bioavailability of polyphenols.

2. Polyphenols: Classification, Structure, and Bioavailability

2.1. Classification and Structure of Polyphenols

Concisely, polyphenols are characterized by at least one aromatic ring and one hydroxyl functional group with evolving structure from a simple molecule to a complex polymer. They are classified into flavonoids and non-flavonoids, according to the intricacy of their structure, number of phenol rings, and carbon skeleton [32,39–41].

Flavonoids present a benzo-γ-pyrone structure containing two aromatic rings (A and B) bound by a 3-carbon bridge (C6–C3–C6). Based on the differences in the C ring, flavonoids can be subdivided into six sub-classes, namely: (1) flavonols (e.g., quercetin, kaempferol); (2) flavones (e.g., luteolin, apigenin); (3) isoflavones (e.g., daidzein, genistein); (4) flavanones (e.g., naringenin, hesperetin); (5) flavanols (e.g., catechins, epigallocatechins); and (6) anthocyanidins (e.g., malvidin, cyanidin). Although polyphenols may be encountered in plants as aglycones, they are generally found as glycoside derivatives, glucoside, galactoside, rhamnoside, xyloside, rutinoside, arabinopyranoside, and finally arabinofuranoside, being the most common [39]. Interestingly, the various tannin functional groups (e.g., hydroxyls) allow them to create non-covalent (hydrophobic and hydrogen) bounds and covalent bonds with proteins and carbohydrates [42–44]. Moreover, tannins are polyphenols that are most involved in binding to proline-rich proteins [45].

2.2. Bioavailability of Polyphenols

2.2.1. Factors Affecting Polyphenol Absorption

Polyphenol bioavailability (e.g., alimentary proportion delivered to blood circulation) is dependent on diverse conditions such as their stability, transport, and metabolic behavior. Several intrinsic and extrinsic factors influence the content of plant polyphenols [35]. It is worth mentioning that their relative composition and levels vary extensively among species and between varieties of the same species in association with their genetic background and state of ripeness, which is related to the time of harvest. For example, the content of olive secoiridoid phenolic derivatives decreased when irrigation and ripening increased [35,46–48]. Furthermore, the concentration and variety of phenolic derivatives depend on storage conditions as illustrated by the variability of the phenolic content and total antioxidant capacity of 14 apple and 6 pear cultivars harvested at different periods [46]. The chemical structure is another important intrinsic factor when considering bioavailability. Polyphenols vary widely in molecular weight, secondary and tertiary structures, as well as glycosylation level [49–51].

For instance, resveratrol and quercetin glycosides are absorbed to a lesser extent than the respective aglycones, which influences their metabolic fate through the digestive tract, their absorption and release in the portal circulation, and excretion in urine or feces [52].

With regards to extrinsic aspects, short time storage at low temperature also affects polyphenol content as exemplified by the broccoli loss of its caffeoyl-quinic and sinapic acid contents [53]. Food processing is yet another external factor when considering polyphenol bioavailability. Thermal processing causes diverging effects on polyphenol content and absorption. For instance, Xu et al. [54] reported that thermal processing significantly decreased bean total phenolic content and antioxidant properties, while Khatun et al. [55] reported the opposite by documenting an increase in the same properties when heating spices. These dissimilarities may be explained both by the difference in polyphenol moieties, matrix, and cooking process. The role of the food matrix on the disposition of polyphenols should not be under-evaluated given the interaction between food components ingested simultaneously (fat, proteins, complex carbohydrates) and different polyphenolic compounds likely influencing their interaction with the microbiota and ultimately their absorption. In a rat model, the combination of lecithin and soya bean oil or emulsifiers (sucrose, fatty ester, polyglycerol fatty acid ester, and sodium taurocholate) increased intestinal absorption efficiency of water-dissolved quercetin when these constituents had no effect given separately [56]. Similarly, the administration of hydroxytyrosol as a sole natural product to humans or rats provides a better bioavailability than when combined with refined oil or yogurt, although its handling differed between the two species [57]. Differing digestive tract possesses such as luminal machinery, bile acid secretion, and enterohepatic cycle may explain the differences. The above examples suffice to warrant precautions when interpreting results obtained from different protocols. The several factors mentioned, and others have to be carefully taken into account when conducting studies using cellular models, animals, or clinical investigations, as they could explain variability in outcomes results from numerous reports. In general, polyphenols exhibit a low bioavailability with maximal plasma concentrations reached within 2 to 4 h post ingestion, and an apparent short elimination time with a return to baseline levels within 8 to 12 h [58]. Hence, 24 h-urine collections generally provide a more accurate evaluation of total polyphenol absorption, metabolism, and excretion [59]. These pharmacokinetic properties suggest that only long-term consumption of a variety of polyphenols will affect the health trajectory and support the role of the "Mediterranean diet" in improving population health perspectives.

2.2.2. Absorption of Polyphenols and Derivatives

Polyphenol intestinal absorption was indirectly estimated by raised antioxidant defense in the circulation in response to their consumption. The basic principles for their transport and local metabolism are partway established. Figure 1 schematizes the digestion and absorption processes as generally accepted today for polyphenols. Distinction must be made between the digestive and absorption processes of low and high molecular weight polyphenols along the digestive tract as the sites and the efficacy may differ. In decreasing order of absorption kinetics, isoflavones, caffeic and gallic acids lead, followed by catechins, flavanones, and quercetin glucosides. High molecular weight polyphenols such as proanthocyanidins, galloylated catechins, and anthocyanins come last [60]. For instance, the degree of polymerization of procyanidins (dimer B3, trimer C2, and polymers), isolated from willow tree catkins, decreased their absorption through the rat gut barrier and can limit their metabolism by the intestinal microbiota when compared to catechins [61].

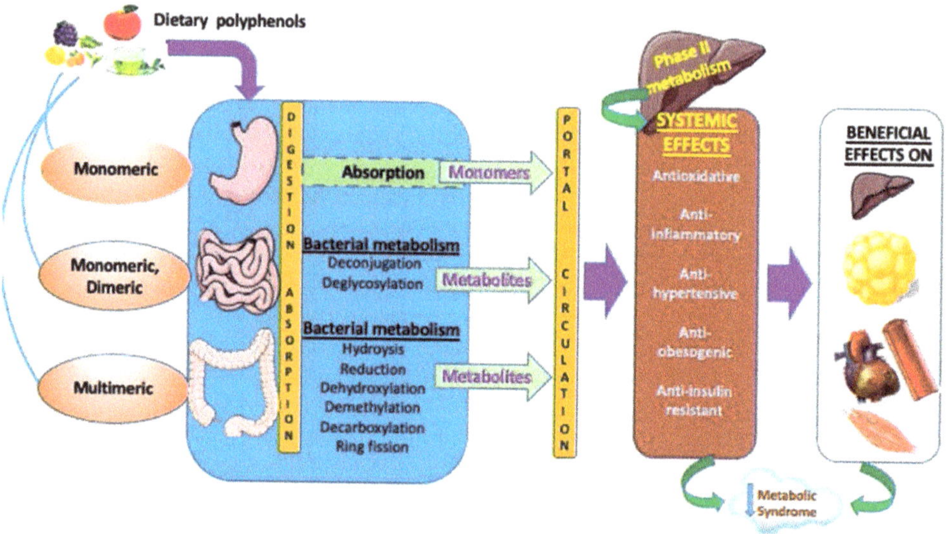

Figure 1. Integrated view of intestinal dietary polyphenol absorption, luminal transformation, and actions of their relevant metabolites on cardiometabolic disorders. As dietary polyphenols have limited absorption in the stomach and the small intestine, the unabsorbed polyphenols continue their transit to the colon where they are hydrolyzed, demethylated, decarboxylated, dehydroxylated, and ring fissioned by microbiota. Following these processes, microbial metabolites are subjected to phase II metabolism in the colon and liver, and enter the bloodstream to exert their biological effects, which extend to peripheral organs. Unabsorbed polyphenols and metabolites are excreted in feces, and absorbed microbial metabolites are mostly excreted in the urine. Noteworthy, whereas polyphenols improve microbiota composition, diversity and functions through their prebiotic actions, gut microbiota transform them into efficient bioactive regulators capable of optimizing cardiometabolic health in healthy individuals, while alleviating and mitigating the metabolic syndrome in patients. *Created with Servier Medical Art (A service to medicine provided by Les Laboratoires Servier, 50, rue Carnot—92284 Suresnes Cedex—France.

2.2.3. Gastric Uptake

It is well established that plant polyphenols predominantly present as glycosides undergo deglycosylation to their respective lipophilic aglycones, thereby enhancing their absorption through the GI [62,63]. Whether the stomach plays a significant role remains uncertain until today. To our knowledge, there is little direct in vivo evidence on the contribution of the human stomach to the assimilation of phenolic compounds. It needs to be underlined that in vitro simulation or animal models are the most common sources of information. Table 1 summarizes the results obtained in in vitro and in vivo models as well as in humans.

Table 1. Gastric handling of polyphenols.

Substrates	Sources	Experimental Conditions and Model	End Products	Main Observations	Conclusions	Ref
Procyanidins polymers (2–6 mers)	Cocoa	Incubation in simulated gastric juice (pH 2.0) at 37 °C for up to 3.5 h	Catechin/epicatechin Monomer and dimer	Time-dependant hydrolysis of oligomers.	Role of stomach in the processing of phenolic compounds.	[64]
Dimeric catechin/ epicatechin	Cocoa	Incubation in simulated gastric juice (pH 1.8) at 37 °C for up to 60 min	Catechin/epicatechin Dimers isomerization	Time-dependant hydrolysis of dimers.	Role of stomach in the processing of phenolic compounds.	[65]
Free and conjugated Hydroxytyrosol and Tyrosol	Olive oil	Incubation in simulated gastric juice (pH 2.0) at 37 °C for up to 4 h	Free hydroxytyrosol and tyrosol	Time-dependent hydrolysis of hydroxytyrosol and tyrosol conjugates.	Stomach hydrolyzes phenolic compounds conjugates.	[66]
Monomeric/ Polymeric Catechin/ epicatechin	Grape seed extract	Incubation in simulated gastric juice (pH 2.0) + pepsin at 37 °C for 2 h	Catechin/epicatechin Oligomers	Stability of catechin/epicat.	Stability of phenolic compounds at the gastric level.	[67]
Hydroxycinnamic acid derivatives, Flavonols, dihydrochlcones monomeric flavans-3-ols, procyanidin B_2	Apple juice	Incubation in simulated gastric juice (pH 2.0) + pepsin at 37 °C for up to 4 h	Hydroxycinnamic acid derivatives, flavonols, dihydrochlcones monomeric, Epicatechin monomer	Stability of Hydroxycinnamic acid derivatives, flavonols, dihydrochalcones monomeric. Hydrolysis of procyanidin B_2.	Stability of phenolic compounds in the stomach dependent on structure.	[68]
Purified Hesperidin 2S	*Citrus sinsensis* peel extract	Digestion in the simulator of human intestinal microbial ecosystem (pH 2.0).	Intact hesperidin 2S	No degradation of Hesperidin 2S.	Hesperidin is resistant to the degradation in the stomach.	[69]
Resveratrol caprylic esters	Synthesis product	Incubation in simulated gastric juice (pH 1.2) + pepsin at 37 °C for up to 2 h.	Intact resveratrol caprylic esters	No hydrolysis of resveratrol caprylic esters.	Resveratrol caprylic esters are not metabolized in the gastric phase.	[70]

Table 1. *Cont.*

Substrates	Sources	Experimental Conditions and Model	End Products	Main Observations	Conclusions	Ref
Polyphenols	Simulated oral digestion of peeled apple tissue	Incubation in simulated gastric juice (pH 1.6) + pepsin at 37 °C for up to 1 h in a dynamic rat stomach wall model	Released from initial material in deceasing order: chlorogenic acid, epicatechin, catechin, procyanidin B_2 flavan-3-ols, hydoxycinnamic acids, dihydrochalcones flavonols.	All polyphenols were stable except for procyanidin B_2 that was hydrolyzed to epicatechin.	Polyphenol resistance to degradation dependent on structure.	[71]
Polyphenols	Simulated oral digestion of Kiwifruit tissue	Incubation in simulated gastric juice (pH 1.2) + pepsin at 37 °C for 2 h	Release of the 16 identified polyphenols from initial material during stomach digestion: catechin, epicatechin, quercetin, rutin, chlorogenic acid, caffeic acid, ferulic acid, p-coumaric acid, gallic acid, salicylic acid, vanillic acid.	All polyphenols were stable.	Polyphenol resistance to degradation.	[72]
Procyanidins oligomers and flavonol monomers	Cocoa	Oral administration of procyanidin oligomers and flavonol monomers to healthy subjects. Gastric contents collected and analyzed at 20 min.	Intact procyanidins oligomers and flavonol monomers	Stomach Procyanidins oligomer and flavonol monomer profiles similar to original product.	Procyanidins oligomers and flavonol monomers are stable in the gastric environment.	[73]
Caffeic acid, gallic acid, chlorogenic acid, ferulic acid, coumaric acid	Purchased purified phenolic acids	In vivo rat ligated pylorus model for in situ gastric digestion at 37 °C for 25 min. Portal vein and abdominal aorta blood collected. Plasma analyzed with and without sulfatase and β-glucuronidase treatment.	Intact and conjugated coumaric acid, ferulic acid, caffeic acid, gallic acid, chlorogenic acid	Rapid appearance of coumaric acid > ferulic acid > caffeic acid > gallic acid > chlorogenic acid in portal vein and abdominal artery. Rapid appearance mainly of coumaric and ferulic acid conjugates in portal vein and abdominal artery.	Differential absorption efficiency of phenolic acids and differential affinity of monocarboxylic acid transporters.	[74]

Table 1. *Cont.*

Substrates	Sources	Experimental Conditions and Model	End Products	Main Observations	Conclusions	Ref
Flavone glycosides (apigenin, luteolin, chrysoeriol) and flavonoid glycosides (kaempferol, quercetin, isorhamnetin)	Parsley	Oral administration of glycoside extracts to rats. Animals sacrificed at 1, 1.5, 2, 4, or 12 h post administration. GI tract segmented (stomach, small intestine, colon, cecum). Stomach wall and lumen content analyzed at 2 h.	At 2 h: flavonoid glycosides in the stomach lumen and wall. Quercetin aglycone in the stomach wall.	At 2 h: flavonoid composition of stomach wall similar to stomach lumen but concentration lower. One aglycone present.	Stomach absorbs intact flavonoid glycosides.	[75]
Isoflavones (Daidzein, daidzin, genistein, and genistin)	Commercial source	In vivo rat ligated pylorus model for in situ gastric digestion at 37 °C for 25 min. Jugular vein blood analyzed for daidzein, daidzin, genistein, and genistin up to 30 min.	Daidzein and genetein (isoflavone aglycones)	Time-dependent absorption and transport of Daidzein and genetein, but not their respective glycosides daidzin and genistin.	Selective absorption and transport of isoflavone aglycones by the stomach.	[76]
Quercetin, rutin, and isoquercetin	Commercial source	In vivo rat ligated pylorus model for in situ gastric digestion at 37 °C for 30 min. Biliary duct cannulation and content analyzed. Aortic blood collected and analyzed.	Biliary quercetin and 3′-O-methyl-quercetin	Querecetin absorbed by the stomach and secreted in bile. No absorption of rutin or isoquercetin.	Limited role of the stomach in flavonoid glycosides. Selective absorption and transport of aglycones.	[77]

Purified cocoa procyanidin oligomers, incubated in simulated gastric juice, led to the appearance of dimers and monomers in a time-dependent fashion [64,65]. Similarly, olive oil polyphenol glycosides, including oleuropein (glycoside of elenolic acid linked to hydroxytyrosol) underwent rapid and time-dependent hydrolysis, yielding appreciable amounts of free hydroxytyrosol and tyrosol, probably through nucleophilic attack [66]. On the contrary, other studies, using the same in vitro model, along with oral administration of polyphenols to humans, showed stability of a variety of polyphenols in the acidic milieu [67–73]. Interestingly, the incubation of apple phloretin and quercetin for 5 min with native saliva resulted in the production of the respective aglycones, likely by oral bacterial flora as the process was blocked by antibiotics [68]. In the same study, hydroxycinnamic acid derivatives, flavonols, dihydrochalcones, and monomeric flavan-3-ols were stable when incubated with simulated gastric juice at pH 1.8, but procyanidin B_2 was degraded. Obviously, the in vitro models suffer from being an inadequate reflection of the gastric fluid, which lacks mucus normally secreted by parietal cells. This is supported by the notable stability of cocoa procyanidin polymers in vivo in the human stomach environment [73]. In terms of in vivo studies, the in situ digestion rat pylorus ligated model shows that polyphenol aglycones are absorbed by the stomach, and suggests the limited role of the stomach in flavonoid glycosides metabolism [74–77]. These results reveal inequality in the behavior of polyphenols in different experimental models and warrant careful analysis to obtain a complete representation.

2.2.4. Small Intestine Uptake

Duodenum

Using the in vitro three-step model simulating the digestive process from the mouth to the small intestine [78], it was shown that the duodenal digestion phase either with static or continuous-flow cellulose membrane-based dialysis containing bile salts and pancreatin resulted in dialyzable (chyme-available for passive absorption into the systemic circulation) and non-dialyzable (digested fraction available for colon) fractions [79]. Profiling the chyme and the non-dialyzable procyanidin content of cocoa liquor post gastric digestion revealed that the high-molecular weight procyanidins (pentamers to nonamers) were hydrolyzed essentially into monomers and dimers. Similar profiles were observed for the duodenal digest. Interestingly, the cocoa liquor duodenal digest contained much higher procyanidin concentrations than the cocoa powder counterpart that the authors attributed to the protective effect of fat micellar structures present in the liquor. More recently, using the same dynamic duodenal model, most of the intact procyanidins isolated from chocolate nibs were retained in the non-dialyzable fraction, which would likely be available for colonic digestion by the microbiota, while smaller molecular weight dimers and trimers were quantified in the dialyzable fraction [67]. On the other hand, the addition of carbohydrate-enriched food resulted in lower duodenal digestibility of procyanidin. Last, with the ex vivo reverted duodenal sac model, the bioavailability of curcumin was found to be significantly enhanced when encapsulated in low- and high-molecular weight polylactic-co-glycolic acid nanoparticles, possibly attributable to faster dissolution following the administration [80]. Overall, these results underpin the effect of the food matrix on the metabolism of polyphenols.

Jejunum and Ileum

It is well known that under physiological conditions, the small intestine, particularly the jejunum and ileum segments, represents the principal location for the digestion and absorption of dietary lipids. This holds for polyphenols, although with differing efficiency according to their molecular weight and type of glycosylation. Upon reaching the jejunum and ileum, glycosylated polyphenols are either hydrolyzed into their respective aglycones by the membrane-bound brush-border lactase phlorizin hydrolase [81] or transported intact by the enterocyte sodium-glucose co-transporter and hydrolyzed by the cytosolic β-glucosidase [82–84]. The nature of carbohydrate moieties affects the absorption of polyphenols through the small intestine. For instance, whereas glucoside conjugates and their aglycones are absorbed in the small intestine, those containing rhamnose molecules, [e.g., the flavonols hesperidin (hesperidin-7-*O*-glucosyl-rhamnose) and rutin (quercetin-3-*O*-glucosyl-rhamnose)], must proceed to the colon where rhamnose will be removed by bacterial rhamnosidase [85]. On the other hand, catechin and epicatechin, monomers of the flavonol family, which are often acylated by gallic acid, are readily absorbed by enterocytes without any deconjugation or hydrolysis [86,87].

Caco-2 cells, a human colon carcinoma derived cell line, which upon confluence develops enterocyte-like characteristics, have extensively been used as a model for studying lipid and drug metabolism [88] and transport [89]. They also have served as a model for studying the absorption, transport, and metabolism of polyphenols as detailed in Table 2.

Caco-2 cells have been shown to absorb major dietary hydroxycinnamates and diferulates after de-esterification (phase 1 transformation) and conversion into glucuronate, methyl, and sulfate conjugates (Phase 2 transformation) [90]. In 2006, Corona et al. [66] reported that Caco-2 cells transported hyroxytyrosol from the apical to the basolateral compartment, with the appearance of 3-*O*-methyl–hydroxytyrosol and glutathionyl-hydroxytyrosol in both compartments. They observed a similar transport for tyrosol, without any evidence of further metabolism or transport for oleuropein. The same year, trans-epithelial transport of trans-piceid (3-β trans-resveratrol glucoside) and deglycosylation in trans-resveratrol were reported using human intestinal Caco-2 cell monolayers [82]. The important question in the context of nutrition is whether polyphenols behave similarly when evaluated as single compounds as when assessed in complex mixtures [93,94]. Su et al. [92] partially answered this question by showing that, in Caco-2 cells, (+)-catechin significantly enhanced the cellular uptake and transport of the isoflavone daidzein-8-C-glucoside, whereas this compound significantly abated that of (+)-catechin, and that inhibition of efflux pumps increased their uptake. Although valuable for screening polyphenol bioavailability, caution should be used when interpreting the data obtained from the Caco-2 cellular model in terms of predictability. Indeed, even if in vitro smaller phenolic derivatives of hesperidin 2S were absorbed, hesperidin conjugates were the main bioavailable moieties in a randomized human controlled study [69].

The intestinal porcine enterocyte cell line (IPEC), a non-transformed non-tumorigenic permanent intestinal cell line derived from the jejunum is another model that merits attention for two main reasons. First, it maintains its differentiated characteristics, and second, there is a close analogy in results obtained in vivo with the porcine model, the closest to the human GI system [95,96]. To our knowledge, it has seldom been used to study polyphenol absorption. However, a good correlation for the uptake of grape pomace polyphenols was observed between IPEC cells and the duodenum and colon of piglets [97]. In addition, Wan et al. [98] proved functionality of the same in vitro model by showing enhanced the secretion of α-defensins-1 and 2 and reduced the E. coli translocation across a confluent IPEC cells by epigallocatechin-3-gallate, the major polyphenol specie in green tea.

Table 2. Experimented Caco-2 cells in polyphenol transport and metabolism.

Substrates	Sources	Experimental Conditions	Main Observations	Conclusions	Ref
hydroxytyrosol, tyrosol, and oleuropein	Olive oil	Polyphenols added to the apical chamber and incubation for 2 h. Apical and basolateral compartments collected and analyzed.	↓ hydroxytyrosol and tyrosol in the apical and ↑ in the basolateral media. Appearance of 3-*O*-methyl-tyrosol and glutathionyl-hydroxytyrosol in the apical and basolateral compartments. Φ transport of oleuropein from the apical to the basolateral compartment.	Hydroxytyrosol transported through the enterocyte apical membrane to the basolateral compartment with formation of conjugates.	[66]
Trans-piceid (Resveratrol 3-β-mono-D-glucoside)	*V. Vinifera*	Incubation with trans-piceid up to 360 min.	Bidirectional (apical to basolateral and inverse) transport of trans-piceid. Detectable trans-resveratrol in both chambers.	Trans-piceid and its aglycone are transported across the apical side and effluxed by the basolateral membrane.	[82]
Trans-piceid (Resveratrol 3-β-mono-D-glucoside)	*V. Vinifera*	Pre-incubated with ± chrysin (5,7-Dihydroxyflavone) and ± D-saccharic lactone in 6-well plates. Incubated with *trans*-piceid or *trans*-resveratrol 24 h.	Appearance of *trans*- Resveratrol after incubation with *trans*-piceid. ↑ *trans*- resveratrol-glucuronides production in chrysin (UDP-glucuronosyl transferase inducer) treated cells.	*Trans*-piceid undergoes hydrolysis to its aglycone and *trans*-resveratrol undergoes phase II metabolism within the enterocyte.	[82]
Quercetin, Quercetin-4-*O*-β-D-glucoside, Quercetin-3-*O*-β-D-glucoside, Quercetin-3,4-di-*O*-β-D-glucoside	Purified polyphenols	Polyphenols added to the apical chamber and incubation up to 2 h. Apical, cellular, and basolateral compartments collected.	Time-dependent ↓ quercetin in apical chamber. Stability of quercetin glycoside in apical chamber. Time-dependent appearance of quercetin glycosides in apical chamber and cellular compartment when quercetin is added. Time-dependent ↑ quercetin glycosides in basolateral compartment and stable low quercetin.	Quercetin aglycone is preferentially transported through the enterocyte apical membrane to the basolateral compartment after intracellular conjugation.	[84]

Table 2. *Cont.*

Substrates	Sources	Experimental Conditions	Main Observations	Conclusions	Ref
Catechin, epigallocatechin gallate encapsulated or not in non-ionic surfactant-based vesicles (niosomes)	Purified polyphenols	Polyphenols added to the apical chamber and incubation up to 6 h at 37 °C or 4 °C. Cell and basolateral compartments collected and analyzed.	Time-, concentration-, and temperature-dependent uptake of polyphenols. ↑ uptake when inserted into niosomes. Time-, concentration-, and temperature-dependent transport of polyphenol to the basolateral compartment, ↑ with niosomes. ↓ transport with ATP inhibitor, ↑transport with EDTA and P-glycoproteins and multidrug resistance proteins inhibitors.	Temperature dependence of uptake suggested energy-driven process. Deactivation of efflux pumps resulted in increased uptake by apical membrane and efflux in basolateral compartment.	[86]
Free and methyl esters of Hydroxycinnamic acids (ferulic, sinapic, *p*-coumaric and caffeic methyl esters), Ethyl esters of 5,5-diferulate, 8-*O*-4-diferulate, 8,5-benzofuran	Purified polyphenols	Medium collected after 24 h incubation and analyzed.	Glucuronides of ferulic, sinapic, *p*-coumaric and caffeic methyl esters, sulfates of ferulic, sinapic, *p*-coumaric and caffeic methyl esters, ferulic, sinapic, *p*-coumaric sulfates.	Metabolites produced either intra-cellularly and excreted in medium or produced in the medium by secreted phase I and phase II enzymes.	[90]
Quercetin, Quercetin-3,7,3,4-*O*-tetra-ethylacetate	Purified polyphenols	Polyphenols added to the apical (A) or basolateral (B) chamber and incubate up to 2 h. Apical, and basolateral compartments collected.	Time- and temperature-dependent bidirectional but preferential B-A transport of quercetin and ethylacetate derivative. Transport more efficient for the ethylacetate derivative. P-glycoproteins and multidrug resistance proteins inhibitors. ↑ quercetin permeation coefficient from A-B and ↓ from B-A. Φ on ethylacetate derivative.	Quercetin might be a substrate of P-glycoproteins and multidrug resistance proteins, causing the ↓ bioavailability of quercetin. Quercetin ethylacetate derivative exhibited better membrane permeation than parent compound.	[91]

Table 2. *Cont.*

Substrates	Sources	Experimental Conditions	Main Observations	Conclusions	Ref
Catechin, puerarin (Daidzein-8-*C*-glucoside)	Purified polyphenols	Incubation up to 2 h.	Time- and concentration-dependent uptake. Catechin enhanced uptake and transcellular transport of puerarin but puerarin inhibited that of catechin. P-glycoproteins and multidrug resistance proteins inhibitors. ↑ polyphenol uptake and transport.	Deactivation of efflux pumps resulted in increased uptake by apical membrane and efflux in basolateral compartment.	[92]

↑: increase; ↓: decrease; Φ: no effect.

Colon

The case of the colon bears some singularity in the sense that there is a reciprocal relationship between polyphenols and colonic microbiota. It has been clearly established that the gut microbiome is a key actor in modulating the production, bioavailability, and biological activities of complex low molecular weight phenolic metabolites grouped under the term polyphenol metabolome [99,100]. Rarely, the transport of polyphenols by the colon has directly been evaluated as most of the studies employed intestinal cell models displaying features of the small intestine [101]. It is, however, worthy to mention that T84 colon carcinoma cell line monolayers, incubated with ferulic, isoferulic, cinnamic, and hydroxycinnamic acids, as well as with flavonoids, showed appreciable transport from the luminal to the basolateral side [102]. Measurable levels of ferulic glucuronide and sulfate were detected in the apical and basolateral compartments when supplied at supra-physiological concentrations. Other reports bear mostly on the polyphenol anticancerogenic properties without addressing the uptake and transport per se [103–106].

The enterohepatic cycle is an important mechanism for the bioavailability and disposition of polyphenol aglycones and their metabolites. This is true whether absorbed by the small intestine or the colon (where they undergo in situ phase II enzymatic conversion to their respective methyl, glucuronide, and sulfate conjugates) or delivered to the liver via the portal circulation for further conjugation (before entering the blood stream and being excreted in urine). Alternatively, they enter the bile duct where mixed with bile salts, re-enter the small intestine for either a second-round absorption by enterocytes and transport to the liver, or excretion in the stools [37,107–109].

3. Interaction between Polyphenols and the Colon Microbiota

Humans are colonized by a wide array of microorganisms referred to as the microbiota that consists of obligate and facultative anaerobes, mainly targeting the colon [110,111]. Their main assignment is to provide assistance to their host in digesting food complex polysaccharides and proteins unprocessed in the upper GI tract [112]. The saccharolytic pathway, mainly active in the proximal colon, yields short chain fatty acids, of which acetic, propionic, and butyric acids are the most abundant, as well as lactic acid, CO_2, methane, and ethanol. The second catabolic pathway responsible for protein fermentation, mainly present in the distal colon, leads to metabolite species such as ammonia, various amines, thiols, phenols, and indoles [113,114]. The healthy gut ecosystem consists of the anaerobes *Bacteroidetes* and *Firmicutes*, the latter contributing to more than 90% of the total bacterial species, and *Proteobacteria*, *Verrucomicrobia*, and *Actinobacteria* accounting for the balance [115].

Diet composition has a definite role in the taxonomic and functional profile of the colon microbiota. For example, several reports have shown that high carbohydrate and fiber intakes are related to higher abundance of 3 enterotypes: *Lachnospiraceae*, *Ruminococcaceae*, and *Bifidobacteria* [116–118]. For their part, omnivorous women were shown to have a higher abundance of fecal butyrate producing taxa (*Clostridium cluster XIVa* and *Roseburia/Eubacterium*) than vegetarians [119]. Intervention studies have also established that changing from a carnivorous diet to a plant-based diet resulted in a gradual decrease in abundance of *Bacteroides*, a bile-tolerant symbiotic microorganism, and an increase in *Firmicutes* that preferentially metabolizes plant polysaccharides [120]. Furthermore, population food tradition-related differences in microbiota constitution have been reported; those consuming meat-based diets had higher abundances of *Bacteroides* than those traditionally consuming plant-based diets [118,121]. A controlled 10-week-diet intervention, involving overweight volunteers, demonstrated a rapid modification in microbiota species composition despite the inter-individual variation in initial composition [122]. Overall, the results indicate that *Bacteroides*, together with *Alistipes* and *Parabacteroides*, may be the primary proteolytic taxon in the human colon [112].

3.1. Polyphenol–Microbiota Interaction

Of interest, although still poorly understood, evidence is accruing on the influence of dietary polyphenols, perceived as xenobiotics, on the modulation of the colonic microflora and health [123]. They interfere with bacterial cell-to-cell communication and coordinate pathogenic behaviors, two clinically important characteristics with regards to virulence control and wound healing [124,125]. They also sensitize bacteria to xenobiotics and alter membrane permeability. This is exemplified by the sensitization of methicillin-resistant *Staphylococcus aureus* to β-lactam through the (−)-epicatechin gallate-mediated alterations of the bacterium cell wall bilayer, and enhanced release of lipotechoic acid from the cytoplasmic membrane [126]. Polyphenols, acting as prebiotics, have also been shown to modulate gut metabolism and immunity as well as inflammatory pathways through the change of T-cell functions, inhibition of mast cell degranulation, and down regulation of inflammatory cytokine responses [127–130]. Studies in animals and in humans have shown modification of colon microflora by polyphenolic compounds, resulting in growth inhibition of certain bacterial groups while permitting others to flourish [131–133]. Valuable effects of polyphenols, such as those present in red wine, include among others, enhanced abundance of beneficial bacterium taxa such as *Bifidobacterium* and *Lactobacillus*, capable of improving gut barrier protection, *Faecalibacterium prausnitzii* having anti-inflammatory properties, and *Roseburia* as a butyrate producer. Interestingly, the proliferation of these bacteria occurred at the expense of the less desirable *Escherichia coli* and *Enterobacter cloacae* [134]. Most of the important biological actions resulting from phenolic compounds and microbiota are depicted in Figure 2.

3.2. Impact of Microbiota on Polyphenol Metabolism

The transformation of polyphenols into bioactive metabolites is tributary to the colonic microflora species. For example, the *Firmicutes*: *Eubacterium ramulus* and *oxidoreducens*, *Clostridium orbiscidens* and *Flavonifractor plautii* metabolize the flavonols Kaempferol, Quercetin, and Myricetin by *O*-deglycosylation and C-ring fission into a series of metabolites comprising protocatechuic, 2-(3,4-dihydroxyphenyl-acetic, 2-(4-hydroxyphenil)-propionic, 3-hydroxyphenylacetic, 3-(3-Hydroxyphenyl)-propionic, 2-(3-hydroxyphenil)-acetic, 2-(3-dihydroxyphenil)-acetic, and 3-(3,4-dihydroxyphenil)-acetic acids as well as short-chain fatty acids (SCFAs) [83,135,136] (Figure 2). Furthermore, *Enterococcus casseliflavus (Firmicutes)* hydrolyzes sugar moieties from quercetin-3-*O*-glucoside, a process releasing the aglycone and ultimately producing lactic, formic, and acetic acids, together with ethanol [137]. The *Actinobacteria*: *Slackia equolifaciens*, *Slackia isoflavoniconvertens*, *Eggerthella lenta*, *Adlercreutzia equolifaciens*, and *Bifidobacterium* spp metabolize the flavonones Naringenin and Hesperidin through C-ring fission yielding 3-(4-hydroxyphenyl)-propionic, 3-phenylpropionic and hydroxyphenylpropionic acids [135,136]. *Adlercreutzia equolifaciens*, *Paraeggerthella hongkongensis*, *Slackia equolifaciens*, *Slackia isoflavoniconvertens*, *Bacteroides ovatus*, *S. intermedius*, *R. productus*, *E. sp. Julong*, *E. faecium EPI1*, *L. mucosae EPI2*, and *F. magna*, have also been shown convert some the isoflavone daidzein to (*S*)-equol, a nonsteroidal estrogen-like molecule [((3*S*)-3-(4-hydroxyphenyl)-7-chromanol)] [108,137].

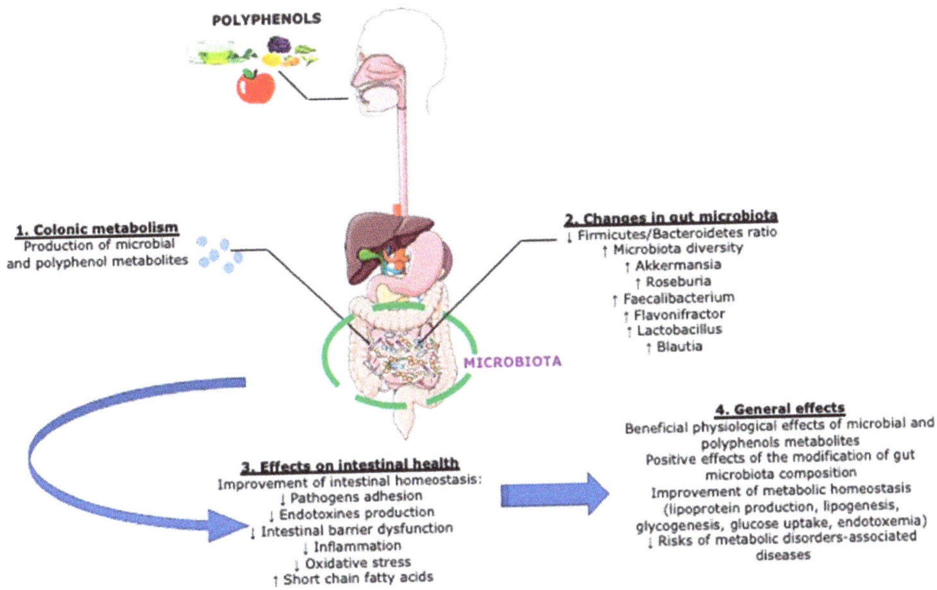

Figure 2. Beneficial actions resulting from the interaction between polyphenols and intestinal microbiota. Polyphenolic compounds exhibit prebiotic ability, which modifies bacterial composition and function. On the other hand, intestinal microbiota ameliorates intestinal homeostasis and cardiometabolic health by producing polyphenolic metabolites. * Created with Servier Medical Art.

4. Polyphenols and Metabolic Syndrome

The polyphenol metabolites, produced by the gut microbiota once absorbed and directed to the target tissues and organs, contribute to the metabolic health, through their antioxidant and anti-inflammatory properties, by preventing or reducing the risk of developing several cardiometabolic disorders (CMD), notably the metabolic syndrome (MetS).

The increasing prevalence of childhood MetS along with overweight and obesity is a major public health concern in both developed and developing countries. MetS is a key risk factor for the development of T2D, non-alcoholic fatty liver disease, and CVD in early adulthood [138–140]. MetS is a cluster of interrelated risk factors, including abdominal obesity, dyslipidemia, hypertension, hyperglycemia, and IR, which is triggered by cellular redox imbalance and inflammation as cardinal features [141–143].

In the last two decades, strategies to limit the deleterious effects of the MetS have been focused on diet regimen containing natural fruits, green vegetables, whole grains, legumes, probiotics, vitamin C, vitamin E, and ω-3 polyunsaturated-fatty acids [20,144–148]. Likewise, prebiotics, including polyphenols with their prebiotic potential, have also demonstrated promising effects on IR, glycemia, lipid profile, and CVD [8,134,149]. Other multiple studies, involving animal models, have consistently shown that administration of polyphenols, extracted from various sources, increased insulin sensitivity (via measurement of homeostatic model assessment for insulin resistance index, glucose intolerance, or insulin tolerance tests) and decreasing circulating free-fatty acid, triglyceride, cholesterol, C-reactive protein, resistin, and leptin concentrations [150–154]. At the clinical level, a parallel, double-blinded controlled and randomized 6-week dietary intervention demonstrated that strawberry and cranberry polyphenols improved insulin sensitivity in overweight and obese nondiabetic, insulin-resistant human subjects [8]. There was not a significant improvement of other cardiometabolic risk factors, which is probably due to the length of the intervention. Nevertheless, these in vivo results show the potential clinical preventive and therapeutic impact of polyphenols on human diseases. Considering that impaired redox potential and inflammatory processes are the basis of MetS, polyphenols, identified for their antioxidant and anti-inflammatory properties [155,156], appear as suitable candidates for preventing its development. The review of the mechanisms linking polyphenols and alleviation of these features follows.

4.1. Antioxidative Effects of Polyphenols

Polyphenols are known as major contributors to the fruit total antioxidant activity [157]. They do so by donation of an electron or hydrogen atom to reactive oxygen, nitrogen, and chlorine species based either on hydrogen atom transfer or single electron transfer by proton transfer [158]. Reacting with the inner side of the plasma membrane hydrophobic compounds, polyphenols impede lipid and protein oxidation, thus protecting the structure, fluidity, and function of the cell membrane [159]. Polyphenols also chelate transition metals as Fe^{2+}, thus directly reducing the Fenton reaction and preventing oxidation by highly reactive hydroxyl radicals [160].

Reactive oxygen species (ROS) play leading roles in provoking pan-cellular inflammation. Their action is mediated by the activation of powerful transcription factors such as nuclear redox factor-2 (Nrf2), nuclear factor-kappa B (NF-κB), and activator protein 1 [161]. Under unstressed conditions, inactive Nrf2 is linked to its cytoplasmic regulator actin-anchored Kelch-like ECH-associating protein 1. Due to reactive species accumulation in cells under OxS, Nrf2 dissociates from Kelch-like ECH-associating protein 1, translocates into the nucleus where it modulates antioxidant-responsive elements-mediated transcription cytoprotective genes [162,163]. Phenolic compounds are able to enhance protective pathways by activating Nrf2. For example, flavonoids induce Nrf2 translocation to the nucleus where it forms heterodimers with the leucine zipper-like transcription factors small musculo-aponeurotic fibrosarcoma proteins, then bind to antioxidant-responsive elements to activate target genes such as Phase II enzymes involved in detoxification [164]. In addition, Zhang et al. [165], using human breast cells, hepatic human liver cancer cells, and mouse Hepa-1 cells, demonstrated cell-dependent agonist/antagonist effects of several flavonoids on the regulation of aryl-hydrocarbon receptor-mediated signalling pathways. Polyphenols may also suppress pro-oxidative response by down-regulating the synthesis of the pro-inflammatory interleukin-(IL)-1β, tumor necrosis factor-alpha (TNFα), and interferon-gamma. They do so by modulating NF-κB and mitogen-activated protein kinase signaling pathways [166]. Moreover, phenolic compounds can display genoprotective effects, thereby preventing/reversing the progression of various CMD diseases [167]. Through the protection of DNA against OxS damages, some types of polyphenols can promote metabolic health [168–171]. For example, epigallocatechin gallate and resveratrol exhibit high capacity to decrease DNA double strand breaks, chromosome loss, and DNA deterioration response in H_2O_2-induced genotoxicity in cells [172,173]. A human supplementation trial has also shown the ability of green tea polyphenolic antioxidants to increase the resistance of DNA to oxidation damages [174].

4.2. Anti-Inflammatory Effects of Polyphenols

The crosstalk between oxidative and inflammation signalling pathways is important for cell homeostasis and survival. Indeed, overproduction of ROS has been shown to stimulate the release of pro-inflammatory cytokines. Reciprocally, NF-κB controlled genes are able to regulate ROS production [175,176]. Various studies, particularly those oriented toward cancer, revealed that polyphenols are able to control chronic inflammatory responses at the level of cytokine production, NF-κB-mediated gene expression, and the release of the tumor-suppressing transforming growth factor-beta [177–179].

Furthermore, closely related to the MetS, our group has more recently demonstrated, in post-confluent Caco-2/15 cells, that polyphenols extracted from dried apple peels (DAPP), containing phenolic acids, flavonol glycosides, flavan-3-ols, and procyanidins prevented iron-ascorbate-mediated lipid peroxidation and counteracted lipopolysaccharide-mediated inflammation as shown by the down regulation of the cytokines TNFα and IL-6, lessening of cyclooxygenase-2 expression, and production of prostaglandin E_2 via the decline of NF-κB. Concomitantly, the alleviation of inflammation was accompanied by the induction of Nrf2 (orchestrating cellular antioxidant defenses and maintaining redox homeostasis) and peroxisome proliferator-activated receptor gamma coactivator 1-alpha (the "master controller" of mitochondrial biogenesis) [180]. We later demonstrated that DAPP prevented and alleviated dextran sodium sulfate-induced intestinal inflammation in mice, stimulated antioxidant transcription factors, and improved mitochondrial dysfunction. They also decreased lipid peroxidation and up-regulated antioxidant enzymes, while decreasing the activity of myeloperoxidase, the expression of cyclooxygenase-2, and the production of prostaglandin E2. Moreover, DAPP partially restored mitochondrial redox homeostasis, fatty acid β-oxidation, ATP synthesis, apoptosis, and regulatory mitochondrial transcription factors [181]. As DAPP in parallel decreased the relative abundance of *Peptostreptococcaceae* and *Enterobacteriaceae* bacteria [182], it is possible that the modifications of intestinal microbiota brought about the amelioration of oxidative and inflammatory processes.

4.3. Epigenetic Control of Polyphenols

Epigenetic modification of gene expression is extremely important in the modulation of the OxS and inflammatory pathways briefly described above. Such alterations have been associated with abnormalities in various metabolic pathways [183]. Epigenetic control includes methylation of CpG rich regions, post-transcriptional modulation of chromatin histone/non-histone, and micro-RNAs (miRNAs) regulating gene expression [184–186]. Importantly, while these alterations may persist for cells lifespan and may have inter-generational effects, they are reversible and have thus become prime targets for interventions in the treatment of MetS [187]. Unfortunately, data on these mechanisms, particularly those bearing on DNA methylation and histone acetylation, are scarce. Table 3 shows that, in both animal models and human studies, polyphenol consumption decreases obesity, IR, and hypertension. These metabolic improvements are associated to epigenetic modifications, including increased DNA methylation, and histone methylation and acetylation.

Interestingly, reports involving miRNAs are more profuse. Table 4 summarizes the major findings obtained regarding the modulation of miRNAs and their effect on the expression of specific genes. It can be appreciated that polyphenols, independently of their origin, modulate miRNAs, especially controlling OxS and inflammation pathways in different cell models. In animal models, they also regulate miRNAs expression involved in the control of inflammation, obesity, lipogenesis, and energy expenditure. One human study shows that polyphenol consumption modulates miRNAs expression related to inflammation processes.

Table 3. Effect of polyphenols on DNA and histone methylation and acetylation.

Polyphenols	Experimental Model	Tx Duration (Week)	Polyphenol Dosage	Epigenetic Modifications	Outcomes in Response to Polyphenols	Ref
			Obesity and Insulin Resistance			
Raspberry extract	HFD-fed mice	16	120 mg/kg/d	↑AT Histone methylation and acetylation	↓Obesity ↓IR ↓Inflammation ↓Liver steatosis	[188]
Quercetin and Q2 derivative	HFD-fed rats	12	0.26 mg/kg/d	↑AT Histone methylation	↓Obesity ↓IR ↓Dyslipidemia ↓Liver steatosis	[189]
Apples	HFHSD-fed rats	8	700 mg/kg/d	↑Methylation Aqp7 ↑PGC genes ↑Methylation leptin gene	↓Obesity ↓IR ↑AT lipolysis	[190]
			Hypertension			
Cocoa	Humans Pre-hypertension or hypercholesterolemia	2	6 g/d	↓Leuk DNA methylation ↓ DNA Mtases methylation	ND	[191]
Resveratrol	Salt-sensitive hypertensive rats	0–12	50 g/L drinking water	↑histone H3K27me3 in renal aorta	Prevention of hypertension ↑Antioxidant defence	[192]

Tx: treatment; HFD: high-fat diet; HFHSD: high-fat high-sucrose diet; miR: micro-RNA; IR: insulin resistance; AT: adipose tissue; Leuk: leukocytes; Mtases: methylases; ND: no data; ↑: increase; ↓: decrease.

Table 4. Effects of polyphenols on the regulation of miRNAs on oxidative and inflammation pathways.

Polyphenols	Experimental Model/Conditions	Regulated miRNAs	Expression Pattern and Function	Ref
		Cellular models		
Quercetin and Isorhamnetin	Pre LPS Tx stimulation of murine RAW 264.7 macrophages. Polyphenols (0, 10–100 μmol/L)	↓miR-155	↓TNFα, ↓ iNOS, ↓IL-1β, ↓IL-6, ↓MIP1α, ↓NF-kB ↑Nrf2 and ↑ARE	[193]
Resveratrol	Human THP-1 cell line HPBMC	↑miR-663 ↓miR-155	↓basal AP-1 and ↓LPS-induced AP-1 ↓ JunB/D mRNA	[194]
EGCG	IL-1β-stimulated human OA chondrocytes	↑hsa-miR-199a-3p	↓COX-2 mRNA/protein expression ↓PGE_2 production	[195]
EVOO oleocanthal (OC) and oleacein (OA) secoiridoids	SGBS adipocytes pretreated with OC or OA before stimulation by TNFα	↓miR-155-5p, ↓miR-34a-5p ↑let-7c-5p	↓IL-1β, ↓COX-2, ↓MMP-2, ↓NF-kB, ↓NADPH oxidase ↑SOD and ↑GPx, ↑PPARγ ↓MCP-1, ↓CXCL-10, ↓M-CSF	[196]
Olive oil hydroxytyrosol (HT)	SGBS adipocytes pretreated with HT before stimulation by TNFα	↓miR-155-5p, ↓miR-34a-5p ↑let-7c-5p	↓MCP-1, ↓CXCL-10, ↓IL-1β, ↓IL-6, ↓vEGF, ↓COX-2, ↓M-CSF, ↓MMP-2, ↓NF-B and ↓ROS production ↑GPX ↑eNOS, ↑PGC-1α	[197]
Propolis extracts	HaCat cell line treated for 24 h with propolis extracts (3.125, 1.56, and 0.78 mg/mL)	↑miR-19a-3p ↑ miR-203a-3p ↑miR-27a-3p ↓miR-17-3p	↓TNFα mRNA ↓NFE2L2 mRNA ↑GPX2, ↑MnSOD and ↑TRXR2 mRNAs	[198]
Curcumin polyphenolic compound	ARPE-19 cells treated with 20 μM curcumin and 200 μM H_2O_2	↑miR-146a ↑miR-155 ↓miR-23b ↓miR-27b ↓miR-26b ↓miR-15b ↓miR-9 ↓miR-30b, miR-30e	↓NF-κB ↑CAT, ↑GPx	[199]

Table 4. *Cont.*

Polyphenols	Experimental Model/Conditions	Regulated miRNAs	Expression Pattern and Function	Ref
Açai and red muscadine grape polyphenols	HUVEC ROS induction by 25 mM glucose for 30 min	↑miR-126 ↑MiR146a	↓IL-6, ↓IL-8, ↓NF-kB, ↓PXR, ↓VCAM-1 ↑CYP1A1, ↑MDRP1, ↑CAT, ↑GST activity	[200]
Resveratrol	LPS-stimulated THP-1 macrophages pretreated with resveratrol	↑miR-Let7A	↓TNFα, ↓IL-6, ↓IL-10, ↓IL-4, ↓SIRT1 mRNAs	[201]
		Animal models		
Quercetin	Ctrl or HFD C57BL/6 J fed mice 0.2 or 2.0 mg/g diet	↑miR-125b	↓IL-6, ↓CRP, ↓MCP-1, ↓AOAH, ↓HO-1, ↓Ref-1, ↓TLR-2 mRNAs	[202]
Grape seed extract	HFD-fed obese Rats 30 mg/kg/d	↓miR-33a, ↓miR-122	↓TC, ↓TAG, ↓LDL-C, ↓TNFα, ↓liver MDA ↑SOD, CAT; ↑liver GSH	[203]
Polydatin (3,4′,5-trihydroxy-stilbene-3-β-D-glucoside)	Crtl, or HFrD, HFrD+Polydatin (7.5, 15, 30 mg/kg), HFrD + PG (4 mg/kg) fed SD rats IG saline, polydatin or PG 7 week	↓ miR200-a	↓ TXNIP, ↓NLRP3, ↓ASC, ↓Casp-1, ↓ SREBP-1 and ↓SCD-1 ↑ PPAR-α and ↑CPT-1	[204]
Tea extract	HFD-fed mice for 12 weeks 500 mg/kg/d	↓miR-335 ↓ miR-155 in AT	↓Obesity, ↓IR, ↓Inflammation, ↑Energy expenditure	[205]
Resveratrol	HFHS-fed rats for 6 weeks 30 mg/kg/d	↑miR-211-3p ↑miR-1224 ↑miR-539-5p ↓AT miR-511-3p	↓Obesity, ↓AT lipogenesis	[206]

Table 4. *Cont.*

Polyphenols	Experimental Model/Conditions	Regulated miRNAs	Expression Pattern and Function	Ref
		Human studies		
Resveratrol	T2D HT patients. 1-year daily intake of grape extract (8.1 mg/d for first 6 months and 16.2 mg/d for last 6 months)	HPBMC ↑miR-21 ↑miR-181b ↑miR-663 ↑miR-30c2	↓IL-6, CCL-3, IL-1β, TNFα, ↑LRRFIP-1	[207]

Tx: Treatment; Ctrl: control, HFD: high-fat diet; HFrD: high-fructose diet; PG: pioglitazone; SD: Sprague-Dawley rat; ARPE-19 cells: adult retinal pigment epithelial cell line-19; HUVEC: human umbilical vascular endothelial cells; THP-1: human monocyte leukemia cells; LPS: lipopolysaccharides; TNFα: tumor necrosis factor alpha; iNOS: inducible nitric oxide synthase; IL-1β: interleukin-1β; IL-6: interleukin-6; MIP-1α: macrophage inflammatory protein-1α; NF-kB: nuclear factor κB; Nrf2: NF-E2-related factor 2; AOAH: acyloxyacyl hydrolase; HO-1: heme oxygenase-1; Ref-1: redox factor-1; TLR-2: toll like receptor-2; ARE: antioxidant response element; CRP: C-reactive protein; MCP-1: monocyte chemo-attractant protein-1; T2D: type-2 diabetes; HT: hypertensive; HPBMCs: human peripheral blood mononuclear cells; LRRFIP-1: leucine-rich repeat flightless-interacting protein-1; HaCat: human keratinocyte cell line; IG: intra-gastric; ASC: apoptosis-associated Speck-like protein Casp1: caspase-1; SGBS: Simpson–Golabi–Behmel syndrome; ALP: alkaline phosphatase; CCL-3: chemokine (C-C motif) ligand-3; TC: total cholesterol, TAG: triacylglycerol; LDL-C: low-density lipoprotein cholesterol; TAC: total antioxidant capacity; MDA: malondialdehyde; SOD: superoxide dismutase; CAT: catalase; GPX: glutathione peroxidase; GSH: glutathion, OA: osteoarthritis; COX2: cyclooxygenase-2; PGE2: prostaglandin-E2; EGCG: epigallocatechin-3-*O*-gallate; MMP-2: matrix-degrading enzyme metalloproteinase; NADPH oxidase: nicotinamide adenine dinucleotide phosphate oxidase; CXCL-10: C-X-C motif ligand 10; M-CSF: macrophage colony-stimulating factor; PPARγ: peroxisome proliferator-activated receptor; NFE2L2: nuclear factor, erythroid 2 like 2; TRXR2: thioredoxin reductase 2; MnSOD: manganese superoxide dismutase; TXNIP: thioredoxin-interacting protein; NLRP3: NOD-like receptor (NLR) family, pyrin domain containing 3; PPARα: peroxisome proliferator activated receptor-α; CPT-1: carnitine palmitoyl transferase-1; SREBP-1: sterol regulatory element binging protein 1; SCD-1: stearoyl-CoA desaturase-1; vEGF: vascular endothelial growth factor; eNOS: endothelial nitric oxide synthase; PGC-1α: peroxisome proliferator-activated receptor coactivator 1α; GLUT-4: glucose transporter-4; VCAM-1: endothelial vascular cell adhesion molecule-1; PXR: pregnane X receptor; GST: glutathione S-transferase; MDRP1: multidrug-resistant protein 1, CYP1A1: cytochrome P450; SIRT1: sirtuin 1; ↑: increase; ↓: decrease.

5. Polyphenol Metabolites

In the first part of the present review, we have defined the polyphenols, described their structure, reported their digestion and absorption, elaborated on their mode of action and potential benefits, emphasized their epigenetic regulation, and pointed out their interaction with intestinal microbiota. The goal of the second part is to focus on the polyphenol metabolites, derived from their colonic microbial metabolism and biotransformation. As mentioned above, parent or native polyphenols have glycosidic linkages, which limit their absorption in the small intestine, forcing them to continue their way to the colon. It is in this part of the large intestine that glycosides are cleaved and further metabolized by microbiota to potentially generate more active and better-absorbed metabolites. Using transporters and passive diffusion, the small molecular weight end-products have easy access to the circulation. Hence, the second part of this review is dedicated to specify the biological activity and consequential functional effects of polyphenol metabolites on CMD. Table 5 provides information on the different metabolites issued from colonic digestion and on their pleiotropic effects.

5.1. Flavonoid Metabolites and Their Antioxidant and Anti-Inflammatory Effects

Bacterial flavonoid catabolism in the colon follows a generic pattern yielding non-specific metabolites such as hydroxylated phenylpropionic acid (HPPA) and phenylacetic acids (HPAA) that ultimately may be subjected to β-oxidation and glycination to yield hippuric acid and its hydroxylated counterpart [34,217]. It can therefore be appreciated that a relatively small number of degradation products emanate from extremely diverse parent flavonoids.

5.1.1. Flavonol Quercetin Metabolites

Quercetin (2-(3,4-dihydroxyphenyl)-3,5,7-trihydroxychromen-4-one) is supplied by apples, tea, red wine, berries, tomatoes, and onions. This polyphenol has attracted a lot of interest from the biomedical milieu in view of its beneficial properties in prevention-diseases (e.g., T2D, CVD) despite its limited bioavailability because of its low solubility, weak stability in the upper gastrointestinal tract, rapid fast metabolism, and short biological half-life [98]. However, quercetin molecules unable to be absorbed in the small intestine reach the colon, and are metabolized into a series of phenolic acids by luminal bacteria [265]. In fact, the health-promoting properties observed with quercetin could account for by microbiota-mediated metabolites.

Quercetin, quercetin 3-*O*-rutinoside (rutin) and their colon-derived flavonoid metabolites 3,4-dihydroxyphenylacetic acid (3,4-HPAA), 4-hydroxyphenylacetic acid (4-HPAA), 3-hydroxyphenylacetic acid (3-HPAA), and hippuric acid (*N*-benzoylglycine) have been shown to possess antioxidant and anti-inflammatory properties. On the other hand, when in vitro studies using primary cultures of rat liver parenchymal cells or the human HepG2 hepatoma cell line were treated with t-butylhydroperoxide to produce OxS, the radical scavenging properties were noted only for the parent molecules quercetin and rutin, as well as for 3,4-dihydroxytoluene (3,4-DHT). The metabolites 3-HPAA, 4-HPAA, and hippuric acid remained almost ineffective [208]. Similarly, quercetin and 3,4-DHT showed the potency to inhibit cholesterol synthesis. Another in vitro study, testing the radical scavenging capacity of 3,4-DHPAA, 3-HPAA, and 4-HPPA in human polymorphonuclear cells (stimulated with opsonized zymosan or N-formyl-methionyl-leucyl phenylalanine) revealed that out of the 3 metabolites investigated, only 3,4-DHPAA was active [209]. This radical scavenging capacity is of utmost importance when considering the cell protection from cytotoxicity. Indeed as an adjunct to its radical scavenging properties, 3,4-DHPAA activates phase 2 cytoprotective enzymes (i.e., hemeoxynegese-1, glutathione *S*-transferase), increases the gene and protein expression of aldehyde dehydrogenase isozymes in mouse hepatoma Hepa1c1c7 cells, and induces nuclear translocation of Nrf2 and aryl hydrocarbon receptor [210,211], all involved in controlling cellular REDOX status and in inhibiting cell cytotoxicity.

Table 5. Polyphenol metabolites and functions.

Polyphenols	Subclasses	Metabolites	Bacterial Catabolism	Metabolites Functions	Ref
Flavonols 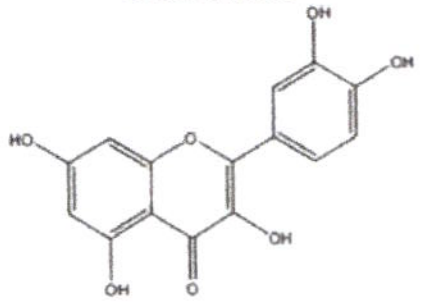	Quercetin	3,4-DHPAA 3-HPAA 4-HPAA	*Clostridium orbiscidens* *Eubacterium oxidoreducens* *Eubacterium ramulus* *Enterococcus casseliflavus*	Oxygen radical scavenging (all the metabolites), SOD- like activities (3,4 DHPAA), ↑glutathione S-transferase (3,4 DHPAA), ↑Nrf2-AhR (3,4 DHPAA) ↓Proinflammatory cytokines (3,4 DHPAA) ↑Glucose induced-insulin secretion (3,4 DHPAA) ↑Function and survival of pancreatic β-cells (3,4 DHPAA) Protective effect against OxS induced-endothelial dysfunction (3,4 DHPAA)	[132,208–220]
Flavones 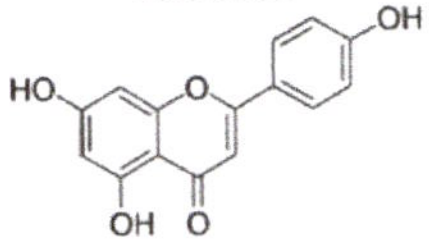	Apigenin	Phloretin 3-HPPA 4-HPPA 4-HCA	*Clostridium orbiscindens*	↓Oxygen radical scavenging (3-HPPA) ↓Proinflammatory cytokines (3-HPPA) ↑Glucose induced-insulin secretion (3-HPPA) ↑Function and survival of pancreatic β-cells (3-HPPA) Protective effect against OxS induced-endothelial dysfunction (3-HPPA)	[132,208,209,212–215,221]

Table 5. *Cont.*

Polyphenols	Subclasses	Metabolites	Bacterial Catabolism	Metabolites Functions	Ref
Flavanones 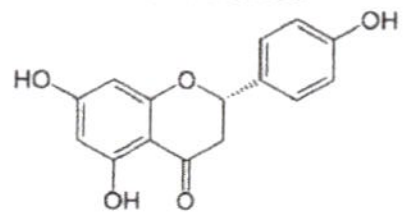	Naringenin	3,4-DHPPA 3-HPPA 4-HPPA	*Clostridium* strains *Eubacterium ramulus*	↓Oxygen radical scavenging (3-HPPA) ↓Proinflammatory; 3,4 DHPPA) ↑Glucose induced-insulin secretion (3-HPPA) Protective effect against OxS induced-endothelial dysfunction (3-HPPA)	[34,60,209,212–215,217,222,223]
Isoflavones 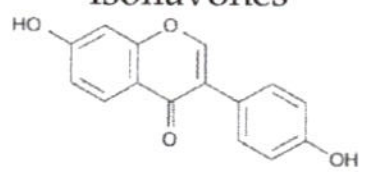	Daidzein	(S)-Equol *O*-DMA	*Bacteroides ovatus*, *Streptococcus intermedius*, *Ruminococcus productus*, *Eggerthella* sp.Julong 732, *Enterococcus faecium* EPI1, *Lactobacillus mucosae* EPI2, *Finegoldia magna* EPI3 *Clostridium* spp. HGHA136	Stimulation of cellular antioxidant systems ↑Catalase and SOD activity Anti-atherogenic effect	[224–226]

Table 5. *Cont.*

Polyphenols	Subclasses	Metabolites	Bacterial Catabolism	Metabolites Functions	Ref
Flavan-3-ols 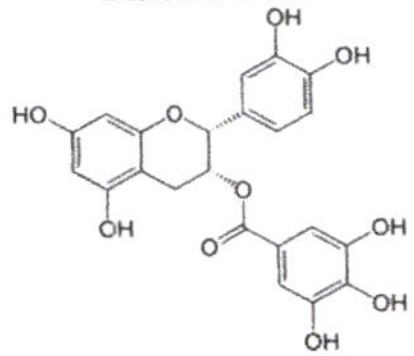	Monomers (catechins, epicatechins) and proanthocyanidins	3-HPPA 3,4-DHPPA 3′,4′-DHPVL 3,4-DHPVA 3′-HPVL 3′,4′,5′-THPVL 3′,5′-DHPVL	*Clostridium coccoides*, *Bifidobacterium* spp. *Eggerthella lenta* *Flavonifractor plautii*	↓Oxygen radical scavenging (3-HPPA) ↓ROS generation (3′-HPVL, 3′,4′-DHPVL) ↓NF-κB transcriptional activity ↓NO synthesis (3′,4′,5′-THPVL; 3′,4′-DHPVL) ↓iNOS expression (3′,4′-DHPVL) Maintenance of endothelial homeostasis and functions (3′,4′-DHPVL): ↓Endothelial adhesion (3′,4′-DHPVL) ↓VCAM1 and MCP1 (3′,4′-DHPVL) ↓Systolic blood pressure (3′,4′,5′-THPVL; 3′,5′-DHPVL)	[87,92,109,131,227–236]
Anthocyanins 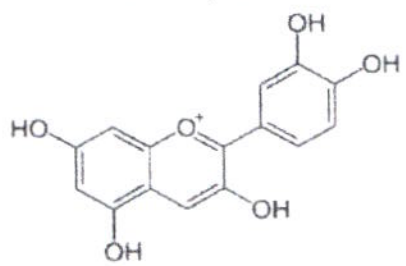	Cyanidin Peonidin Pelargonidin Malvidin Delphinidin	Protocatechuic acid Vanillic acid 4-Hydroxybenzoic acid Syringic acid Gallic acid	*Lactobacillus plantarum*, *Lactobacillus casei* *Lactobacillus acidophilus* LA-5 *Bifidobacterium lactis* BB-12	Antidiabetic activities due to their antioxidant capacity ↓DNA damages, ↓ROS production ↑Cellular glutathione level, ↑glucose uptake by HepG2 and human skeletal cells, ↑glycogen production by HepG2 cells, ↑Mitochondria homeostasis	[237–242]

Table 5. *Cont.*

Polyphenols	Subclasses	Metabolites	Bacterial Catabolism	Metabolites Functions	Ref
Hydroxycinnamic acids 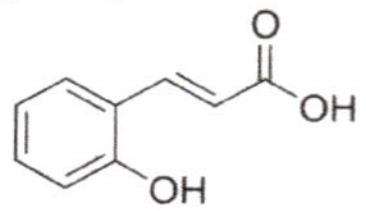	Chlorogenic acids	3-HPPA 3,4-DHPPA Caffeic acid	*Escherichia coli, Bifidobacterium lactis, Lactobacillus gasseri*	↓Oxygen radical scavenging(3-HPPA) ↓Proinflammatory cytokines (3-HPPA; 3,4 DHPPA) Antidiabetic activities due to its antioxidant capacity (caffeic acid): ↑Cellular glutathione level ↓DNA damages ↓Cytotoxicity, ↓ROS production ↑Glucose consumption ↑Glycogen production	[209,212,214,215,243,244]
Hydrolyzables tannins 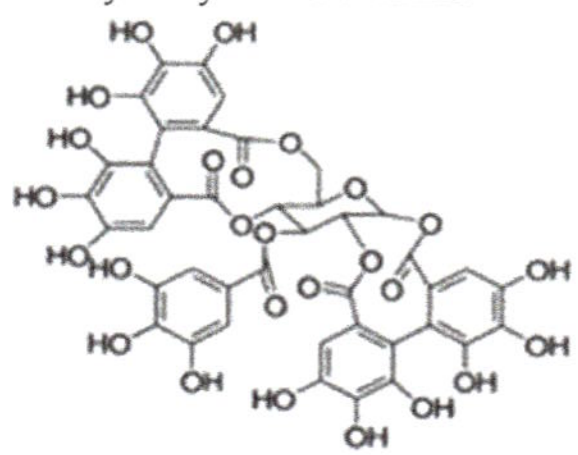	Ellagitannins	Ellagic acid Urolithin A Urolithin B	*Butyrivibrio* spp.	↓Intracellular ROS accumulation (Urolithin A) ↓Cellular injury by ROS ↓Proinflammatory mediators (Ellagic acid and Urolithin A) ↓NADPH oxidase activation (Urolithin A) ↓PGE2 production (Urolithin A and B) ↓mPGES-1 and COX-2 expression (Urolithin A and B) ↓Proteins glycation (Urolithin A and B) ↓Triglycerides accumulation (Ellagic acid and Urolithin A) ↓Expression of adipogenic protein and gene (Urolithin A) ↑Fatty acid β-oxidation (Urolithin A) Alleviation of myocardial ischemia/reperfusion injury (Urolithin A)	[245–257]

Table 5. *Cont.*

Polyphenols	Subclasses	Metabolites	Bacterial Catabolism	Metabolites Functions	Ref
Lignans 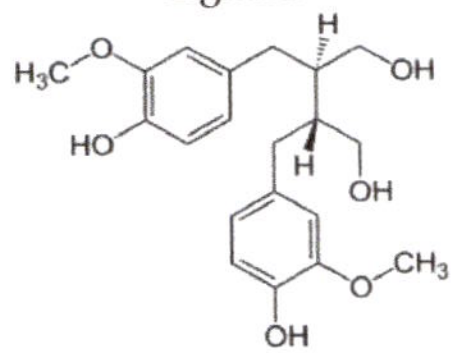	Secoisolariciresinol	Enterodiol Enterolactone	*Bacteroides distasonis, Bacteroides fragilis, Bacteroides ovatus, Clostridium cocleatum, Butyribacterium methylotrophicum, Eubacterium callanderi, Eubacterium limosum, Peptostreptococcus productus, Clostridium scindens, Eggerthella lenta*	Antioxidant capacity OH-scavenging activity Immunomodulatory effects in human cells ↓NF-κB transcriptional activity ↓Proinflammatory cytokines expression	[258–261]
Stilbenes 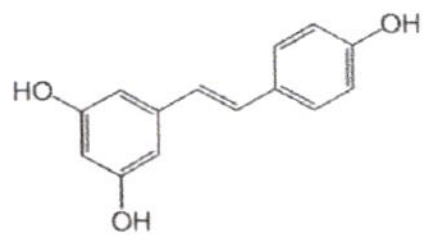	Trans-resveratrol	DHR 3,4′-dihydroxy-trans-stilbene 3,4′-dihydroxybibenzyl (lunularin)	*Slackia equolifaciens Adlercreutzia equolifaciens*	Antioxidant activity Free radical scavenging (DHR) ↓NO production (DHR)	[262–264]

3,4-DHPPA, 3,4-dihydroxyphenylpropionic acid; 3-HPPA, 3-hydroxyphenylpropionic acid; 4-HPPA, 4-hydroxyphenylpropionic acid; 4-HCA, 4-hydroxycinnamic acid; 3,4-DHPAA, 3,4-dihydroxyphenylacetic acid; 3-HPAA, hydroxyphenylacetic acid; 4-HPAA,4-hydroxyphenylacetic acid. *O*-DMA, O-demethylangolensin, 3,4-DHPVA, 3,4-dihydroxyphenyl-γ-valeric acid; 3′,4′,5′-THPVL, 3′,4′,5′-trihydroxyphenyl-γ valerolactone; 3′,4′-DHPVL, 3′,4′-dihydroxyphenyl-γ-valerolactone; 3′,5′-DHPVL, 3′,5′-Dihydroxyphenyl-γ-valerolactone 3′,-HPVL, 3′-hydroxyphenyl-γ-valerolactone; DHR, dihydroresveratrol, NO: nitric oxide; ↑: increase; ↓: decrease.

In vitro studies have also established the anti-inflammatory properties of quercetin metabolites. For example, experiments with lipopolysaccharide-stimulated human peripheral blood mononuclear cells showed the modulatory effect of 3,4-DHPPA, 3-HPPA, 3,4-DHPAA, 3-HPAA, and 4-hydroxybenzoic acid on the expression of the central pro-inflammatory cytokines; TNFα, IL-1β, and IL-6. TNFα expression was also significantly decreased by the metabolites 3,4-DHPPA, 3,4-DHPAA, and 4-hydroxy-hippuric acid (4-HHA), whereas 3,4-DHPPA and 3,4-DHPAA only suppressed secretion of IL-1β and IL-6 [212]. In another study, in which lipopolysaccharide was used to induce inflammation in RAW 264.7 murine macrophage cell line, 3,4-DHT disclosed an anti-inflammatory capacity by modulating the I κB/NF-κB signaling pathway [266].

Surprisingly, very little research has highlighted the potential anti-diabetic properties of colonic metabolites derived flavonoids. However, when the ability of DHPAA and HPPA was examined on β cell function, the findings clearly emphasized an elevated glucose-stimulated insulin secretion, high protection against tert-butyl hydroperoxide-induced β cell toxicity, hence better survival and function of β cells [215]. The last functions were mediated by the activation of protein kinase C and the extracellular regulated kinases (ERKs) pathways. Collectively, these interesting data suggest that flavonol quercetin/rutin metabolites exert anti-diabetic actions.

5.1.2. Flavones and Flavanones Metabolites

Flavones (e.g., apigenin) are present in foods as cereals, parsley, thyme, celery, and citrus fruits under their *O*-glycosidic or C-glycosides derivatives. Once glucosides are hydrolyzed at the intestinal level by digestive enzymes, unabsorbed aglycons undergo further reactions in the large intestine by specific micro-organisms (*Clostridium orbiscindens*, *Enterococcus avium*) causing C-ring fission. Following the metabolism of apigenin, metabolites [e.g., phloretin chalcon, 3-(4-hydroxyphenyl)-propionic, 3-(3-hydroxyphenyl)-propionic, and 4-hydroxycinnamic acids] are acquired [221].

Flavanones (e.g., naringenin) are abundant in tomatoes and citrus fruits, and seem to have higher bioavailability in comparison to flavonols and flavan-3-ols [267]. Flavanone deglycosylation and further degradation by colonic microbiota pathways follow the similar fate of flavonols [267]. *Clostridium* species and *Eubacterium ramulus* are largely involved in these transformations in the colon [217,268]. Metabolites, including 3-(3,4-dihydroxyphenyl)-propionic, 3-(4-hydroxyphenyl)-propionic, and 3-(3-hydroxyphenyl)-propionic acids are produced following naringenin metabolism. However, as for quercetin metabolites, hydroxyphenyl propionic acids (derived from apigenin and naringenin metabolism) display antioxidant and anti-inflammatory capacities (Table 5).

5.1.3. Isoflavone Daidzein and Daidzin Metabolites

Isoflavone, as a phytoestrogen precursor, is another flavonoid class of interest. Like many types of polyphenols, isoflavones exert advantageous effects on intestinal health, menopause symptoms, hormone-mediated syndromes, and CVD [269]. They are also transformed into their active metabolites by enzymes from the gut microbiota to generate compounds endowed with elevated estrogenic activity (e.g., 4′,7-isoflavandiol; equol) or inactive compounds (e.g., *O*-desmethylangolensin; *O*-DMA) [270]. Importantly, the majority of humans are competent to yield *O*-DMA, but only 25–50% among them engender equol [271,272].

The potential health properties of equol and *O*-DMA, both derived from daidzein and its 7-*O*-glycoside daidzin, are the subject of continuing debate. Some studies claim their beneficial effects on the risk of coronary heart disease events [273,274] and others have shown a limited action [275]. The diverging conclusions stems in part from the inter-individual variability in the gut microbiota [276] or from a characteristic that is influenced by ethnicity [277]. Interestingly enough, prehypertensive postmenopausal women producing equol/*O*-DMA had more favorable cardiovascular risk profiles than non-producers [278]. Furthermore, equol/*O*-DMA producers exhibited lower serum uric acid, triglycerides, and waist/hip ratio, as well as a tendency to have higher high density lipoprotein-cholesterol compared to equol non-producers [279]. These effects are probably exerted through the control of cell redox potential as both equol and *O*-DMA have antioxidant capacity, exemplified by the stimulation of catalase and total superoxide dismutase activities [226]. In a population-based, cross-sectional investigation, urinary concentrations of isoflavones and equol/*O*-DMA were related to well-disposed insulin sensitivity and cardiometabolic markers in pregnant women [280]. Other studies emphasized the beneficial role of *O*-DMA in obesity [281]. However, caution should be exerted when concluding on the effects of isoflavone and related metabolite interventions, as sufficiently randomized controlled trials with statistical are needed [275].

5.1.4. Flavanol Catechin Metabolites

Flavanols or flavan-3-ols are among the most consumed phenolic compounds in Western populations [282]. They are composed of monomers (e.g., catechin, epicatechin, epigallocatechin) and oligomers/polymers (known as proanthocyanidins) [283]. Several groups have reported their ability to prevent chronic diseases (e.g., T2D and CVD) [283–285]. Hydroxy-phenyl-γ-valerolactones and their related hydroxy-phenylvaleric acids are the main metabolites of dietary flavan-3-ols [87,286]. Although a number of studies based on different models aiming at assessing hydroxy-phenyl-γ-valerolactones bioactivity have been reported in the past decade, those involving human subjects are scarce. Inflammatory and OxS processes were particularly studied. For example, (3′4′-dihydroxyphenyl)-γ-valerolactone significantly decreased TNFα-induced NF-kB transcriptional activity in HepG2 cells [92]. Furthermore, a decrease was noted in nitrous oxide production and inducible nitrous oxide synthase expression in the murine macrophage RAW264.7 cells and freshly isolated human monocytes, both treated with (δ-(3,4-dihydroxyphenyl)-γ-valerolactone) a metabolite of the maritime pine bark Pycnogenol [233].

The accumulation of pro-inflammatory cytokine-producing T-lymphocytes and macrophages with the concomitant secretion of the vascular cell-adhesion molecule-1 by the endothelium is key to the early development of the atherosclerotic plaque [287]. Inhibition of this process was observed following the incubation of cultured human umbilical vein endothelial cells with 3′,4′-dihydroxyphenyl)-γ-valerolactone [234], thereby providing the evidence that this compound is a potential therapeutic target. Transferability of the concentrations (quantities) of natural compounds to in vivo or clinical environment (cytotoxicity) represents a major limit of these in vitro experiments. The polyphenolic metabolites 3′,4′,5′-trihydroxyphenyl)-γ-valerolactone and 5-(3′,5′-dihydroxyphenyl)-γ-valerolactone also displayed, in spontaneously hypertensive rats, the capacity to lower angiotensin-1 converting enzyme activity and blood pressure, both contributing to improving endothelial function and arterial elasticity [235]. Paradoxically, the flavan-3-ol metabolites 5-(3′,4′,5′-trihydroxyphenyl)-γ-valerolactone, 5-(3′,4′-dihydroxyphenyl)-γ-valerolactone, and 5-(3′-methoxy, 4′-hydroxyphenyl)-γ-valerolactone did not improve vascular endothelial plasticity of freshly isolated mouse saphenous arteries [236]. The dissimilarity in data could partially be explained by the different models used, especially that isolated saphenous artery endothelium is less reactive than the vessels of the spontaneously hypertensive rat. The in vivo versus the in vitro situations and the experimental time frame are also factors that have to be considered.

The decrease in obesity, a major element in the development of MetS, in response to the consumption of foods rich in flavan-3-ols (e.g., green tea), has been attributed to the role of their content on polyphenolic metabolites in increasing energy expenditure via the modulation of cell differentiation and thermogenesis in adipose tissues [231]. However, when tested in vitro, the flavan-3-ol metabolites [5-(3′,4′-dihydroxyphenyl)-γ-valerolactone, 5-(3′-hydroxyphenyl)-γ-valerolactone-4′-O-sulfate, and 5-phenyl-γ-valerolactone-3′,4′-di-O-sulfate] did not stimulate cell differentiation and the activation of immortalized pre-adipocytes. In addition, none of the metabolites regulated the expression of the uncouple protein 1, nor the main transcription factors implicated in brown adipocyte genesis. Nevertheless, both 5-(3′,4′-dihydroxyphenyl)-γ-valerolactone and 5-(3′-hydroxyphenyl)-γ-valerolactone-4′-O-sulfate protected immortalized pre-adipocytes from

H_2O_2 generated OxS [231]. In a recent clinical trial, Rodriguez-Mateos et al. [288] showed that cranberry juice given to healthy volunteers improved, in a concentration- and time-dependent manner, the flow-mediated vasodilatation, an early clinical sign in the development of atherosclerosis. Interestingly, the plasma level of the flavan-3-ols metabolite, 5-(3′-hydroxyphenyl)-γ-valerolactone-4′-sulfate, was significantly positively correlated to flow-mediated vasodilatation. Noteworthy, mice on high-fat diet improved non-alcoholic fatty liver disease by lessening IR and endotoxin-toll-like receptor 4/NF-κB pathway in response to raised hepatic catechin metabolites (namely phenyl-γ-valerolactones) following catechin-rich green tea extract [289].

Overall, these findings offer promising beneficial health effects of polyphenols metabolites. However, additional clinical studies are necessary to validate the data in humans and to ensure their transferability from in vitro and animal models to in vivo human situations.

5.1.5. Tannin Ellagitannin and Urolithin Metabolites

Hydrolysable tannins, named ellagitannins, encompass one or more gallic acid units and one or more hexahydroxydiphenoic acid units, ester-connected with a sugar residue. In view of their difficulty to be absorbed in the small intestine, ellagitannins reach the colon where they are hydrolyzed into mainly ellagic acid and urolithin (3,4-benzocoumarin derivatives). Supporting data propose a stringent relationship between the consumption of ellagitannins-rich foods and healthy effects based on animal and human models.

The rat under high carbohydrate high-fat diet-induced MetS model has demonstrated the ability of ellagic acid to improve hepatic and cardiometabolic functions, while normalizing glucose tolerance, blood lipid components, central obesity, and body weight [290]. The mechanisms involved in the modulation of OxS and inflammation are through the protein level regulation of Nrf2 and NF-κB, with an impact on metabolic pathways such as fatty acid β-oxidation [290]. Other studies confirm the health benefit properties of ellagic acid, for example, by showing that it dose-dependently repressed the IL-1β-induced production of ROS [248]. This was associated with a significant reduction of the attachment of the established human monocyte cell line to IL-1β-treated human umbilical vein endothelial cells through the suppression of NF-κB, thereby downregulating the expression of vascular cell-adhesion molecule-1 and E-selectin and resulting in decreased monocyte adhesion. Along the same line of evidence, ellagic acid significantly inhibited oxidized low-density lipoprotein and platelet-derived growth factor-BB-induced proliferation of rat aortic smooth muscle cells via inactivation of ERK and marked decrease of proliferating cell nuclear antigen. [249]. More evidence for ellagic acid effects in preventing the development of atherosclerotic events has been provided by showing its capacity of reducing platelet growth factor (PDGF) receptor-β-phosphorylation, diminishing ROS production, and lessening downstream activation of ERK, and finally blocking platelet-derived growth factor-induced expression of cyclin D1 in primary cultures of rat aortic smooth muscle cells [291]. Moreover, these in vitro results were further supported by the ability of ellagic acid to block T2D-induced medial thickness, and lipid and collagen deposition in the arch of aorta in streptozotocin-induced diabetic rats [291]. Collectively, these data suggest that ellagic acid can inhibit OxS and inflammation response, which thus favorably contributes to the regulation of vascular homeostasis.

Urolithins, formed primarily from the oxidative linkage of galloyl groups, have extensively emerged as novel natural bioactive compounds, being currently the focus of extensive investigations. The type of urolithin produced depends on the colon microbiome. Three ellagitannins-metabolizing metabotypes have been identified: 'metabotype A' (producers of the dihydoxylated urolithin-A conjugates), 'metabotype B' (producers of urolithin-A/isourolithin-A conjugates and/or monohydoxylated urolithin-B conjugates), and 'metabotype 0' (no urolithins produced) [292]. Contrary to the parent compounds, which bloodstream concentrations are negligible, these ellagic acid derivatives may reach micromolar concentrations and thus act as surrogate biomarkers of the inter-individual variation in the colon metabotype [245]. Among the numerous metabolites, urolithin-A has been particularly studied in terms of its anti-inflammatory and anti-oxidative properties both in vitro and in vivo [250,293,294], as well as anti-atherosclerotic and anti-angiogenic activities [293,295]. In line with these features, adhesion model involving established human aortic endothelial cells and human leukemia monocytic THP-1 cells showed that urolithin-A inhibited TNFα-stimulated adhesiveness and endothelial cell migration [257]. To delve deeper and uncover the mechanisms, the authors could further demonstrate that these effects were concomitant to a down regulation of the levels of chemokine ligand 2, plasminogen activator inhibitor-1, and IL-8. Even more interesting, a clinical trial provided evidence that overweight-obese individuals consuming ellagitannin-containing food with the urolithin metabotype producing urolithin-A were protected against CMD when compared to those with a urolithin-B metabotype [296]. Despite the small number of participants, this clinical trial pointed out the important inter-individual disparity in metabolizing polyphenols, the regulation by microbiota, and indicated that urine metabolite profiling is a valuable tool in assessing potential effects. Fecal polyphenol metabolite profiling could also shed some light on the mechanisms of action. In this context, the profile of fecal phenolic-derived metabolites correlated to blood biomarkers of OxS and inflammation, thus providing support to this avenue [297].

Urolithin may constitute a central regulator of CMD as IR decreased in response to urolithin administration to mice fed high-fat, high-sucrose diet [298]. Concomitantly, urolithin lowered glucose, fatty acids, and triglycerides, while ameliorating adiponectin and mitochondrial function. Furthermore, in the same animal model, through thyroid hormone modulation, urolithin increased energy expenditure by enhancing thermogenesis in brown adipose tissue and inducing browning of white adipose tissue [299]. Therefore, this first-in-class natural compound contributed to metabolic homeostasis via its capacity to normalize OxS, inflammation, obesity, and IR, which overall contribute to MetS.

5.2. Non-Flavonoids

5.2.1. Lignan Metabolites

Plant lignans are non-flavonoids, but polyphenolic, phytoestrogenic compounds. They exhibit several biological functions with potent protective benefits. Anti-estrogenic, antioxidant, anti-inflammatory, anti-atherosclerotic, and anti-carcinogenic effects are among their properties, thereby conferring therapeutic values [300,301]. These properties explain the impact of lignan-rich diets on the prevention of hyperlipidemia, IR and hyperglycemia, T2D, and CMD [302–304]. The most frequent forms of dietary lignans are secoisolariciresinol, lariciresinol, pinoresinol, and matairesinol, which are metabolized by gut microflora enzymes into enterolignans such as enterodiol and enterolactone undergoing enterohepatic circulation [305,306]. At least, 28 bacterial species belonging to 12 different genera (e.g., *Bacteroides*, *Clostridium*, *Bifidobacterium*, and *Ruminococcus*) have been identified to be involved in the transformation of lignans to enterolignans [307]. Several groups have quantified the concentrations of the enterolactone metabolite in blood and urine, and reported a protective association in view of the reduction (30–45%) of cardiovascular mortality [308]. Similarly, in a prospective investigation, enterolactone (more than enterodiol) was associated with a lower risk of T2D in U.S. women [309]. A cross sectional trial of the general U.S. population found that subjects with high excretion of enteractone had lower odds of having MetS [310]. Moreover, in a representative sample of U.S. adults, the cross-sectional analysis of data detected that urinary enterolignan concentrations are associated with lower serum triglyceride and greater high-density lipoprotein-cholesterol concentrations in U.S. adults [311]. Additional efforts are obviously needed to confirm these interesting cardiometabolic findings and to explore potential mechanisms of action behind diverse enterolignan metabolites.

5.2.2. Resveratrol Metabolites

Resveratrol (3,5,4′-trihydroxystilbene) belongs to the stilbene derivative and is a dietary polyphenolic compound accessible in 70 plant species such as grapes, mulberries, and peanuts [312].

Its pleiotropic effects have made it so famous that it is called a star molecule. It displays numerous biological activities, including antioxidant, anti-inflammatory, cardioprotective, anti-aging, and cancer-chemopreventive features [313,314]. Accordingly, the wealth of scientific reports revealed the effectiveness and beneficial actions of resveratrol in MetS and related disorders such as obesity, IR, and T2D [315,316]). For example, a meta-analysis evaluating 11 studies involving 388 T2D subjects mentioned that resveratrol markedly lessened fasting glucose, insulin, haemoglobin A1c, and IR [278]. Resveratrol was also able to improve hypertension by acting on vascular nitric oxide production, reducing endothelial dysfunction and arteriolar remodeling [317,318]. However, other studies could not record advantageous data on MetS in response to resveratrol [319].

Today, it is well established that resveratrol is easily absorbed by the intestine and promptly metabolized to produce glucuronide and sulfate conjugates [320,321]. Although many gaps are still persistent in the metabolic route of resveratrol, intestinal bacteria transform resveratrol into dihydroresveratrol, known to be partially absorbed and subsequently metabolized to conjugated forms, which can be excreted in urine [322,323]. Besides dihydroresveratrol, two additional bacterial *trans*-resveratrol metabolites were identified: 3,4′-dihydroxy-*trans*-stilbene and 3,4′-dihydroxybibenzyl (lunularin) [264]. Large inter-individual variation was noticed in different subjects: some of them were lunularin producers, others were dihydroresveratrol producers, and the last category was mixed producers, according to levels of these metabolites [264]. Clearly, there was a close relationship between lunularin producers and the abundance of *Bacteroidetes*, *Actinobacteria*, *Verrucomicrobia*, and *Cyanobacteria*, along with lower abundance of *Firmicutes* compared to dihydroresveratrol or mixed producers [264]. Whether resveratrol metabolites are physiologically relevant to tackle metabolic disorders still remains an enigma.

6. Conclusions

Consuming a diet rich in fruits and vegetables is associated with a reduced risk of several noncommunicable lifestyle-related diseases, including CVD. They contribute several essential and bioactive micronutrients such as polyphenols. The first goal of this review was to focus on the available and pertinent literature, which assessed the health-promoting potential of polyphenols to modulate OxS and inflammation. These two major components are powerful inducers of IR, leading to CMD, especially MetS and related complications such as T2D and atherosclerosis.

For the reader's benefit, our review first summarized the most widely distributed classes, sources, and properties of the naturally occurring polyphenols, separated into two main groups, flavonoids and non-flavonoid compounds. Second, attention was paid to their transformation in the intestinal lumen, absorption in different regions of the GI tract, and their bioavailability in the bloodstream. Clearly, most of polyphenols undergo complex intraluminal transformation during digestive and absorptive processes. Data obtained from critical literature revision underline the poor bioavailability, aqueous solubility, permeability, and instability of especially polyphenols with high molecular weights. Third, considerable emphasis was devoted to the essential polyphenol-microbiota interaction as polyphenols directly exert their regulatory effects on intestinal bacteria with concomitant biotransformation of polyphenols into metabolites by gut microbiota. This fundamental interplay provides the central explanation as to the stupefying pleiotropic biological/clinical actions and functions of "parent" polyphenols whereas their systemic bioavailability is incredibly limited, which highlights the prominent contribution of polyphenol metabolites. Fourth, the multifaceted actions and functions of "parent" polyphenols were summed up. Not only do polyphenols show the great ability to regulate cellular, molecular, and physiological pathways in relation with OxS and inflammation, but they also exhibit a remarkable effectiveness in improving cardiometabolic biomarkers to the advantage of human health. Thanks to their anti-obesogenic, -hyperlipidemic, -hypertensive, -IR, -atherosclerotic, as well as -diabetic effects, they may help to prevent and treat CMD as global public health problems with vast mortality and economic consequences. Fifth, although polyphenol metabolites were also shown to play a similar beneficial role in metabolic complications, there are various enigmatic issues to deal with: inter-individual differences in the microbial conversion of parent polyphenols, largely unknown microbial metabolites from distinct polyphenols, unidentifiable bacterial strains involved, metabolite health-promoting effects of a multitude of metabolites, elusive metabolite doses, the underlying mechanisms of their specific actions, and the lack of clinical trials.

Author Contributions: Conceptualization, E.L. and E.D.; methodology, E.L.; validation, E.L. and E.D.; resources, E.L.; data curation, M.K., F.F. and E.D.; writing—original draft preparation, E.D. and E.L.; writing—review and editing, S.S., Y.D. and E.L.; supervision, E.L.; project administration, S.S.; funding acquisition, E.L. and Y.D. All authors have read and agreed to the published version of the manuscript.

Funding: This work was supported by a grant from the JA deSève Research Chair in nutrition (E.L.) and the NSERC-Diana Food Industrial Chair on prebiotic effects of polyphenols (401240871) (E.L. & Y.D.).

Conflicts of Interest: The authors declare no conflict of interest.

Abbreviations

CMD: cardiometabolic disorders; CVD: cardiovascular disease; DAPP: dried apple peels; ERK: extracellular regulated kinases; GI: gastrointestinal; HPAA: phenylacetic acids; HPPA: phenylpropionic acid; IL: interleukin; IPEC: intestinal porcine enterocyte cell line; IR: insulin resistance; MetS: Metabolic syndrome; miRNAs: micro-RNAs; NF- κB: nuclear factor-kappa B; Nrf2: nuclear redox factor-2; OxS: oxidative stress; ROS: reactive oxygen species; TNFα: tumor necrosis factor-alpha.

References

1. Chen, Y.; Huang, L.; Liang, X.; Dai, P.; Zhang, Y.; Li, B.; Lin, X.; Sun, C. Enhancement of polyphenolic metabolism as an adaptive response of lettuce (Lactuca sativa) roots to aluminum stress. *Environ. Pollut.* **2020**, *261*, 114230. [CrossRef]
2. Czerniewicz, P.; Sytykiewicz, H.; Durak, R.; Borowiak-Sobkowiak, B.; Chrzanowski, G. Role of phenolic compounds during antioxidative responses of winter triticale to aphid and beetle attack. *Plant Physiol. Biochem.* **2017**, *118*, 529–540. [CrossRef]
3. Mestar, N.G.; Boudiaf, M.N.; Lahcene, S.; Abbaci, H.; Aiche, G.I.; Metna, B.; Saadoun, N.S.; Taibi, F.; Houali, K. Bio-insecticidal effects of Oleaster leaves aqueous extracts against Psylla larvae (Euphyllura olivina (Costa)), a primary pest of *Olea europaea* L. *Cell Mol. Biol.* **2018**, *64*, 35–40. [CrossRef]
4. Antoniolli, A.; Fontana, A.R.; Piccoli, P.; Bottini, R. Characterization of polyphenols and evaluation of antioxidant capacity in grape pomace of the cv. Malbec. *Food Chem.* **2015**, *178*, 172–178. [CrossRef]
5. Anhe, F.F.; Roy, D.; Pilon, G.; Dudonne, S.; Matamoros, S.; Varin, T.V.; Garofalo, C.; Moine, Q.; Desjardins, Y.; Levy, E.; et al. A polyphenol-rich cranberry extract protects from diet-induced obesity, insulin resistance and intestinal inflammation in association with increased Akkermansia spp. population in the gut microbiota of mice. *Gut* **2015**, *64*, 872–883. [CrossRef]
6. D'Amore, S.; Vacca, M.; Cariello, M.; Graziano, G.; D'Orazio, A.; Salvia, R.; Sasso, R.C.; Sabba, C.; Palasciano, G.; Moschetta, A. Genes and miRNA expression signatures in peripheral blood mononuclear cells in healthy subjects and patients with metabolic syndrome after acute intake of extra virgin olive oil. *Biochim. Biophys. Acta* **2016**, *1861*, 1671–1680. [CrossRef] [PubMed]
7. Kantartzis, K.; Fritsche, L.; Bombrich, M.; Machann, J.; Schick, F.; Staiger, H.; Kunz, I.; Schoop, R.; Lehn-Stefan, A.; Heni, M.; et al. Effects of resveratrol supplementation on liver fat content in overweight and insulin-resistant subjects: A randomized, double-blind, placebo-controlled clinical trial. *Diabet. Obes. Metab.* **2018**, *20*, 1793–1797. [CrossRef]
8. Paquette, M.; Medina Larque, A.S.; Weisnagel, S.J.; Desjardins, Y.; Marois, J.; Pilon, G.; Dudonne, S.; Marette, A.; Jacques, H. Strawberry and cranberry polyphenols improve insulin sensitivity in insulin-resistant, non-diabetic adults: A parallel, double-blind, controlled and randomised clinical trial. *Br. J. Nutr.* **2017**, *117*, 519–531. [CrossRef]
9. Babu, P.V.A.; Liu, D.; Gilbert, E.R. Recent advances in understanding the anti-diabetic actions of dietary flavonoids. *J. Nutr. Biochem.* **2013**, *24*, 1777–1789. [CrossRef]
10. Dominguez Avila, J.A.; Rodrigo Garcia, J.; Gonzalez Aguilar, G.A.; de la Rosa, L.A. The Antidiabetic Mechanisms of Polyphenols Related to Increased Glucagon-Like Peptide-1 (GLP1) and Insulin Signaling. *Molecules* **2017**, *22*, 903. [CrossRef]

11. Les, F.; Casedas, G.; Gomez, C.; Moliner, C.; Valero, M.S.; Lopez, V. The role of anthocyanins as antidiabetic agents: From molecular mechanisms to in vivo and human studies. *J. Physiol. Biochem.* **2020**. [CrossRef] [PubMed]
12. Macho-Gonzalez, A.; Lopez-Oliva, M.E.; Merino, J.J.; Garcia-Fernandez, R.A.; Garcimartin, A.; Redondo-Castillejo, R.; Bastida, S.; Sanchez-Muniz, F.J.; Benedi, J. Carob fruit extract-enriched meat improves pancreatic beta-cell dysfunction, hepatic insulin signaling and lipogenesis in late-stage type 2 diabetes mellitus model. *J. Nutr. Biochem.* **2020**, *84*, 108461. [CrossRef] [PubMed]
13. Wang, S.; Moustaid-Moussa, N.; Chen, L.; Mo, H.; Shastri, A.; Su, R.; Bapat, P.; Kwun, I.; Shen, C.L. Novel insights of dietary polyphenols and obesity. *J. Nutr. Biochem.* **2014**, *25*, 1–18. [CrossRef] [PubMed]
14. Cho, Y.J.; Lee, H.G.; Seo, K.H.; Yokoyama, W.; Kim, H. Antiobesity Effect of Prebiotic Polyphenol-Rich Grape Seed Flour Supplemented with Probiotic Kefir-Derived Lactic Acid Bacteria. *J. Agric. Food. Chem.* **2018**, *66*, 12498–12511. [CrossRef]
15. Sandoval, V.; Sanz-Lamora, H.; Arias, G.; Marrero, P.F.; Haro, D.; Relat, J. Metabolic Impact of Flavonoids Consumption in Obesity: From Central to Peripheral. *Nutrients* **2020**, *12*, 2393. [CrossRef]
16. Anhe, F.F.; Nachbar, R.T.; Varin, T.V.; Trottier, J.; Dudonne, S.; Le Barz, M.; Feutry, P.; Pilon, G.; Barbier, O.; Desjardins, Y.; et al. Treatment with camu camu (Myrciaria dubia) prevents obesity by altering the gut microbiota and increasing energy expenditure in diet-induced obese mice. *Gut* **2019**, *68*, 453–464. [CrossRef]
17. McKay, D.L.; Blumberg, J.B. Cranberries (*Vaccinium macrocarpon*) and cardiovascular disease risk factors. *Nutr. Rev.* **2007**, *65*, 490–502. [CrossRef]
18. Ruel, G.; Couillard, C. Evidences of the cardioprotective potential of fruits: The case of cranberries. *Mol. Nutr. Food Res.* **2007**, *51*, 692–701. [CrossRef]
19. Ruel, G.; Lapointe, A.; Pomerleau, S.; Couture, P.; Lemieux, S.; Lamarche, B.; Couillard, C. Evidence that cranberry juice may improve augmentation index in overweight men. *Nutr. Res.* **2013**, *33*, 41–49. [CrossRef]
20. Martinez-Gonzalez, M.A.; Gea, A.; Ruiz-Canela, M. The Mediterranean Diet and Cardiovascular Health. *Circ. Res.* **2019**, *124*, 779–798. [CrossRef]
21. Ferrari, C.K.B. Anti-atherosclerotic and cardiovascular protective benefits of Brazilian nuts. *Front. Biosci. Schol Ed.* **2020**, *12*, 38–56. [CrossRef]
22. Perez de Vega, M.J.; Moreno-Fernandez, S.; Pontes-Quero, G.M.; Gonzalez-Amor, M.; Vazquez-Lasa, B.; Sabater-Munoz, B.; Briones, A.M.; Aguilar, M.R.; Miguel, M.; Gonzalez-Muniz, R. Characterization of Novel Synthetic Polyphenols: Validation of Antioxidant and Vasculoprotective Activities. *Antioxidants* **2020**, *9*, 787. [CrossRef]
23. Devi, S.A.; Chamoli, A. Polyphenols as an Effective Therapeutic Intervention Against Cognitive Decline During Normal and Pathological Brain Aging. *Adv. Exp. Med. Biol.* **2020**, *1260*, 159–174.
24. Oppedisano, F.; Maiuolo, J.; Gliozzi, M.; Musolino, V.; Carresi, C.; Nucera, S.; Scicchitano, M.; Scarano, F.; Bosco, F.; Macri, R.; et al. The Potential for Natural Antioxidant Supplementation in the Early Stages of Neurodegenerative Disorders. *Int. J. Mol. Sci.* **2020**, *21*, 2618. [CrossRef]
25. Silva, R.F.M.; Pogacnik, L. Polyphenols from Food and Natural Products: Neuroprotection and Safety. *Antioxidants* **2020**, *9*, 61. [CrossRef]
26. Bensalem, J.; Dudonne, S.; Gaudout, D.; Servant, L.; Calon, F.; Desjardins, Y.; Laye, S.; Lafenetre, P.; Pallet, V. Polyphenol-rich extract from grape and blueberry attenuates cognitive decline and improves neuronal function in aged mice. *J. Nutr. Sci.* **2018**, *7*, e19. [CrossRef]
27. Mao, X.; Xiao, X.; Chen, D.; Yu, B.; He, J. Tea and Its Components Prevent Cancer: A Review of the Redox-Related Mechanism. *Int. J. Mol. Sci.* **2019**, *20*, 5249. [CrossRef]
28. Toric, J.; Markovic, A.K.; Brala, C.J.; Barbaric, M. Anticancer effects of olive oil polyphenols and their combinations with anticancer drugs. *Acta Pharm.* **2019**, *69*, 461–482. [CrossRef]
29. Abdelgawad, I.Y.; Grant, M.K.O.; Zordoky, B.N. Leveraging the Cardio-Protective and Anticancer Properties of Resveratrol in Cardio-Oncology. *Nutrients* **2019**, *11*, 627. [CrossRef] [PubMed]
30. Horne, J.R.; Vohl, M.C. Biological plausibility for interactions between dietary fat, resveratrol, ACE2, and SARS-CoV illness severity. *Am. J. Physiol. Endocrinol. Metab.* **2020**, *318*, e830–e833. [CrossRef]
31. Mani, J.S.; Johnson, J.B.; Steel, J.C.; Broszczak, D.A.; Neilsen, P.M.; Walsh, K.B.; Naiker, M. Natural product-derived phytochemicals as potential agents against coronaviruses: A review. *Virus Res.* **2020**, *284*, 197989. [CrossRef]

32. Levy, E.; Delvin, E.; Marcil, V.; Spahis, S. May phytotherapy with polyphenols serve as a powerful approach for the prevention and therapy tool of novel coronavirus disease 2019 (Covid-19)? *Am. J. Physiol. Endocrinol. Metab.* **2020**, *319*. [CrossRef]
33. Dabeek, W.M.; Marra, M.V. Dietary Quercetin and Kaempferol: Bioavailability and Potential Cardiovascular-Related Bioactivity in Humans. *Nutrients* **2019**, *11*, 2288. [CrossRef]
34. Rechner, A.R.; Kuhnle, G.; Bremner, P.; Hubbard, G.P.; Moore, K.P.; Rice-Evans, C.A. The metabolic fate of dietary polyphenols in humans. *Free Radic. Biol. Med.* **2002**, *33*, 220–235. [CrossRef]
35. D'Archivio, M.; Filesi, C.; Vari, R.; Scazzocchio, B.; Masella, R. Bioavailability of the polyphenols: Status and controversies. *Int. J. Mol. Sci.* **2010**, *11*, 1321–1342. [CrossRef]
36. Brglez Mojzer, E.; Knez Hrncic, M.; Skerget, M.; Knez, Z.; Bren, U. Polyphenols: Extraction Methods, Antioxidative Action, Bioavailability and Anticarcinogenic Effects. *Molecules* **2016**, *21*, 901. [CrossRef]
37. Kawabata, K.; Yoshioka, Y.; Terao, J. Role of Intestinal Microbiota in the Bioavailability and Physiological Functions of Dietary Polyphenols. *Molecules* **2019**, *24*, 370. [CrossRef]
38. Cardona, F.; Andrés-Lacueva, C.; Tulipani, S.; Tinahones, F.J.; Queipo-Ortuño, M.I. Benefits of polyphenols on gut microbiota and implications in human health. *J. Nutr. Biochem.* **2013**, *24*, 1415–1422. [CrossRef]
39. Bravo, L. Polyphenols: Chemistry, dietary sources, metabolism, and nutritional significance. *Nutr. Rev.* **1998**, *56*, 317–333. [CrossRef]
40. Crozier, A.; Jaganath, I.B.; Clifford, M.N. Dietary phenolics: Chemistry, bioavailability and effects on health. *Nat. Prod. Rep.* **2009**, *26*, 1001–1043. [CrossRef]
41. Tsao, R. Chemistry and biochemistry of dietary polyphenols. *Nutrients* **2010**, *2*, 1231–1246. [CrossRef]
42. Smeriglio, A.; Barreca, D.; Bellocco, E.; Trombetta, D. Proanthocyanidins and hydrolysable tannins: Occurrence, dietary intake and pharmacological effects. *Br. J. Pharmacol.* **2017**, *174*, 1244–1262. [CrossRef]
43. Prigent, S.V.; Gruppen, H.; Visser, A.J.; Van Koningsveld, G.A.; De Jong, G.A.; Voragen, A.G. Effects of non-covalent interactions with 5-O-caffeoylquinic acid (chlorogenic acid) on the heat denaturation and solubility of globular proteins. *J. Agric. Food. Chem.* **2003**, *51*, 5088–5095. [CrossRef]
44. Prigent, S.V.; Voragen, A.G.; Li, F.; Visser, A.J.; van Koningsveld, G.A.; Gruppen, H. Covalent interactions between amino acid side chains and oxidation products of caffeoylquinic acid (chlorogenic acid). *J. Sci. Food Agric.* **2008**, *88*, 1748–1754. [CrossRef]
45. Le Bourvellec, C.; Renard, C.M. Interactions between polyphenols and macromolecules: Quantification methods and mechanisms. *Crit. Rev. Food Sci. Nutr.* **2012**, *52*, 213–248. [CrossRef]
46. Kevers, C.; Pincemail, J.; Tabart, J.; Defraigne, J.O.; Dommes, J. Influence of cultivar, harvest time, storage conditions, and peeling on the antioxidant capacity and phenolic and ascorbic acid contents of apples and pears. *J. Agric. Food Chem.* **2011**, *59*, 6165–6171. [CrossRef] [PubMed]
47. Gomez-Rico, A.; Salvador, M.D.; La Greca, M.; Fregapane, G. Phenolic and volatile compounds of extra virgin olive oil (Olea europaea L. Cv. Cornicabra) with regard to fruit ripening and irrigation management. *J. Agric. Food Chem.* **2006**, *54*, 7130–7136. [CrossRef]
48. Bonoli, M.; Bendini, A.; Cerretani, L.; Lercker, G.; Toschi, T.G. Qualitative and semiquantitative analysis of phenolic compounds in extra virgin olive oils as a function of the ripening degree of olive fruits by different analytical techniques. *J. Agric. Food Chem.* **2004**, *52*, 7026–7032. [CrossRef]
49. Yao, L.H.; Jiang, Y.M.; Shi, J.; Tomas-Barberan, F.A.; Datta, N.; Singanusong, R.; Chen, S.S. Flavonoids in food and their health benefits. *Plant Foods Hum. Nutr.* **2004**, *59*, 113–122. [CrossRef]
50. Marti, M.P.; Pantaleon, A.; Rozek, A.; Soler, A.; Valls, J.; Macia, A.; Romero, M.P.; Motilva, M.J. Rapid analysis of procyanidins and anthocyanins in plasma by microelution SPE and ultra-HPLC. *J. Sep. Sci.* **2010**, *33*, 2841–2853. [CrossRef]
51. Denis, M.C.; Dubé, P.; Dudonné, S.; Desjardins, Y.; Matei, C.; Delvin, E.; Levy, E.; Furtos, A. Characterization of bioactive cranberry fractions by mass spectrometry. *Can. J. Chem.* **2020**, in press. [CrossRef]
52. Meng, X.; Maliakal, P.; Lu, H.; Lee, M.J.; Yang, C.S. Urinary and plasma levels of resveratrol and quercetin in humans, mice, and rats after ingestion of pure compounds and grape juice. *J. Agric. Food Chem.* **2004**, *52*, 935–942. [CrossRef]
53. Vallejo, F.; Tomas-Barberan, F.; Garcia-Viguera, C. Health-promoting compounds in broccoli as influenced by refrigerated transport and retail sale period. *J. Agric. Food Chem.* **2003**, *51*, 3029–3034. [CrossRef]

54. Xu, B.; Chang, S.K. Total Phenolic, Phenolic Acid, Anthocyanin, Flavan-3-ol, and Flavonol Profiles and Antioxidant Properties of Pinto and Black Beans (Phaseolus vulgaris L.) as Affected by Thermal Processing. *J. Agric. Food Chem.* **2009**, *57*, 4754–4764. [CrossRef] [PubMed]
55. Khatun, M.; Eguchi, S.; Yamaguchi, T.; Takamura, H.; Matoba, T. Effect of Thermal Treatment on Radical-scavenging Activity of Some Spices. *Food Sci. Technol. Res.* **2006**, *12*, 178–185. [CrossRef]
56. Azuma, K.; Ippoushi, K.; Ito, H.; Higashio, H.; Terao, J. Combination of lipids and emulsifiers enhances the absorption of orally administered quercetin in rats. *J. Agric. Food Chem.* **2002**, *50*, 1706–1712. [CrossRef]
57. Visioli, F.; Galli, C.; Grande, S.; Colonnelli, K.; Patelli, C.; Galli, G.; Caruso, D. Hydroxytyrosol excretion differs between rats and humans and depends on the vehicle of administration. *J. Nutr.* **2003**, *133*, 2612–2615. [CrossRef] [PubMed]
58. Williamson, G.; Kay, C.D.; Crozier, A. The Bioavailability, Transport, and Bioactivity of Dietary Flavonoids: A Review from a Historical Perspective. *Compr. Rev. Food Sci. Food Saf.* **2018**, *17*, 1054–1112. [CrossRef]
59. Pérez-Jiménez, J.; Hubert, J.; Hooper, L.; Cassidy, A.; Manach, C.; Williamson, G.; Scalbert, A. Urinary metabolites as biomarkers of polyphenol intake in humans: A systematic review. *Am. J. Clin. Nutr.* **2010**, *92*, 801–809. [CrossRef]
60. Manach, C.; Scalbert, A.; Morand, C.; Remesy, C.; Jimenez, L. Polyphenols: Food sources and bioavailability. *Am. J. Clin. Nutr.* **2004**, *79*, 727–747. [CrossRef]
61. Gonthier, M.P.; Donovan, J.L.; Texier, O.; Felgines, C.; Remesy, C.; Scalbert, A. Metabolism of dietary procyanidins in rats. *Free Radic. Biol. Med.* **2003**, *35*, 837–844. [CrossRef]
62. Chen, Z.; Zheng, S.; Li, L.; Jiang, H. Metabolism of flavonoids in human: A comprehensive review. *Curr. Drug Metab.* **2014**, *15*, 48–61. [CrossRef]
63. Villa-Rodriguez, J.A.; Ifie, I.; Gonzalez-Aguilar, G.A.; Roopchand, D.E. The Gastrointestinal Tract as Prime Site for Cardiometabolic Protection by Dietary Polyphenols. *Adv. Nutr.* **2019**, *10*, 999–1011. [CrossRef]
64. Spencer, J.P.; Chaudry, F.; Pannala, A.S.; Srai, S.K.; Debnam, E.; Rice-Evans, C. Decomposition of cocoa procyanidins in the gastric milieu. *Biochem. Biophys. Res. Commun.* **2000**, *272*, 236–241. [CrossRef] [PubMed]
65. Zhu, Q.Y.; Holt, R.R.; Lazarus, S.A.; Ensunsa, J.L.; Hammerstone, J.F.; Schmitz, H.H.; Keen, C.L. Stability of the flavan-3-ols epicatechin and catechin and related dimeric procyanidins derived from cocoa. *J. Agric. Food Chem.* **2002**, *50*, 1700–1705. [CrossRef] [PubMed]
66. Corona, G.; Tzounis, X.; Assunta Dessi, M.; Deiana, M.; Debnam, E.S.; Visioli, F.; Spencer, J.P. The fate of olive oil polyphenols in the gastrointestinal tract: Implications of gastric and colonic microflora-dependent biotransformation. *Free Radic. Res.* **2006**, *40*, 647–658. [CrossRef] [PubMed]
67. Serra, A.; Macia, A.; Romero, M.P.; Valls, J.; Blade, C.; Arola, L.; Motilva, M.J. Bioavailability of procyanidin dimers and trimers and matrix food effects in in vitro and in vivo models. *Br. J. Nutr.* **2010**, *103*, 944–952. [CrossRef]
68. Kahle, K.; Kempf, M.; Schreier, P.; Scheppach, W.; Schrenk, D.; Kautenburger, T.; Hecker, D.; Huemmer, W.; Ackermann, M.; Richling, E. Intestinal transit and systemic metabolism of apple polyphenols. *Eur. J. Nutr.* **2011**, *50*, 507–522. [CrossRef]
69. Van Rymenant, E.; Salden, B.; Voorspoels, S.; Jacobs, G.; Noten, B.; Pitart, J.; Possemiers, S.; Smagghe, G.; Grootaert, C.; Van Camp, J. A Critical Evaluation of In Vitro Hesperidin 2S Bioavailability in a Model Combining Luminal (Microbial) Digestion and Caco-2 Cell Absorption in Comparison to a Randomized Controlled Human Trial. *Mol. Nutr. Food Res.* **2018**, *62*, e1700881. [CrossRef]
70. Hu, X.P.; Yin, F.W.; Zhou, D.Y.; Xie, H.K.; Zhu, B.W.; Ma, X.C.; Tian, X.G.; Wang, C.; Shahidi, F. Stability of resveratrol esters with caprylic acid during simulated in vitro gastrointestinal digestion. *Food Chem.* **2019**, *276*, 675–679. [CrossRef]
71. Liu, D.; Dhital, S.; Wu, P.; Chen, X.D.; Gidley, M.J. In Vitro Digestion of Apple Tissue Using a Dynamic Stomach Model: Grinding and Crushing Effects on Polyphenol Bioaccessibility. *J. Agric. Food Chem.* **2020**, *68*, 574–583. [CrossRef] [PubMed]
72. Ma, T.; Lan, T.; Geng, T.; Ju, Y.; Cheng, G.; Que, Z.; Gao, G.; Fang, Y.; Sun, X. Nutritional properties and biological activities of kiwifruit (Actinidia) and kiwifruit products under simulated gastrointestinal in vitro digestion. *Food Nutr. Res.* **2019**, *63*. [CrossRef] [PubMed]
73. Rios, L.Y.; Bennett, R.N.; Lazarus, S.A.; Remesy, C.; Scalbert, A.; Williamson, G. Cocoa procyanidins are stable during gastric transit in humans. *Am. J. Clin. Nutr.* **2002**, *76*, 1106–1110. [CrossRef] [PubMed]

74. Konishi, Y.; Zhao, Z.; Shimizu, M. Phenolic acids are absorbed from the rat stomach with different absorption rates. *J. Agric. Food Chem.* **2006**, *54*, 7539–7543. [CrossRef]
75. Pforte, H.; Hempel, J.; Jacobasch, G. Distribution pattern of a flavonoid extract in the gastrointestinal lumen and wall of rats. *Nahrung* **1999**, *43*, 205–208. [CrossRef]
76. Piskula, M.K.; Yamakoshi, J.; Iwai, Y. Daidzein and genistein but not their glucosides are absorbed from the rat stomach. *FEBS Lett.* **1999**, *447*, 287–291. [CrossRef]
77. Crespy, V.; Morand, C.; Besson, C.; Manach, C.; Demigne, C.; Remesy, C. Quercetin, but not its glycosides, is absorbed from the rat stomach. *J. Agric. Food Chem.* **2002**, *50*, 618–621. [CrossRef]
78. Gil-Izquierdo, A.; Zafrilla, P.; Tomás-Barberán, F.A. An in vitro method to simulate phenolic compound release from the food matrix in the gastrointestinal tract. *Eur. Food Res. Technol.* **2002**, *214*, 155–159. [CrossRef]
79. Ortega, N.; Reguant, J.; Romero, M.P.; Macia, A.; Motilva, M.J. Effect of fat content on the digestibility and bioaccessibility of cocoa polyphenol by an in vitro digestion model. *J. Agric. Food Chem.* **2009**, *57*, 5743–5749. [CrossRef]
80. Tsai, Y.M.; Chang-Liao, W.L.; Chien, C.F.; Lin, L.C.; Tsai, T.H. Effects of polymer molecular weight on relative oral bioavailability of curcumin. *Int. J. Nanomed.* **2012**, *7*, 2957–2966. [CrossRef]
81. Day, A.J.; Gee, J.M.; DuPont, M.S.; Johnson, I.T.; Williamson, G. Absorption of quercetin-3-glucoside and quercetin-4′-glucoside in the rat small intestine: The role of lactase phlorizin hydrolase and the sodium-dependent glucose transporter. *Biochem. Pharmacol.* **2003**, *65*, 1199–1206. [CrossRef]
82. Henry-Vitrac, C.; Desmouliere, A.; Girard, D.; Merillon, J.M.; Krisa, S. Transport, deglycosylation, and metabolism of trans-piceid by small intestinal epithelial cells. *Eur. J. Nutr.* **2006**, *45*, 376–382. [CrossRef]
83. Murota, K.; Nakamura, Y.; Uehara, M. Flavonoid metabolism: The interaction of metabolites and gut microbiota. *Biosci. Biotechnol. Biochem.* **2018**, *82*, 600–610. [CrossRef] [PubMed]
84. Murota, K.; Shimizu, S.; Chujo, H.; Moon, J.-H.; Terao, J. Efficiency of Absorption and Metabolic Conversion of Quercetin and Its Glucosides in Human Intestinal Cell Line Caco-2. *Arch. Biochem. Biophys.* **2000**, *384*, 391–397. [CrossRef] [PubMed]
85. Amaretti, A.; Raimondi, S.; Leonardi, A.; Quartieri, A.; Rossi, M. Hydrolysis of the rutinose-conjugates flavonoids rutin and hesperidin by the gut microbiota and bifidobacteria. *Nutrients* **2015**, *7*, 2788–2800. [CrossRef] [PubMed]
86. Song, Q.; Li, D.; Zhou, Y.; Yang, J.; Yang, W.; Zhou, G.; Wen, J. Enhanced uptake and transport of (+)-catechin and (−)-epigallocatechin gallate in niosomal formulation by human intestinal Caco-2 cells. *Int. J. Nanomed.* **2014**, *9*, 2157–2165. [CrossRef]
87. Borges, G.; Ottaviani, J.I.; van der Hooft, J.J.J.; Schroeter, H.; Crozier, A. Absorption, metabolism, distribution and excretion of (−)-epicatechin: A review of recent findings. *Mol. Asp. Med.* **2018**, *61*, 18–30. [CrossRef]
88. Levy, E.; Yotov, W.; Seidman, E.G.; Garofalo, C.; Delvin, E.; Menard, D. Caco-2 cells and human fetal colon: A comparative analysis of their lipid transport. *Biochim. Biophys. Acta* **1999**, *1439*, 353–362. [CrossRef]
89. Nauli, A.M.; Nauli, S.M. Intestinal transport as a potential determinant of drug bioavailability. *Curr. Clin. Pharmacol.* **2013**, *8*, 247–255. [CrossRef]
90. Kern, S.M.; Bennett, R.N.; Needs, P.W.; Mellon, F.A.; Kroon, P.A.; Garcia-Conesa, M.T. Characterization of metabolites of hydroxycinnamates in the in vitro model of human small intestinal epithelium caco-2 cells. *J. Agric. Food Chem.* **2003**, *51*, 7884–7891. [CrossRef]
91. Hu, J.N.; Zou, X.G.; He, Y.; Chen, F.; Deng, Z.Y. Esterification of Quercetin Increases Its Transport across Human Caco-2 Cells. *J. Food Sci.* **2016**, *81*, H1825–H1832. [CrossRef] [PubMed]
92. Su, H.F.; Lin, Q.; Wang, X.Y.; Fu, Y.; Gong, T.; Sun, X.; Zhang, Z.R. Absorptive interactions of concurrent oral administration of (+)-catechin and puerarin in rats and the underlying mechanisms. *Acta Pharmacol. Sin.* **2016**, *37*, 545–554. [CrossRef] [PubMed]
93. Dudonne, S.; Dal-Pan, A.; Dube, P.; Varin, T.V.; Calon, F.; Desjardins, Y. Potentiation of the bioavailability of blueberry phenolic compounds by co-ingested grape phenolic compounds in mice, revealed by targeted metabolomic profiling in plasma and feces. *Food Funct.* **2016**, *7*, 3421–3430. [CrossRef] [PubMed]

94. Dudonne, S.; Dube, P.; Pilon, G.; Marette, A.; Jacques, H.; Weisnagel, J.; Desjardins, Y. Modulation of Strawberry/Cranberry Phenolic Compounds Glucuronidation by Co-Supplementation with Onion: Characterization of Phenolic Metabolites in Rat Plasma Using an Optimized muSPE-UHPLC-MS/MS Method. *J. Agric. Food Chem.* **2014**, *62*, 3244–3256. [CrossRef]
95. Deglaire, A.; Moughan, P.J. Animal models for determining amino acid digestibility in humans—A review. *Br. J. Nutr.* **2012**, *108* (Suppl. S2), S273–S281. [CrossRef]
96. Vergauwen, H. The IPEC-J2 Cell Line. In *The Impact of Food Bioactives on Health: In Vitro and Ex Vivo Models*; Verhoeckx, K., Cotter, P., López-Expósito, I., Kleiveland, C., Lea, T., Mackie, A., Requena, T., Swiatecka, D., Wichers, H., Eds.; Springer International Publishing: Cham, Switzerland, 2015; pp. 125–134.
97. Chedea, V.S.; Palade, L.M.; Marin, D.E.; Pelmus, R.S.; Habeanu, M.; Rotar, M.C.; Gras, M.A.; Pistol, G.C.; Taranu, I. Intestinal Absorption and Antioxidant Activity of Grape Pomace Polyphenols. *Nutrients* **2018**, *10*, 588. [CrossRef]
98. Wang, W.; Sun, C.; Mao, L.; Ma, P.; Liu, F.; Yang, J.; Gao, Y. The biological activities, chemical stability, metabolism and delivery systems of quercetin: A review. *Trends Food Sci. Tech.* **2016**, *56*, 21–38. [CrossRef]
99. Zhang, L.; Wang, Y.; Li, D.; Ho, C.T.; Li, J.; Wan, X. The absorption, distribution, metabolism and excretion of procyanidins. *Food Funct.* **2016**, *7*, 1273–1281. [CrossRef]
100. Rothwell, J.A.; Urpi-Sarda, M.; Boto-Ordonez, M.; Llorach, R.; Farran-Codina, A.; Barupal, D.K.; Neveu, V.; Manach, C.; Andres-Lacueva, C.; Scalbert, A. Systematic analysis of the polyphenol metabolome using the Phenol-Explorer database. *Mol. Nutr. Food Res.* **2016**, *60*, 203–211. [CrossRef]
101. Martínez-Maqueda, D.; Miralles, B.; Recio, I. HT29 Cell Line. In *The Impact of Food Bioactives on Health: In Vitro and Ex Vivo Models*; Verhoeckx, K., Cotter, P., López-Expósito, I., Kleiveland, C., Lea, T., Mackie, A., Requena, T., Swiatecka, D., Wichers, H., Eds.; Springer International Publishing: Cham, Switzerland, 2015; pp. 113–124.
102. Bergmann, H.; Rogoll, D.; Scheppach, W.; Melcher, R.; Richling, E. The Ussing type chamber model to study the intestinal transport and modulation of specific tight-junction genes using a colonic cell line. *Mol. Nutr. Food Res.* **2009**, *53*, 1211–1225. [CrossRef]
103. Alsolmei, F.A.; Li, H.; Pereira, S.L.; Krishnan, P.; Johns, P.W.; Siddiqui, R.A. Polyphenol-Enriched Plum Extract Enhances Myotubule Formation and Anabolism while Attenuating Colon Cancer-induced Cellular Damage in C2C12 Cells. *Nutrients* **2019**, *11*, 1077. [CrossRef]
104. Blanquer-Rossello, M.D.; Hernandez-Lopez, R.; Roca, P.; Oliver, J.; Valle, A. Resveratrol induces mitochondrial respiration and apoptosis in SW620 colon cancer cells. *Biochim. Biophys. Acta Gen. Subj.* **2017**, *1861*, 431–440. [CrossRef]
105. Saunier, E.; Antonio, S.; Regazzetti, A.; Auzeil, N.; Laprevote, O.; Shay, J.W.; Coumoul, X.; Barouki, R.; Benelli, C.; Huc, L.; et al. Resveratrol reverses the Warburg effect by targeting the pyruvate dehydrogenase complex in colon cancer cells. *Sci. Rep.* **2017**, *7*, 6945. [CrossRef]
106. Valdes, A.; Garcia-Canas, V.; Simo, C.; Ibanez, C.; Micol, V.; Ferragut, J.A.; Cifuentes, A. Comprehensive foodomics study on the mechanisms operating at various molecular levels in cancer cells in response to individual rosemary polyphenols. *Anal. Chem.* **2014**, *86*, 9807–9815. [CrossRef]
107. Williamson, G.; Clifford, M.N. Role of the small intestine, colon and microbiota in determining the metabolic fate of polyphenols. *Biochem. Pharmacol.* **2017**, *139*, 24–39. [CrossRef]
108. Marin, L.; Miguelez, E.M.; Villar, C.J.; Lombo, F. Bioavailability of dietary polyphenols and gut microbiota metabolism: Antimicrobial properties. *Biomed. Res. Int.* **2015**, *2015*, 905215. [CrossRef]
109. Monagas, M.; Urpi-Sarda, M.; Sanchez-Patan, F.; Llorach, R.; Garrido, I.; Gomez-Cordoves, C.; Andres-Lacueva, C.; Bartolome, B. Insights into the metabolism and microbial biotransformation of dietary flavan-3-ols and the bioactivity of their metabolites. *Food Funct.* **2010**, *1*, 233–253. [CrossRef]
110. Neish, A.S. Microbes in gastrointestinal health and disease. *Gastroenterology* **2009**, *136*, 65–80. [CrossRef]
111. Arora, T.; Backhed, F. The gut microbiota and metabolic disease: Current understanding and future perspectives. *J. Intern. Med.* **2016**, *280*, 339–349. [CrossRef]
112. Korpela, K. Diet, Microbiota, and Metabolic Health: Trade-Off between Saccharolytic and Proteolytic Fermentation. *Annu. Rev. Food Sci. Technol.* **2018**, *9*, 65–84. [CrossRef]
113. Ma, N.; Tian, Y.; Wu, Y.; Ma, X. Contributions of the Interaction Between Dietary Protein and Gut Microbiota to Intestinal Health. *Curr. Protein Pept. Sci.* **2017**, *18*, 795–808. [CrossRef]

114. Zhao, J.; Zhang, X.; Liu, H.; Brown, M.A.; Qiao, S. Dietary Protein and Gut Microbiota Composition and Function. *Curr. Protein Pept. Sci.* **2019**, *20*, 145–154. [CrossRef]
115. Qin, J.; Li, R.; Raes, J.; Arumugam, M.; Burgdorf, K.S.; Manichanh, C.; Nielsen, T.; Pons, N.; Levenez, F.; Yamada, T.; et al. A human gut microbial gene catalogue established by metagenomic sequencing. *Nature* **2010**, *464*, 59–65. [CrossRef]
116. Ou, J.; Carbonero, F.; Zoetendal, E.G.; DeLany, J.P.; Wang, M.; Newton, K.; Gaskins, H.R.; O'Keefe, S.J. Diet, microbiota, and microbial metabolites in colon cancer risk in rural Africans and African Americans. *Am. J. Clin. Nutr.* **2013**, *98*, 111–120. [CrossRef]
117. Wu, G.D.; Chen, J.; Hoffmann, C.; Bittinger, K.; Chen, Y.Y.; Keilbaugh, S.A.; Bewtra, M.; Knights, D.; Walters, W.A.; Knight, R.; et al. Linking long-term dietary patterns with gut microbial enterotypes. *Science* **2011**, *334*, 105–108. [CrossRef]
118. Zhernakova, A.; Kurilshikov, A.; Bonder, M.J.; Tigchelaar, E.F.; Schirmer, M.; Vatanen, T.; Mujagic, Z.; Vila, A.V.; Falony, G.; Vieira-Silva, S.; et al. Population-based metagenomics analysis reveals markers for gut microbiome composition and diversity. *Science* **2016**, *352*, 565–569. [CrossRef]
119. Kabeerdoss, J.; Devi, R.S.; Mary, R.R.; Ramakrishna, B.S. Faecal microbiota composition in vegetarians: Comparison with omnivores in a cohort of young women in southern India. *Br. J. Nutr.* **2012**, *108*, 953–957. [CrossRef]
120. David, L.A.; Maurice, C.F.; Carmody, R.N.; Gootenberg, D.B.; Button, J.E.; Wolfe, B.E.; Ling, A.V.; Devlin, A.S.; Varma, Y.; Fischbach, M.A.; et al. Diet rapidly and reproducibly alters the human gut microbiome. *Nature* **2014**, *505*, 559–563. [CrossRef]
121. De Filippo, C.; Cavalieri, D.; Di Paola, M.; Ramazzotti, M.; Poullet, J.B.; Massart, S.; Collini, S.; Pieraccini, G.; Lionetti, P. Impact of diet in shaping gut microbiota revealed by a comparative study in children from Europe and rural Africa. *Proc. Natl. Acad. Sci. USA* **2010**, *107*, 14691–14696. [CrossRef]
122. Walker, A.W.; Ince, J.; Duncan, S.H.; Webster, L.M.; Holtrop, G.; Ze, X.; Brown, D.; Stares, M.D.; Scott, P.; Bergerat, A.; et al. Dominant and diet-responsive groups of bacteria within the human colonic microbiota. *ISME J.* **2011**, *5*, 220–230. [CrossRef]
123. Kumar Singh, A.; Cabral, C.; Kumar, R.; Ganguly, R.; Kumar Rana, H.; Gupta, A.; Rosaria Lauro, M.; Carbone, C.; Reis, F.; Pandey, A.K. Beneficial Effects of Dietary Polyphenols on Gut Microbiota and Strategies to Improve Delivery Efficiency. *Nutrients* **2019**, *11*, 2216. [CrossRef]
124. Yin, H.; Deng, Y.; Wang, H.; Liu, W.; Zhuang, X.; Chu, W. Tea polyphenols as an antivirulence compound Disrupt Quorum-Sensing Regulated Pathogenicity of Pseudomonas aeruginosa. *Sci. Rep.* **2015**, *5*, 16158. [CrossRef]
125. Stenvang, M.; Dueholm, M.S.; Vad, B.S.; Seviour, T.; Zeng, G.; Geifman-Shochat, S.; Søndergaard, M.T.; Christiansen, G.; Meyer, R.L.; Kjelleberg, S.; et al. Epigallocatechin Gallate Remodels Overexpressed Functional Amyloids in Pseudomonas aeruginosa and Increases Biofilm Susceptibility to Antibiotic Treatment. *J. Biol. Chem.* **2016**, *291*, 26540–26553. [CrossRef]
126. Stapleton, P.D.; Shah, S.; Ehlert, K.; Hara, Y.; Taylor, P.W. The beta-lactam-resistance modifier (−)-epicatechin gallate alters the architecture of the cell wall of Staphylococcus aureus. *Microbiology* **2007**, *153*, 2093–2103. [CrossRef]
127. Hachimura, S.; Totsuka, M.; Hosono, A. Immunomodulation by food: Impact on gut immunity and immune cell function. *Biosci. Biotechnol. Biochem.* **2018**, *82*, 584–599. [CrossRef]
128. Vitetta, L.; Vitetta, G.; Hall, S. Immunological Tolerance and Function: Associations Between Intestinal Bacteria, Probiotics, Prebiotics, and Phages. *Front. Immunol.* **2018**, *9*, 2240. [CrossRef]
129. Stan, M.S.; Voicu, S.N.; Caruntu, S.; Nica, I.C.; Olah, N.K.; Burtescu, R.; Balta, C.; Rosu, M.; Herman, H.; Hermenean, A.; et al. Antioxidant and Anti-Inflammatory Properties of a Thuja occidentalis Mother Tincture for the Treatment of Ulcerative Colitis. *Antioxidants* **2019**, *8*, 416. [CrossRef]
130. Mileo, A.M.; Nisticò, P.; Miccadei, S. Polyphenols: Immunomodulatory and Therapeutic Implication in Colorectal Cancer. *Front. Immunol.* **2019**, *10*, 729. [CrossRef]
131. Yamakoshi, J.; Tokutake, S.; Kikuchi, M.; Kubota, Y.; Konishi, H.; Mitsuoka, T. Effect of Proanthocyanidin-Rich Extract from Grape Seeds on Human Fecal Flora and Fecal Odor. *Microb. Ecol. Health Dis.* **2001**, *13*, 25–31.
132. Larrosa, M.; Luceri, C.; Vivoli, E.; Pagliuca, C.; Lodovici, M.; Moneti, G.; Dolara, P. Polyphenol metabolites from colonic microbiota exert anti-inflammatory activity on different inflammation models. *Mol. Nutr. Food Res.* **2009**, *53*, 1044–1054. [CrossRef]

133. Tzounis, X.; Rodriguez-Mateos, A.; Vulevic, J.; Gibson, G.R.; Kwik-Uribe, C.; Spencer, J.P. Prebiotic evaluation of cocoa-derived flavanols in healthy humans by using a randomized, controlled, double-blind, crossover intervention study. *Am. J. Clin. Nutr.* **2011**, *93*, 62–72. [CrossRef]
134. Moreno-Indias, I.; Sánchez-Alcoholado, L.; Pérez-Martínez, P.; Andrés-Lacueva, C.; Cardona, F.; Tinahones, F.; Queipo-Ortuño, M.I. Red wine polyphenols modulate fecal microbiota and reduce markers of the metabolic syndrome in obese patients. *Food Funct.* **2016**, *7*, 1775–1787. [CrossRef]
135. Rowland, I.; Gibson, G.; Heinken, A.; Scott, K.; Swann, J.; Thiele, I.; Tuohy, K. Gut microbiota functions: Metabolism of nutrients and other food components. *Eur. J. Nutr.* **2018**, *57*, 1–24. [CrossRef] [PubMed]
136. Braune, A.; Blaut, M. Bacterial species involved in the conversion of dietary flavonoids in the human gut. *Gut Microbes* **2016**, *7*, 216–234. [CrossRef]
137. Corrêa, T.A.F.; Rogero, M.M.; Hassimotto, N.M.A.; Lajolo, F.M. The Two-Way Polyphenols-Microbiota Interactions and Their Effects on Obesity and Related Metabolic Diseases. *Front. Nutr.* **2019**, *6*, 188. [CrossRef]
138. Cote, A.T.; Harris, K.C.; Panagiotopoulos, C.; Sandor, G.G.; Devlin, A.M. Childhood obesity and cardiovascular dysfunction. *J. Am.Coll. Cardiol.* **2013**, *62*, 1309–1319. [CrossRef]
139. Levy, E.; Saenger, A.K.; Steffes, M.W.; Delvin, E. Pediatric Obesity and Cardiometabolic Disorders: Risk Factors and Biomarkers. *Ejifcc* **2017**, *28*, 6–24. [PubMed]
140. Nehus, E.; Mitsnefes, M. Childhood Obesity and the Metabolic Syndrome. *Pediatr. Clin. N. Am.* **2019**, *66*, 31–43. [CrossRef]
141. Vona, R.; Gambardella, L.; Cittadini, C.; Straface, E.; Pietraforte, D. Biomarkers of Oxidative Stress in Metabolic Syndrome and Associated Diseases. *Oxid. Med. Cell. Longev.* **2019**, *2019*, 8267234. [CrossRef] [PubMed]
142. Reinehr, T.; Roth, C.L. Inflammation Markers in Type 2 Diabetes and the Metabolic Syndrome in the Pediatric Population. *Curr. Diab. Rep.* **2018**, *18*, 131. [CrossRef]
143. Rani, V.; Deep, G.; Singh, R.K.; Palle, K.; Yadav, U.C.S. Oxidative stress and metabolic disorders: Pathogenesis and therapeutic strategies. *Life Sci.* **2016**, *148*, 183–193. [CrossRef] [PubMed]
144. McKeown, N.M.; Meigs, J.B.; Liu, S.; Saltzman, E.; Wilson, P.W.; Jacques, P.F. Carbohydrate nutrition, insulin resistance, and the prevalence of the metabolic syndrome in the Framingham Offspring Cohort. *Diabet. Care* **2004**, *27*, 538–546. [CrossRef] [PubMed]
145. Spahis, S.; Alvarez, F.; Ahmed, N.; Dubois, J.; Jalbout, R.; Paganelli, M.; Grzywacz, K.; Delvin, E.; Peretti, N.; Levy, E. Non-alcoholic fatty liver disease severity and metabolic complications in obese children: Impact of omega-3 fatty acids. *J. Nutr. Biochem.* **2018**, *58*, 28–36. [CrossRef]
146. Sabaté, J.; Wien, M. A perspective on vegetarian dietary patterns and risk of metabolic syndrome. *Br. J. Nutr.* **2015**, *113* (Suppl. S2), S136–S143. [CrossRef] [PubMed]
147. Esmaillzadeh, A.; Mirmiran, P.; Azizi, F. Whole-grain consumption and the metabolic syndrome: A favorable association in Tehranian adults. *Eur. J. Clin. Nutr.* **2005**, *59*, 353–362. [CrossRef]
148. Mallappa, R.H.; Rokana, N.; Duary, R.K.; Panwar, H.; Batish, V.K.; Grover, S. Management of metabolic syndrome through probiotic and prebiotic interventions. *Indian J. Endocrinol. Metab.* **2012**, *16*, 20–27.
149. Joseph, S.V.; Edirisinghe, I.; Burton-Freeman, B.M. Fruit Polyphenols: A Review of Anti-inflammatory Effects in Humans. *Crit. Rev. Food Sci. Nutr.* **2016**, *56*, 419–444. [CrossRef]
150. Gourineni, V.; Shay, N.F.; Chung, S.; Sandhu, A.K.; Gu, L. Muscadine grape (*Vitis rotundifolia*) and wine phytochemicals prevented obesity-associated metabolic complications in C57BL/6J mice. *J. Agric. Food Chem.* **2012**, *60*, 7674–7681. [CrossRef]
151. Makino-Wakagi, Y.; Yoshimura, Y.; Uzawa, Y.; Zaima, N.; Moriyama, T.; Kawamura, Y. Ellagic acid in pomegranate suppresses resistin secretion by a novel regulatory mechanism involving the degradation of intracellular resistin protein in adipocytes. *Biochem. Biophys. Res. Commun.* **2012**, *417*, 880–885. [CrossRef]
152. Lei, F.; Zhang, X.N.; Wang, W.; Xing, D.M.; Xie, W.D.; Su, H.; Du, L.J. Evidence of anti-obesity effects of the pomegranate leaf extract in high-fat diet induced obese mice. *Int. J. Obes.* **2007**, *31*, 1023–1029. [CrossRef]
153. Monika, P.; Geetha, A. The modulating effect of Persea americana fruit extract on the level of expression of fatty acid synthase complex, lipoprotein lipase, fibroblast growth factor-21 and leptin—A biochemical study in rats subjected to experimental hyperlipidemia and obesity. *Phytomedicine* **2015**, *22*, 939–945. [CrossRef] [PubMed]
154. Kang, I.; Espín, J.C.; Carr, T.P.; Tomás-Barberán, F.A.; Chung, S. Raspberry seed flour attenuates high-sucrose diet-mediated hepatic stress and adipose tissue inflammation. *J. Nutr. Biochem.* **2016**, *32*, 64–72. [CrossRef] [PubMed]

155. Upadhyay, S.; Dixit, M. Role of Polyphenols and Other Phytochemicals on Molecular Signaling. *Oxid. Med. Cell. Longev.* **2015**, *2015*, 504253. [CrossRef] [PubMed]
156. Gothai, S.; Ganesan, P.; Park, S.Y.; Fakurazi, S.; Choi, D.K.; Arulselvan, P. Natural Phyto-Bioactive Compounds for the Treatment of Type 2 Diabetes: Inflammation as a Target. *Nutrients* **2016**, *8*, 461. [CrossRef] [PubMed]
157. Wang, H.; Guo, X.; Hu, X.; Li, T.; Fu, X.; Liu, R.H. Comparison of phytochemical profiles, antioxidant and cellular antioxidant activities of different varieties of blueberry (Vaccinium spp.). *Food Chem.* **2017**, *217*, 773–781. [CrossRef]
158. Leopoldini, M.; Marino, T.; Russo, N.; Toscano, M. Antioxidant properties of phenolic compounds: H-Atom versus Elecgron Transfer Mechanism. *J. Phys. Chem.* **2004**, *108*, 4916–4922. [CrossRef]
159. Oteiza, P.I.; Erlejman, A.G.; Verstraeten, S.V.; Keen, C.L.; Fraga, C.G. Flavonoid-membrane interactions: A protective role of flavonoids at the membrane surface? *Clin. Dev. Immunol.* **2005**, *12*, 19–25. [CrossRef]
160. Perron, N.R.; Brumaghim, J.L. A review of the antioxidant mechanisms of polyphenol compounds related to iron binding. *Cell Biochem. Biophys.* **2009**, *53*, 75–100. [CrossRef]
161. Spahis, S.; Borys, J.M.; Levy, E. Metabolic Syndrome as a Multifaceted Risk Factor for Oxidative Stress. *Antioxid. Redox Signal.* **2017**, *26*, 445–461. [CrossRef]
162. Kensler, T.W.; Wakabayashi, N.; Biswal, S. Cell survival responses to environmental stresses via the Keap1-Nrf2-ARE pathway. *Annu. Rev. Pharmacol. Toxicol.* **2007**, *47*, 89–116. [CrossRef]
163. Kansanen, E.; Kuosmanen, S.M.; Leinonen, H.; Levonen, A.L. The Keap1-Nrf2 pathway: Mechanisms of activation and dysregulation in cancer. *Redox Biol.* **2013**, *1*, 45–49. [CrossRef] [PubMed]
164. Köhle, C.; Bock, K.W. Activation of coupled Ah receptor and Nrf2 gene batteries by dietary phytochemicals in relation to chemoprevention. *Biochem. Pharmacol.* **2006**, *72*, 795–805. [CrossRef] [PubMed]
165. Zhang, S.; Qin, C.; Safe, S.H. Flavonoids as aryl hydrocarbon receptor agonists/antagonists: Effects of structure and cell context. *Environ. Health Perspect.* **2003**, *111*, 1877–1882. [CrossRef] [PubMed]
166. Chuang, C.C.; McIntosh, M.K. Potential mechanisms by which polyphenol-rich grapes prevent obesity-mediated inflammation and metabolic diseases. *Annu. Rev. Nutr.* **2011**, *31*, 155–176. [CrossRef]
167. Diotallevi, C.; Fava, F.; Gobbetti, M.; Tuohy, K. Healthy dietary patterns to reduce obesity-related metabolic disease: Polyphenol: Microbiome interactions unifying health effects across geography. *Curr. Opin. Clin. Nutr. Metab. Care* **2020**, *23*, 437–444. [CrossRef]
168. Sanchez-Marzo, N.; Perez-Sanchez, A.; Barrajon-Catalan, E.; Castillo, J.; Herranz-Lopez, M.; Micol, V. Rosemary Diterpenes and Flavanone Aglycones Provide Improved Genoprotection against UV-Induced DNA Damage in a Human Skin Cell Model. *Antioxidants* **2020**, *9*, 255. [CrossRef]
169. Matsuno, Y.; Atsumi, Y.; Alauddin, M.; Rana, M.M.; Fujimori, H.; Hyodo, M.; Shimizu, A.; Ikuta, T.; Tani, H.; Torigoe, H.; et al. Resveratrol and its Related Polyphenols Contribute to the Maintenance of Genome Stability. *Sci. Rep.* **2020**, *10*, 5388. [CrossRef]
170. Huarte, E.; Cid, C.; Azqueta, A.; de Pena, M.P. DNA damage and DNA protection from digested raw and griddled green pepper (poly)phenols in human colorectal adenocarcinoma cells (HT-29). *Eur. J. Nutr.* **2020**. [CrossRef]
171. Groh, I.A.M.; Bakuradze, T.; Pahlke, G.; Richling, E.; Marko, D. Consumption of anthocyanin-rich beverages affects Nrf2 and Nrf2-dependent gene transcription in peripheral lymphocytes and DNA integrity of healthy volunteers. *BMC Chem.* **2020**, *14*, 39. [CrossRef]
172. Abib, R.T.; Quincozes-Santos, A.; Zanotto, C.; Zeidán-Chuliá, F.; Lunardi, P.S.; Gonçalves, C.A.; Gottfried, C. Genoprotective effects of the green tea-derived polyphenol/epicatechin gallate in C6 astroglial cells. *J. Med. Food* **2010**, *13*, 1111–1115. [CrossRef]
173. Quincozes-Santos, A.; Andreazza, A.C.; Nardin, P.; Funchal, C.; Gonçalves, C.A.; Gottfried, C. Resveratrol attenuates oxidative-induced DNA damage in C6 Glioma cells. *Neurotoxicology* **2007**, *28*, 886–891. [CrossRef]
174. Han, K.C.; Wong, W.C.; Benzie, I.F. Genoprotective effects of green tea (Camellia sinensis) in human subjects: Results of a controlled supplementation trial. *Br. J. Nutr.* **2011**, *105*, 171–179. [CrossRef]
175. Morgan, M.J.; Liu, Z.G. Crosstalk of reactive oxygen species and NF-κB signaling. *Cell Res.* **2011**, *21*, 103–115. [CrossRef]
176. Blaser, H.; Dostert, C.; Mak, T.W.; Brenner, D. TNF and ROS Crosstalk in Inflammation. *Trends Cell. Biol.* **2016**, *26*, 249–261. [CrossRef] [PubMed]
177. Kim, Y.S.; Young, M.R.; Bobe, G.; Colburn, N.H.; Milner, J.A. Bioactive food components, inflammatory targets, and cancer prevention. *Cancer Prev. Res.* **2009**, *2*, 200–208. [CrossRef] [PubMed]

178. Yu, Z.; Tang, Y.; Hu, D.; Li, J. Inhibitory effect of genistein on mouse colon cancer MC-26 cells involved TGF-beta1/Smad pathway. *Biochem. Biophys. Res. Commun.* **2005**, *333*, 827–832. [CrossRef] [PubMed]
179. Nguyen, K.A.; Cao, Y.; Chen, J.R.; Townsend, C.M., Jr.; Ko, T.C. Dietary fiber enhances a tumor suppressor signaling pathway in the gut. *Ann. Surg.* **2006**, *243*, 619–625. [CrossRef]
180. Denis, M.C.; Furtos, A.; Dudonne, S.; Montoudis, A.; Garofalo, C.; Desjardins, Y.; Delvin, E.; Levy, E. Apple peel polyphenols and their beneficial actions on oxidative stress and inflammation. *PLoS ONE* **2013**, *8*, e53725. [CrossRef]
181. Yeganeh, P.R.; Leahy, J.; Spahis, S.; Patey, N.; Desjardins, Y.; Roy, D.; Delvin, E.; Garofalo, C.; Leduc-Gaudet, J.P.; St-Pierre, D.; et al. Apple peel polyphenols reduce mitochondrial dysfunction in mice with DSS-induced ulcerative colitis. *J. Nutr. Biochem.* **2018**, *57*, 56–66. [CrossRef]
182. Denis, M.C.; Roy, D.; Yeganeh, P.R.; Desjardins, Y.; Varin, T.; Haddad, N.; Amre, D.; Sane, A.T.; Garofalo, C.; Furtos, A.; et al. Apple peel polyphenols: A key player in the prevention and treatment of experimental inflammatory bowel disease. *Clin. Sci.* **2016**, *130*, 2217–2237. [CrossRef]
183. Cheng, Z.; Zheng, L.; Almeida, F.A. Epigenetic reprogramming in metabolic disorders: Nutritional factors and beyond. *J. Nutr. Biochem.* **2018**, *54*, 1–10. [CrossRef]
184. Lim, W.-J.; Kim, K.H.; Kim, J.-Y.; Jeong, S.; Kim, N. Identification of DNA-Methylated CpG Islands Associated With Gene Silencing in the Adult Body Tissues of the Ogye Chicken Using RNA-Seq and Reduced Representation Bisulfite Sequencing. *Front. Genet.* **2019**, *10*, 346. [CrossRef]
185. Bannister, A.J.; Kouzarides, T. Regulation of chromatin by histone modifications. *Cell Res.* **2011**, *21*, 381–395. [CrossRef] [PubMed]
186. Moutinho, C.; Esteller, M. Chapter Seven—MicroRNAs and Epigenetics. In *Advances in Cancer Research*; Croce, C.M., Fisher, P.B., Eds.; Academic Press: Cambridge, MA, USA, 2017; Volume 135, pp. 189–220.
187. Bruce, K.D.; Cagampang, F.R. Epigenetic priming of the metabolic syndrome. *Toxicol. Mech. Methods* **2011**, *21*, 353–361. [CrossRef] [PubMed]
188. Fan, R.; You, M.; Toney, A.M.; Kim, J.; Giraud, D.; Xian, Y.; Ye, F.; Gu, L.; Ramer-Tait, A.E.; Chung, S. Red Raspberry Polyphenols Attenuate High-Fat Diet-Driven Activation of NLRP3 Inflammasome and its Paracrine Suppression of Adipogenesis via Histone Modifications. *Mol. Nutr. Food Res.* **2019**, *64*, e1900995. [CrossRef] [PubMed]
189. Nettore, I.C.; Rocca, C.; Mancino, G.; Albano, L.; Amelio, D.; Grande, F.; Puoci, F.; Pasqua, T.; Desiderio, S.; Mazza, R.; et al. Quercetin and its derivative Q2 modulate chromatin dynamics in adipogenesis and Q2 prevents obesity and metabolic disorders in rats. *J. Nutr. Biochem.* **2019**, *69*, 151–162. [CrossRef] [PubMed]
190. Boqué, N.; de la Iglesia, R.; de la Garza, A.L.; Milagro, F.I.; Olivares, M.; Bañuelos, O.; Soria, A.C.; Rodríguez-Sánchez, S.; Martínez, J.A.; Campión, J. Prevention of diet-induced obesity by apple polyphenols in Wistar rats through regulation of adipocyte gene expression and DNA methylation patterns. *Mol. Nutr. Food Res.* **2013**, *57*, 1473–1478. [CrossRef] [PubMed]
191. Crescenti, A.; Solà, R.; Valls, R.M.; Caimari, A.; Del Bas, J.M.; Anguera, A.; Anglés, N.; Arola, L. Cocoa Consumption Alters the Global DNA Methylation of Peripheral Leukocytes in Humans with Cardiovascular Disease Risk Factors: A Randomized Controlled Trial. *PLoS ONE* **2013**, *8*, e65744. [CrossRef]
192. Han, S.; Uludag, M.O.; Usanmaz, S.E.; Ayaloglu-Butun, F.; Akcali, K.C.; Demirel-Yilmaz, E. Resveratrol affects histone 3 lysine 27 methylation of vessels and blood biomarkers in DOCA salt-induced hypertension. *Mol. Biol. Rep.* **2015**, *42*, 35–42. [CrossRef]
193. Boesch-Saadatmandi, C.; Loboda, A.; Wagner, A.E.; Stachurska, A.; Jozkowicz, A.; Dulak, J.; Döring, F.; Wolffram, S.; Rimbach, G. Effect of quercetin and its metabolites isorhamnetin and quercetin-3-glucuronide on inflammatory gene expression: Role of miR-155. *J. Nutr. Biochem.* **2011**, *22*, 293–299. [CrossRef]
194. Tili, E.; Michaille, J.J.; Adair, B.; Alder, H.; Limagne, E.; Taccioli, C.; Ferracin, M.; Delmas, D.; Latruffe, N.; Croce, C.M. Resveratrol decreases the levels of miR-155 by upregulating miR-663, a microRNA targeting JunB and JunD. *Carcinogenesis* **2010**, *31*, 1561–1566. [CrossRef] [PubMed]
195. Rasheed, Z.; Rasheed, N.; Al-Shobaili, H.A. Epigallocatechin-3-O-gallate up-regulates microRNA-199a-3p expression by down-regulating the expression of cyclooxygenase-2 in stimulated human osteoarthritis chondrocytes. *J. Cell. Mol. Med.* **2016**, *20*, 2241–2248. [CrossRef]

196. Carpi, S.; Scoditti, E.; Massaro, M.; Polini, B.; Manera, C.; Digiacomo, M.; Esposito Salsano, J.; Poli, G.; Tuccinardi, T.; Doccini, S.; et al. The Extra-Virgin Olive Oil Polyphenols Oleocanthal and Oleacein Counteract Inflammation-Related Gene and miRNA Expression in Adipocytes by Attenuating NF-κB Activation. *Nutrients* **2019**, *11*, 2855. [CrossRef] [PubMed]
197. Scoditti, E.; Carpi, S.; Massaro, M.; Pellegrino, M.; Polini, B.; Carluccio, M.A.; Wabitsch, M.; Verri, T.; Nieri, P.; De Caterina, R. Hydroxytyrosol Modulates Adipocyte Gene and miRNA Expression Under Inflammatory Condition. *Nutrients* **2019**, *11*, 2493. [CrossRef] [PubMed]
198. Zaccaria, V.; Curti, V.; Di Lorenzo, A.; Baldi, A.; Maccario, C.; Sommatis, S.; Mocchi, R.; Daglia, M. Effect of Green and Brown Propolis Extracts on the Expression Levels of microRNAs, mRNAs and Proteins, Related to Oxidative Stress and Inflammation. *Nutrients* **2017**, *9*, 1090. [CrossRef]
199. Howell, J.C.; Chun, E.; Farrell, A.N.; Hur, E.Y.; Caroti, C.M.; Iuvone, P.M.; Haque, R. Global microRNA expression profiling: Curcumin (diferuloylmethane) alters oxidative stress-responsive microRNAs in human ARPE-19 cells. *Mol. Vis.* **2013**, *19*, 544–560.
200. Noratto, G.D.; Angel-Morales, G.; Talcott, S.T.; Mertens-Talcott, S.U. Polyphenolics from açaí (*Euterpe oleracea* Mart.) and red muscadine grape (*Vitis rotundifolia*) protect human umbilical vascular Endothelial cells (HUVEC) from glucose- and lipopolysaccharide (LPS)-induced inflammation and target microRNA-126. *J. Agric. Food Chem.* **2011**, *59*, 7999–8012. [CrossRef]
201. Song, J.; Jun, M.; Ahn, M.R.; Kim, O.Y. Involvement of miR-Let7A in inflammatory response and cell survival/apoptosis regulated by resveratrol in THP-1 macrophage. *Nutr. Res. Pract.* **2016**, *10*, 377–384. [CrossRef]
202. Boesch-Saadatmandi, C.; Wagner, A.E.; Wolffram, S.; Rimbach, G. Effect of quercetin on inflammatory gene expression in mice liver in vivo—Role of redox factor 1, miRNA-122 and miRNA-125b. *Pharmacol. Res.* **2012**, *65*, 523–530. [CrossRef]
203. Mohamed, H.E.; Abo-ELmatty, D.M.; Salah, S.M.; Sakr, A.T. Ameliorative effect of grapre seed extract on metabolic disorders caused by high fat diet induced obesity in rats by reversing the indrease inhepatic miR-33a and miR-122. *Afr. J. Pharm. Pharmacol.* **2016**, *10*, 699–708.
204. Zhao, X.J.; Yu, H.W.; Yang, Y.Z.; Wu, W.Y.; Chen, T.Y.; Jia, K.K.; Kang, L.L.; Jiao, R.Q.; Kong, L.D. Polydatin prevents fructose-induced liver inflammation and lipid deposition through increasing miR-200a to regulate Keap1/Nrf2 pathway. *Redox Biol.* **2018**, *18*, 124–137. [CrossRef] [PubMed]
205. Otton, R.; Bolin, A.P.; Ferreira, L.T.; Marinovic, M.P.; Rocha, A.L.S.; Mori, M.A. Polyphenol-rich green tea extract improves adipose tissue metabolism by down-regulating miR-335 expression and mitigating insulin resistance and inflammation. *J. Nutr. Biochem.* **2018**, *57*, 170–179. [CrossRef] [PubMed]
206. Gracia, A.; Miranda, J.; Fernández-Quintela, A.; Eseberri, I.; Garcia-Lacarte, M.; Milagro, F.I.; Martínez, J.A.; Aguirre, L.; Portillo, M.P. Involvement of miR-539-5p in the inhibition of de novo lipogenesis induced by resveratrol in white adipose tissue. *Food Funct.* **2016**, *7*, 1680–1688. [CrossRef] [PubMed]
207. Tomé-Carneiro, J.; Larrosa, M.; Yáñez-Gascón, M.J.; Dávalos, A.; Gil-Zamorano, J.; Gonzálvez, M.; García-Almagro, F.J.; Ruiz Ros, J.A.; Tomás-Barberán, F.A.; Espín, J.C.; et al. One-year supplementation with a grape extract containing resveratrol modulates inflammatory-related microRNAs and cytokines expression in peripheral blood mononuclear cells of type 2 diabetes and hypertensive patients with coronary artery disease. *Pharmacol. Res.* **2013**, *72*, 69–82. [CrossRef]
208. Glässer, G.; Graefe, E.U.; Struck, F.; Veit, M.; Gebhardt, R. Comparison of antioxidative capacities and inhibitory effects on cholesterol biosynthesis of quercetin and potential metabolites. *Phytomedicine* **2002**, *9*, 33–40. [CrossRef]
209. Merfort, I.; Heilmann, J.; Weiss, M.; Pietta, P.; Gardana, C. Radical scavenger activity of three flavonoid metabolites studied by inhibition of chemiluminescence in human PMNs. *Planta Med.* **1996**, *62*, 289–292. [CrossRef]
210. Tang, Y.; Nakashima, S.; Saiki, S.; Myoi, Y.; Abe, N.; Kuwazuru, S.; Zhu, B.; Ashida, H.; Murata, Y.; Nakamura, Y. 3,4-Dihydroxyphenylacetic acid is a predominant biologically-active catabolite of quercetin glycosides. *Food Res. Int.* **2016**, *89*, 716–723. [CrossRef]
211. Liu, Y.; Kurita, A.; Nakashima, S.; Zhu, B.; Munemasa, S.; Nakamura, T.; Murata, Y.; Nakamura, Y. 3,4-Dihydroxyphenylacetic acid is a potential aldehyde dehydrogenase inducer in murine hepatoma Hepa1c1c7 cells. *Biosci. Biotechnol. Biochem.* **2017**, *81*, 1978–1983. [CrossRef]

212. Monagas, M.; Khan, N.; Andres-Lacueva, C.; Urpi-Sarda, M.; Vazquez-Agell, M.; Lamuela-Raventos, R.M.; Estruch, R. Dihydroxylated phenolic acids derived from microbial metabolism reduce lipopolysaccharide-stimulated cytokine secretion by human peripheral blood mononuclear cells. *Br. J. Nutr.* **2009**, *102*, 201–206. [CrossRef]
213. Rechner, A.R.; Kroner, C. Anthocyanins and colonic metabolites of dietary polyphenols inhibit platelet function. *Thromb. Res.* **2005**, *116*, 327–334. [CrossRef] [PubMed]
214. Álvarez-Cilleros, D.; Ramos, S.; Goya, L.; Martín, M.Á. Colonic metabolites from flavanols stimulate nitric oxide production in human endothelial cells and protect against oxidative stress-induced toxicity and endothelial dysfunction. *Food Chem. Toxicol.* **2018**, *115*, 88–97. [CrossRef] [PubMed]
215. Fernandez-Millan, E.; Ramos, S.; Alvarez, C.; Bravo, L.; Goya, L.; Martin, M.A. Microbial phenolic metabolites improve glucose-stimulated insulin secretion and protect pancreatic beta cells against tert-butyl hydroperoxide-induced toxicity via ERKs and PKC pathways. *Food Chem. Toxicol.* **2014**, *66*, 245–253. [CrossRef] [PubMed]
216. Carrasco-Pozo, C.; Gotteland, M.; Castillo, R.L.; Chen, C. 3,4-Dihydroxyphenylacetic acid, a microbiota-derived metabolite of quercetin, protects against pancreatic beta-cells dysfunction induced by high cholesterol. *Exp. Cell Res.* **2015**, *334*, 270–282. [CrossRef] [PubMed]
217. Rechner, A.R.; Smith, M.A.; Kuhnle, G.; Gibson, G.R.; Debnam, E.S.; Srai, S.K.S.; Moore, K.P.; Rice-Evans, C.A. Colonic metabolism of dietary polyphenols: Influence of structure on microbial fermentation products. *Free Radic. Biol. Med.* **2004**, *36*, 212–225. [CrossRef] [PubMed]
218. Winter, J.; Moore, L.H.; Dowell, V.R., Jr.; Bokkenheuser, V.D. C-ring cleavage of flavonoids by human intestinal bacteria. *Appl. Environ. Microbiol.* **1989**, *55*, 1203–1208. [CrossRef]
219. Schneider, H.; Simmering, R.; Hartmann, L.; Pforte, H.; Blaut, M. Degradation of quercetin-3-glucoside in gnotobiotic rats associated with human intestinal bacteria. *J. Appl. Microbiol.* **2000**, *89*, 1027–1037. [CrossRef]
220. Feng, X.; Li, Y.; Brobbey Oppong, M.; Qiu, F. Insights into the intestinal bacterial metabolism of flavonoids and the bioactivities of their microbe-derived ring cleavage metabolites. *Drug Metab. Rev.* **2018**, *50*, 343–356. [CrossRef]
221. Hanske, L.; Loh, G.; Sczesny, S.; Blaut, M.; Braune, A. The Bioavailability of Apigenin-7-Glucoside Is Influenced by Human Intestinal Microbiota in Rats. *J. Nutr.* **2009**, *139*, 1095–1102. [CrossRef]
222. Labib, S.; Erb, A.; Kraus, M.; Wickert, T.; Richling, E. The pig caecum model: A suitable tool to study the intestinal metabolism of flavonoids. *Mol. Nutr. Food Res.* **2004**, *48*, 326–332. [CrossRef]
223. Zou, W.; Luo, Y.; Liu, M.; Chen, S.; Wang, S.; Nie, Y.; Cheng, G.; Su, W.; Zhang, K. Human intestinal microbial metabolism of naringin. *Eur. J. Drug Metab.Pharmacokinet.* **2015**, *40*, 363–367. [CrossRef]
224. Yuan, J.-P.; Wang, J.-H.; Liu, X. Metabolism of dietary soy isoflavones to equol by human intestinal microflora—Implications for health. *Mol. Nutr. Food Res.* **2007**, *51*, 765–781. [CrossRef] [PubMed]
225. Sekikawa, A.; Ihara, M.; Lopez, O.; Kakuta, C.; Lopresti, B.; Higashiyama, A.; Aizenstein, H.; Chang, Y.F.; Mathis, C.; Miyamoto, Y.; et al. Effect of S-equol and Soy Isoflavones on Heart and Brain. *Curr. Cardiol. Rev.* **2019**, *15*, 114–135. [CrossRef] [PubMed]
226. Choi, E.J.; Kim, G.H. The antioxidant activity of daidzein metabolites, O-desmethylangolensin and equol, in HepG2 cells. *Mol. Med. Rep.* **2014**, *9*, 328–332. [CrossRef] [PubMed]
227. Schneider, H.; Blaut, M. Anaerobic degradation of flavonoids by Eubacterium ramulus. *Arch. Microbiol.* **2000**, *173*, 71–75. [CrossRef] [PubMed]
228. Tzounis, X.; Vulevic, J.; Kuhnle, G.G.C.; George, T.; Leonczak, J.; Gibson, G.R.; Kwik-Uribe, C.; Spencer, J.P.E. Flavanol monomer-induced changes to the human faecal microflora. *Br. J. Nutr.* **2008**, *99*, 782–792. [CrossRef] [PubMed]
229. Mena, P.; Bresciani, L.; Brindani, N.; Ludwig, I.; Pereira Caro, G.; Angelino, D.; Llorach, R.; Calani, L.; Brighenti, F.; Clifford, M.; et al. Phenyl-γ-valerolactones and phenylvaleric acids, the main colonic metabolites of flavan-3-ols: Synthesis, analysis, bioavailability, and bioactivity. *Nat. Prod. Rep.* **2019**, *36*, 714–752. [CrossRef]
230. Ottaviani, J.I.; Borges, G.; Momma, T.Y.; Spencer, J.P.E.; Keen, C.L.; Crozier, A.; Schroeter, H. The metabolome of [2-14C](−)-epicatechin in humans: Implications for the assessment of efficacy, safety, and mechanisms of action of polyphenolic bioactives. *Sci. Rep.* **2016**, *6*, 29034. [CrossRef]

231. Mele, L.; Carobbio, S.; Brindani, N.; Curti, C.; Rodriguez-Cuenca, S.; Bidault, G.; Mena, P.; Zanotti, I.; Vacca, M.; Vidal-Puig, A.; et al. Phenyl-γ-valerolactones, flavan-3-ol colonic metabolites, protect brown adipocytes from oxidative stress without affecting their differentiation or function. *Mol. Nutr. Food Res.* **2017**, *61*, 1700074. [CrossRef]
232. Lambert, J.D.; Rice, J.E.; Hong, J.; Hou, Z.; Yang, C.S. Synthesis and biological activity of the tea catechin metabolites, M4 and M6 and their methoxy-derivatives. *Bioorg. Med. Chem. Lett.* **2005**, *15*, 873–876. [CrossRef]
233. Uhlenhut, K.; Högger, P. Facilitated cellular uptake and suppression of inducible nitric oxide synthase by a metabolite of maritime pine bark extract (Pycnogenol). *Free Radic. Biol. Med.* **2012**, *53*, 305–313. [CrossRef]
234. Lee, C.C.; Kim, J.H.; Kim, J.S.; Oh, Y.S.; Han, S.M.; Park, J.H.Y.; Lee, K.W.; Lee, C.Y. 5-(3′,4′-Dihydroxyphenyl-gamma-valerolactone), a Major Microbial Metabolite of Proanthocyanidin, Attenuates THP-1 Monocyte-Endothelial Adhesion. *Int. J. Mol. Sci.* **2017**, *18*, 1363. [CrossRef] [PubMed]
235. Takagaki, A.; Nanjo, F. Effects of Metabolites Produced from (−)-Epigallocatechin Gallate by Rat Intestinal Bacteria on Angiotensin I-Converting Enzyme Activity and Blood Pressure in Spontaneously Hypertensive Rats. *J. Agric. Food Chem.* **2015**, *63*, 8262–8266. [CrossRef] [PubMed]
236. Van Rymenant, E.; Grootaert, C.; Beerens, K.; Needs, P.W.; Kroon, P.A.; Kerimi, A.; Williamson, G.; Garcia-Villalba, R.; Gonzalez-Sarrias, A.; Tomas-Barberan, F.; et al. Vasorelaxant activity of twenty-one physiologically relevant (poly)phenolic metabolites on isolated mouse arteries. *Food Funct.* **2017**, *8*, 4331–4335. [CrossRef] [PubMed]
237. Miladinović, B.; Kostić, M.; Šavikin, K.; Đorđević, B.; Mihajilov-Krstev, T.; Živanović, S.; Kitić, D. Chemical Profile and Antioxidative and Antimicrobial Activity of Juices and Extracts of 4 Black Currants Varieties (*Ribes nigrum* L.). *J. Food Sci.* **2014**, *79*, C301–C309. [CrossRef]
238. Aura, A.M.; Martin-Lopez, P.; O'Leary, K.A.; Williamson, G.; Oksman-Caldentey, K.M.; Poutanen, K.; Santos-Buelga, C. In vitro metabolism of anthocyanins by human gut microflora. *Eur. J. Nutr.* **2005**, *44*, 133–142. [CrossRef]
239. Vitaglione, P.; Donnarumma, G.; Napolitano, A.; Galvano, F.; Gallo, A.; Scalfi, L.; Fogliano, V. Protocatechuic acid is the major human metabolite of cyanidin-glucosides. *J. Nutr.* **2007**, *137*, 2043–2048. [CrossRef] [PubMed]
240. Williamson, G.; Clifford, M.N. Colonic metabolites of berry polyphenols: The missing link to biological activity? *Br. J. Nutr.* **2010**, *104* (Suppl. S3), S48–S66. [CrossRef]
241. Gowd, V.; Bao, T.; Wang, L.; Huang, Y.; Chen, S.; Zheng, X.; Cui, S.; Chen, W. Antioxidant and antidiabetic activity of blackberry after gastrointestinal digestion and human gut microbiota fermentation. *Food Chem.* **2018**, *269*, 618–627. [CrossRef]
242. Ho, G.T.; Wangensteen, H.; Barsett, H. Elderberry and Elderflower Extracts, Phenolic Compounds, and Metabolites and Their Effect on Complement, RAW 264.7 Macrophages and Dendritic Cells. *Int. J. Mol. Sci.* **2017**, *18*, 584. [CrossRef]
243. Andreasen, M.F.; Kroon, P.A.; Williamson, G.; Garcia-Conesa, M.T. Esterase activity able to hydrolyze dietary antioxidant hydroxycinnamates is distributed along the intestine of mammals. *J. Agric. Food Chem.* **2001**, *49*, 5679–5684. [CrossRef]
244. Couteau, D.; McCartney, A.L.; Gibson, G.R.; Williamson, G.; Faulds, C.B. Isolation and characterization of human colonic bacteria able to hydrolyse chlorogenic acid. *J. Appl. Microbiol.* **2001**, *90*, 873–881. [CrossRef] [PubMed]
245. Cerdá, B.; Periago, P.; Espín, J.C.; Tomás-Barberán, F.A. Identification of Urolithin A as a Metabolite Produced by Human Colon Microflora from Ellagic Acid and Related Compounds. *J. Agric. Food Chem.* **2005**, *53*, 5571–5576. [CrossRef] [PubMed]
246. González-Barrio, R.; Borges, G.; Mullen, W.; Crozier, A. Bioavailability of anthocyanins and ellagitannins following consumption of raspberries by healthy humans and subjects with an ileostomy. *J. Agric. Food Chem.* **2010**, *58*, 3933–3939. [CrossRef]
247. Iino, T.; Tashima, K.; Umeda, M.; Ogawa, Y.; Takeeda, M.; Takata, K.; Takeuchi, K. Effect of ellagic acid on gastric damage induced in ischemic rat stomachs following ammonia or reperfusion. *Life Sci.* **2002**, *70*, 1139–1150. [CrossRef]
248. Yu, Y.M.; Wang, Z.H.; Liu, C.H.; Chen, C.S. Ellagic acid inhibits IL-1beta-induced cell adhesion molecule expression in human umbilical vein endothelial cells. *Br. J. Nutr.* **2007**, *97*, 692–698. [CrossRef]

249. Chang, W.C.; Yu, Y.M.; Chiang, S.Y.; Tseng, C.Y. Ellagic acid suppresses oxidised low-density lipoprotein-induced aortic smooth muscle cell proliferation: Studies on the activation of extracellular signal-regulated kinase 1/2 and proliferating cell nuclear antigen expression. *Br. J. Nutr.* **2008**, *99*, 709–714. [CrossRef]
250. Ishimoto, H.; Tai, A.; Yoshimura, M.; Amakura, Y.; Yoshida, T.; Hatano, T.; Ito, H. Antioxidative properties of functional polyphenols and their metabolites assessed by an ORAC assay. *Biosci. Biotechnol. Biochem.* **2012**, *76*, 395–399. [CrossRef]
251. Bialonska, D.; Kasimsetty, S.G.; Khan, S.I.; Ferreira, D. Urolithins, intestinal microbial metabolites of Pomegranate ellagitannins, exhibit potent antioxidant activity in a cell-based assay. *J. Agric. Food Chem.* **2009**, *57*, 10181–10186. [CrossRef]
252. Verzelloni, E.; Pellacani, C.; Tagliazucchi, D.; Tagliaferri, S.; Calani, L.; Costa, L.G.; Brighenti, F.; Borges, G.; Crozier, A.; Conte, A.; et al. Antiglycative and neuroprotective activity of colon-derived polyphenol catabolites. *Mol. Nutr. Food Res.* **2011**, *55* (Suppl. S1), S35–S43. [CrossRef]
253. Komatsu, W.; Kishi, H.; Yagasaki, K.; Ohhira, S. Urolithin A attenuates pro-inflammatory mediator production by suppressing PI3-K/Akt/NF-κB and JNK/AP-1 signaling pathways in lipopolysaccharide-stimulated RAW264 macrophages: Possible involvement of NADPH oxidase-derived reactive oxygen species. *Eur. J. Pharmacol.* **2018**, *833*, 411–424. [CrossRef]
254. Singh, R.; Chandrashekharappa, S.; Bodduluri, S.R.; Baby, B.V.; Hegde, B.; Kotla, N.G.; Hiwale, A.A.; Saiyed, T.; Patel, P.; Vijay-Kumar, M.; et al. Enhancement of the gut barrier integrity by a microbial metabolite through the Nrf2 pathway. *Nat. Commun.* **2019**, *10*, 89. [CrossRef] [PubMed]
255. Tang, L.; Mo, Y.; Li, Y.; Zhong, Y.; He, S.; Zhang, Y.; Tang, Y.; Fu, S.; Wang, X.; Chen, A. Urolithin A alleviates myocardial ischemia/reperfusion injury via PI3K/Akt pathway. *Biochem. Biophys. Res. Commun.* **2017**, *486*, 774–780. [CrossRef] [PubMed]
256. González-Sarrías, A.; Larrosa, M.; Tomás-Barberán, F.A.; Dolara, P.; Espín, J.C. NF-kappaB-dependent anti-inflammatory activity of urolithins, gut microbiota ellagic acid-derived metabolites, in human colonic fibroblasts. *Br. J. Nutr.* **2010**, *104*, 503–512. [CrossRef] [PubMed]
257. Giménez-Bastida, J.A.; González-Sarrías, A.; Larrosa, M.; Tomás-Barberán, F.; Espín, J.C.; García-Conesa, M.T. Ellagitannin metabolites, urolithin A glucuronide and its aglycone urolithin A, ameliorate TNF-α-induced inflammation and associated molecular markers in human aortic endothelial cells. *Mol. Nutr. Food Res.* **2012**, *56*, 784–796. [CrossRef] [PubMed]
258. Corsini, E.; Dell'Agli, M.; Facchi, A.; De Fabiani, E.; Lucchi, L.; Boraso, M.S.; Marinovich, M.; Galli, C.L. Enterodiol and enterolactone modulate the immune response by acting on nuclear factor-kappaB (NF-kappaB) signaling. *J. Agric. Food Chem.* **2010**, *58*, 6678–6684. [CrossRef] [PubMed]
259. Prasad, K. Antioxidant Activity of Secoisolariciresinol Diglucoside-derived Metabolites, Secoisolariciresinol, Enterodiol, and Enterolactone. *Int. J. Angiol.* **2000**, *9*, 220–225. [CrossRef] [PubMed]
260. Kitts, D.D.; Yuan, Y.V.; Wijewickreme, A.N.; Thompson, L.U. Antioxidant activity of the flaxseed lignan secoisolariciresinol diglycoside and its mammalian lignan metabolites enterodiol and enterolactone. *Mol. Cell. Biochem.* **1999**, *202*, 91–100. [CrossRef]
261. Heinonen, S.; Nurmi, T.; Liukkonen, K.; Poutanen, K.; Wähälä, K.; Deyama, T.; Nishibe, S.; Adlercreutz, H. In Vitro Metabolism of Plant Lignans: New Precursors of Mammalian Lignans Enterolactone and Enterodiol. *J. Agric. Food Chem.* **2001**, *49*, 3178–3186. [CrossRef]
262. Springer, M.; Moco, S. Resveratrol and Its Human Metabolites-Effects on Metabolic Health and Obesity. *Nutrients* **2019**, *11*, 143. [CrossRef]
263. Lu, D.L.; Ding, D.J.; Yan, W.J.; Li, R.R.; Dai, F.; Wang, Q.; Yu, S.S.; Li, Y.; Jin, X.L.; Zhou, B. Influence of glucuronidation and reduction modifications of resveratrol on its biological activities. *Chembiochem* **2013**, *14*, 1094–1104. [CrossRef]
264. Bode, L.M.; Bunzel, D.; Huch, M.; Cho, G.S.; Ruhland, D.; Bunzel, M.; Bub, A.; Franz, C.M.; Kulling, S.E. In vivo and in vitro metabolism of trans-resveratrol by human gut microbiota. *Am. J. Clin. Nutr.* **2013**, *97*, 295–309. [CrossRef] [PubMed]
265. Rodriguez-Mateos, A.; Heiss, C.; Borges, G.; Crozier, A. Berry (poly)phenols and cardiovascular health. *J. Agric. Food Chem.* **2014**, *62*, 3842–3851. [CrossRef] [PubMed]

266. Su, K.-Y.; Yu, C.Y.; Chen, Y.-P.; Hua, K.-F.; Chen, Y.-L.S. 3,4-Dihydroxytoluene, a metabolite of rutin, inhibits inflammatory responses in lipopolysaccharide-activated macrophages by reducing the activation of NF-κB signaling. *BMC Complement Altern. Med.* **2014**, *14*, 1–9. [CrossRef] [PubMed]

267. Ozdal, T.; Sela, D.A.; Xiao, J.; Boyacioglu, D.; Chen, F.; Capanoglu, E. The Reciprocal Interactions between Polyphenols and Gut Microbiota and Effects on Bioaccessibility. *Nutrients* **2016**, *8*, 78. [CrossRef] [PubMed]

268. Manach, C.; Williamson, G.; Morand, C.; Scalbert, A.; Remesy, C. Bioavailability and bioefficacy of polyphenols in humans. I. Review of 97 bioavailability studies. *Am. J. Clin. Nutr.* **2005**, *81*, 230S–242S. [CrossRef] [PubMed]

269. Rietjens, I.M.C.M.; Louisse, J.; Beekmann, K. The potential health effects of dietary phytoestrogens. *Br. J. Pharmacol.* **2017**, *174*, 1263–1280. [CrossRef]

270. Sánchez-Calvo, J.M.; Rodríguez-Iglesias, M.A.; Molinillo, J.M.G.; Macías, F.A. Soy isoflavones and their relationship with microflora: Beneficial effects on human health in equol producers. *Phytochem. Rev.* **2013**, *12*, 979–1000. [CrossRef]

271. Tousen, Y.; Uehara, M.; Abe, F.; Kimira, Y.; Ishimi, Y. Effects of short-term fructooligosaccharide intake on equol production in Japanese postmenopausal women consuming soy isoflavone supplements: A pilot study. *Nutr. J.* **2013**, *12*, 127. [CrossRef]

272. Setchell, K.D.; Brown, N.M.; Summer, S.; King, E.C.; Heubi, J.E.; Cole, S.; Guy, T.; Hokin, B. Dietary factors influence production of the soy isoflavone metabolite s-(−)equol in healthy adults. *J. Nutr.* **2013**, *143*, 1950–1958. [CrossRef]

273. Usui, T.; Tochiya, M.; Sasaki, Y.; Muranaka, K.; Yamakage, H.; Himeno, A.; Shimatsu, A.; Inaguma, A.; Ueno, T.; Uchiyama, S.; et al. Effects of natural S-equol supplements on overweight or obesity and metabolic syndrome in the Japanese, based on sex and equol status. *Clin. Endocrinol.* **2013**, *78*, 365–372. [CrossRef]

274. Yoshikata, R.; Myint, K.Z.Y.; Ohta, H. Effects of Equol Supplement on Bone and Cardiovascular Parameters in Middle-Aged Japanese Women: A Prospective Observational Study. *J. Altern. Complement. Med.* **2018**, *24*, 701–708. [CrossRef] [PubMed]

275. Birru, R.L.; Ahuja, V.; Vishnu, A.; Evans, R.W.; Miyamoto, Y.; Miura, K.; Usui, T.; Sekikawa, A. The impact of equol-producing status in modifying the effect of soya isoflavones on risk factors for CHD: A systematic review of randomised controlled trials. *J. Nutr. Sci.* **2016**, *5*, e30. [CrossRef] [PubMed]

276. Atkinson, C.; Frankenfeld, C.L.; Lampe, J.W. Gut bacterial metabolism of the soy isoflavone daidzein: Exploring the relevance to human health. *Exp. Biol. Med.* **2005**, *230*, 155–170. [CrossRef] [PubMed]

277. Atkinson, C.; Newton, K.M.; Bowles, E.J.; Yong, M.; Lampe, J.W. Demographic, anthropometric, and lifestyle factors and dietary intakes in relation to daidzein-metabolizing phenotypes among premenopausal women in the United States. *Am. J. Clin. Nutr.* **2008**, *87*, 679–687. [CrossRef]

278. Liu, Z.M.; Ho, S.C.; Chen, Y.M.; Liu, J.; Woo, J. Cardiovascular risks in relation to daidzein metabolizing phenotypes among Chinese postmenopausal women. *PLoS ONE* **2014**, *9*, e87861. [CrossRef]

279. Guo, K.; Zhang, B.; Chen, C.; Uchiyama, S.; Ueno, T.; Chen, Y.; Su, Y. Daidzein-metabolising phenotypes in relation to serum lipids and uric acid in adults in Guangzhou, China. *Br. J. Nutr.* **2010**, *104*, 118–124. [CrossRef]

280. Shi, L.; Ryan, H.H.; Jones, E.; Simas, T.A.M.; Lichtenstein, A.H.; Sun, Q.; Hayman, L.L. Urinary isoflavone concentrations are inversely associated with cardiometabolic risk markers in pregnant U.S. women. *J. Nutr.* **2014**, *144*, 344–351. [CrossRef]

281. Frankenfeld, C.L.; Atkinson, C.; Wähälä, K.; Lampe, J.W. Obesity prevalence in relation to gut microbial environments capable of producing equol or O-desmethylangolensin from the isoflavone daidzein. *Eur. J. Clin. Nutr.* **2014**, *68*, 526–530. [CrossRef]

282. Pérez-Jiménez, J.; Fezeu, L.; Touvier, M.; Arnault, N.; Manach, C.; Hercberg, S.; Galan, P.; Scalbert, A. Dietary intake of 337 polyphenols in French adults. *Am. J. Clin. Nutr.* **2011**, *93*, 1220–1228. [CrossRef]

283. Del Rio, D.; Rodriguez-Mateos, A.; Spencer, J.P.E.; Tognolini, M.; Borges, G.; Crozier, A. Dietary (poly)phenolics in human health: Structures, bioavailability, and evidence of protective effects against chronic diseases. *Antioxid. Redox Signal.* **2013**, *18*, 1818–1892. [CrossRef]

284. Sansone, R.; Rodriguez-Mateos, A.; Heuel, J.; Falk, D.; Schuler, D.; Wagstaff, R.; Kuhnle, G.G.C.; Spencer, J.P.E.; Schroeter, H.; Merx, M.W.; et al. Cocoa flavanol intake improves endothelial function and Framingham Risk Score in healthy men and women: A randomised, controlled, double-masked trial: The Flaviola Health Study. *Br. J. Nutr.* **2015**, *114*, 1246–1255. [CrossRef] [PubMed]

285. Stote, K.S.; Clevidence, B.A.; Novotny, J.A.; Henderson, T.; Radecki, S.V.; Baer, D.J. Effect of cocoa and green tea on biomarkers of glucose regulation, oxidative stress, inflammation and hemostasis in obese adults at risk for insulin resistance. *Eur. J. Clin. Nutr.* **2012**, *66*, 1153–1159. [CrossRef] [PubMed]
286. Castello, F.; Costabile, G.; Bresciani, L.; Tassotti, M.; Naviglio, D.; Luongo, D.; Ciciola, P.; Vitale, M.; Vetrani, C.; Galaverna, G.; et al. Bioavailability and pharmacokinetic profile of grape pomace phenolic compounds in humans. *Arch. Biochem. Biophys.* **2018**, *646*, 1–9. [CrossRef] [PubMed]
287. Hansson, G.K.; Libby, P. The immune response in atherosclerosis: A double-edged sword. *Nat. Rev. Immunol.* **2006**, *6*, 508–519. [CrossRef] [PubMed]
288. Rodriguez-Mateos, A.; Feliciano, R.P.; Boeres, A.; Weber, T.; Dos Santos, C.N.; Ventura, M.R.; Heiss, C. Cranberry (poly)phenol metabolites correlate with improvements in vascular function: A double-blind, randomized, controlled, dose-response, crossover study. *Mol. Nutr. Food Res.* **2016**, *60*, 2130–2140. [CrossRef]
289. Sasaki, G.Y.; Li, J.; Cichon, M.J.; Riedl, K.M.; Kopec, R.E.; Bruno, R.S. Green Tea Extract Treatment in Obese Mice with Nonalcoholic Steatohepatitis Restores the Hepatic Metabolome in Association with Limiting Endotoxemia-TLR4-NFκB-Mediated Inflammation. *Mol. Nutr. Food Res.* **2019**, *63*, e1900811. [CrossRef]
290. Panchal, S.K.; Ward, L.; Brown, L. Ellagic acid attenuates high-carbohydrate, high-fat diet-induced metabolic syndrome in rats. *Eur. J. Nutr.* **2013**, *52*, 559–568. [CrossRef]
291. Rani, U.P.; Kesavan, R.; Ganugula, R.; Avaneesh, T.; Kumar, U.P.; Reddy, G.B.; Dixit, M. Ellagic acid inhibits PDGF-BB-induced vascular smooth muscle cell proliferation and prevents atheroma formation in streptozotocin-induced diabetic rats. *J. Nutr. Biochem.* **2013**, *24*, 1830–1839. [CrossRef]
292. Tomás-Barberán, F.A.; García-Villalba, R.; González-Sarrías, A.; Selma, M.V.; Espín, J.C. Ellagic acid metabolism by human gut microbiota: Consistent observation of three urolithin phenotypes in intervention trials, independent of food source, age, and health status. *J. Agric. Food Chem.* **2014**, *62*, 6535–6538. [CrossRef]
293. Larrosa, M.; González-Sarrías, A.; Yáñez-Gascón, M.J.; Selma, M.V.; Azorín-Ortuño, M.; Toti, S.; Tomás-Barberán, F.; Dolara, P.; Espín, J.C. Anti-inflammatory properties of a pomegranate extract and its metabolite urolithin-A in a colitis rat model and the effect of colon inflammation on phenolic metabolism. *J. Nutr. Biochem.* **2010**, *21*, 717–725. [CrossRef]
294. Ishimoto, H.; Shibata, M.; Myojin, Y.; Ito, H.; Sugimoto, Y.; Tai, A.; Hatano, T. In vivo anti-inflammatory and antioxidant properties of ellagitannin metabolite urolithin A. *Bioorg. Med. Chem. Lett.* **2011**, *21*, 5901–5904. [CrossRef] [PubMed]
295. Kawai, Y. Immunochemical detection of food-derived polyphenols in the aorta: Macrophages as a major target underlying the anti-atherosclerotic activity of polyphenols. *Biosci. Biotechnol. Biochem.* **2011**, *75*, 609–617. [CrossRef] [PubMed]
296. Selma, M.V.; González-Sarrías, A.; Salas-Salvadó, J.; Andrés-Lacueva, C.; Alasalvar, C.; Örem, A.; Tomás-Barberán, F.A.; Espín, J.C. The gut microbiota metabolism of pomegranate or walnut ellagitannins yields two urolithin-metabotypes that correlate with cardiometabolic risk biomarkers: Comparison between normoweight, overweight-obesity and metabolic syndrome. *Clin. Nutr.* **2018**, *37*, 897–905. [CrossRef]
297. Gutiérrez-Díaz, I.; Fernández-Navarro, T.; Salazar, N.; Bartolomé, B.; Moreno-Arribas, M.V.; López, P.; Suárez, A.; de Los Reyes-Gavilán, C.G.; Gueimonde, M.; González, S. Could Fecal Phenylacetic and Phenylpropionic Acids Be Used as Indicators of Health Status? *J. Agric. Food Chem.* **2018**, *66*, 10438–10446. [CrossRef] [PubMed]
298. Yang, J.; Guo, Y.; Henning, S.M.; Chan, B.; Long, J.; Zhong, J.; Acin-Perez, R.; Petcherski, A.; Shirihai, O.; Heber, D.; et al. Ellagic Acid and Its Microbial Metabolite Urolithin A Alleviate Diet-Induced Insulin Resistance in Mice. *Mol. Nutr. Food Res.* **2020**, *64*, e2000091. [CrossRef]
299. Xia, B.; Shi, X.C.; Xie, B.C.; Zhu, M.Q.; Chen, Y.; Chu, X.Y.; Cai, G.H.; Liu, M.; Yang, S.Z.; Mitchell, G.A.; et al. Urolithin A exerts antiobesity effects through enhancing adipose tissue thermogenesis in mice. *PLoS Biol.* **2020**, *18*, e3000688. [CrossRef]
300. Adolphe, J.L.; Whiting, S.J.; Juurlink, B.H.; Thorpe, L.U.; Alcorn, J. Health effects with consumption of the flax lignan secoisolariciresinol diglucoside. *Br. J. Nutr.* **2010**, *103*, 929–938. [CrossRef]
301. Imran, M.; Ahmad, N.; Anjum, F.M.; Khan, M.K.; Mushtaq, Z.; Nadeem, M.; Hussain, S. Potential protective properties of flax lignan secoisolariciresinol diglucoside. *Nutr. J.* **2015**, *14*, 71. [CrossRef]
302. Zhang, W.; Wang, X.; Liu, Y.; Tian, H.; Flickinger, B.; Empie, M.W.; Sun, S.Z. Dietary flaxseed lignan extract lowers plasma cholesterol and glucose concentrations in hypercholesterolaemic subjects. *Br. J. Nutr.* **2008**, *99*, 1301–1309. [CrossRef]

303. Prasad, K. Secoisolariciresinol diglucoside from flaxseed delays the development of type 2 diabetes in Zucker rat. *J. Lab. Clin. Med.* **2001**, *138*, 32–39. [CrossRef]
304. Kottke, T.E.; Ogwang, Z.; Smith, J.C. Reasons for not meeting coronary artery disease targets of care in ambulatory practice. *Perm. J.* **2010**, *14*, 12–16. [CrossRef] [PubMed]
305. Durazzo, A.; Lucarini, M.; Camilli, E.; Marconi, S.; Gabrielli, P.; Lisciani, S.; Gambelli, L.; Aguzzi, A.; Novellino, E.; Santini, A.; et al. Dietary Lignans: Definition, Description and Research Trends in Databases Development. *Molecules* **2018**, *23*, 3251. [CrossRef]
306. Corona, G.; Kreimes, A.; Barone, M.; Turroni, S.; Brigidi, P.; Keleszade, E.; Costabile, A. Impact of lignans in oilseed mix on gut microbiome composition and enterolignan production in younger healthy and premenopausal women: An in vitro pilot study. *Microb. Cell Fact.* **2020**, *19*, 82. [CrossRef] [PubMed]
307. Brito, A.F.; Zang, Y. A Review of Lignan Metabolism, Milk Enterolactone Concentration, and Antioxidant Status of Dairy Cows Fed Flaxseed. *Molecules* **2018**, *24*, 41. [CrossRef] [PubMed]
308. Rienks, J.; Barbaresko, J.; Nöthlings, U. Association of Polyphenol Biomarkers with Cardiovascular Disease and Mortality Risk: A Systematic Review and Meta-Analysis of Observational Studies. *Nutrients* **2017**, *9*, 415. [CrossRef]
309. Sun, Q.; Wedick, N.M.; Pan, A.; Townsend, M.K.; Cassidy, A.; Franke, A.A.; Rimm, E.B.; Hu, F.B.; van Dam, R.M. Gut microbiota metabolites of dietary lignans and risk of type 2 diabetes: A prospective investigation in two cohorts of U.S. women. *Diabet. Care* **2014**, *37*, 1287–1295. [CrossRef]
310. Struja, T.; Richard, A.; Linseisen, J.; Eichholzer, M.; Rohrmann, S. The association between urinary phytoestrogen excretion and components of the metabolic syndrome in NHANES. *Eur. J. Nutr.* **2014**, *53*, 1371–1381. [CrossRef]
311. Peñalvo, J.L.; López-Romero, P. Urinary enterolignan concentrations are positively associated with serum HDL cholesterol and negatively associated with serum triglycerides in U.S. adults. *J. Nutr.* **2012**, *142*, 751–756. [CrossRef]
312. Baur, J.A.; Sinclair, D.A. Therapeutic potential of resveratrol: The in vivo evidence. *Nat. Rev. Drug Discov.* **2006**, *5*, 493–506. [CrossRef]
313. Alberdi, G.; Rodríguez, V.M.; Miranda, J.; Macarulla, M.T.; Churruca, I.; Portillo, M.P. Thermogenesis is involved in the body-fat lowering effects of resveratrol in rats. *Food Chem.* **2013**, *141*, 1530–1535. [CrossRef]
314. Szkudelski, T.; Szkudelska, K. Resveratrol and diabetes: From animal to human studies. *Biochim. Biophys. Acta* **2015**, *1852*, 1145–1154. [CrossRef] [PubMed]
315. Méndez-del Villar, M.; González-Ortiz, M.; Martínez-Abundis, E.; Pérez-Rubio, K.G.; Lizárraga-Valdez, R. Effect of resveratrol administration on metabolic syndrome, insulin sensitivity, and insulin secretion. *Metab. Syndr. Relat. Disord.* **2014**, *12*, 497–501. [CrossRef] [PubMed]
316. Timmers, S.; Konings, E.; Bilet, L.; Houtkooper, R.H.; van de Weijer, T.; Goossens, G.H.; Hoeks, J.; van der Krieken, S.; Ryu, D.; Kersten, S.; et al. Calorie restriction-like effects of 30 days of resveratrol supplementation on energy metabolism and metabolic profile in obese humans. *Cell Metab.* **2011**, *14*, 612–622. [CrossRef] [PubMed]
317. Javkhedkar, A.A.; Quiroz, Y.; Rodriguez-Iturbe, B.; Vaziri, N.D.; Lokhandwala, M.F.; Banday, A.A. Resveratrol restored Nrf2 function, reduced renal inflammation, and mitigated hypertension in spontaneously hypertensive rats. *Am. J. Physiol. Regul. Integr. Comp. Physiol.* **2015**, *308*, R840–R846. [CrossRef]
318. Dolinsky, V.W.; Chakrabarti, S.; Pereira, T.J.; Oka, T.; Levasseur, J.; Beker, D.; Zordoky, B.N.; Morton, J.S.; Nagendran, J.; Lopaschuk, G.D.; et al. Resveratrol prevents hypertension and cardiac hypertrophy in hypertensive rats and mice. *Biochim. Biophys. Acta* **2013**, *1832*, 1723–1733. [CrossRef] [PubMed]
319. Walker, J.M.; Eckardt, P.; Aleman, J.O.; da Rosa, J.C.; Liang, Y.; Iizumi, T.; Etheve, S.; Blaser, M.J.; Breslow, J.L.; Holt, P.R. The effects of trans-resveratrol on insulin resistance, inflammation, and microbiota in men with the metabolic syndrome: A pilot randomized, placebo-controlled clinical trial. *J. Clin. Transl. Res.* **2019**, *4*, 122–135. [PubMed]
320. Wenzel, E.; Soldo, T.; Erbersdobler, H.; Somoza, V. Bioactivity and metabolism of trans-resveratrol orally administered to Wistar rats. *Mol. Nutr. Food Res.* **2005**, *49*, 482–494. [CrossRef]
321. Boocock, D.J.; Patel, K.R.; Faust, G.E.; Normolle, D.P.; Marczylo, T.H.; Crowell, J.A.; Brenner, D.E.; Booth, T.D.; Gescher, A.; Steward, W.P. Quantitation of trans-resveratrol and detection of its metabolites in human plasma and urine by high performance liquid chromatography. *J. Chromatogr. B Analyt. Technol. Biomed. Life Sci.* **2007**, *848*, 182–187. [CrossRef]

322. Walle, T.; Hsieh, F.; DeLegge, M.H.; Oatis, J.E., Jr.; Walle, U.K. High absorption but very low bioavailability of oral resveratrol in humans. *Drug. Metab. Dispos.* **2004**, *32*, 1377–1382. [CrossRef]

323. Azorín-Ortuño, M.; Yáñez-Gascón, M.J.; Vallejo, F.; Pallarés, F.J.; Larrosa, M.; Lucas, R.; Morales, J.C.; Tomás-Barberán, F.A.; García-Conesa, M.T.; Espín, J.C. Metabolites and tissue distribution of resveratrol in the pig. *Mol. Nutr. Food Res.* **2011**, *55*, 1154–1168. [CrossRef]

antioxidants

MDPI

Article

Effects of Selenium- and Zinc-Enriched *Lactobacillus plantarum* SeZi on Antioxidant Capacities and Gut Microbiome in an ICR Mouse Model

Sini Kang [1], Rui Li [1], Hui Jin [1], Hyun Ju You [2,*] and Geun Eog Ji [1,3,*]

[1] Department of Food and Nutrition, Research Institute of Human Ecology, Seoul National University, Seoul 08826, Korea; kangsini@snu.ac.kr (S.K.); lirui@ribolia.com (R.L.); huijin1030@hotmail.com (H.J.)

[2] Institute of Health and Environment, Graduate School of Public Health, Seoul National University, Seoul 08826, Korea

[3] Research Center, BIFIDO Co., Ltd., Hongcheon 25117, Korea

* Correspondence: dhlover1@snu.ac.kr (H.J.Y.); geji@snu.ac.kr (G.E.J.); Tel.: +82-2-880-2790 (H.J.Y.); +82-2-880-6282 (G.E.J.)

Received: 28 September 2020; Accepted: 19 October 2020; Published: 21 October 2020

Abstract: Selenium and zinc are essential trace minerals for humans with various biological functions. In this study, selenium- and zinc-tolerant lactic acid bacteria (LAB) isolates were screened out from human fecal samples. Amongst three hundred LAB isolates, the *Lactobacillus plantarum* SeZi strain displayed the tolerance against selenium and zinc with the greatest biomass production and bioaccumulation of selenium and zinc. To further assess the characteristics of this strain, the lyophilized *L. plantarum* SeZi were prepared and administered to Institute of Cancer Research (ICR) mice. The mice were divided into four groups, provided with normal chow (Con), or normal chow supplemented with Na_2SeO_3 and $ZnSO_4{\cdot}7H_2O$ (SZ), *L. plantarum* SeZi (Lp), or selenium- and zinc-enriched *L. plantarum* SeZi (SZ + Lp), respectively. After 4 weeks of oral administration, the concentrations of selenium and zinc in blood were significantly increased in the SZ + Lp group when compared to the control or SZ group ($p < 0.05$). The increased selenium level led to an enhanced glutathione peroxidase activity and decreased blood malondialdehyde level in the SZ + Lp group ($p < 0.05$). Meanwhile, the results of bacterial community and microbial metabolic pathway analysis via 16S rRNA gene amplicon sequencing showed that *L. plantarum* SeZi significantly promoted the utilization of selenocysteine, seleno-cystathionine and seleno-methionine in the selenocompounds metabolism. Here, the in vivo antioxidant capacities of the selenium- and zinc-enriched lactobacillus strain showed us the utilization of a unique probiotic as a Se/Zn supplement with high availability, low toxicity, and additional probiotic advantages.

Keywords: selenium; zinc; bioaccumulation; antioxidant capacities; gut microbiota

1. Introduction

Micronutrient deficiencies, known as "hidden hunger", have affected more than 50% of the world's population [1]. Selenium (Se) is a vital trace element, contributing to modulation of growth, regulation of antiviral capacity, and prevention of disease, especially cancer and cardiovascular disease [2–4]. The antioxidant activity of selenium is exhibited as a form of selenoenzymes, including glutathione peroxidase (GSH-Px), selenoprotein P, thioredoxin reductase, and methionine sulfoxide reductase [5]. Selenium deficiency can trigger serious health issues such as poor growth, muscle pain, decreased immune responses, and hypofunction of glandula thyreoidea [6,7]. Besides, the deficit of selenium is associated with a cardiomyopathy named as Keshan disease (KD) and an osteoarthropathy named as Kashin-Beck disease (KBD) [8]. Although KD and KBD are just local problems primarily in China and east Serbia due to environmentally low selenium [9], the hypothyreose or insuline secretion impairments associated with the lack of selenium have a global impact. Zinc (Zn) is another essential micronutrient for humans. It is a key component of many metalloenzymes (i.e., superoxide dismutase (SOD), carbonic anhydrase, alcohol dehydrogenase) associated with human growth, immunity, fertility, and reproduction [10]. Additionally, zinc is significant for the correct secretion of hormone isoline by pancreas [11]. The chronic deficiency of zinc can lead to glucose intolerance and pre-diabetic syndromes [12,13]. On the other hand, zinc chronic overdose, which can be caused by some nutraceuticals, might be responsible for some neurodegenerations such as nervus opticus inflammation [14,15].

Inorganic selenium, such as selenate (SeO_4^{2-}) and selenite (SeO_3^{2-}), are toxic and poorly bioavailable [16]. The reduction of selenium oxyanions largely depends on biotic processes by microorganisms [17]. The utilization of microorganisms as the natural adsorbent for metal ions (i.e., selenium and zinc) is eco-friendly and cost-effective [18]. The bioabsorption capacities are attributed to their intrinsic biochemical and structural properties of the cellular membrane [19]. Lactic acid bacteria (LAB), as important food-grade bacteria with probiotic advantages, have been extensively studied in this field. The selenium concentration in the medium is highly linked to bacterial selenium level, but the growth of most bacterial isolates from the human gut can be inhibited by the addition of inorganic selenium into the medium [20]. However, some LAB strains have been reported to be capable of resisting selenium oxyanions at high concentrations during cultivation [21–23]. Especially, *Lactobacillus plantarum* has been suggested as Se-enriched lactobacilli for food applications [24]. Few studies about Zn-enriched LAB have been conducted, but it has been found that the bacterial growth and probiotic effect of *L. plantarum* can be enhanced by zinc in the gut [25].

Although the resistances of LAB to selenium and zinc have been reported, the in vivo antioxidant capacities of the SeZn-enriched probiotic products have not been reported. In this study, SeZn-tolerant LAB strains isolated from human feces were screened out to further investigate the effects on selenium and zinc bioaccumulation and related metabolism, antioxidant activities, and compositional changes of intestinal microbiota in vivo in an Institute of Cancer Research (ICR) mouse model.

2. Materials and Methods

2.1. Isolation of Probiotic Strains from Human Feces

According to the protocol approved by the Institutional Review Board of Seoul National University (IRB No. 1702/002-013), fresh fecal samples were obtained from five children (1–6 years old) in Korea and stored at 4 °C during transportation. Each fecal sample (1 g) was serially diluted with a sterilized phosphate buffered saline solution (pH 7.4). The suspension was plated on *Lactobacillus* Selection (LBS) agar (Difco, Sparks, MD, USA) to isolate *Lactobacillus* spp. The plates were incubated anaerobically at 37 °C for 48 h [26]. Three hundreds of morphologically different microbial colonies were collected and cultured for further tests. The isolated LAB strains were then cultured in De Man, Rogosa and Sharpe (MRS) medium (Becton Dickinson, Cockeysville, MD, USA) containing 0.05% L-cysteine hydrochloride anaerobically at 37 °C. The bacterial stocks were stored at −80 °C with 17% glycerol utilized as a cryoprotectant.

2.2. Screening of SeZn-Tolerant LAB Isolates from Human Feces

To identify selenium-tolerant strains, the isolates were plated on the MRS agar in the presence of 60 mM Na_2SeO_3 (Sigma-Aldrich, St. Louis, MO, USA) at 37 °C for 24 h under anaerobic condition. When the concentration of selenium in the medium is high, strains convert inorganic selenium into element of selenium (red color) in the medium [27]. Thus, the strains with selenium resistance were selected based on the results of bacteria growth and color changes.

Thereafter, the screened strains were further tested for zinc-tolerant abilities by culturing in the MRS agar with 100 mM $ZnSO_4{\cdot}7H_2O$ (Sigma-Aldrich) at 37 °C for 24 h under anaerobic condition. Strains with strong zinc tolerance were selected by observing the bacterial growth. The final screened SeZn-tolerant bacteria were identified by phylogenetic analysis of 16S rRNA gene sequence.

2.3. *Assessing Bioaccumulation of Selenium and Zinc in LAB Strains During the Cultivation*

Considering the application for food and feed additive, the initial concentrations of Na_2SeO_3 and $ZnSO_4{\cdot}7H_2O$ were set at 0.01 mM and 3.5 mM, respectively. After 24 h anaerobic culture in MRS broth, the LAB strains were centrifuged (15,600× *g*, 5 min) to gain pellets. The bacterial pellets were washed three times with phosphate buffered saline (PBS) and frozen at −80 °C for lyophilization. One liter of PBS buffer (pH 7.4) was prepared by dissolving NaCl (8 g), KCl (200 mg), Na_2HPO_4 (1.44 g) and KH_2PO_4 (245 mg) in the distilled water and autoclaved at 121 °C for 15 min.

The concentrations of zinc and selenium in the bacterial biomass were measured using an inductively coupled plasma-atomic emission spectrometer (ICP-AES, Optima-4300 DV, Perkin Elmer, Waltham, MA, USA). The lyophilized sample (400 mg) was digested with HNO_3 (5 mL) and $HClO_4$ (0.5 mL) by heating in a Multiwave 3000 microwave. After the cool-down to room temperature, the solution was diluted with deionized water to reach a final volume of 20 mL, and mineral levels were assessed by the Inductively coupled plasma atomic emission (ICP-AES). The bioconversion rates of Se and Zn were calculated by dividing the Se or Zn content in dry cell mass by the total Se or Zn content added in the broth. The strain with the highest levels of selenium and zinc bioaccumulation was selected as the experimental strain for in vivo study.

$$\text{Bioconversion rate of Se } (\%) = (\text{Se content in dry cell mass/total Se content added in broth}) \times 100\%$$

$$\text{Bioconversion rate of Zn } (\%) = (\text{Zn content in dry cell mass/total Zn content added in broth}) \times 100\%$$

2.4. *Gene Analysis of Se/Zn Uptake and Resistance in L. plantarum SeZi*

The genomic DNA of pure cultured *L. plantarum* SeZi isolate was extracted by using MG™ Cell Genomic DNA Extraction SV kit (MGmed, Seoul, Korea), following the manufacturer's instructions. Whole genome sequencing was carried out by using a Nextera XT Library Preparation kit (Illumina, San Diego, CA, USA) and sequenced at a read length of 300 bp with paired-end library via an Illumina MiSeq sequencer (Illumina, San Diego, CA, USA). The Illumina sequencing raw data in the FASTQ format were assembled with SPAdes 3.9.0. Gene-finding and functional annotation pipeline of whole genome assemblies used in EzBioCloud genome database (http://www.ezbiocloud.net, ChunLab Co., Ltd., Seoul, Korea) [28]. The tRNA genes were investigated via tRNAscan-SE 1.3.1 [29]. The rRNA and other non-coding RNAs were explored by using a Rfam covariance model version 12.0 [30]. Protein-coding sequences (CDSs) were predicted via Prodigal 2.6.2 [31], and classified into different functional groups (EggNOG 4.5; http://eggnogdb.embl.de) [32]. In order to obtain more functional annotation, the UBLAST program [33] was utilized to search and compare the predicted CDSs in the protein databases, including Swissprot [34], Kyoto Encyclopedia of Genes and Genomes (KEGG) [35] and subsystems-based annotations (SEED) [36]. The comparative genomics analysis was conducted by using the genome sequences of closely related *Lactobacillus plantarum* strains from the EzBioCloud database and analyzed via ChunLab's comparative genomics tool (http://www.ezbiocloud.net/contents/cg).

2.5. *Effect of SeZn-Enriched L. plantarum SeZi in an ICR Mouse Model*

2.5.1. Preparation of SeZn-Enriched *L. plantarum* SeZi for Mouse Study

The *L. plantarum* SeZi strain was anaerobically grown at 37 °C for 24 h in MRS medium with the addition of 0.01 mM Na_2SeO_3 and 3.5 mM $ZnSO_4{\cdot}7H_2O$. For harvesting probiotic powder, the bacterial pellets were collected by centrifugation (15,600× *g*, 5 min) after SeZn enrichment, thoroughly washed with the PBS buffer, and frozen at −80 °C for lyophilization.

2.5.2. Animals and Diets

Seven-week old male ICR mice were purchased from Central Lab Animal (Seoul, Korea). The animal breeding environment was adjusted to a dark cycle of 12 h light/12 h dark at a temperature of 23 ± 1 °C and a humidity of 40–60%. The mice were acclimatized in the laboratory room for one week and then randomly divided into four groups (n = 8/group). The control group was provided with a normal chow diet AIN-93G purchased from Doo Yeol Biotech (Seoul, Korea). The treatment groups (SZ, Lp, SZ + Lp) were fed the same normal chow diet mixed with Na_2SeO_3 (1.2 μg/g Se^{4+}) and $ZnSO_4{\cdot}7H_2O$ (5 μg/g Zn^{2+}), 10^{12} CFU/mouse *Lactobacillus plantarum* SeZi, and 10^{12} CFU/mouse SeZn-enriched *L. plantarum* SeZi, respectively. The daily administration was conducted for 4 weeks. The protocols and facilities utilized in this animal experiment were approved by the Institutional Animal Care and Use Committee of Seoul National University (SNU-180403-2-2).

2.5.3. Blood Analysis

To assess Se and Zn concentrations and oxidative stress-related parameters in mouse blood, blood samples were collected into a 1.5 mL heparinized tube from the mouse heart by cardiac puncture. Approximately 0.9 mL of the blood samples were centrifuged (2500 rpm, 10 min) to separate serum. The serum and whole blood samples were stored at −80 °C. The concentrations of selenium and zinc in the whole blood were measured using ICP-AES. GSH-Px activity, SOD activity, and malondialdehyde (MDA) level in serum were assessed via antioxidant enzyme detection kits purchased from Jiancheng Bioengineering Institute, Nanjing, China.

2.5.4. Bacterial Community Analysis by 16S rRNA Gene Amplicon Sequencing

Fecal DNA was extracted using a QIAamp DNA Stool Mini Kit (Qiagen, Manchester, UK). The V3-V4 hypervariable regions of the 16S rRNA genes in the stool DNA samples were targeted and amplified using interest-specific primers. A pooled library was constituted by attaching specific barcode sequences to the 16S rRNA amplicons. The denatured and diluted pooled library and PhiX control (Phix control v3, 30%, *v*/*v*) library were mixed and loaded onto a MiSeq v2 (500 cycle) reagent cartridge (Illumina, San Diego, CA, USA). The primers and methods were as described in our previous study [37]. After the metagenomic sequencing, paired-end FASTQ files were collected and imported into Quantitative Insights Into Microbial Ecology 2 (QIIME2) (ver. 2020.6, https://qiime2.org) for analysis. Operational taxonomic unit (OTU) taxonomy and related analysis were performed using QIIME2 as described in our previous study [37]. KEGG associated with selenocompounds metabolism pathways were assessed by conducting phylogenetic investigation of the community by reconstruction of unobserved states (PICRUSt) with the entire picrust2 pipeline command [38].

2.6. Statistic Analysis

Differential abundance analyses were performed by non-parametric one-way analysis of variance (ANOVA) using the Kruskal-Wallis test, or non-parametric t-test with Mann-Whitney test. Other analyses were conducted by one-way ANOVA with Tukey's multiple comparisons test or paired t-test analysis. All statistical analyses were carried out via Graph-Pad Prism 8. Statistically significant difference was accepted at $p < 0.05$.

3. Results

3.1. Screening and Selection of SeZn-Tolerant LAB Strains

Amongst the three hundreds of isolated strains, only four LAB species grew in the presence of 60 mM selenite, including *L. plantarum*, *L. pentosus*, *L. fermentum*, and *L. rhamnosus*. All of these four strains are able to resist 100 mM $ZnSO_4{\cdot}7H_2O$. Among these four SeZn-tolerant strains, *L. plantarum* SeZi yielded the greatest dry cell mass with the best selenium and zinc bioaccumulation capability as shown in Table 1. Based on the results of whole genome sequencing analysis in Tables 2 and 3, the genes coding DedA and CysA proteins related to Na_2SeO_3 uptake and detoxification, and zinc uptake regulation protein ZUR (zinc uptake regulation) and zinc resistance protein MerR (Mercury resistance) were found in *L. plantarum* SeZi genome. Thus, *L. plantarum* SeZi was selected for in vivo mouse study. The abundances of these microbial selenium/zinc metabolism-related genes in genomes from other LAB strains were investigated using publicly deposited genome databases (NCBI genome datasets, https://www.ncbi.nlm.nih.gov/genome/). Among the 1912 genome assemblies available, only a few *Lactobacillus* genomes contain genes encoding DedA, CysA, ZUR, or MerR proteins (Table 4).

Table 1. Generation of biomass and bioconvertion rates of selenium and zinc in the selected SeZn-tolerant probiotic strains in vitro.

Strains	Biomass (g/L)	Bioconversion Rate (%)	
		Selenium	Zinc
Lactobacillus plantarum SeZi	2.82 ± 0.28 [a]	19.47	0.35
Lactobacillus pentosus SeZi	2.33 ± 0.11 [ab]	8.93	0.36
Lactobacillus fermentum SeZi	1.94 ± 0.34 [bc]	18.04	0.33
Lactobacillus rhamnosus SeZi	1.78 ± 0.34 [c]	6.90	0.20

Data are expressed as mean ± SD ($n = 5$). Treatments with different letters are significantly different at $p < 0.05$ ($n = 5$).

Table 2. Gene products of selenium resistance gene cluster in *Lactobacillus plantarum* SeZi.

Coding Region [a]	Length (aa)	Product	Function
264974–265621 (−)	648	DedA protein	Detoxification and uptake of selenate
193218–193874 (−)	657	DedA protein	Detoxification and uptake of selenate
19823–20716	894	Sulfate permease-CysA	Detoxification and uptake of selenate

[a] Genes encoded on the minus strand are indicated with (−).

Table 3. Gene products of zinc resistance gene cluster in *Lactobacillus plantarum* SeZi.

Coding Region [a]	Length (aa)	Product	Function
237659–238102	444	ZUR	Zinc uptake regulation
151–966	816	Multidrug efflux transporter 1 regulator	Zinc resistance
31131–31565	435	Uncharacterized HTH-type transcriptional regulator	Zinc resistance
174455–174916	462	MerR family	Zinc resistance
22703–23146	444	MerR family	Zinc resistance
115660–116040	381	MerR family	Zinc resistance
27565–27957 (−)	393	MerR family	Zinc resistance
4022–4399	378	MerR family	Zinc resistance
227–619 (−)	393	MerR family	Zinc resistance
342254–342718	465	Hypothetical protein	Zinc resistance
54502–54756 (−)	255	Hypothetical protein	Zinc resistance
54351–54800	450	Hypothetical protein	Zinc resistance

[a] Genes encoded on the minus strand are indicated with (−). ZUR: zinc uptake regulation.

Table 4. Other *Lactobacillus* strains with selenium and/or zinc resistance gene clusters. (From NCBI genome databases).

DedA	CysA	ZUR	MerR Family
L. acidophilus La-14	-	*L.rhamnosus* LOCK908	*L. acidophilus* La-14
L. gasseri ATCC 33323		*L.rhamnosus* LOCK900	*L. curvatus* JCM 1096
L. dekbrueckii subsp. *delbrueckii*		*L.rhamnosus* LOCK919	*L. gasseri* ATCC 33323
L. salivarius str. Ren			*L. ruminis* ATCC 25644
L. buchneri subsp. *silagei* CD034			*L. buchneri* subsp. *silagei* CD034
			L. rhamnosus GG
			L. paracasei subsp. *paracasei* 8700:2

3.2. Increased Concentrations of Selenium and Zinc in Blood after L. plantarum SeZi Administration

The selenium and zinc contents in mouse blood are presented in Figure 1. The blood selenium and zinc levels in the SZ + Lp group were significantly higher than that of the control group and the SZ group, respectively ($p < 0.05$). The differences between the control, SZ and Lp groups were not significant ($p > 0.05$).

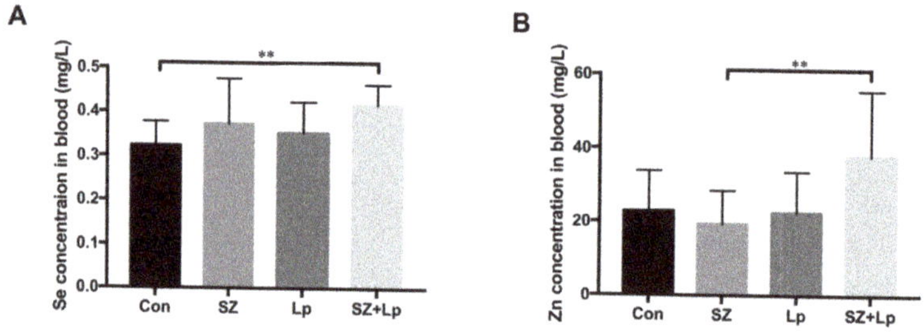

Figure 1. The concentrations of blood selenium (**A**) and zinc (**B**) in mice after 4 weeks of administration. Data were analyzed by unpaired t-test analysis and expressed as mean ± SD. ** $p < 0.01$ ($n = 8$). Con, control; SZ, selenium and zinc supplemented; Lp, *Lactobacillus plantarum* SeZi; SZ + Lp, selenium- and zinc-enriched *L. plantarum* SeZi.

3.3. Increased Antioxidant Activities in Mice after L. plantarum SeZi Administration

GSH-Px and SOD are imperative antioxidant defenses against oxidative stress [39,40], and MDA is the most commonly utilized biomarker of oxidative stress [41]. As shown in Figure 2, the GSH-Px activity was highest in the SZ + Lp group, followed by the SZ group when compared with other groups ($p < 0.05$). The activity of SOD was significantly increased in the SZ + Lp group compared to the Lp group ($p < 0.05$). Meanwhile, significant decreases in the MDA level were observed in the SZ group and SZ + Lp group compared to other two groups ($p < 0.05$).

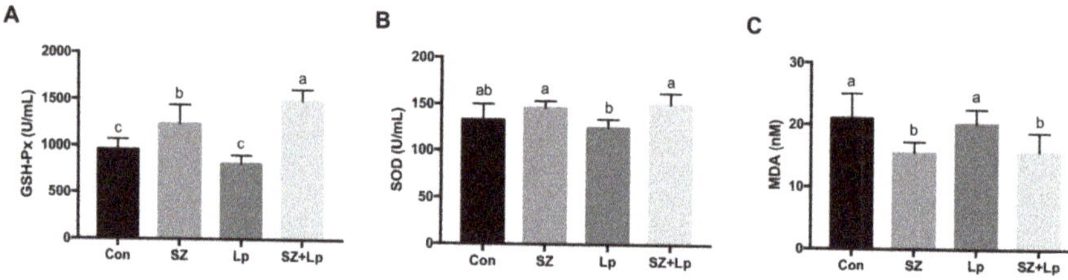

Figure 2. Glutathione peroxidase (GSH-Px) activity (**A**), superoxide dismutase (SOD) activity (**B**), and lipid oxidation product malondialdehyde (MDA) level (**C**) in serum of mice at the final day. Treatments with different letters (a, b, c) are significantly different at $p < 0.05$ ($n = 8$). Con, control; SZ, selenium and zinc supplemented; Lp, *Lactobacillus plantarum* SeZi; SZ + Lp, selenium- and zinc-enriched *L. plantarum* SeZi.

3.4. Changes in the Gut Microbiota after L. plantarum SeZi Administration

Gut microbiota alpha diversities were assessed by richness (Faith-pd) and Pielou's evenness analyses, which represent the number of species and the degree of species homogeneity, respectively. No significant difference was observed between the groups in the alpha diversity of richness (Figure 3A), while the evenness in the SZ group was significantly larger than Lp and SZ + Lp groups (Figure 3B). The results of beta diversity (Bray-Curtis dissimilarity) indicated that the clustering in microbial communities in the Lp and SZ + Lp group was distinct from that in the control and SZ group (Figure 3C).

The average relative abundances of the final day fecal samples at the phylum level (Figure 3D) and the genus level (Figure 3E) suggested the different microbial compositions amongst the groups after the oral administration of SZ and SZ + Lp. To further evaluate the effects of treatments on microbial compositional changes, the three significantly different genera between the groups were identified and are displayed in Figure 4A–C. The relative abundance of *Lactobacillus* in the SZ + Lp group was significantly higher than that of the control and SZ groups, and the *Lactobacillus* level in the Lp group was also significantly higher than the SZ group. *Adlercreutzia* was significantly abundant in the SZ group compared to the SZ and SZ + Lp groups. Interestingly, *Lactococcus* was highly enriched only in the SZ group with the relative abundance at 4.15%. In addition, the relative abundance of *Allobaculum* in the SZ group was much larger than the SZ + Lp group, although the difference was not statistically significant (Figure 4D, $p > 0.05$).

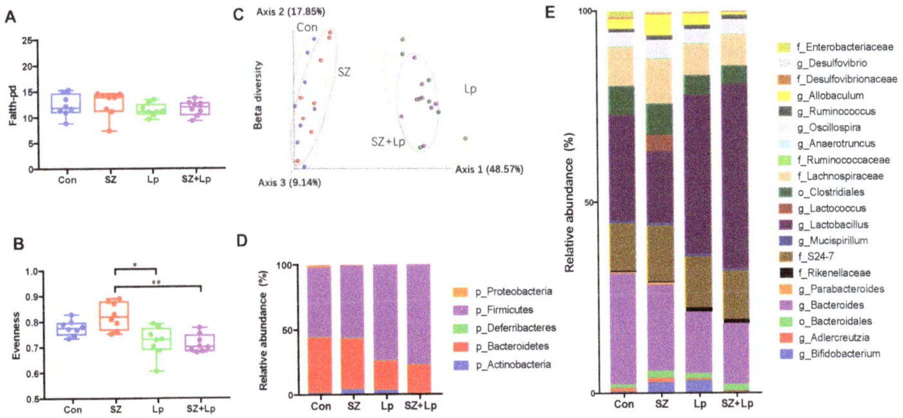

Figure 3. Comparison of diversity indices and microbial compositions amongst groups after 4 weeks of oral administration. Alpha-diversities of microbial communities are shown as (**A**) richness and (**B**) evenness. (**C**) Principal Coordinates Analysis (PCoA) plot represents beta-diversity based on Bray-Curtis dissimilarity. (**D**) Taxonomic profiles at the phylum level (**D**) and the genus level (**E**). Relative Abundance of taxa below 0.1% were excluded prior to analyses. Significance was accepted at * $p < 0.05$, ** $p < 0.01$ ($n = 8$). Con, control; SZ, selenium and zinc supplemented; Lp, *Lactobacillus plantarum* SeZi; SZ + Lp, selenium- and zinc-enriched *L. plantarum* SeZi.

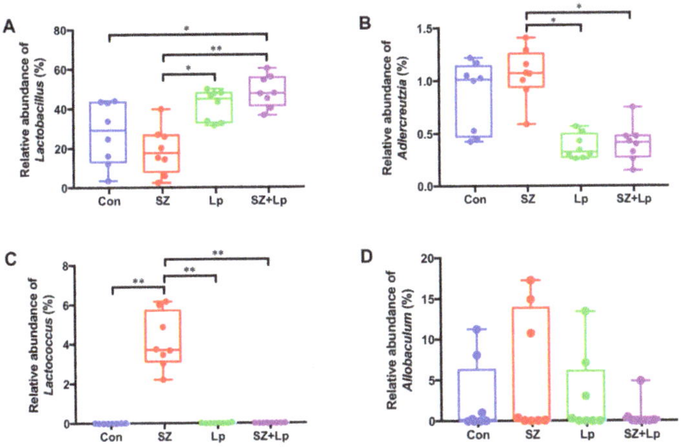

Figure 4. Relative abundances of *Lactobacillus* (**A**), *Adlercreutzia* (**B**), *Lactococcus* (**C**), and *Allobaculum* (**D**) in fecal samples after 4 weeks of oral administration. Data are expressed as mean ± SD. Significance was accepted at * $p < 0.05$, ** $p < 0.01$ ($n = 8$). Con, control; SZ, selenium and zinc supplemented; Lp, *Lactobacillus plantarum* SeZi; SZ + Lp, selenium- and zinc-enriched *L. plantarum* SeZi.

3.5. Microbial Function Analysis Related to Selenocompounds Metabolism

To investigate the functional changes in selenocompounds metabolism of the gut microbiome, KEGG analysis was performed by phylogenetic investigation of the community by reconstruction of unobserved states (PICRUSt).

As presented in Figure 5, *L. plantarum* SeZi significantly increased the relative abundances of the SCLY gene coding for selenocysteine lyase (EC: 4.4.1.16), CCBL gene coding for cysteine-S-conjugate beta-lyase (EC: 4.4.1.13), and MARS gene coding for methionyl-tRNA synthetase (EC: 6.1.1.10). These selenocompounds metabolism-related genes were responsible for the utilization of selenocysteine, seleno-cystathionine and seleno-methionine, respectively.

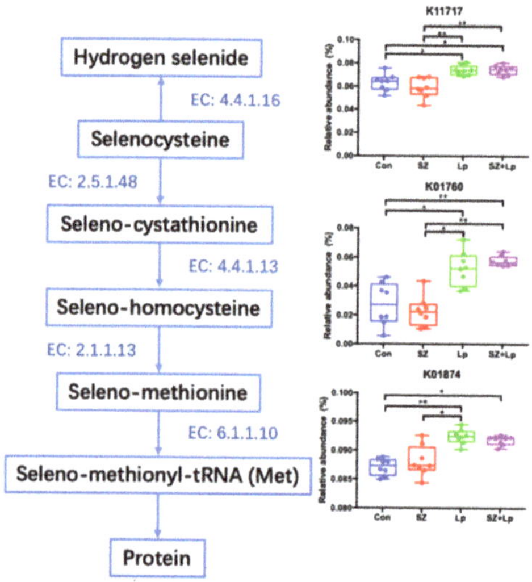

Figure 5. Selenocompounds metabolism pathway. Related enzymes include selenocysteine lyase (EC: 4.4.1.16), cystathionine gamma-synthase (EC: 2.5.1.48), cysteine-S-conjugate beta-lyase (EC: 4.4.1.13), homocysteine methyltransferase (EC: 2.1.1.13), and methionyl-tRNA synthetase (EC: 6.1.1.10). Arrows indicate the related genes that are involved in the corresponding pathway. Error bars represent means ± SD. Significance was accepted at * $p < 0.05$, ** $p < 0.01$ ($n = 8$). Con, control; SZ, selenium and zinc supplemented; Lp, *Lactobacillus plantarum* SeZi; SZ + Lp, selenium- and zinc-enriched *L. plantarum* SeZi. K11717, cysteine desulfurase/selenocysteine lyase; K01760, cystathionine beta-lyase; K01874, methionyl-tRNA synthetase.

To detoxify selenite and selenate in the selenocompounds metabolism pathway, selenite is converted to selenate directly or via an intermediate, and selenate is further metabolized into hydrogen selenide. As shown in Figure 6, the relative abundances of enzymes related to oxidation of selenite to selenate (EC: 2.7.7.4, EC: 1.97.1.9) were significantly reduced in the Lp and SZ + Lp groups.

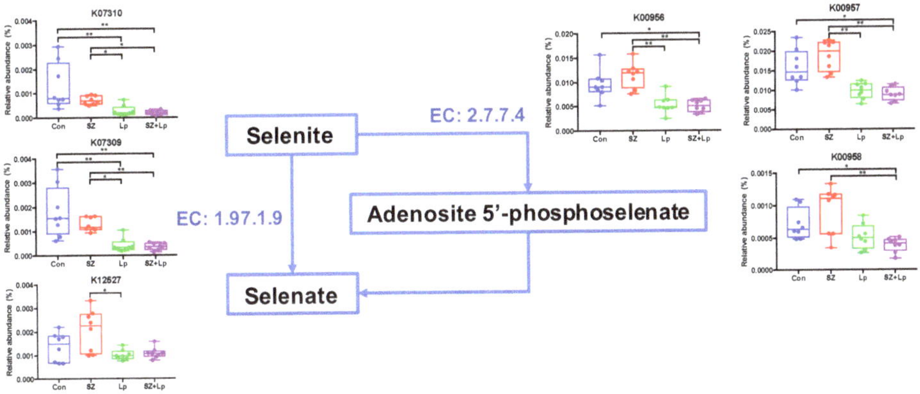

Figure 6. Detoxification process of inorganic selenium. Related enzymes include 3′-phosphoadenosine 5′-phosphosulfate synthase (EC: 2.7.7.4) and selenate reductase subunit alpha (EC: 1.97.1.9). Arrows indicate the related genes that are involved in the corresponding pathway. Error bars represent means ± SD. Significance was accepted at * $p < 0.05$, ** $p < 0.01$ ($n = 8$). Con, control; SZ, selenium and zinc supplemented; Lp, *Lactobacillus plantarum* SeZi; SZ + Lp, selenium- and zinc-enriched *L. plantarum* SeZi. K07310, Tat-targeted selenate reductase subunit YnfF; K07309, Tat-targeted selenate reductase subunit YnfE; K12527, putative selenate reductase; K00956, sulfate adenylyltransferase subunit 1; K00957, sulfate adenylyltransferase subunit 2; K00958, sulfate adenylyltransferase.

4. Discussion

In this study, *L. plantarum* SeZi isolate was screened out from human fecal bacteria by the selenium- and zinc-tolerant abilities. The selenium- and zinc-enriched *L. plantarum* SeZi strain increased the levels of selenium and zinc and presented antioxidative properties in an ICR mouse model. A thorough search of the literature reporting the bioavailability and functionality of Se- and Zn-enriched microorganisms using a mouse model yielded only one related article [42]. To determine appropriate concentrations of selenium and zinc for mice, we referenced this article and 300 μg of selenium and 1.5 mg of zinc (/kg body weight/day) were used in our study. Yan et al. reported the antioxidant and antitumor activities were significantly increased by supplementation with Se/Zn-enriched mushrooms. However, the in vivo antioxidant activity of Se/Zn-enriched LAB has not been reported yet.

Selenium is an essential element that must be exogenously provided to reach the requirement of human and animal health [43]. The toxicity order of selenium species from high to low is selenate, selenite, nano-selenium, and lactomicro-selenium [44]. The accumulation of selenium in bacteria is processed by extracellular binding via active groups in the cell-membrane conjunction or intracellular binding via ion transportation on the membrane [45]. Many *Lactobacillus* strains are well-known to accumulate and biotransform toxic selenite into non-toxic seleno-amino acids (i.e., selenocysteine and selenomethionine) and selenoprotein [46,47] The utilization of Se-enriched *Lactobacillus* possesses unique advantages, including low toxicity, low cost production and additional probiotic effects.

In this study, we focused on the changes in the gut microbial composition as well as functional metabolism associated with Se/Zn uptake and utilization. Interestingly, Se/Zn supplementation greatly induced the enrichment of specific genus, *Lactococcus* (belonging to LAB), which confirmed the previous reports regarding in vitro tolerance of LAB to selenium [46,47]. However, this indigenously enriched LAB by inorganic Se/Zn supplement showed significantly different patterns in the microbial selenocompounds metabolism compared with Se/Zn-bioaccumulated *L. plantarum* SeZi strain.

To further investigate strain-specific functions, whole genome sequencing of *L. plantarum* SeZi strain was conducted and analyzed based on public databases. The genes coding *cysA* and *dedA* were observed in the whole genome sequencing of *L. plantarum* SeZi. Both of the *cysA* and *dedA* genes are associated with selenite uptake and detoxification. These genes were abundantly present in *L. plantarum* SeZi strain, but not observed in *Lactococcus* or other *Lactobacillus* spp. According to previous studies, selenite may enter the cells of *E. coli* through the sulfate permease CysA [48]. DedA can uptake selenite into cells as a direct transporter or a cofactor. Additionally, the *dedA* gene-contained mutant *E. coli* displayed selenite resistance [17]. Based on the results of targeted metagenome sequencing, the *L. plantarum* SeZi strain promoted the utilization of selenocysteine, seleno-cystathionine and seleno-methionine in selenocompounds metabolism pathway of gut microbiome. Besides, the transformation between the toxic inorganic selenium was reduced by *L. plantarum* SeZi in the detoxification process. These are consistent with the selenium-related function detected in the genome sequencing of *L. plantarum* SeZi.

Zinc plays an essential role in catalytic, structural, and regulatory functions in enzymes and protein domains [49]. Similar to the other trace elements (i.e., selenium), zinc in the organic forms have more bioavailability its inorganic forms [49]. Pharmacological zinc supplements often have low bioavailability and are easily overdosed [50]. Although internalizing zinc by *Lactobacillus* has not been studied in depth, the utilization of certain *Lactobacillus* species can be a promising alternative to deliver zinc in a highly organic form [51].

The selenium and zinc levels in the blood can be affected by dietary supplementation and related metabolism. In this study, the concentrations of selenium and zinc in the Se/Zn-enriched *Lactobacillus* group were significantly higher than the control or SZ group. This is consistent with a recently published study which indicates that feeding a diet supplemented with Se/Zn-enriched probiotics, 0.3 mg/L selenium, and 100 mg/L zinc significantly enhanced the blood selenium and zinc concentrations in Wistar rats [52]. It is also the only published study referred to Se/Zn-enriched probiotics in a murine model to date. However, the assessment of antioxidant capability and its potential mechanisms was not reported yet.

In this study, the enhancement of antioxidant ability in the SZ + Lp group was probably triggered by the increased selenium level. When selenium is incorporated into the selenoenzymes (i.e., GSH-Px), it enhances the antioxidant activities by suppressing the nuclear factor-kappa B (NF-κB) signal pathway [53]. Most selenoproteins take part in the defense against oxidative stress, protecting tissues and cells from oxidative damages [47]. Although the SZ group was administered with inorganic forms of Se/Zn, the SZ + Lp group was administered with bio-accumulated Se/Zn in bacteria. To understand the mechanisms of bioavailability and functionality related to inorganic/organic forms of trace nutrients, it is necessary to investigate the changes in selenium and zinc metabolites (e.g., selenoproteins, zincproteins, etc.) from blood and fecal samples.

Alteration of microbial communities was evaluated by 16S rRNA community analysis in this study. Up to now, the 16S metagenomic technique is still rarely used in the selenium or zinc-related animal studies. Obviously, the significant surge of *Lactobacillus* levels in the Lp and SZ + Lp groups was caused by the oral administration of *L. plantarum* SeZi. The relative abundance of *Adlercreutzia* was significantly higher in the SZ + Lp group compared to the control and SZ groups. According to the previous studies, a decreased level of *Adlercreutzia* was observed in multiple sclerosis patients and Alzheimer's disease patients [54,55]. Amongst *Adlercreutzia* species, *A. equolifaciens* is an equol-producing bacteria, promoting intestinal health [56]. Besides, a high relative abundance of *Lactococcus* (approximately 4.15%) was found in the SZ group, while this bacterium was almost undetectable in the other groups. It might be caused by utilization of selenium and zinc by *Lactococcus* spp. According to the previous studies, *Lactococcus lactis* is capable of selenium biotransformation and zinc uptake [57,58]. The relative abundance of *Allobaculum* spp., which was markedly reduced in the SZ + Lp group compared with other groups, was reported to be adversely associated with mRNA expression levels of tight junction protein genes (*zo-1* and *occludin*) and anti-inflammatory genes (*foxp3* and *Il-10*) in the colon of rats [59]. In some inflammatory bowel disease (IBD) patients, *Allobaculum* spp. was one of the uniquely observed species [60].

To the best of our knowledge, this study is the first attempt to evaluate the effects of Se/Zn-enriched LAB on in vivo gut microbiome changes. The microbiome analysis suggests a microbial aspect of selenocompound metabolism, however, lacks information on host's Se/Zn metabolic process.

Future studies on evaluating functionality of *L. plantarum* SeZi in a disease-induced mouse model should include host's metabolites analysis as well.

5. Conclusions

In conclusion, the selected strain *L. plantarum* SeZi is able to resist and biotransform inorganic selenium into organic selenium. In the in vivo study, the selenium and zinc-enriched *L. plantarum* SeZi increased blood selenium level, antioxidant capability and the utilization of seleno-amino acids. Therefore, the *L. plantarum* SeZi strain is a potential selenium and zinc-enriched probiotic for application as functional food ingredients in the future.

Author Contributions: Conceptualization, S.K., R.L., H.J.Y., and G.E.J.; Investigation, S.K., R.L., and H.J.; Data curation, S.K. and R.L.; Methodology, S.K. and R.L.; Formal analysis, S.K.; Software, S.K.; Visualization, S.K.; writing-original draft preparation, S.K. and R.L.; writing—review and editing, S.K., H.J.Y. and G.E.J.; supervising, H.J.Y. and G.E.J.; funding acquisition, H.J.Y. and G.E.J. All authors have read and agreed to the published version of the manuscript.

Funding: This work was carried out with the support of the Ministry of Small and Medium-sized Enterprises (SMEs) and Startups (MSS), Korea, under the "Regional Specialized Industry Development Program (R&D, Project number S2848321)" supervised by the Korea Institute for Advancement of Technology (KIAT). This work was also supported by the "K-BIO KIURI Center program (Project number 2020M3H1A1073304)".

Conflicts of Interest: G.E.J. holds BIFIDO Ltd. stocks. Other authors declare no conflict of interest.

References

1. Mrvčić, J.; Stanzer, D.; Šolić, E.; Stehlik-Tomas, V. Interaction of lactic acid bacteria with metal ions: Opportunities for improving food safety and quality. *World J. Microb. Biot.* **2012**, *28*, 2771–2782. [CrossRef]
2. Rayman, M.P. Selenium and adverse health conditions of human pregnancy. In *Selenium*; Springer: New York, NY, USA, 2011; pp. 531–544.
3. Flores-Mateo, G.; Navas-Acien, A.; Pastor-Barriuso, R.; Guallar, E. Selenium and coronary heart disease: A meta-analysis. *Am. J. Clin. Nutr.* **2006**, *84*, 762–773. [CrossRef] [PubMed]
4. Brenneisen, P.; Steinbrenner, H.; Sies, H. Selenium, oxidative stress, and health aspects. *Mol. Asp. Med.* **2005**, *26*, 256–267. [CrossRef] [PubMed]
5. Ferguson, L.R.; Karunasinghe, N.; Zhu, S.; Wang, A.H. Selenium and its' role in the maintenance of genomic stability. *Mutat. Res. Fundam. Mol. Mech. Mutagenesis* **2012**, 733, 100–110. [CrossRef]
6. Rayman, M.P. Selenium and human health. *Lancet* **2012**, *379*, 1256–1268. [CrossRef]
7. Arthur, J.R.; Nicol, F.; Beckett, G.J. The role of selenium in thyroid hormone metabolism and effects of selenium deficiency on thyroid hormone and iodine metabolism. *Biol. Trace Elem. Res.* **1992**, *33*, 37–42. [CrossRef]
8. Burke, M.P.; Opeskin, K. Fulminant heart failure due to selenium deficiency cardiomyopathy (Keshan disease). *Med. Sci. Law* **2002**, *42*, 10–13. [CrossRef]
9. Maksimović, Z.J. Selenium deficiency and Balkan endemic nephropathy. *Kidney Int. Suppl.* **1991**, *34*, S12–S14. [PubMed]
10. Gropper, S.S.; Smith, J.L. Advanced Nutrition and Human Metabolism. Wadsworth Cengage Learning: Belmot, CA, USA, 2012.
11. Chimienti, F. Zinc, pancreatic islet cell function and diabetes: New insights into an old story. *Nutr. Res. Rev.* **2013**, *26*, 1–11. [CrossRef] [PubMed]
12. Cruz, K.J.C.; de Oliveira, A.R.S.; Morais, J.B.S.; Severo, J.S.; Mendes, P.M.V.; de Sousa Melo, S.R.; de Sousa, G.S.; do Nascimento Marreiro, D. Zinc and insulin resistance: Biochemical and molecular aspects. *Biol. Trace Elem. Res.* **2018**, *186*, 407–412. [CrossRef] [PubMed]
13. Islam, M.R.; Attia, J.; Ali, L.; McEvoy, M.; Selim, S.; Sibbritt, D.; Akhter, A.; Akter, S.; Peel, R.; Faruque, O. Zinc supplementation for improving glucose handling in pre-diabetes: A double blind randomized placebo controlled pilot study. *Diabetes Res. Clin. Pract.* **2016**, *115*, 39–46. [CrossRef] [PubMed]
14. Afolabi, O.B.; Balogun, B.D.; Oloyede, O.I.; Akinyemi, A.J. Zinc and neurodegenerative disorders. In *Handbook of Research on Critical Examinations of Neurodegenerative Disorders*; IGI Global: Harrisburg, PA, USA, 2019; pp. 176–193.

15. Gomolak, J.R. Zinc accumulation in the midbrain following methamphetamine exposure as a potential biomarker for neurodegeneration. In Proceedings of the 25th UNG Annual Research Conference, Dahlonega, GA, USA, 11 November 2016.
16. Kieliszek, M.; Błażejak, S. Selenium: Significance, and outlook for supplementation. *Nutrition* **2013**, *29*, 713–718. [CrossRef] [PubMed]
17. Ledgham, F.; Quest, B.; Vallaeys, T.; Mergeay, M.; Covès, J. A probable link between the DedA protein and resistance to selenite. *Res. Microbiol.* **2005**, *156*, 367–374. [CrossRef]
18. Ayangbenro, A.S.; Babalola, O.O. A new strategy for heavy metal polluted environments: A review of microbial biosorbents. *Int. J. Environ. Res. Pub. He.* **2017**, *14*, 94. [CrossRef]
19. Blackwell, K.; Singleton, I.; Tobin, J.M. Metal cation uptake by yeast: A review. *Appl. Microbiol. Biot.* **1995**, *43*, 579–584. [CrossRef]
20. Andreoni, V.; Luischi, M.M.; Cavalca, L.; Erba, D.; Ciappellano, S. Selenite tolerance and accumulation in the Lactobacillus species. *Ann. Microbiol.* **2000**, *50*, 77–88.
21. Saini, K.; Tomar, S.K. In vitro evaluation of probiotic potential of Lactobacillus cultures of human origin capable of selenium bioaccumulation. *LWT* **2017**, *84*, 497–504. [CrossRef]
22. Shu, G.; Mei, S.; Chen, L.; Zhang, B.; Guo, M.; Cui, X.; Chen, H. Screening, identification, and application of selenium–enriched *Lactobacillus* in goat milk powder and tablet. *J. Food Process. Pres.* **2020**, e14470. [CrossRef]
23. Xia, S.K.; Chen, L.; Liang, J.Q. Enriched selenium and its effects on growth and biochemical composition in Lactobacillus bulgaricus. *J. Agric. Food. Chem.* **2007**, *55*, 2413–2417. [CrossRef]
24. Diowksz, A.; Ambroziak, W.; Wlodarczyk, M. Investigation of the ability of selenium accumulation by lactic acid bacteria of *Lactobacillus species* and *yeast Saccharomyces cerevisiae. Pol. J. Food Nutr. Sci.* **1999**, *1*, 17–22.
25. Mudroňová, D.; Gancarčíková, S.; Nemcová, R. Influence of Zinc Sulphate on the Probiotic Properties of Lactobacillus plantarum CCM 7102. *Folia Veterinaria* **2019**, *63*, 45–54. [CrossRef]
26. Jin, H.; Jeong, Y.; Yoo, S.H.; Johnston, T.V.; Ku, S.; Ji, G.E. Isolation and characterization of high exopolysaccharide-producing Weissella confusa VP30 from young children's feces. *Microb. Cell Fact.* **2019**, *18*, 110. [CrossRef]
27. Chaney, A.L.; Marbach, E.P. Modified reagents for determination of urea and ammonia. *Clin. Chem.* **1962**, *8*, 130–132. [CrossRef]
28. Yoon, S.H.; Ha, S.M.; Kwon, S.; Lim, J.; Kim, Y.; Seo, H.; Chun, J. Introducing EzBioCloud: A taxonomically united database of 16S rRNA gene sequences and whole-genome assemblies. *Int. J. Syst. Evol. Microbiol.* **2017**, *67*, 1613. [CrossRef]
29. Schattner, P.; Brooks, A.N.; Lowe, T.M. The tRNAscan-SE, snoscan and snoGPS web servers for the detection of tRNAs and snoRNAs. *Nucleic Acids Res.* **2005**, *33* (Suppl. 2), W686–W689. [CrossRef]
30. Nawrocki, E.P.; Eddy, S.R. Computational identification of functional RNA homologs in metagenomic data. *RNA Biol.* **2013**, *10*, 1170–1179. [CrossRef]
31. Hyatt, D.; Chen, G.L.; LoCascio, P.F.; Land, M.L.; Larimer, F.W.; Hauser, L.J. Prodigal: Prokaryotic gene recognition and translation initiation site identification. *BMC Bioinform.* **2010**, *11*, 119. [CrossRef]
32. Powell, S.; Forslund, K.; Szklarczyk, D.; Trachana, K.; Roth, A.; Huerta-Cepas, J.; Gabaldon, T.; Rattei, T.; Creevey, C.; Kuhn, M. eggNOG v4. 0: Nested orthology inference across 3686 organisms. *Nucleic Acids Res.* **2014**, *42*, D231–D239. [CrossRef]
33. Edgar, R.C. Search and clustering orders of magnitude faster than BLAST. *Bioinformatics* **2010**, *26*, 2460–2461. [CrossRef]
34. Consortium, U. UniProt: A hub for protein information. *Nucleic Acids Res.* **2015**, *43*, D204–D212. [CrossRef]
35. Kanehisa, M.; Goto, S.; Sato, Y.; Kawashima, M.; Furumichi, M.; Tanabe, M. Data, information, knowledge and principle: Back to metabolism in KEGG. *Nucleic Acids Res.* **2014**, *42*, D199–D205. [CrossRef]
36. Overbeek, R.; Begley, T.; Butler, R.M.; Choudhuri, J.V.; Chuang, H.Y.; Cohoon, M.; de Crécy-Lagard, V.; Diaz, N.; Disz, T.; Edwards, R. The subsystems approach to genome annotation and its use in the project to annotate 1000 genomes. *Nucleic Acids Res.* **2005**, *33*, 5691–5702. [CrossRef]
37. Kang, S.; You, H.J.; Lee, Y.G.; Jeong, Y.; Johnston, T.V.; Baek, N.I.; Ku, S.; Ji, G.E. Production, structural characterization, and In Vitro assessment of the prebiotic potential of butyl-fructooligosaccharides. *Int. J. Mol. Sci.* **2020**, *21*, 445. [CrossRef]
38. Douglas, G.M.; Maffei, V.J.; Zaneveld, J.R.; Yurgel, S.N.; Brown, J.R.; Taylor, C.M.; Huttenhower, C.; Langille, M.G. PICRUSt2 for prediction of metagenome functions. *Nat. Biotechnol.* **2020**, 1–5. [CrossRef]

39. Younus, H. Therapeutic potentials of superoxide dismutase. *Int. J. Health Sci.* **2018**, *12*, 88.
40. Blankenberg, S.; Rupprecht, H.J.; Bickel, C.; Torzewski, M.; Hafner, G.; Tiret, L.; Smieja, M.; Cambien, F.; Meyer, J.; Lackner, K.J. Glutathione peroxidase 1 activity and cardiovascular events in patients with coronary artery disease. *N. Engl. J. Med.* **2003**, *349*, 1605–1613. [CrossRef]
41. Khoubnasabjafari, M.; Ansarin, K.; Jouyban, A. Reliability of malondialdehyde as a biomarker of oxidative stress in psychological disorders. *BioImpacts* **2015**, *5*, 123. [PubMed]
42. Yang, H.; Chang, H. Antioxidant and antitumor activities of selenium and zinc-enriched oyster mushroom in mice. *Biol. Trace Elem. Res.* **2012**, *150*, 236–241.
43. Xu, C.; Guo, Y.; Qiao, L.; Ma, L.; Cheng, Y.; Roman, A. Biogenic synthesis of novel functionalized selenium nanoparticles by Lactobacillus casei ATCC 393 and its protective effects on intestinal barrier dysfunction caused by enterotoxigenic Escherichia coli K88. *Front. Microbiol.* **2018**, *9*, 1129. [CrossRef] [PubMed]
44. Nagy, G.; Pinczes, G.; Pinter, G.; Pocsi, I.; Prokisch, J.; Banfalvi, G. In situ electron microscopy of lactomicroselenium particles in probiotic bacteria. *Int. J. Mol. Sci.* **2016**, *17*, 1047. [CrossRef]
45. Kurek, E.; Ruszczynska, A.; Wojciechowski, M.; Luciuk, A.; Michalska-Kacymirow, M.; Motyl, I.; Bulska, E. Bio-transformation of selenium in Se-enriched bacterial strains of *Lactobacillus casei. Roczniki Państwowego Zakładu Higieny* **2016**, *67*, 3.
46. Yazdi, M.H.; Mahdavi, M.; Setayesh, N.; Esfandyar, M.; Shahverdi, A.R. Selenium n6noparticle-enriched *Lactobacillus brevis* causes more efficient immune responses in vivo and reduces the liver metastasis in metastatic form of mouse breast cancer. *DARU* **2013**, *21*, 33. [CrossRef] [PubMed]
47. Qiao, L.; Dou, X.; Yan, S.; Zhang, B.; Xu, C. Biogenic selenium nanoparticles synthesized by Lactobacillus casei ATCC 393 alleviate diquat-induced intestinal barrier dysfunction in C57BL/6 mice through their antioxidant activity. *Food Funct.* **2020**, *11*, 3020–3031. [CrossRef] [PubMed]
48. Turner, R.J.; Weiner, J.H.; Taylor, D.E. Selenium metabolism in Escherichia coli. *Biometals* **1998**, *11*, 223–227. [CrossRef] [PubMed]
49. Leonardi, A.; Zanoni, S.; de Lucia, M.; Amaretti, A.; Raimondi, S.; Rossi, M. Zinc uptake by lactic acid bacteria. *ISRN Biotechnol.* **2013**, *2013*. [CrossRef]
50. Góral, M.; Pankiewicz, U.; Sujka, M.; Kowalski, R. Bioaccumulation of zinc ions in Lactobacillus rhamnosus B 442 cells under treatment of the culture with pulsed electric field. *Eur. Food Res. Technol.* **2019**, *245*, 817–824. [CrossRef]
51. Lule, V.K.; Tomar, S.K.; Chawla, P.; Pophaly, S.; Kapila, S.; Arora, S. Bioavailability assessment of zinc enriched lactobacillus biomass in a human colon carcinoma cell line (Caco-2). *Food Chem.* **2020**, *309*, 125583. [CrossRef]
52. Malyar, R.M.; Li, H.; Liu, D.; Abdulrahim, Y.; Farid, R.A.; Gan, F.; Ali, W.; Enayatullah, H.; Banuree, S.A.H.; Huang, K. Selenium/Zinc-Enriched probiotics improve serum enzyme activity, antioxidant ability, inflammatory factors and related gene expression of Wistar rats inflated under heat stress. *Life Sci.* **2020**, *248*, 117464. [CrossRef]
53. Benstoem, C.; Goetzenich, A.; Kraemer, S.; Borosch, S.; Manzanares, W.; Hardy, G.; Stoppe, C. Selenium and its supplementation in cardiovascular disease—what do we know? *Nutrients* **2015**, *7*, 3094–3118. [CrossRef]
54. Fan, Y.; Zhang, J. Dietary modulation of intestinal microbiota: Future opportunities in experimental autoimmune encephalomyelitis and multiple sclerosis. *Front. Microbiol.* **2019**, *10*, 740. [CrossRef]
55. Vogt, N.M.; Kerby, R.L.; Dill-McFarland, K.A.; Harding, S.J.; Merluzzi, A.P.; Johnson, S.C.; Carlsson, C.M.; Asthana, S.; Zetterberg, H.; Blennow, K. Gut microbiome alterations in Alzheimer's disease. *Sci. Rep.* **2017**, *7*, 1–11. [CrossRef]
56. Maruo, T.; Sakamoto, M.; Ito, C.; Toda, T.; Benno, Y. Adlercreutzia equolifaciens gen. nov., sp. nov., an equol-producing bacterium isolated from human faeces, and emended description of the genus Eggerthella. *Int. J. Syst. Evol. Micr.* **2008**, *58*, 1221–1227. [CrossRef]
57. Martínez, F.G.; Moreno-Martin, G.; Pescuma, M.; Madrid-Albarrán, Y.; Mozzi, F. Biotransformation of selenium by lactic acid bacteria: Formation of seleno-nanoparticles and seleno-amino acids. *Front. Bioeng. Biotechnol.* **2020**, *8*, 506. [CrossRef]
58. Llull, D.; Poquet, I. New expression system tightly controlled by zinc availability in Lactococcus lactis. *Appl. Environ. Microbiol.* **2004**, *70*, 5398–5406. [CrossRef]

59. Lee, S.M.; Han, H.W.; Yim, S.Y. Beneficial effects of soymilk and fiber on high cholesterol diet-induced alteration of gut microbiota and inflammatory gene expression in rats. *Food Funct.* **2015**, *6*, 492–500. [CrossRef]
60. Palm, N.W.; de Zoete, M.R.; Cullen, T.W.; Barry, N.A.; Stefanowski, J.; Hao, L.; Degnan, P.H.; Hu, J.; Peter, I.; Zhang, W. Immunoglobulin A coating identifies colitogenic bacteria in inflammatory bowel disease. *Cell* **2014**, *158*, 1000–1010. [CrossRef]

antioxidants

MDPI

Review

The Interactions between Polyphenols and Microorganisms, Especially Gut Microbiota

Małgorzata Makarewicz, Iwona Drożdż, Tomasz Tarko and Aleksandra Duda-Chodak *

Department of Fermentation Technology and Microbiology, Faculty of Food Technology, University of Agriculture in Krakow, 30-149 Kraków, Poland; malgorzata.makarewicz@urk.edu.pl (M.M.); iwona.drozdz@urk.edu.pl (I.D.); tomasz.tarko@urk.edu.pl (T.T.)

* Correspondence: aleksandra.duda-chodak@urk.edu.pl; Tel.: +48-12-662-4792

Abstract: This review presents the comprehensive knowledge about the bidirectional relationship between polyphenols and the gut microbiome. The first part is related to polyphenols' impacts on various microorganisms, especially bacteria, and their influence on intestinal pathogens. The research data on the mechanisms of polyphenol action were collected together and organized. The impact of various polyphenols groups on intestinal bacteria both on the whole "microbiota" and on particular species, including probiotics, are presented. Moreover, the impact of polyphenols present in food (bound to the matrix) was compared with the purified polyphenols (such as in dietary supplements) as well as polyphenols in the form of derivatives (such as glycosides) with those in the form of aglycones. The second part of the paper discusses in detail the mechanisms (pathways) and the role of bacterial biotransformation of the most important groups of polyphenols, including the production of bioactive metabolites with a significant impact on the human organism (both positive and negative).

Keywords: intestinal microbiota; inhibition; metabolism; catabolism; biotransformation; bioactive compounds; health; metabolites; diversity

Citation: Makarewicz, M.; Drożdż, I.; Tarko, T.; Duda-Chodak, A. The Interactions between Polyphenols and Microorganisms, Especially Gut Microbiota. *Antioxidants* **2021**, *10*, 188. https://doi.org/10.3390/antiox10020188

Received: 22 December 2020
Accepted: 25 January 2021
Published: 28 January 2021

1. Introduction

The intestinal microbiome plays an important, if not crucial, role in the metabolism of chemical compounds delivered with food, especially those that are undigested in the upper digestive tract. The enormous number of bacterial cells inhabiting the large intestine forms a complex ecosystem called the "intestinal microbiome". The word microbiome was introduced for the first time in 2001 to define the collective genomes of the microbiota [1]. Since then, much research and many projects were dedicated to assessing the impact of intestinal microbiota on a host's health, especially by determining its role in food metabolism, xenobiotics biotransformation and various disease development.

It is estimated that the microbiota of an adult is composed of ~10^{14} bacteria cells [2] belonging to 1000–1150 species, with each individual harboring at least 160 species (usually about 500 species) [3]. Based on the large-scale 16S rRNA or metagenomic studies, scientists stated that ~80% of the bacteria identified by molecular tools in the human gut are uncultured and hence can be characterized only by metagenomic studies [4,5]. There are significant interindividual differences in the bacterial species found in the gastrointestinal tract. The composition, as well as the ratio of different species that form the intestinal microbiome, is very diverse within the human population, and each individual has his or her own unique profile of microbial species, which can be compared to a fingerprint. The differentiation of gut microbiota composition and profile is caused by the influence of multiple and diverse factors, such as age, origin, geographical location, environment, dietary habits (including probiotics), health, the application of antibiotics or even in the way an individual is born [6–8]. However, despite the great diversity of bacterial species, the majority of them belong to only four bacterial phyla: Firmicutes (64%), Bacteroidetes (23%), Proteobacteria

(8%) and Actinobacteria (3%), whereas other taxons highly diverse [2]. Among the key functions of microbiota is occupying the intestinal surfaces and production of antimicrobial compounds that prevent the invasion of pathogens. Both commensal bacteria and gut pathogens (such as *Salmonella*, *Shigella*, *Helicobacter*, *Vibrio*, *Campylobacter*, *Yersinia*, *Clostridia*, *Aeromonas*, *Listeria*, *Streptococcus*, and *Staphylococcus*, as well as pathogenic strains of *Escherichia coli*, *Klebsiella pneumoniae*, *Enterococcus faecalis*) require similar ecological niches to colonize and proliferate in the intestine [9,10]. Therefore, various mechanisms designed to compete with each other have evolved. Commensal bacteria produce bacteriocins that specifically inhibit members of the same or similar bacterial species (e.g., *E. coli* versus pathogen enterohemorrhagic *E. coli*). Commensal bacteria produce short-chain fatty acids and cause pH reduction, thereby preventing the colonization by pathogens whose optimal pH for growth is neutral [11]. An altered bacterial community structure may facilitate the gut colonization by enteric pathogens but can also favor the overgrowth of potentially harmful subsets of indigenous bacteria, like virulent *E. coli* or *Clostridium difficile*.

The great diversity of bacterial species forming the gut microbiota implicates the large number of genes which they contain [2] and the enormous metabolic capacity of the intestinal microbiome, which is approximately 100-fold greater than that of the human liver [2,12]. The intestinal microbiota is equipped with a large set of different enzymes able to hydrolyze glycosides, glucuronides, sulfates, amides, esters and lactones through the action of enzymes such as α-rhamnosidase, β-glucuronidase, β-glucosidase, sulfatase and various esterases. Other reactions catalyzed by the gut microbial enzymes are aromatic ring cleavage, reductions (reductases, hydrogenases), decarboxylation (decarboxylase), demethylation (demethylase), isomerization (isomerase), and dehydroxylation (dehydroxylase) [13,14].

Intestinal bacteria contribute to the breakdown of polysaccharides and polyphenols as well as participate in the synthesis of vitamins (K, B12) and amino acids [8]. Many metabolites produced by gut microbiota are involved in various important physiological processes of the host, including energy metabolism and immunity. For example, the essential aromatic amino acid tryptophan can be metabolized by—among others—*Peptostreptococcus russellii*, *Clostridium sporogenes*, and *Lactobacillus* spp. to indole derivatives, which are ligands for aryl hydrocarbon receptor (AhR). This transcription factor plays an important role in the human immunological response, and via modulating T cell differentiation, Th17 development and IL-22 production may inhibit inflammation [15]. Branched-chain amino acids (BCAAs) (such as leucine, isoleucine, and valine) are essential amino acids that possess an aliphatic side chain with a branch, and that cannot be synthesized by humans. Therefore, they are provided by diet or synthesized by gut microbiota. The main species that contribute to the BCAAs production are *Prevotella copri* and *Bacteroides vulgatus* [15].

To date, thousands of microbial metabolites with known and unknown functions have been identified as components of the human metabolome. Well-known are short-chain fatty acids (SCFAs) with the acetate, propionate, and butyrate, being the metabolite of resistant starch and dietary fiber fermentation [15]. SCFAs are generally considered to have beneficial effects on host health, and they modulate metabolism, inflammation, hormone production, lipogenesis, and gut homeostasis. Gut microbiota can also metabolize dietary L-carnitine, choline, and lecithin into trimethylamine (TMA), which is then converted to trimethylamine-N-oxide (TMAO) in the liver of a host.

Owing to the multitude of direct and indirect interactions with the host organism, the intestinal microbiome is hence closely linked to the health of a host [16,17]. Dysbiosis, an imbalanced or disturbed microbiota composition, may play a significant role in etiology or the development of various gastrointestinal diseases such as inflammatory bowel disease (IBD), irritable bowel syndrome (IBS), colon cancer, and antibiotic-associated diarrhea [12,18]. The gut microbiota also plays a critical role in the transformation of dietary polyphenols into absorbable biologically active compounds. It is estimated that about 90–95% of the total polyphenol intake remains unabsorbed and colonic bacteria act enzymatically on their backbone, producing metabolites with a different physiological significance [19].

2. The Structure and Role of Polyphenols

Polyphenols are secondary metabolites playing an important role in plant tissues. They provide the color to flowers and fruits (mainly anthocyanins), which attracts pollinators and seed dispersers; they are also responsible for flavor in fruit and vegetables, as well as protecting plant tissues against herbivores and other biotic and abiotic stressors, like UV radiation, cold, heat or salinity [20]. Flavonoids take part in energy transfers, the regulation of photosynthesis and morphogenesis, regulation of growth factors, and sex determination and—due to antimicrobial activity—protect against the spread of pathogens in plant tissues [21]. Polyphenols also influence human health. Because of antioxidant properties and free-radical scavenging activity, they are believed to protect against various diseases, e.g., cancer, stroke and myocardial infarction, cardiovascular diseases, and some immunological and neurological disorders; they are also thought to have a beneficial impact on humans with diabetes and obesity [22–31]. Several in vitro and in vivo animal studies have demonstrated the antioxidant and anti-inflammatory effects of polyphenols in the brain–liver–gut axis [32], and polyphenols have been shown to target different stages of the inflammatory cascade to reduce the severity of inflammation. Polyphenols can also modulate various signal pathways, for example, through interaction with AMP-activated protein kinase (AMPK), CCAAT/enhancer-binding protein α (C/EBPα), peroxisome proliferator-activated receptor γ (PPARγ), and peroxisome proliferator-activated receptor-gamma coactivator 1-alpha (PGC-1α), sirtuin 1, and sterol regulatory element-binding protein-1c (SREBP1c) involved mainly in cellular energy metabolism and adipogenesis, as well as uncoupling proteins 1 and 2 (UCP1 and UCP2), and NF-κB that regulate antioxidant and anti-inflammatory responses [28].

Polyphenols are a large group of compounds that comprise phenolic acids, flavonoids, tannins, lignans, stilbens and coumarins (Figure 1 presents the chemical structure of flavonoids, Figure 2—non-flavonoids). In a human diet, they are provided mainly by plant food such as fruit, vegetables, tea, wine, coffee, and cocoa. However, even when phenolic compounds occur in the human diet in large quantities, they do not always show high biological activity after consumption.

Polyphenols' influence on human health depends both on their amount of food and on their bioavailability, bioaccessibility and the biological activity of metabolites produced in the human body. Some phenolic compounds have limited absorption in the digestive tract, while the others undergo an intensive metabolism to derivatives with a lower activity or they undergo a rapid elimination (degradation). The uniqueness of the gut microbiota composition causes that in one individual, a given polyphenol will undergo bacterial metabolism and will have an effect (beneficial or negative), whereas in another human being, the metabolism of the same polyphenol will follow a different path, and there will be no effect.

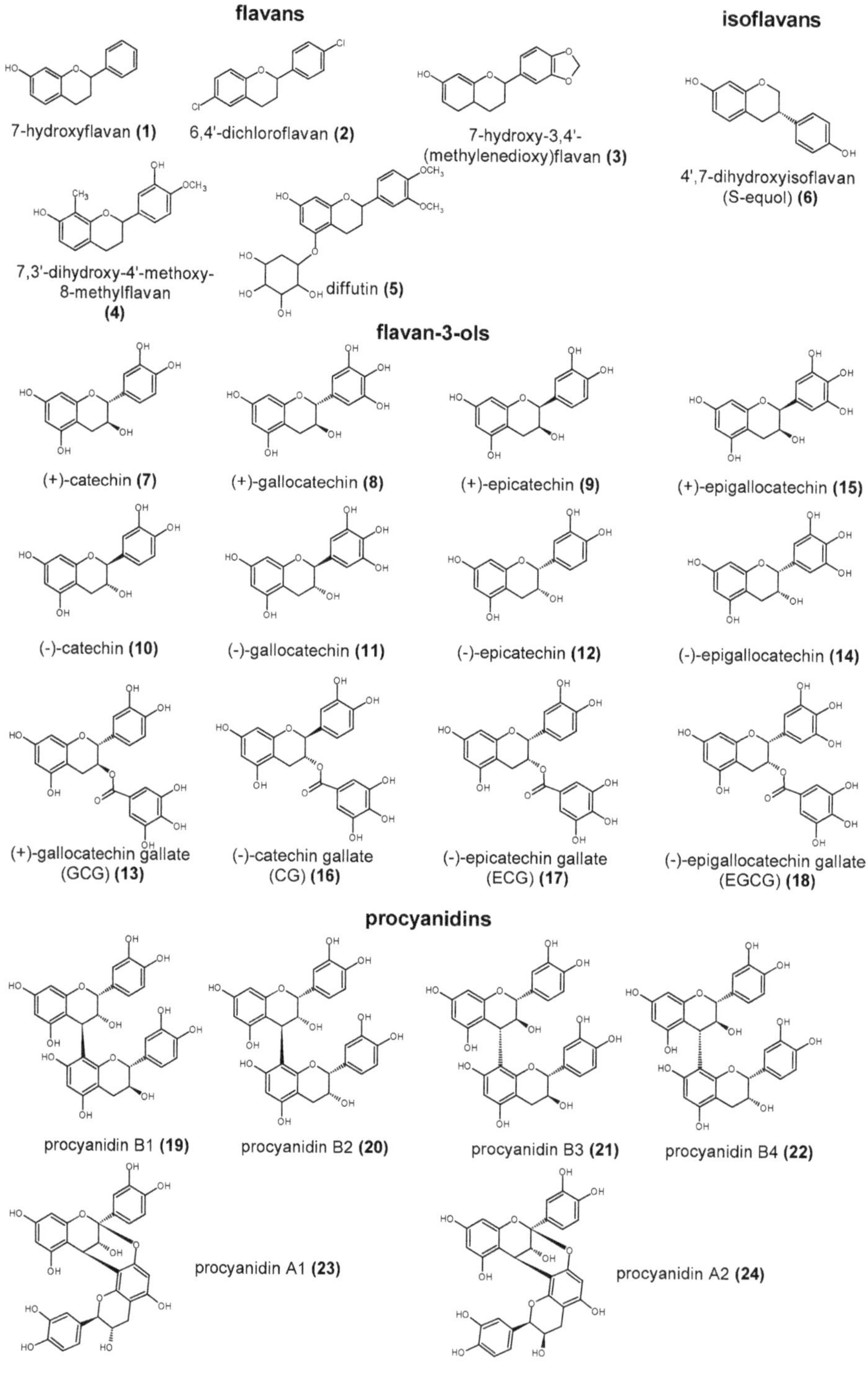

Figure 1. *Cont.*

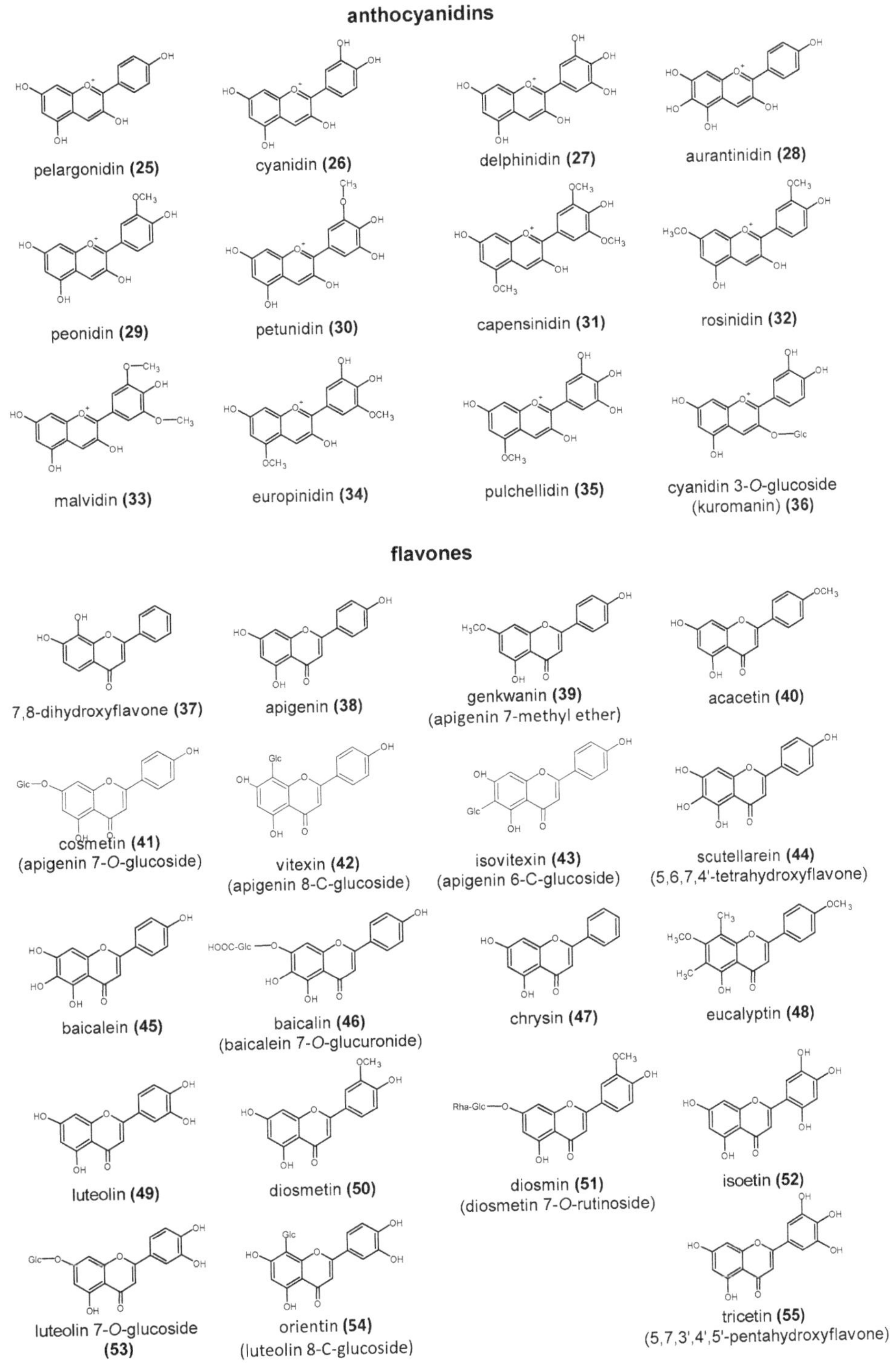

Figure 1. *Cont.*

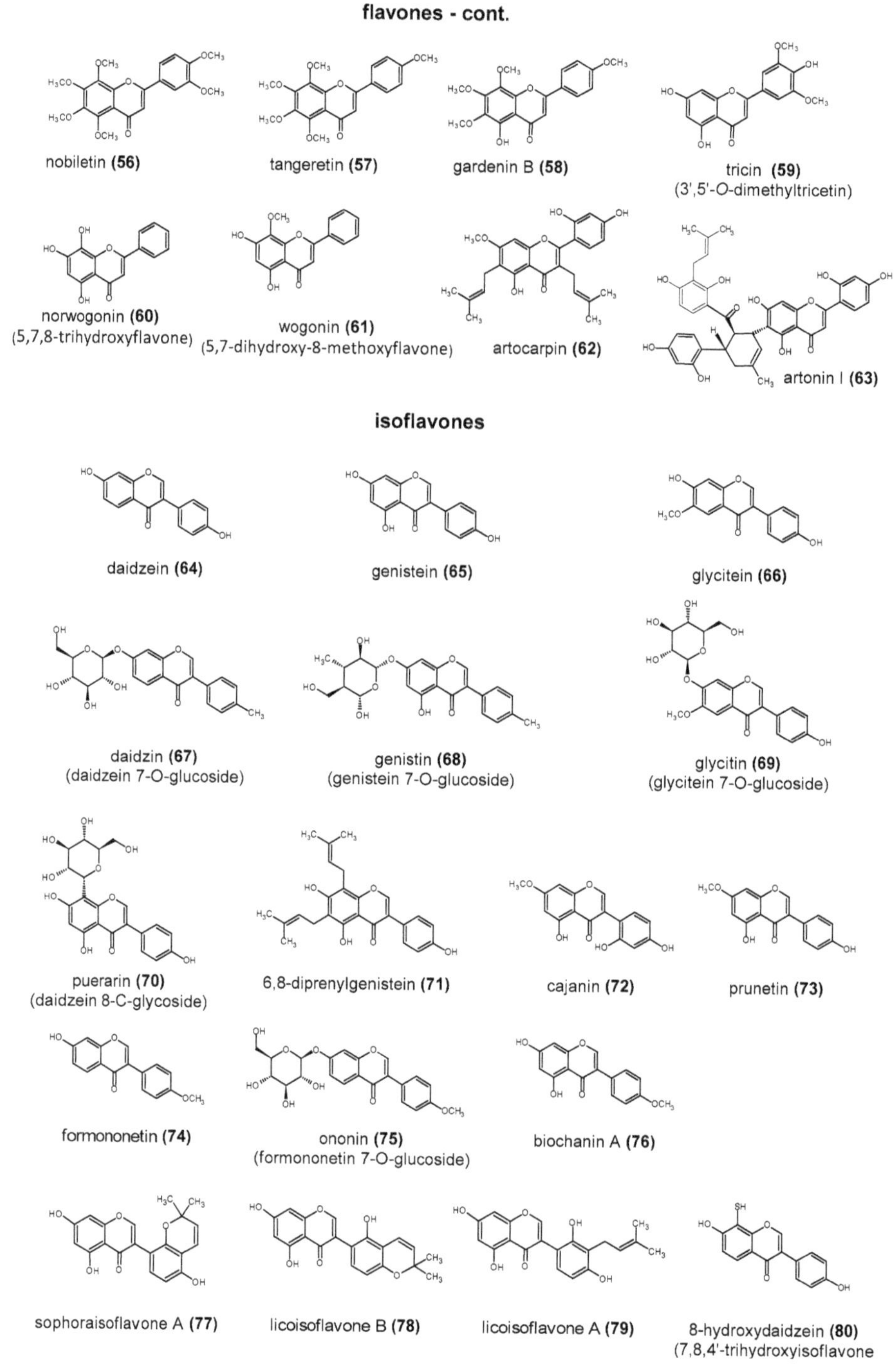

Figure 1. *Cont.*

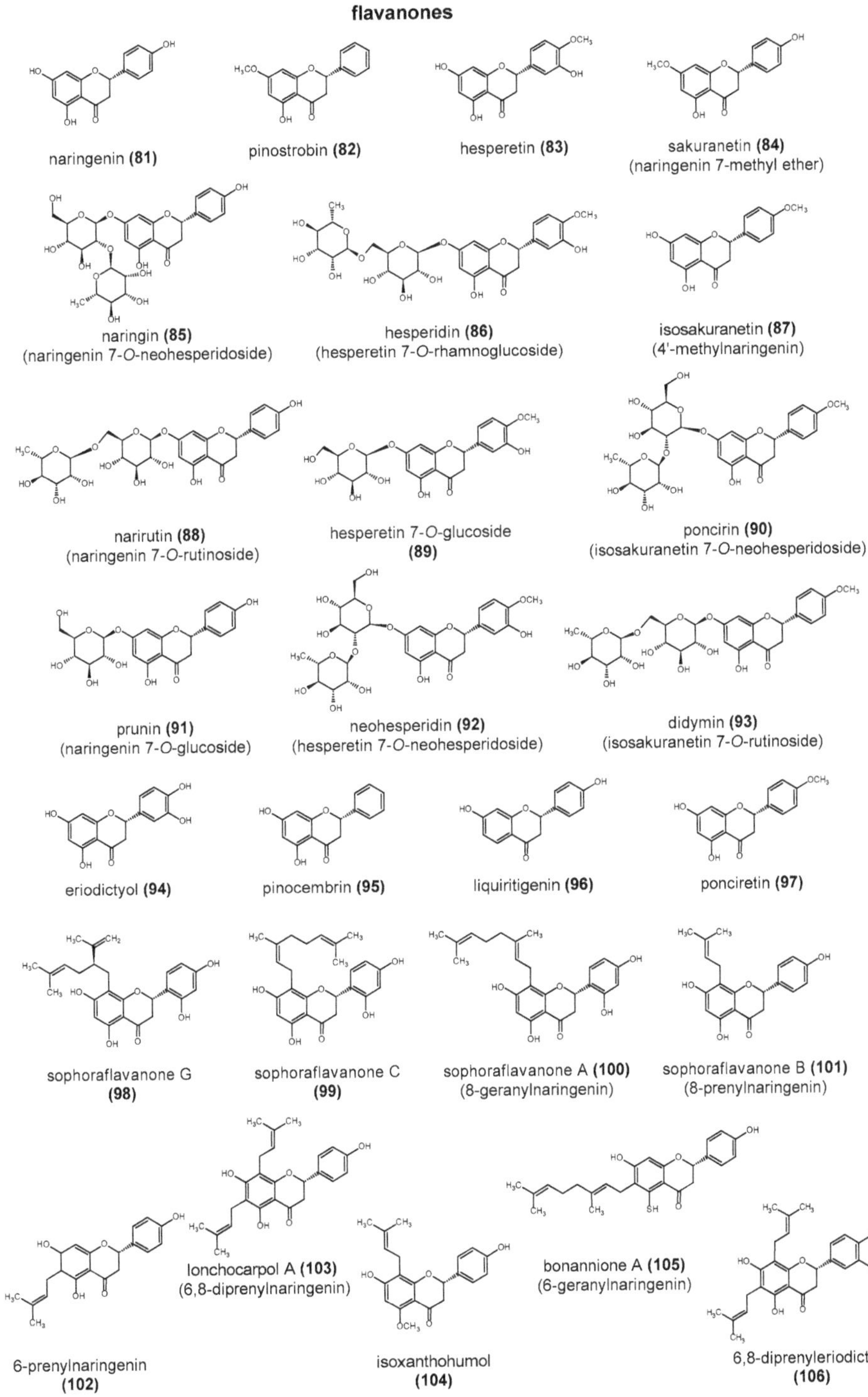

Figure 1. *Cont.*

flavonols

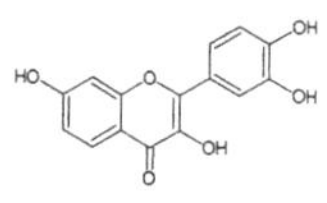
fisetin **(107)**

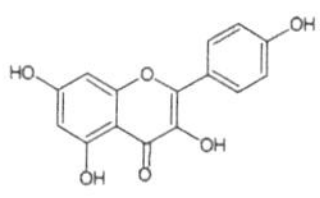
kaempferol (**108**)

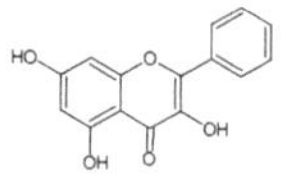
galangin **(109)**

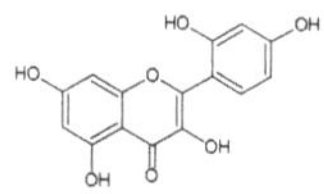
morin **(110)**

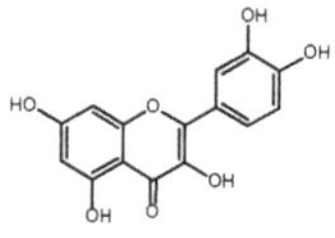
quercetin **(111)**

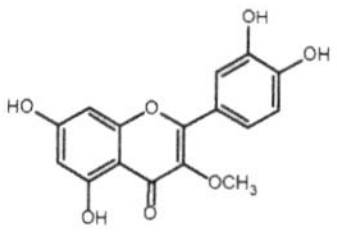
3-*O*-methylquercetin **(112)**

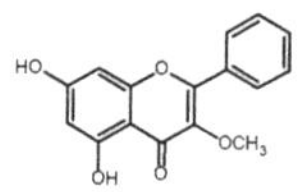
3-*O*-methylgalangin **(113)**

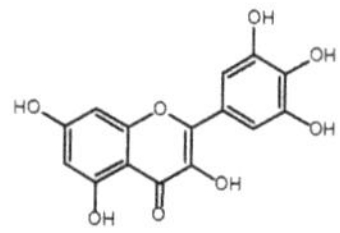
myricetin **(114)**

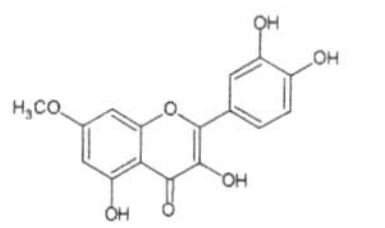
rhamnetin **(115)**
(7-*O*-methylquercetin)

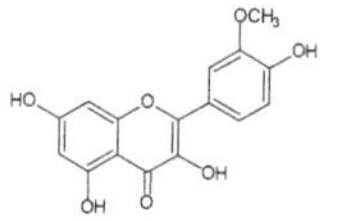
isorhamnetin **(116)**
(3'-*O*-methylquercetin)

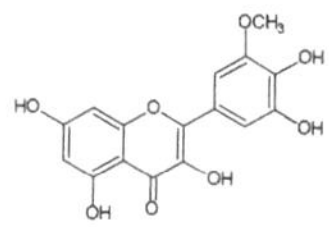
laricitrin **(117)**
(3'-*O*-methylmyricetin)

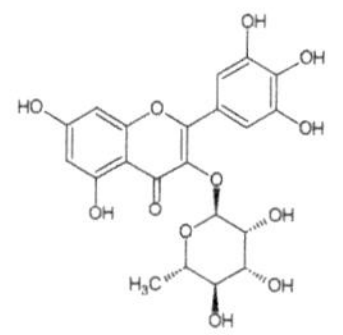
myricitrin **(118)**
(myricetin 3-*O*-rhamnoside)

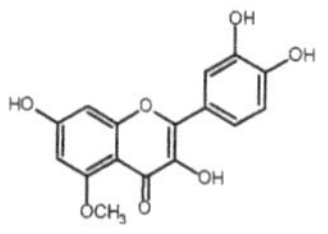
azaleatin **(119)**
(quercetin 5-methyl ether)

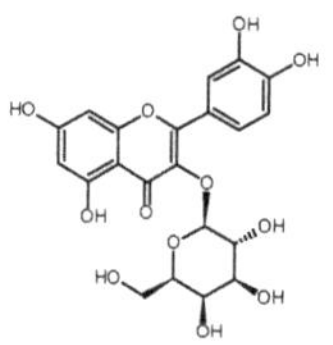
hyperoside **(120)**
(quercetin 3-*O*-galactoside)

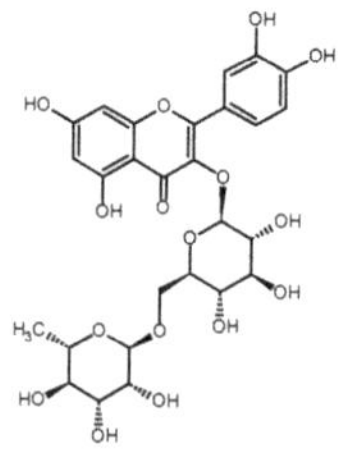
rutin **(121)**
(quercetin 3-*O*-rutinoside)

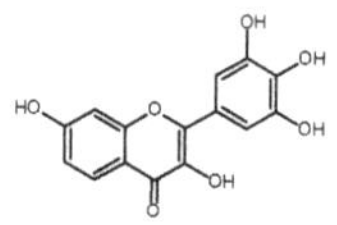
robinetin **(122)**
(5-deoxymyricetin)

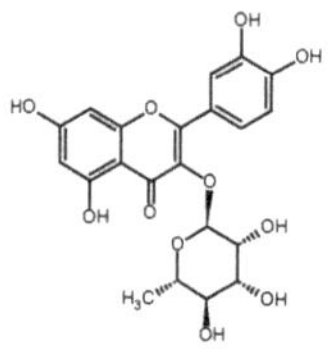
quercitrin **(123)**
(quercetin 3-*O*-rhamnoside)

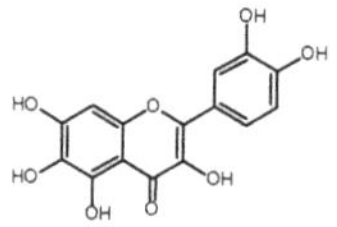
quercetagetin **(124)**
(6-hydroxyquercetin)

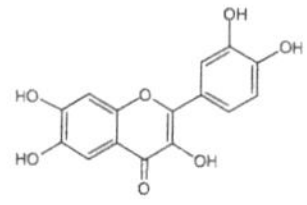
3,3',4',6',7-pentahydroxyflavone
(125)

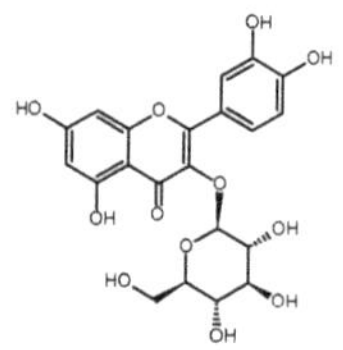
isoquercetin **(126)**
(quercetin 3-*O*-glucoside)

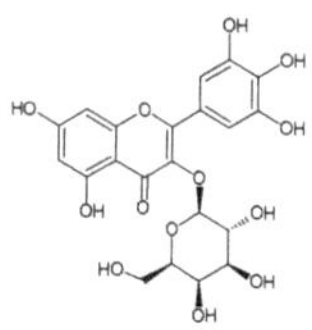
myricetin 3-*O*-galactoside
(127)

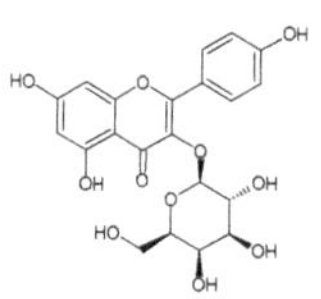
trifolin **(128)**
(kaempferol 3-*O*-galactoside)

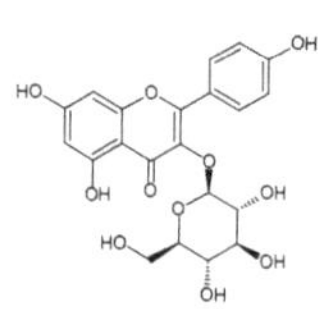
astragalin **(129)**
(kaempferol 3-*O*-glucoside)

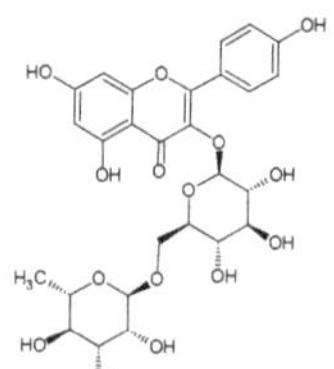
nicotiflorin **(130)**
(kaempferol 3-*O*-rutinoside)

Figure 1. *Cont.*

dihydroflavonols

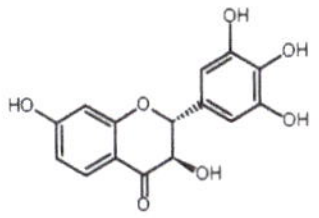

fustin **(131)**
2,3-dihydrorobinetin

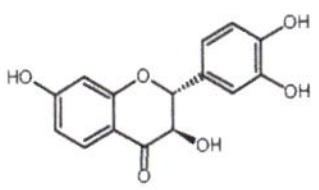

2,3-dihydrofisetin
(132)

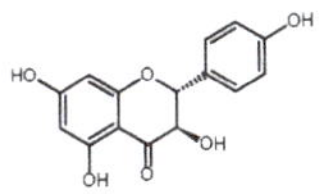

aromadendrin **(133)**
2,3-dihydrokaempferol

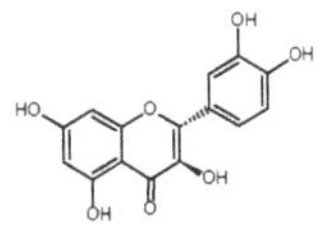

taxifolin **(134)**
(2,3-dihydroquercetin)

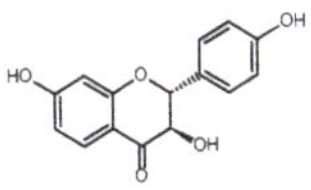

garbanzol **(135)**
(3,7,4'-trihydroxyflavanone)

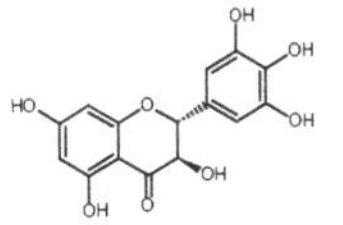

ampelopsin **(136)**
(2,3-dihydromyricetin)

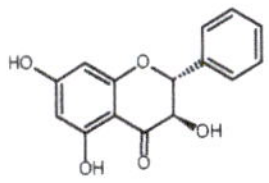

pinobanksin **(137)**
(3,5,7-trihydroxyflavanone)

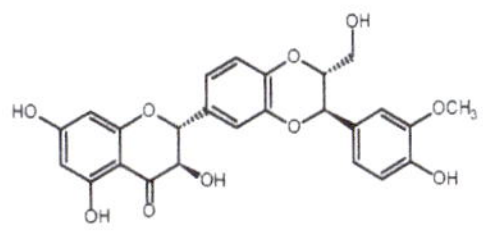

silibinin **(138)**
(sylimarin)

chalcones

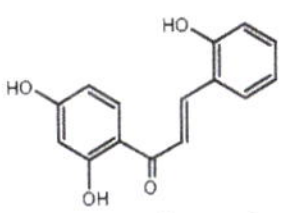

2,2',4'-trihydroxychalcone
(139)

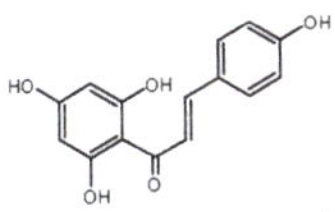

naringenin chalcone **(140)**
2',4,4',6'-tetrahydroxychalcone

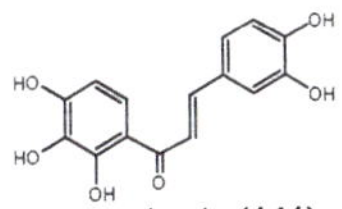

okanin **(141)**

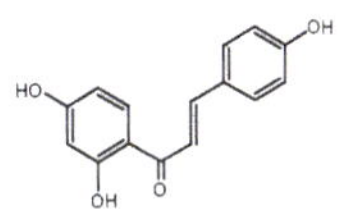

isoliquiritigenin **(142)**
2',4,4'-trihydroxychalcone

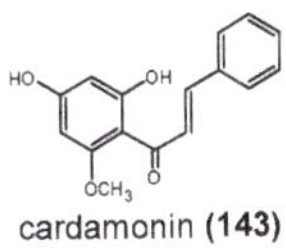

cardamonin **(143)**

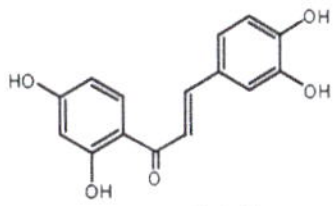

butein **(144)**
2',3,4,4'-tetrahydroxychalcone

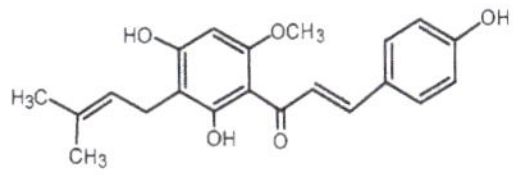

xanthohumol **(145)**

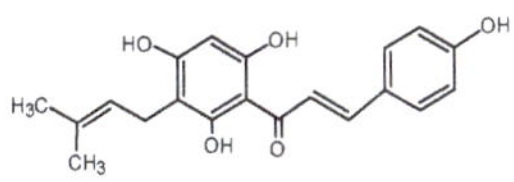

desmethylxanthohumol **(146)**

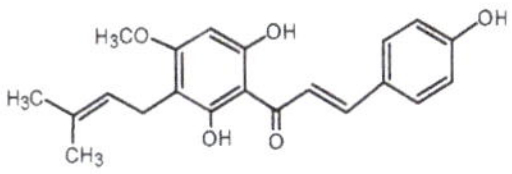

xanthogalenol **(147)**

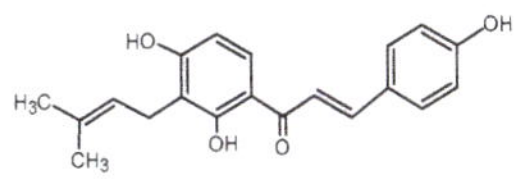

isobavachalcone **(148)**

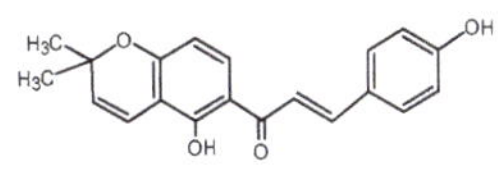

4-hydroxylonchocarpin **(149)**

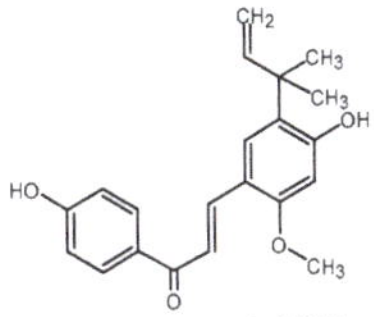

licochalcone A **(150)**

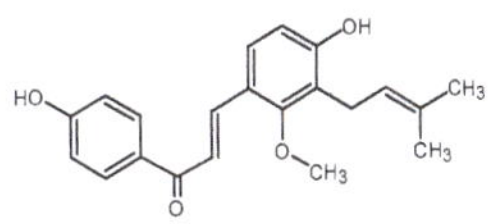

licochalcone C **(151)**

dihydrochalcones

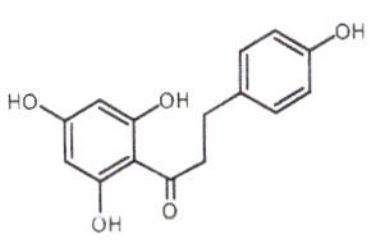

phloretin **(152)**

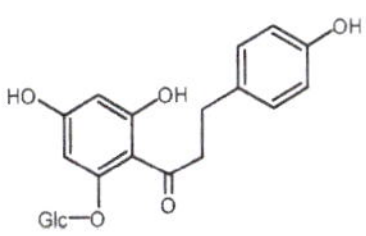

phloridzin **(153)**
(phloretin 2'-glucoside)

Figure 1. The chemical structure of various classes of flavonoids. Based on [16,24,33–40].

hydroxybenzoic acid derivatives

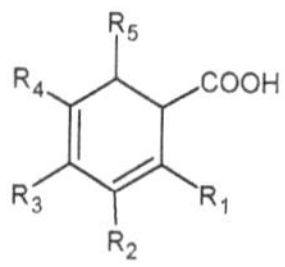

Benzoic acid **(154)**	R_1 = H, R_2 = H, R_3 = H, R_4 = H, R_5 = H
3-Hydroxybenzoic acid **(155)**	R_1 = H, R_2 = OH, R_3 = H, R_4 = H, R_5 = H
4-Hydroxybenzoic acid **(156)**	R_1 = H, R_2 = H, R_3 = OH, R_4 = H, R_5 = H
Protocatechuic acid **(157)**	R_1 = H, R_2 = OH, R_3 = OH, R_4 = H, R_5 = H
Gentisic acid **(158)**	R_1 = OH, R_2 = H, R_3 = H, R_4 = OH, R_5 = H
Gallic acid **(159)**	R_1 = H, R_2 = OH, R_3 = OH, R_4 = OH, R_5 = H
Salicylic acid **(160)**	R_1 = OH, R_2 = H, R_3 = H, R_4 = H, R_5 = H
Vanillic acid **(161)**	R_1 = H, R_2 = OCH_3, R_3 = OH, R_4 = H, R_5 = H
Syringic acid **(162)**	R_1 = H, R_2 = OCH_3, R_3 = OH, R_4 = OCH_3, R_5 = H
Veratric acid **(163)**	R_1 = H, R_2 = OCH_3, R_3 = OCH_3, R_4 = H, R_5 = H

hydroxycinnamic acid derivatives

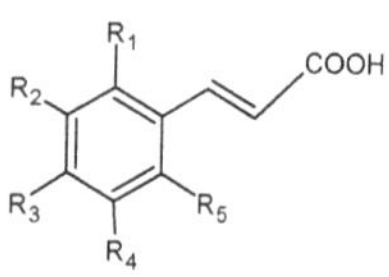

Cinnamic acid* **(164)**	R_1 = H, R_2 = H, R_3 = H, R_4 = H, R_5 = H
o-Coumaric acid **(165)**	R_1 = OH, R_2 = H, R_3 = H, R_4 = H, R_5 = H
p-Coumaric acid **(166)**	R_1 = H, R_2 = H, R_3 = OH, R_4 = H, R_5 = H
Caffeic acid **(167)**	R_1 = H, R_2 = OH, R_3 = OH, R_4 = H, R_5 = H
Ferulic acid **(168)**	R_1 = H, R_2 = OCH_3, R_3 = OH, R_4 = H, R_5 = H
Sinapic acid **(169)**	R_1 = OH, R_2 = OCH_3, R_3 = OH, R_4 = OCH_3, R_5 = H

* *nonphenolic*

chlorogenic acids and other phenolic acids derivatives

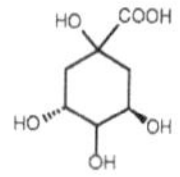

quinic acid **(170)**

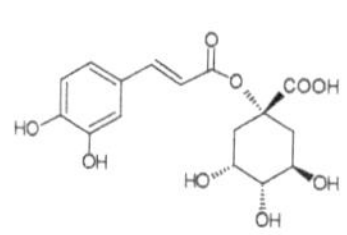

1-*O*-caffeoylquinic acid **(171)**

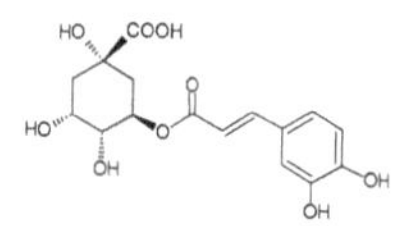

chlorogenic acid **(172)** (3-*O*-caffeoylquinic acid)

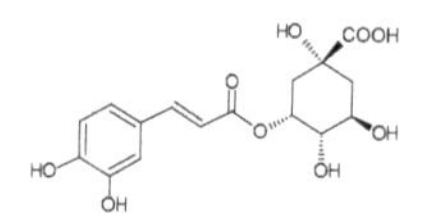

5-*O*-caffeoylquinic acid **(173)** (neochlorogenic acid)

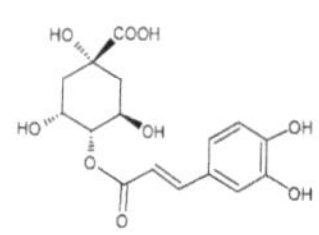

cryptochlorogenic acid **(174)** (4-*O*-caffeoylquinic acid)

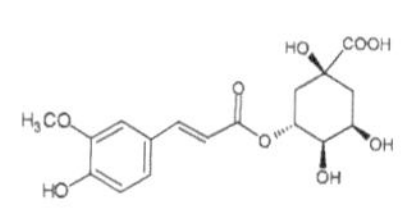

3-*O*-feruloylquinic acid **(175)**

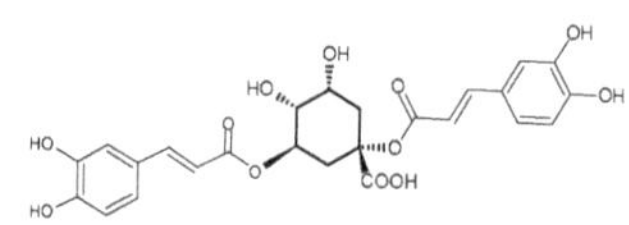

1,3-*O*-dicaffeoylquinic acid **(176)**

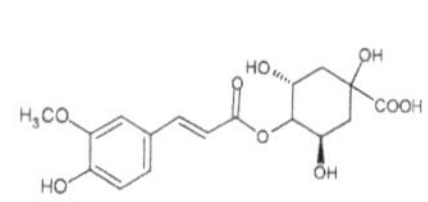

4-*O*-feruloylquinic acid **(177)**

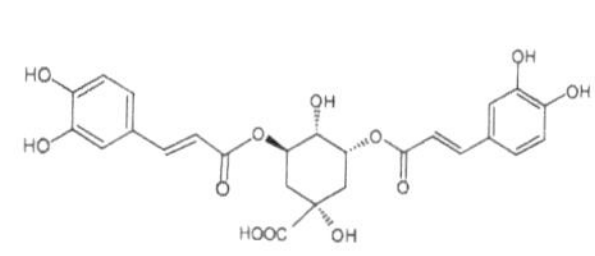

3,5-*O*-dicaffeoylquinic acid **(178)**

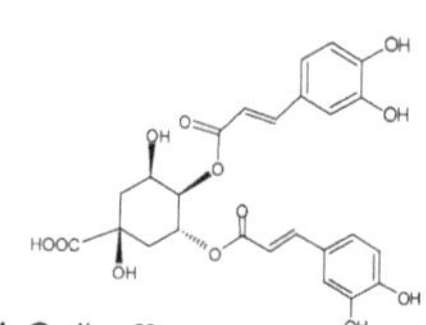

3,4-*O*-dicaffeoylquinic acid **(179)**

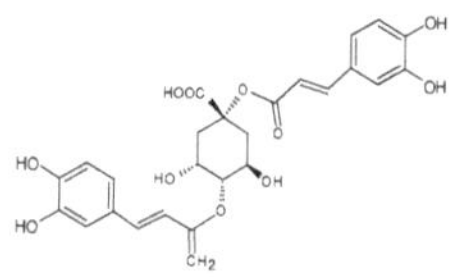

1,4-*O*-dicaffeoylquinic acid **(180)**

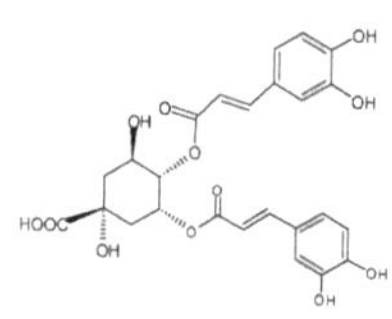

4,5-*O*-dicaffeoylquinic acid **(181)**

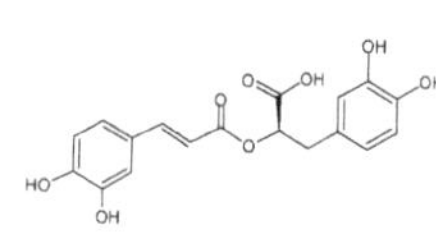

rosmarinic acid **(182)**

Figure 2. *Cont.*

carboxylic acids

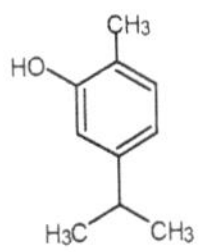

carvacrol **(183)**

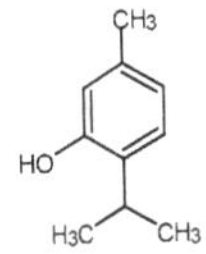

thymol **(184)**

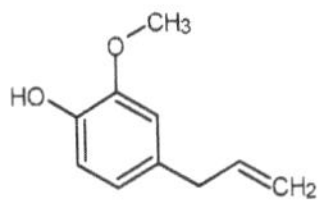

eugenol **(185)**

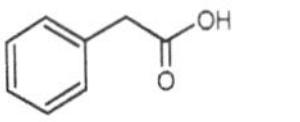

phenyacetic acid **(186)**

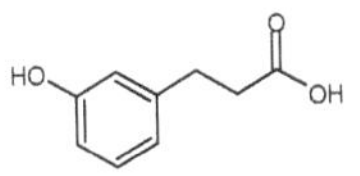

dihydro-3-coumaric acid **(187)**
(3-(3'-hydroxyphenyl)propionic acid)

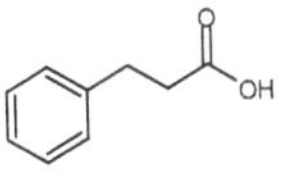

hydroxycinnamic acid **(188)**
(3-phenylpropionic acid)

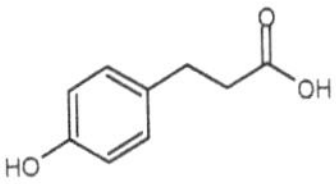

phloretic acid **(189)**
(3-(4'-hydroxyphenyl)propionic acid)

lignans

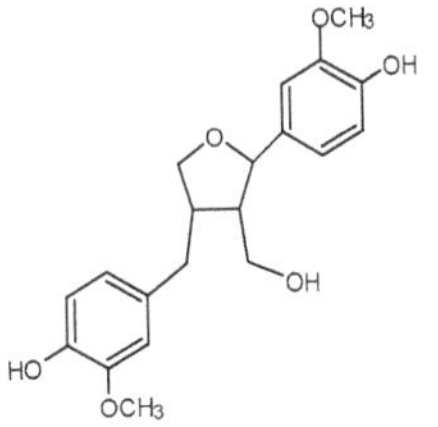

lariciresinol **(190)**

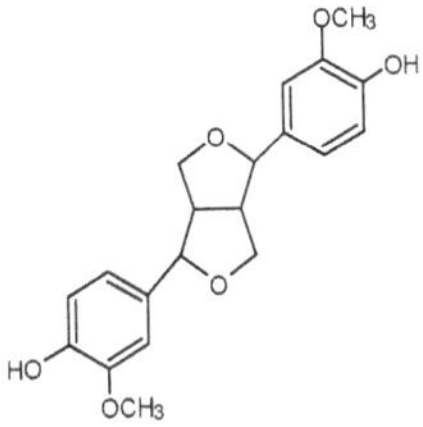

pinoresinol **(191)**

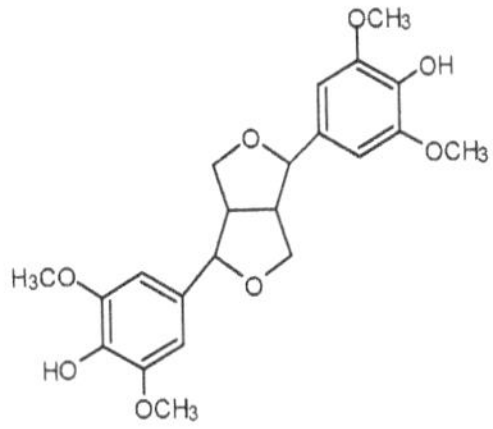

syringaresinol **(192)**

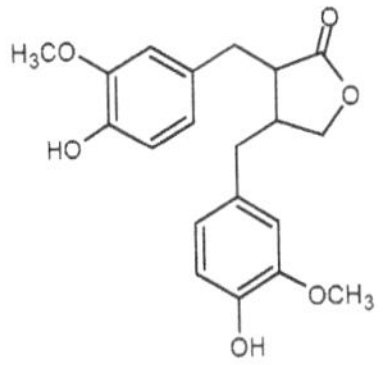

matairesinol **(193)**

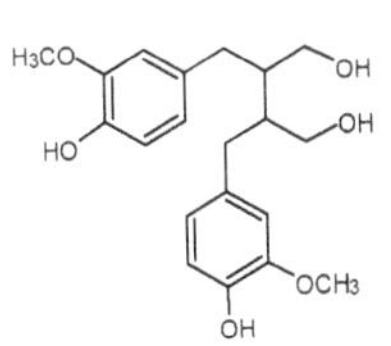

secoisolariciresinol **(194)**

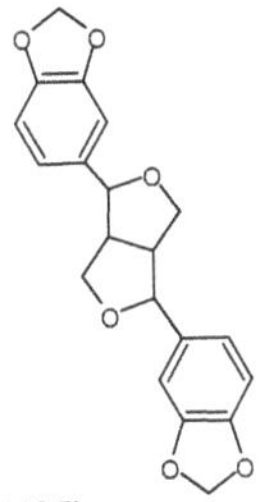

sesamin **(195)**

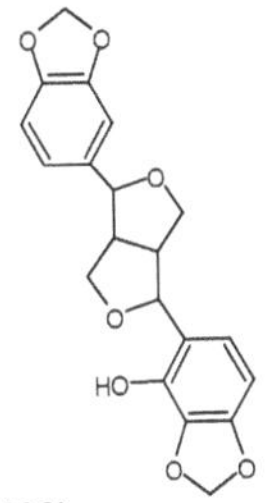

sesaminol **(196)**

coumarins

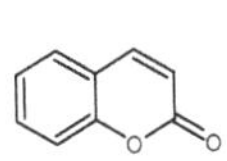

coumarin **(197)**

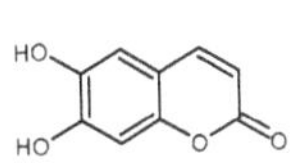

esculetin **(198)**
(6,7-dihydroxycoumarin)

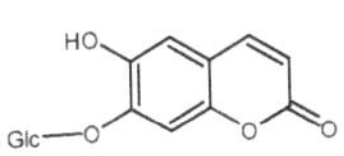

esculin **(199)**
(escultein 6-*O*-glucoside)

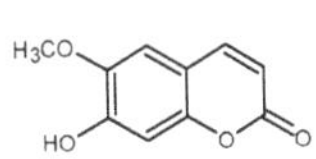

scopoletin **(200)**

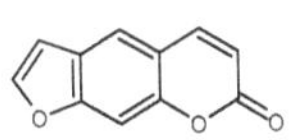

psolaren **(201)**

Figure 2. *Cont.*

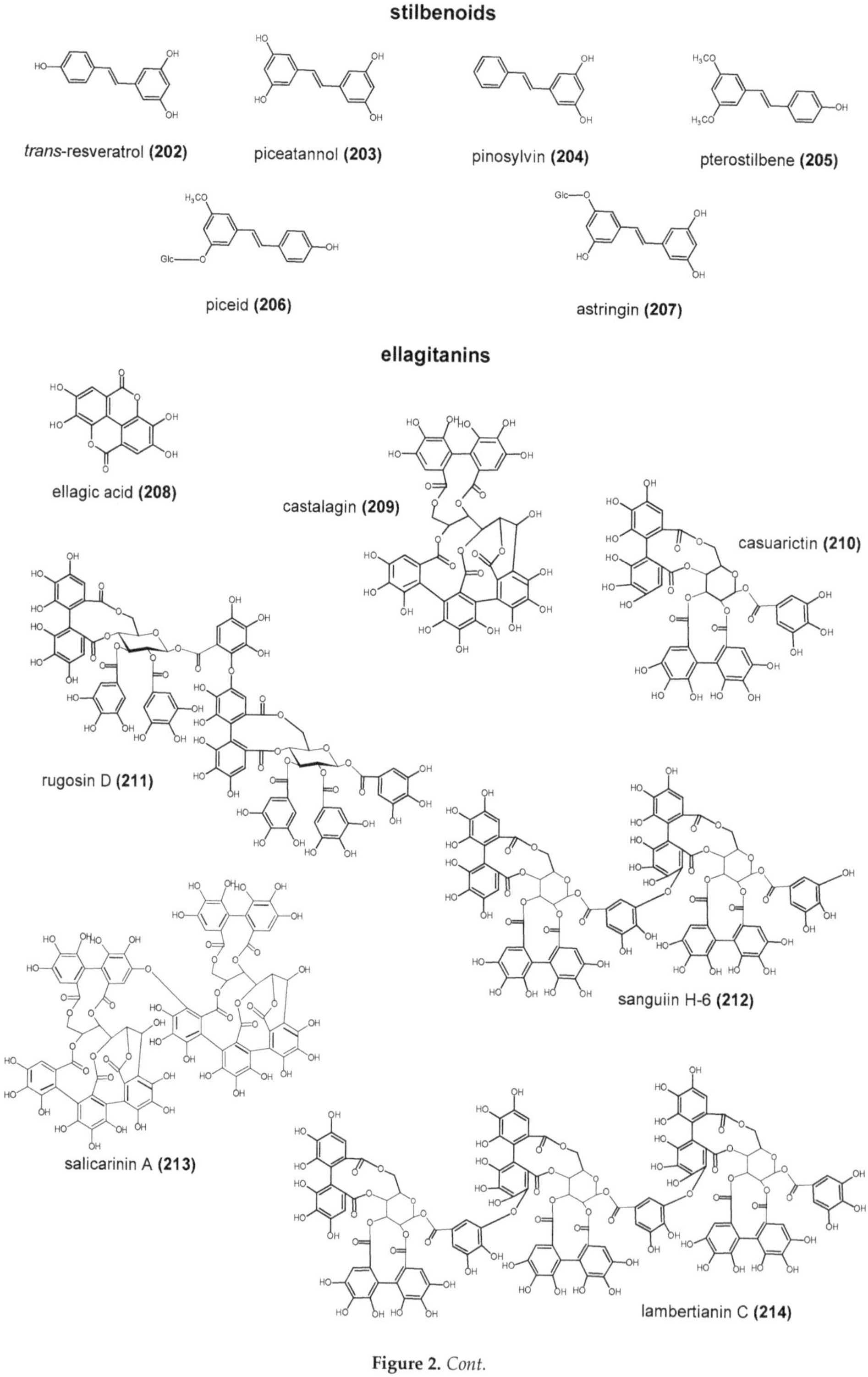

Figure 2. *Cont.*

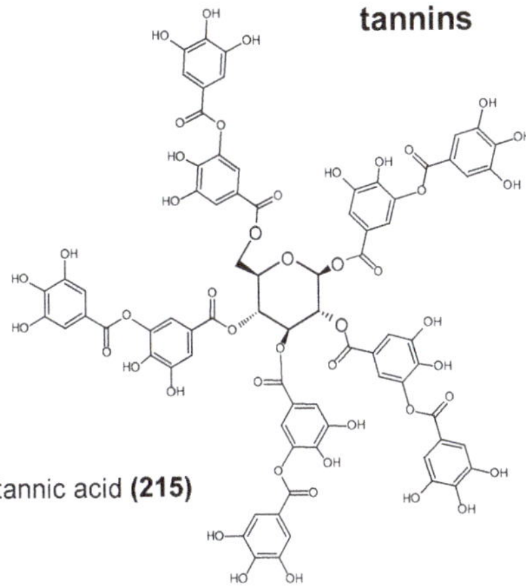

tannic acid **(215)**

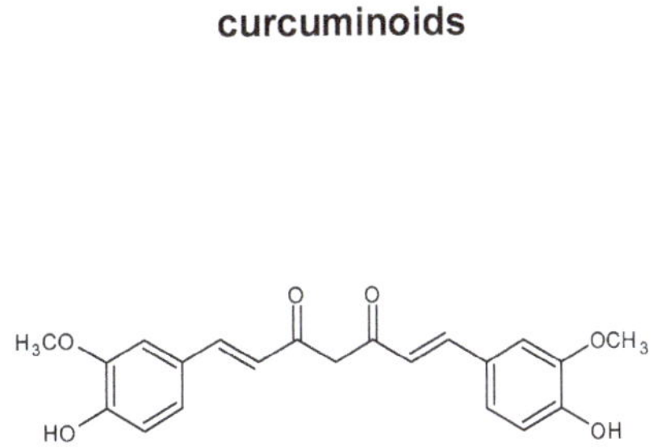

curcumin **(216)**

Figure 2. The chemical structure of various groups of non-flavonoid polyphenols. Based on [16,24,33–40].

3. The Impact of Polyphenols on Microorganisms and the Mechanism of Their Action

Herbs and spices have a long history of being used as natural food preservatives and within folk medicine. It is due to substances with the antimicrobial activity they contain, such as flavonoids, anthocyanins, alkaloids, glycosides, saponins, coumarins, tannins, vitamins, phenolic acids and many more. Essential oils (a complex mixture of various bioactive compounds, inter alia, polyphenols), plant extracts and pure polyphenols are large groups of compounds with strong antibacterial properties. It has been proven many times that thyme, oregano, rosemary, sage, mint and other herbs and spices can inhibit Gram-positive and Gram-negative bacteria, including pathogens [20]. There are many scientific reports and publications demonstrating that pure polyphenols or bioactive compounds present in various types of plant preparations (e.g., aqueous, ethanolic or methanolic extracts, essential oils, enriched extracts) can exert both negative and positive impact on microorganisms. Some examples of such impact are presented in Table 1.

Table 1. Examples of the negative and positive impact of pure polyphenols or plant preparations (aqueous, ethanolic, methanolic or other extracts, essential oils, enriched extracts) on food-associated bacteria or the representatives of the human gastrointestinal tract microbiota.

Plant Material, Preparation or Source	Pure Bioactive Compounds or Polyphenols Identified in Plant Material	Impact on Microorganisms	Reference
Ethanolic extracts of: rosemary, pomegranate peel, grape seed		Antimicrobial activity of rosemary extract against *S. aureus*, *E. coli*, *P. aeruginosa*, *K. pneumonia*, *B. subtilis*, *M. luteus* and *C. albicans*. Pomegranate peel extract inhibited all mentioned microorganism except *P. aeruginosa*. Inhibitory impact on *S. aureus*, *E. coli*, *B. subtilis*, *M. luteus* and *C. albicans*.	[41]
Ethanolic extract of sage (*Salvia officinalis*) flower, leaf and stem and essential oils	Manool, α-pinene, camphene, camphor, limonene, bornyl, 1,8-cineole, linalool, cis- and trans-thujone, acetate, α-humulene	Antibacterial activity against *B. subtilis*, *S. aureus*, *S. enteritidis*, *P. aeruginosa*, *E. coli*. Extracts had stronger antibacterial activity that the essential oils.	[42]

Table 1. *Cont.*

Plant Material, Preparation or Source	Pure Bioactive Compounds or Polyphenols Identified in Plant Material	Impact on Microorganisms	Reference
Ethanolic extracts of Heliotropum europaeum		Antimicrobial activity against *B. subtilis*, *E. coli*, *P. aeruginosa*, *P. fluorescens*, *S. aureus*, *MRSA*, *S. epidermidis*, *S. odorifera P. vulgaris*, *K. pneumoniae*, *E. faecium* and clinic isolate of *E. faecalis*. No impact on *E. aerogenes*, multidrug resistant strains of *E. coli* and *K. pneumoniae*, *E. faecalis* ATCC 29212, *L. monocytogenes*, *S. haemolyticus*, *S. boydi*, *A. baumannii*, and *S. flexneri*.	[43]
Ethanolic extracts from thyme (*Thymus vulgaris*) and sage (*Salvia officinalis*) leaves	Flavonoids tannins rosmarinic acid **(182)**, caffeic acid **(167)**, chlorogenic acid **(172)**, carnosol	Inhibited growth of *S. aureus*, *Vibrio tubiashii*, *L. pneumophila*, *M. luteus*, *Streptococcus* sp., *B. cereus*.	[44]
Aqueous and methanol extracts of *Salvadora persica* L.		Aqueous extract inhibited in dose-dependent manner all tested microorganisms, especially *Streptococcus mutans*, *S. faecalis*, *S. pyogenes*, as well as *S. aureus*, *L. acidophilus*, *P. aeruginosa*, and *C. albicans*. Methanol extract was inactive against *L. acidophilus* and *P. aeruginosa*.	[45]
Hydrodistilled essential oils from *Mentha piperita* and *Rosmarinus officinalis*	Mint: α-terpinene, piperitenone oxide, isomenthone, trans-carveol and β-caryophyllene. Rosemary: piperitone, α-pinene, linalool, camphor, 1,8-cineole	Inhibition of the growth of *Streptococcus mutans* and *S. pyogenes*. Inhibition of biofilm formation by *S. mutans*.	[46]
Phenolic extracts from blueberry, lingonberry, blackcurrant, raspberry, cloudberry, cranberry and strawberry Pure polyphenols	Extracts contained anthocyanins, flavonols, flavan-3-ols, hydroxycinnamate. myricetin **(114)**, luteolin **(49)**, apigenin **(38)**, kaempferol **(202)**, quercetin **(111)** rutin **(121)** isoquercitrin **(126)** (+)-catechin **(7)** phenolic acids: trans-cinnamic acid **(164)**, *m*-coumaric acid, caffeic acid **(167)**, ferulic acid **(168)**, chlorogenic acid **(172)**	Strong antimicrobial activity against *Salmonella enterica* ser. Thyphimurium SH5014, *E. coli* CM871 and *E. coli* 50. Inhibition of *Lactobacillus rhamnosus* and *L. rhamnosus* GG growth by cloudberry, raspberry and strawberry extracts at higher concentrations. Strawberry extract was effective inhibitor against *E. faecalis* and *Bifidobacterium lactis*. Strong inhibition of the growth of all lactic acid bacteria derived from the human gut, and *E. coli* growth by myricetin; but no impact on *Salmonella* Typhimurium and *Lactobacillus plantarum* from beer. Bacteriostatic impact of luteolin on some *Lactobacillus* species, *Bifidobacterium lactis* and *Enterococcus faecalis*, no impact on Gram-negative bacteria. No impact of (+)-catechin, kaempferol, apigenin, isoquercitrin, and rutin. Phenolic acids at concentration 0.5 mg/well phenolic acids inhibited the growth of *E. coli*, and *S. enterica*.	[47]

Table 1. *Cont.*

Plant Material, Preparation or Source	Pure Bioactive Compounds or Polyphenols Identified in Plant Material	Impact on Microorganisms	Reference
Methanolic extracts of six species of Hypericum	tannin, flavonoid and phenolic acids, among them quercitrin **(123)**, hyperoside **(120)**, isoquercitrin **(126)**, chlorogenic acid **(172)** were identified	*H. caprifoliatum* – the most active, inhibition of *S. aureus*. *H. polyanthemum* and *H. ternum* extracts antibacterial activity against *B. subtilis*. No activity of extracts against activity against *S. epidermidis* and *E. coli*.	[48]
Extracts of meadowsweet, willow herb, cloudberry, raspberry, bilberry and crowberry Extracts of white birch, pine, potato	Extract contained flavonoids and phenolic acid Pure compounds: quercetin **(111)**, morin **(110)**, rutin **(121)**, naringenin **(81)**, naringin **(85)**, kaempferol **(202)**	*E. coli* and *S. aureus* growth inhibition. Only *S. aureus* growth inhibition. Quercetin, morin, naringenin inhibited *S. aureus*, *S. epidermidis*, *E. coli*, *B. subtilis*, *M. luteous*, *P. aeruginosa*. Glycosides rutin and naringin were inactive. Kaempferol inhibited only *S. aureus*.	[49]
Peppermint essential oil and various extracts		Strong inhibition of *S. aureus* and *S. pyogenes* growth, less impact on of *E. coli* and *Klebsiella pneumoniae*. Petroleum ether, chloroform and ethyl acetate extracts were more effective antibacterial agents than ethanol and aqueous extracts. The strength of inhibition by extracts: *S. aureus* > *K. pneumoniae* > *S. pyogenes* > *E. coli*.	[50]
Peppermint oil Green tea polyphenols (GTP)	53 constituents of oil, e.g., menthol, menthone, neomenthol, menthofuran, (+)-limonene, piperiton, 3-octanol, *cis*-jasmone, mint lactone, (−)-myrtenol, piperitol, eugenol **(185)**, carvacrol **(183)**, 2-ethylfuran, ocimene (−)-epigallocatechin **(14)**, (−)-epigallocatechin-3-gallate **(18)**	Mentioned compounds had strong antibacterial activity against non-pathogenic *E. coli*. Oil, menthol, menthone and neomenthol killed the enterohemorrhagic strain *E. coli* O157:H7 at concentrations 400 µg/mL within 1 h. GTP inhibited *E. coli* O157:O7 growth at concentration of 800 µg/mL within 18 h. The synergistic effect was reported for peppermint oil + GTP and menthol + GTP.	[51]
Cranberry, blueberry and grape seed extracts and their synergy mixtures		Dose-dependent inhibitory activity against *Helicobacter pylori*. The synergy mixtures with higher concentrations of polymeric phenolics (procyanidins and tannins) had higher antimicrobial activity against *H. pylori*.	[52]
Chinese green tea extract	(−)-epicatechin gallate (ECG) **(17)**, (−)-epigallocatechin-3-gallate (EGCG) **(18)**	Crude green tea extract caused growth inhibition (by 44–100%) of food-borne pathogens *Escherichia coli* O157:H7, *Salmonella* Typhimurium DT104, *L. monocytogenes*, *S. aureus*, a diarrhoea food-poisoning pathogen *Bacillus cereus*. The lowest MIC90 values had EGCG: against *S. aureus* MSSA was 58 µg/mL, while against MRSA 37 µg/mL. ECG had MIC 309 and 105 µg/mL, respectively.	[53–59]

Table 1. *Cont.*

Plant Material, Preparation or Source	Pure Bioactive Compounds or Polyphenols Identified in Plant Material	Impact on Microorganisms	Reference
Green tea, Chung tea, Black tea	(−)-epicatechin (EC) **(12)**, (−)-epigallocatechin (EGC) **(14)**, (−)-epigallocatechin-3-gallate **(EGCG)** (18), (−)-epicatechin gallate (ECG) **(17)**, Teaflavins, quercetin **(111)**	EC and EGC strongly inhibited *S. aureus* ATCC 29213 (MIC 12.5 μg/mL), while ECG had MIC 50 μg/mL, ECG and EGCG strongly inhibited *E. coli* ATCC 25922 (MIC 12.5 μg/mL). ECG inhibited *Enterobacter cloacae* 1321E (MIC 12.5 μg/mL) and *S. pyogenes* (MIC 25 μg/mL), EC inhibited the growth of *P. aeruginosa* (MIC 25 μg/mL)	[54]
9 Hot water extracts of various tea (oolong, green, black, white)	gallic acid **(159)**, quercetin **(111)**, caffeine, (+)-catechin **(7)**, (−)-epicatechin **(12)**, (−)-epigallocatechin **(14)**	5-min extracts inhibited the growth of Helicobacter pylori, but growth of probiotics Bifidobacterium longum and Lactobacillus acidophilus was not affected.	[55]
	(−)-epigallocatechin-3-gallate **(18)**	EGCG inhibited intracellular groth of Listeria monocytogenes in macrophages	[56]
	(−)-epigallocatechin-3-gallate **(18)**	EGCG inhibits major functions of cellular and surface proteins, leading to growth inhibition of Bacillus subtilis	[57]
	(−)-epigallocatechin-3-gallate **(18)**, (−)-epigallocatechin (EGC) **(14)**	EGCG inhibits *S. aureus*, MRSA, *S. mutans*, *E. coli*, *P. aeruginosa*, *K. pneumoniae*. EGC inhibits *E. coli*	[58]
	(+)-catechin **(7)**, (−)-epicatechin **(12)**	Catechin caused significant decrease in the growth of the *Clostridium histolyticum* group and a marked increase in the growth of the beneficial bacterial group of *C. coccoides–Eubacterium rectale*, *Lactobacillus* spp. and *Bifidobacterium* spp. Epicatechin caused a significant increase in the growth of the *Eubacterium rectale–C. coccoides*.	[59]
	Pure polyphenols, inter alia: baicalein **(45)**, quercetin **(111)**, myricetin **(114)**, naringenin **(81)**, naringin **(85)**, hesperetin **(83)**, hesperidin **(86)**, resveratrol **(202)**, gallic acid **(159)**	Aglycones quercetin, naringenin, hesperetin inhibited growth of tested bacteria, while their glycoside did not. The lowest MIC were reported for: baicalein, myricetin, hesperetin and kaempferol against *E. coli* O157; baicalein nad myricetin against *S. aureus;* and baicalein against *Salmonella* Typhimurium, *Enterobacter sakazakii* and *Vibrio parahemolyticus*.	[60]
Citrus fruit	hesperetin **(83)**, naringenin **(81)**, poncirin **(90)**, diosmetin **(50)**	Inhibition of the growth of *H. pylori*.	[61]
	rutin **(121)**, quercetin **(111)**	Inhibition of *L. monocytogenes* growth.	[62]

Table 1. *Cont.*

Plant Material, Preparation or Source	Pure Bioactive Compounds or Polyphenols Identified in Plant Material	Impact on Microorganisms	Reference
	(+)-catechin **(7)**, quercetin **(111)**, naringenin **(81)**, hesperetin **(83)**, rutin **(121)**, naringin **(85)**, hesperidin **(86)**	Aglycones naringenin and hesperetin, inhibited growth of almost all analysed bacteria (MIC $\geq$ 250 µg/mL). Catechin, and glycosides naringin, hesperidin, and rutin had no impact on tested intestinal bacteria. Quercetin had strong inhibitory impact (MIC 20–50 µg/mL) on *Ruminococcus gauvreauii, Bacteroides galacturonicus* and *Lactobacillus* sp. growth.	[63]
	13 phenolic acids: benzoic acid **(154),** 3-hydroxybenzoic acid **(155)**, 4-hydroxybenzoic acid **(156)**, 4-hydroxy-3-methoxy-benzoic acid, 3,4-dihydroxybenzoic acid, phenylpropionic acid **(188)**, 3-hydroxyphenylpropionic acid **(187)**, 4-hydroxyphenylpropionic acid **(189)**, 3,4-dihydroxyphenylpropionic acids, phenylacetic acid **(186)**, 3-hydroxyphenylacetic acid, 4-hydroxyphenylacetic acid, 3,4-dihydroxyphenylacetic acid	No impact on *P. aeruginosa* PAO1. Impact on *E. coli* was strain dependent. *E. coli* ATCC 25922 inhibition by benzoic and 4-hydroxy-3-methoxy-benzoic, phenylacetic and phenylpropionic acids at a concentration of 1000 mg/mL. *E. coli* O157:H7 (CECT 5947) was susceptible to benzoic, 3-hydroxybenzoic, 4-hydroxybenzoic, 4-hydroxy-3-methoxy-benzoic acids, phenylacetic, 3-hydroxy-phenylacetic, 4-hydroxyphenylacetic and 3,4-dihydroxyphenylacetic acids, phenylpropionic, 3-hydroxyphenylpropionic acid and 4-hydroxyphenylpropionic acids. *Lactobacillus paraplantarum* LCH7 was the most susceptible to the action of phenolic acids, while *L. fermentum* LPH1 was the most resistant. The susceptibility was strain-dependent. The most active compound were: 4-hydroxybenzoic acid for *L. fermentum* CECT 5716, *L. fermentum* LPH1, *L. brevis* LCH23, and *L. plantarum* LCH17; 4-hydroxybenzoic acid.and phenylpropionic acid for *L. paraplantarum* LCH7 and *L. coryniformis* CECT 5711, while 3-hydroxyphenylpropionic acid for *L. fermentum* CECT 5716.	[64]
Extracts from 3 *Eucalyptus* species	2′,6′-dihydroxy-3′-methyl-4′-methoxy-dihydrochalcone, eucalyptin **(48)**, 8-desmethyl-eucalyptin	Significantly inhibited growth of Gram-positive bacteria: *S. aureus, MRSA, B. cereus, E. faecalis, Alicyclobacillus acidoterrestris, Propionibacterium acnes.* Not show strong antibacterial activity against Gram-negative *E. coli* and *P. putida.*	[65]
Essential oils from *Origanum vulgare* and *Thymus vulgaris Ocinum basilicum* oil		Minimal Bactericidal Concentration /MBC/ $\leq$ 5 mg/mL against pathogens (*E. coli, Salmonella Enteritidis*, and *S.* Typhimurium), and beneficial bacteria (*Lactobacillus acidophilus* and *Bifidobacterium breve*). Higher activity against pathogenic bacteria (MBCs $\leq$ 10 mg/mL) than beneficial bacteria (MBCs of 80 mg/mL).	[66]

Table 1. *Cont.*

Plant Material, Preparation or Source	Pure Bioactive Compounds or Polyphenols Identified in Plant Material	Impact on Microorganisms	Reference
	13 common flavonoids (flavones, flavonols, flavanones) and 6 organic acids (aliphatic and aromatic acids) kaempferol **(108)**, quercetin **(111)**, chlorogenic acid **(172)**, salicylic acid **(160)**	Antimicrobial activity against Gram-negative bacteria *E. coli* and *P. aeruginosa*, higher than impact on Gram-positive ones: *E. faecalis* and *S. aureus*. Kaempferol, quercetin and chlorogenic acid had no significant influence on *P. aeruginosa*. Salicylic acid – the highest inhibitory activity against all tested bacterial species (MIC = 250–500 µg/mL).	[67]
A flavan-3-ol enriched grape seed extract		Inhibited growth of *Streptococcus thermophilus*, *Bifidobacterium lactis* BB12, *Lactobacillus fermentum*, *L. acidophilus* and *L. vaginalis*. Stimulated growth of some *Lactobacillus plantarum*, *L. casei*, and *L. bulgaricus* strains. No impact on *Bifidobacterium breve* 26M2 and *B. bifidum* HDD541 growth	[68]
	naringenin **(85)**, hes-peridin **(86)**, rutin **(121)**, quercetin **(111)**, gal-lic acid **(159)**, caffeic acid **(167)**, *p*-coumaric acid **(166)**, ferulic acid **(168)**, chlorogenic acid **(172)**, vanillic acid **(161)**, sinapic acid **(169)**, hesperidin **(86)**, quercetin **(111)**	All polyphenols influenced the growth of *Bifidobacterium adolescentis* and *B. bifidum* was assessed. Both the stimulatory and inhibitory effects of polyphenols were observed. Coumaric acid had strongest stimulatory effect on *B. bifidum*, while vanillic and caffeic acid stimulated *B. adolescentis*. Inhibitory dose-dependent impact of hesperidin and quercetin, on *B. bifidum* and *B. adolescentis*.	[69]
Tea phenolics and their derivatives	3-phenylpropionic acid **(188)**, caffeic acid **(167)**, gallic acid **(159)**	Inhibition of pathogenic *Clostridium perfringens*, *C. difficile* and *Bacteroides* spp. Less effect on commensal *Clostridium* spp., *Bifidobacterium* spp. and probiotic *Lactobacillus* sp. The growth of *Lactobacillus* ssp. and *Bifidobacterium* spp. inhibited by caffeic acid, 3-phenylpropionic acid and to the lesser extent by gallic acid, while *Lactobacillus casei* Shirota growth inhibited only by 3-phenylpropionic acid.	[70]
Red wine polyphenols		Stimulated growth of *Bifidobacterium* and *Lactobacillus*.	[71]
Tannin-rich diet		Significantly decreased growth of *Clostridium* spp. with a corresponding shift toward Entero-bacteriaceae and Bacteroides	[72]
Pomegranate extract	ellagitannins	Enhancing the growth of beneficial bacteria *Akkermansia muciniphila*.	[73]
	resveratrol **(202)**	Enhancing the growth of beneficial bacteria *Akkermansia muciniphila*, *Lactobacillus* and *Bifidobacterium*.	[74] [75]
	polymeric procyanidins	Enhancing the growth of beneficial bacteria *Akkermansia muciniphila*. Markedly decreased the Firmicutes/Bacteroidetes ratio	[76]

Table 1. *Cont.*

Plant Material, Preparation or Source	Pure Bioactive Compounds or Polyphenols Identified in Plant Material	Impact on Microorganisms	Reference
Grape polyphenols		Enhancing the growth of beneficial bacteria *Akkermansia muciniphila*, and decreasing the proportion of Firmicutes to Bacteroidetes.	[77]
	rutin **(121)**	The beneficial impact on *Lactobacillus* and *Bifidobacterium*	[63]
Cocoa	polyphenols	The beneficial impact on *Lactobacillus* and *Bifidobacterium*	[78]
Red wine polyphenols		Significantly increased the number of Enterococcus, *Prevotella*, *Bacteroides*, *Bifidobacterium*, *Bacteroides uniformis*, *Eggerthella lenta*, and *Blautia coccoides–Eubacterium rectale* groups	[79]
Blueberry	anthocyanidins	The beneficial impact on *Lactobacillus acidophilus* and *Bifidobacterium* spp	[80]
Red wine	polyphenols	Significantly increase of the number of faecal bifidobacteria and *Lactobacillus* (intestinal barrier protectors), *Faecalibacterium prausnitzii* and *Roseburia* (butyrate-producing bacteria)	[81]
	flavan-3-ols epigallocatechin gallate **(18)**, (−)-epigallocatechin **(14)**, procyanidin B1 **(19)**, procyanidin B2 **(20)**	Significant inhibition of *L. acidophilus* LA-5 and *L. plantarum* IFPL379 adhesion, except 4 compounds. Enhanced *L. acidophilus* LA-5 adhesion to Caco-2 cells. Increased *L. casei* LC115 adhesion to Caco-2 cells. Increased the adhesion of *Lactobacillus casei* LC115 to HT-29 cells.	[82]
Extract from apples	procyanidin B2 **(20)**, chlorogenic acid **(172)**	The increase of adhesion of *Lactobacillus gasseri* and *Lactobacillus casei* to intestinal epithelial cells.	[83]
Cranberry	proanthocyanidins with A-type linkages	Inhibition of the adhesion of both antibiotic-susceptible and antibiotic-resistant strains of uropathogenic P-fimbriated *E. coli*.	[84,85]
	anthocyanidins: pelargonidin **(25)**, cyanidin **(26)**, delphinidin **(27)**, cyanidin-3-glucoside **(36)**	Inhibition of the growth of *E. coli* strain CM871. No effect on *E. faecalis* strain E-203, *S. enterica* SH-5014 and strains of the *Lactobacillus* and *Bifidobacterium*.	[47]
Wine extract	quercetin **(111)**, flavan-3-ols, anthocyanins	No impact on species belonging to the *Lactobacillus*, *Enterococcus*, *Bacteroides*, and *Bifidobacterium* genera.	[86]
Grape seeds and pomace	tannic acid **(215)**	Potent growth-promoting effects on *L. acidophilus*.	[87]
	gallic acid **(159)**, and free anthocyanins, vanillic acid **(161)**, protocatechuic acid **(157)**	Activated cell growth and the rate malolactic fermentation of *Leuconostoc oenos*. Vanillic acid showed a slight inhibiting effect, while protocatechuic acid had no effect.	[88]

As can be seen in the table, not only pathogenic bacteria growth can be inhibited by polyphenols. It was proved that some beneficial microorganisms, inter alia lactic acid bacteria and probiotics strains, can also be inhibited. However, the bacteriostatic or bactericidal effect depends both on the polyphenol structure and bacteria species. Susceptibility to some polyphenols was also proved to be strain-dependent [43,47,50,64]. It was proved that some polyphenols exert a beneficial influence on bacteria. They can stimulate growth or at least change the composition of the microbiome in favor of beneficial bacteria such as *Bifidobacterium* and *Lactobacillus*, which both contribute to gut barrier protection; *Akkermansia muciniphila* and *Faecalibacterium prausnitzii* that possess anti-inflammatory effect by inhibiting the activation of NF-kB; and *Roseburia* sp.—the butyrate-producer [15]. *Akkermansia muciniphila* is an anaerobic, mucin-degrading bacterium residing in the healthy intestinal tract of a host that is believed to have several health benefits in humans [89]. Many studies show that various diseases, e.g., obesity, type II diabetes and inflammatory bowel diseases have an association with reducing *A. muciniphila* abundance [90]. Taking into account that polyphenols have been proven to increase the growth of *Akkermansia muciniphila* [72,73,75–77], the beneficial impact of polyphenols on human health may result from other than antioxidant activity. All these findings (Table 1) suggest that polyphenols reaching the large intestine may not only be catabolized to small phenolic acids but also elicit potentially beneficial effects of intestinal probiotic bacteria. Taken together, polyphenols appear to be able to alter gut microecology and, by affecting the total number of beneficial species in the gut, may confer positive gut health benefits.

4. Mechanism of Antibacterial Activity of Polyphenols

As was reported above, the influence of pure polyphenols and plant extract, as well as the strength of that impact on bacteria, differs depending on the kind of both phenolic compounds and bacteria strain. The mechanism of antimicrobial activity of polyphenols against bacteria can differ and also depends both on the polyphenol type and bacteria species. Among the most important mechanisms of the antibacterial action of polyphenols are [58,91–98]:

- Reactions with proteins;
- Inhibition of nucleic acid synthesis by bacterial cells or DNA damage;
- Interaction with the bacterial cell wall or inhibition of cell wall formation;
- Alteration of cytoplasmic membrane function, such as modifications of the membrane permeability or fluidity, cytoplasmic membrane damage and—in the result—the membrane disruption;
- Inhibition of energy metabolism;
- Changes in cell attachment and inhibition of biofilm formation;
- Substrate and metal deprivation.

4.1. Reactions with Proteins

The antibacterial activity of flavonoids may result from their ability to form complexes with proteins through nonspecific forces such as hydrogen bonding and hydrophobic effects, as well as by covalent bond formation [99]. Due to protein binding by polyphenols, they are sequestered into soluble or insoluble complexes, which affects the function of both polyphenol and protein [100]. Proteins modified by polyphenols binding have some amino acids blocked or undergo conformation transitions, which can cause changes in protein structure, solubility, hydrophobicity, thermal stability, and the isoelectric point. In consequence, protein–phenolic complexation leads to changes in their physicochemical and biological properties, including the digestibility and utilization of food proteins as well as the activity of digestive enzymes [101]. It has been demonstrated that naturally occurring polyphenols, e.g., condensed tannins, can inhibit a number of digestive enzymes, including α-glycosidase, α-amylase, lipase, pepsin, trypsin, and chymotrypsin, changing the availability of nutrients and hence modulating the microbiota composition [102–106]. Furthermore, polyphenols can bind to important bacterial proteins such as adhesins, enzymes, cell envelope transport proteins and—by inactivating them—exert an antimicrobial impact. On the other hand, polyphenols complexation with proteins may influence the bioaccessibility and activity of phenolic compounds.

Quinones are known to complex irreversibly with nucleophilic amino acids in proteins, which leads to inactivation and loss of function in the proteins. They possibly interact with cell wall polypeptides, membrane-bound enzymes and surface-exposed adhesins of pathogenic bacteria [107].

Kaempferol-3-rutinoside (nicotiflorin) **(130)** was demonstrated to inhibit *Streptococcus mutans*. The affected protein was sortase A, a membrane enzyme that actively plays a crucial role in bacteria adhesion and the invasion of host cells [108]. A purified *S. mutans* sortase A was inhibited by curcumin **(216)** at a half-maximal inhibitory concentration; curcumin was also found to release the Pac protein to the supernatant and reduce *S. mutans* biofilm formation [109]. The sortase enzymes (cysteine transpeptidases) are used by Gram-positive bacteria to display proteins in a cell surface (e.g., glycoproteins), and they can attach to proteins in the cross-bridge peptide of the cell wall. Hence they are an important virulence factor. Morin **(110)**, myricetin **(114)**, and quercetin **(111)** exhibited strong inhibitory activity against sortase A and B from *S. aureus* [110].

Some flavones were active against *Escherichia coli* by forming complexes with extracellular and soluble proteins [111]. Chen et al. [112] proved that baicalein **(45)** decreased the expression of intracellular adhesin in *S. aureus*. EGCG can bind to porins, so probably this way, it affects the permeability of the outer membrane of Gram-negative bacteria via porin pores [113].

Nakayama et al. [114] demonstrated with two-dimensional electrophoresis that epigallocatechin gallate **(18)** strongly interacted with one of the outer membrane porin proteins of *E. coli*, especially with basic amino acids such as Arg, Lys and His. The docking simulation revealed that EGCG enters into the porin pore and binds to Arg residues present on the inner surface of the pore channel through hydrogen bonding, resulting in inhibition of the porin function.

It also has been shown that tea catechins [115] and various polyphenols [20] have the capacity to sensitize strains of methicillin-resistant *Staphylococcus aureus* to antibiotics. Taylor et al. [115] postulated that catechin gallates (compounds **13–18**) intercalate into phospholipid bilayers, and probably they affect both virulence and antibiotic resistance by perturbing the function of the key processes associated with the bacterial cytoplasmic membrane.

The inhibitory impact of polyphenols on the action of the bacterial efflux pump, which changes transport through the cell wall and cytoplasmic membrane, is also taken into account [91,116]. It has been demonstrated that quinones and chalcones are substrates of bacterial efflux pumps and could be used in combination with the efflux inhibitors in order to improve the accumulation of the drug in the cells to fight against MRSA infections [117]. Kaempferol **(108)** and galangin **(109)** were found to be effective efflux pump inhibitors in *S. aureus* [118].

Flavonoids can modulate the activity of bacterial enzymes, which are crucial for cell life, such are those catalyzing the synthesis of cell wall elements, cell membrane fatty acids or ATP. Fatty acid synthase II (FAS-II) is a key enzyme for the synthesis of fatty acids building the bacterial membranes. It catalyzes fatty acid chain elongation, from 16–24 carbons obtained de novo by FAS-I to long-chain fatty acids of 36–48 carbons as well as mycolic acids [92]. Flavonoids such as isoliquiritigenin **(142)**, butein **(144)**, fisetin **(107)** and 2,2′,4′-trihydroxychalcone **(139)** inhibited FAS-II, thus preventing the growth of *Mycobacterium smegmatis* [119].

Epigallocatechin gallate (EGCG) **(18)** and the related tea catechins potently inhibited both the FabG and FabI reductase steps in the fatty acid elongation cycle [120]. The authors suggested that the presence of the galloyl moiety was essential for inhibitory activity, and EGCG was a competitive inhibitor of FabI and a mixed-type inhibitor of FabG, demonstrating that EGCG interfered with cofactor binding in both enzymes. Furthermore, EGCG inhibited acetate incorporation into fatty acids in vivo. Molecular docking studies conducted by Xiao et al. [121] revealed the importance of the 3-*O*-galloyl or 3-*O*-glycosides side chain at the flavonoid pyran ring in the mechanism of the inhibition of reductase flavoprotein, dihydroorotate dehydrogenase (PyrD), dihydrofolate reductase (DYR), NADH-dependent enoyl-ACP reductase, and the DNA gyrase subunit in *E. coli*. Results obtained in the study demonstrated that EGCG has the strongest binding with NADH-dependent enoyl-ACP reductase (FabI) in comparison with other flavonoids, while quercitrin **(111)** was also the strongest inhibitor of DNA gyrase subunit B (GyrB) among tested 19 flavonoids. The results also indicated that flavonoids that have galloyl moieties, such as EGCG **(18)**, (−)-catechin gallate **(16)**, (−)-epicatechin gallate **(17)**, and (−)-gallocatechin gallate **(13)**, exhibited higher binding affinities to PyrD, FabI, and DYR than their cognates lacking the galloyl group, i.e., (−)-epigallocatechin **(14)**, (+)-catechin **(7)**, (−)-epicatechin **(12)**, and (−)-gallocatechin **(11)**, respectively.

Quercetin **(111)**, apigenin **(38)**, and sakuranetin **(84)** inhibited the activity of β-hydroxyacyl-acyl carrier protein dehydratase from *Helicobacter pylori* (HpFabZ), which is necessary for bacterial fatty acid biosynthesis. These three flavonoids are all competitive inhibitors against HpFabZ by binding to the substrate tunnel and preventing the substrate from accessing the active site [122]. Similarly, β-ketoacyl acyl carrier protein synthase (KAS) III is a key catalyst in bacterial fatty acid biosynthesis. The docking studies between *Enterococcus faecalis* KAS III (efKAS III), and flavonoids proved that naringenin **(81)**, eriodictyol **(94)**, and taxifolin **(134)**, with high-scoring functions and good binding affinities, docked well with efKAS III, causing the *E. faecalis* growth inhibition. Hydrogen bonds between the 5- and 4′-hydroxy groups and the side-chain of Arg38 and the backbone carbonyl of Phe308 were the key interactions for efKAS III inhibition [123].

Both Gram-positive and Gram-negative bacteria produce hyaluronidases, which are an important virulence factor. They enable the bacteria to avoid the immune system and host defense mechanisms. Terpenes (e.g., glycyrrhizin) have been identified as hyaluronic acid lyases (Hyal B from *Streptococcus agalactiae*, Hyal S from *Streptomyces hyalurolyticus*, and Hay C form *Streptococcus equisimilis*). Compounds with many hydroxyl groups inhibited hyaluronate lyase stronger than those with only a few [124].

Flavonoid 5,6-dihydroxy-4′,7,8-trimethoxyflavone, isolated from *Limnophila heterophylla* Benth, was found to effectively kill *Bacillus subtilis* by cell lysis. Moreover, they enhanced the activity of gluconeogenic fructose 1,6-bisphosphatase, but the decreased activity of phosphofructokinase and isocitrate dehydrogenase, the key enzymes of the Embden–Meyerhof–Parnas pathway and the tricarboxylic acid cycle, respectively, was demonstrated [125].

Isoflavones (4-(p-hydroxyphenethyl) pyrogallol and 7,8,4′-trihydroxyisoflavone **(80)** are potent inhibitors of urease, an enzyme produced by *Helicobacter pylori*, which catalyzes the hydrolysis of urea to produce ammonia and carbon dioxide and to protect the bacteria in the acidic environment of the stomach [126]. The structure–activity relationship of these polyphenols revealed that the two *o*-hydroxyl groups were essential for the inhibitory activity of polyphenol. When the C-ring of isoflavone was broken, the inhibitory activity markedly decreased.

It was observed [127] that treatment *Pseudomonas aeruginosa* with cranberry type-A proanthocyanidins **(23,24)** caused downregulation of a wide variety of proteins, including those related to ATP synthesis (likely cytochrome C PA2482), purine, carbohydrate, amino-acid and fatty acid metabolism (HmgA, GuaB, FdhE, FoaB, LdcA, PurU1) and involved in nucleic acid synthesis and repair (e.g., TopA, Rne, RplC, and Mfd). In addition, several citric acid cycle proteins, such as subunits of the acetyl-CoA carboxylase, aconitate hydratase and fumarase, were found to be significantly reduced. However, more than 30 proteins, mainly related to metal cation utilization, were upregulated.

4.2. Inhibition of Bacterial DNA Synthesis and Interaction with Nucleic Acids

Flavonoids from *Elaeagnus glabra* were tested for their antibacterial activity against *Proteus vulgaris* and *Staphylococcus aureus*. A free 3′,4′,5′-trihydroxy B-ring and a free 3-OH group were necessary for antibacterial activity. DNA synthesis was predominantly inhabited by the active flavonoids in *P. vulgaris*, whereas RNA synthesis was inhibited in *S. aureus* [128]. The most active inhibitors of DNA synthesis were robinetin **(122)**, myricetin

(114), and (−)-epigallocatechin **(14)**. It is probable that the B ring of the flavonoids could intercalate or form a hydrogen bond with the stacking of nucleic acid bases and further lead to the inhibition of nucleic acid synthesis in bacteria. The results of Lou et al. [129] demonstrated that *p*-coumaric acid **(166)** had dual mechanisms of bactericidal activity: disrupting bacterial cell membranes and binding to bacterial genomic DNA leading to inhibition of cellular functions, and ultimately to cell death.

Depolarization of membrane and inhibition of DNA, RNA, and proteins synthesis was observed in *S. aureus* and—in higher concentrations—cell lysis, when treated with flavonoids from *Dorstenia* sp., such as 6,8-diprenyleriodictyol **(106)**, isobavachalcone **(148)**, and 4-hydroxylonchocarpin **(149)** [130].

The synthesis of nucleic acid can be inhibited by polyphenols also through topoisomerase inhibition. Flavonoids are inhibitors of topoisomerases, and it plays an important role in their antimycobacterial activity. Docking studies have proved that quercetin **(111)** effectively binds to the subunit B of DNA gyrase through interaction with residues that are in the Toprim domain of the protein. Due to this activity, it inhibited the growth of *Mycobacterium smegmatis* and *Mycobacterium tuberculosis* [131].

Bandele et al. [132,133] have found that polyphenols may act against topoisomerase II in different a manner; (−)-epigallocatechin gallate **(18)** and (−)-epigallocatechin **(14)** were redox-dependent topoisomerase II poisons, kaempferol **(108)** and quercetin **(111)** were topoisomerase II "poisons", myricetin **(114)** utilized both mechanisms, while (−)-epicatechin gallate **(18)**, and (−)-epicatechin **(12)** displayed no significant activity. Based on the observation, a set of rules has been formed to predict the mechanism of bioflavonoid action against topoisomerase II: while the C4′-OH in B ring is critical for the compound to act as a traditional poison, the addition of –OH groups at C3′ and C5′ increases the redox activity of the B ring and allows the compound to act as a redox-dependent poison. The second rule is that the aromatic and planar structure of the C ring in the flavonols that includes a C4-keto group allows the formation of a proposed pseudo ring with the C5-OH. Disruption of these elements abrogated enzyme binding and precluded the ability to function as a traditional topoisomerase II poison [132,133].

Although the above studies were conducted with human cells, and flavonoids are assumed to be poisons of human topoisomerase IIα and IIβ, there are some data about flavonoids as inhibitors of bacterial type II topoisomerases: DNA gyrase and topoisomerase IIA (also called topoisomerase IV) [92]. Gyrases are enzymes that modify the DNA topology, and they are present only in prokaryotes, making them an attractive target for antibacterial drugs. DNA gyrase consists of two catalytic subunits; GyrA is responsible for DNA breakage and reunion, while the subunit GyrB contains the ATP-binding site. Coumarins and cyclothialidines are natural products that inhibit the ATPase activity of DNA gyrase by blocking the binding of ATP to subunit GyrB [134]. Plaper et al. [135] demonstrated that quercetin **(111)** inhibits the supercoiling activity of the bacterial gyrase and induces DNA cleavage, and the mechanism is probably based on interaction with DNA. They showed that quercetin **(111)** binds to the 24 kDa fragment of gyrase B of *Escherichia coli* with a K(D) value of 15 μM and inhibits ATPase activity of gyrase B. Its binding site overlaps with the ATP binding pocket and could be competitively replaced by either ATP or novobiocin. The proposed mechanism is that quercetin **(111)** inhibits gyrases through either the interaction with DNA or with the ATP binding site of gyrase [135]. Other polyphenols that can inhibit bacterial DNA gyrase by binding to the ATP binding site of the gyrase B subunit are catechins, with epigallocatechin gallate **(18)** being the most active, followed by epicatechin gallate **(17)** and epigallocatechin **(14)** [136]. Furthermore, quercetin **(111)**, apigenin **(38)**, and 3,3′,4′,6,7-pentahydroxyflavone **(167)** demonstrated inhibitory activity against *Escherichia coli* DNA gyrase [137].

The quantitative structure–activity relationship (QSAR) and molecular docking of flavonoids were analyzed in the study of Fang et al. [138]. The QSAR models demonstrated that hydrophobicity, H-bond donor, steric and electronic properties are key factors for the antibacterial activity of flavonoids. Structure requirements including hydroxyl group at

C-3, C-5, C-7 and C-3′, C2-C3 unsaturated double bond and the carbonyl group at C-4 are essential, while the presence of hydroxyl group at C-6, methoxyl group at C-8 and C-3′ could decrease the antibacterial activity. Docking results indicated that half of the tested flavonoids inhibited GyrB by interacting with ATP pocket in the same orientation. Polymethoxyl flavones, flavonoid glycosides, and isoflavonoids changed their orientation, resulting in a decrease in inhibitory activity. Hydroxyl group at C-3, C-5, C-7 and C-4′, carbonyl group at C-4 are key active substituents of flavonoids for inhibiting GyrB by interacting with its key residues. Structure changes, including glycosylation, polymethoxylation or isoflavonoids, will change the action mode and result in a decrease in inhibitory activity [138].

Three flavonoids isolated from cottonseed flour which promoted *Escherichia coli* topoisomerase IV-dependent DNA cleavage were identified as rutin **(121)**, quercetin 3-*O*-rhamnogalactoside, and isoquercetin **(126)**. Moreover, rutin **(121)** also inhibited topoisomerase IV-dependent decatenation activity and induced the SOS response of a permeable *E. coli* strain [139].

Arima et al. [140] observed that morin alone, at a concentration of 50 μg/mL inhibited the synthesis of DNA in the cells of *Salmonella enteritidis*, while its concentration equaled to a concentration of 12.5 μg/mL was enough if rutin **(121)** was added to the medium at a concentration of 12.5 μg/mL. Morin **(110)** alone also inhibited RNA and protein synthesis, but the rutin added did not influence the inhibition process.

Tannic acid **(215)** is strongly bound to DNA, which possibly had led to the covalent modification of DNA bases. Furthermore, tannic acid in the presence of Cu(II) caused strand cleavage in supercoiled plasmid DNA [141].

4.3. *Interaction with the Bacterial Cell Wall or Inhibition of Cell Wall Formation*

Various strengths of antimicrobial activity against bacteria may be caused by differences in cell surface structures between Gram-negative and Gram-positive species [20,64]. The major function of the cell wall is to provide shape and cell integrity and to act as an osmotic barrier. Gram-negative bacteria were reported to be resistant toward many antibacterial substances due to the hydrophilic surface of their outer membrane and associated enzymes in the periplasmic space, which is capable of breaking down many molecules introduced from outside [142,143]. Moreover, the negatively charged lipopolysaccharide (LPS) of the outer membrane protects the bacterial cell against catechins [144]. Gram-positive bacteria seem to be more susceptible to the action of phenolic acids than Gram-negative bacteria [64]. One of the explanations is that the Gram-positive bacterium lacks an outer membrane, which would facilitate diffusion of the phenolic acids through the cell wall and intracellular acidification. Vattem et al. [145] postulated the hyperacidification at the plasma membrane interphase, being a consequence of dissociation of phenolic acids, as one of the possible mechanisms of the antimicrobial action of phenolic acids. This hyperacidification would alter cell membrane potential, making it more permeable, and cause irreversible alterations in the sodium-potassium ATPase pump, therefore leading to cell death.

Wu et al. [146] have demonstrated that quercetin **(111)** and apigenin **(38)** influence the synthesis of the bacterial cell walls by the inhibition of D-alanine:D-alanine ligase (an essential enzyme that catalyzes the ligation of d-Ala–d-Ala in the assembly of peptidoglycan precursors). Moreover, these two flavonoids could inhibit the FabZ enzyme from *H. pylori* [122]. Tasdemir et al. [38] found that quercetin **(111)** could inhibit three consecutive enzymes, β-ketoacyl-ACP reductase (FabG), β-hydroxyacyl-ACP dehydrase (FabZ) and enoyl-ACP reductase (FabI), in the FAS II pathway of *Plasmodium falciparum*, whereas apigenin **(38)** could only inhibit FabI.

Flavones form a complex with cell wall components and consequently inhibit further adhesions and microbial growth as well. The inhibition of bacterial enzymes (such as tyrosyl-tRNA synthetase) was observed for C-7-modified flavonoids containing the naringenin **(81)** core [147]. It was also demonstrated that they were also inhibitors of *S. aureus*,

E. coli, and *Pseudomonas aeruginosa* growth. Baicalein **(45)** was an effective bactericide, and when combined with cefotaxime, the synergistic effects were observed by inhibiting extended-spectrum β-lactamase CTX-M-1 mRNA expression [148]. Inhibition of the bacterial efflux pump and increase in the susceptibility of existing antibiotics (by inducing depolarization of the cell membrane) is another possible mechanism of antibacterial activity. Artonin I **(63)**, from *Morus mesozygia*, was effective against *S. aureus* due to blocking the efflux mechanism and causing depolarization of the cell membrane [149]. Artonin I **(63)** reversed multidrug resistance and increased the susceptibility of existing antibiotics by lowering their minimum inhibitory concentrations.

Many researchers have proved that polyphenols can interact with the cell wall or outer membrane and with their components such as peptidoglycan, lipopolysaccharide. Zhao et al. [150] demonstrated that unlike dextran and lipopolysaccharide, peptidoglycan from *S. aureus* blocked both the antibacterial activity of EGCG **(18)** and the synergism between EGCG and oxacillin, suggesting EGCG may directly bind to the cell wall of *S. aureus* and interfere with its integrity. These results were confirmed by Yoda et al. [113]. As the bactericidal activity of EGCG **(18)** for *S. aureus* was blocked, dose-dependently by purified peptidoglycan, but not by lipopolysaccharide or dextran, it was suggested that EGCG binds directly to the peptidoglycan in the cell wall. These results are consistent with the opinion that the structure of the bacterial cell wall is responsible for the different susceptibilities of Gram-positive and Gram-negative cells to polyphenols.

Gram-negative bacteria (e.g., *Escherichia coli, Salmonella, Shigella*) have in their outer cell membrane a strong "endotoxin"—lipopolysaccharide (LPS). The fraction of procyanidins from cranberries composed of polymers with an average degree of polymerization of 21 can efficiently bind lipopolysaccharide and prevent the interaction of LPS with receptors on the surface of mammalian target cells [151]. On the other side, phenolic extracts of cloudberry and raspberry rich in ellagitannins disintegrated the outer membrane of examined *Salmonella* sp. [152] and released LPS from bacteria cells.

4.4. Alteration of Cytoplasmic Membrane Function

The (inner) bacterial cell membrane is responsible for many essential functions: osmoregulation and respiration processes, transport, biosynthesis and the cross-linking of peptidoglycan and synthesis of lipids [153]. Any disturbance in its structure or functionality can result in metabolic dysfunction and cell death; hence the membrane disruption is postulated to be one of the mechanisms of the antibacterial activity of polyphenols. For example, catechins were shown to rupture the bacterial membrane by binding to the lipid bilayer and by inactivating or inhibiting the synthesis of intracellular and extracellular enzymes [154].

Apigenin **(38)** induced fungal membrane dysfunction and increased cell permeability [155], which caused the release of small intracellular constituents such as ions and sugars, but not proteins. Epicatechin-3-gallate **(17)** and caffeic acid **(167)** targeted both the cell wall and cytoplasmic membrane of *P. aeruginosa* [156]. The cellular membrane destruction and ensuing membrane permeability perturbation of *P. aeruginosa* had led to the ascending access of hydrophobic antibiotics, a release of potassium ions, and leakage of nucleotides. Phenolic acids, due to their partially lipophilic nature, pass through the cell membrane by passive diffusion and cause an increase in membrane permeability. They possibly reduce the intracellular pH and induce protein denaturation [157]. Methanol extract of *Coriolus versicolor* rich in polyphenols disabled *S. aureus* cell division (i.e., the formation of septa) and led to the accumulation of peptidoglycan and teichoic acid precursors in the cytoplasm [157]. In this case, the extract acted directly on the cytoplasmatic membrane, whereas in Gram-negative *Salmonella* Enteritidis, the cell envelope was damaged. On the other side, at high concentrations, catechins were found to generate an oxidative burst by the generation of reactive oxygen species (ROS) that cause alteration in the membrane permeability and membrane damage [158]. Purified flavonoids from *Graptophyllum glandulosum* possessed antimicrobial activities against multidrug-resistant

Vibrio cholerae and caused cell lysis and disruption of the cytoplasmic membrane upon membrane permeability [159].

Flavonoids (acacetin **(40)** and apigenin **(38)**) and flavonols (morin **(110)** and rhamnetin **(115)**) caused destabilization of the membrane structure by disordering and the disorientation of the membrane lipids and induced leakage from the vesicle [160]. The inverse correlation between the number of hydroxyl groups in the flavonoids and their capacity to leakage induction was noted. Studies of Chabot et al. [161] suggested that flavonoids lacking hydroxyl groups on their B rings (genistein **(65)**, hesperetin **(83)**, chrysin **(47)**, galangin **(109)**) were more potent inhibitors of microbial growth than those with the –OH groups. On the other hand, Adamczak et al. [67] have demonstrated that the presence of hydroxyl groups in the phenyl rings A and B usually did not influence the level of the antibacterial activity of flavones. A significant increase in the activity of the hydroxy derivatives of flavone was observed only for *S. aureus*. What is interesting, in contrary to other studies, the compounds tested in the study were generally more active against Gram-negative bacteria: *Escherichia coli* and *Pseudomonas aeruginosa* than Gram-positive ones: *Enterococcus faecalis* and *Staphylococcus aureus*.

Tsuchiya [162] reported that the catechin impact on membrane fluidity also depended on the stereospecificity. (−)-Epicatechin **(12)**, (+)-epicatechin **(9)**, (−)-catechin **(10)** and (+)-catechin **(7)** reduced membrane fluidity in increasing order of intensity; it means that epicatechins in a *cis* form were more effective for reducing membrane fluidity than catechins in a *trans* form. Stereospecificity in the membrane effects of catechin stereoisomers may be induced by the different hydrophobicity of geometrical isomers and the chirality of membrane lipid components. Lipophilic flavonoids may also disrupt microbial membranes [107]. It was suggested that the mode of action of terpenes and their related alcohols involves disruption of microbial membranes by their lipophilic components [163]. According to Tsuchiya [164], bioactive components with amphiphilic or hydrophobic structures interact with biological membranes resulting in the modification of membrane fluidity, microviscosity, order, elasticity, and permeability. The author postulated that interactions of flavonoids with lipid bilayers involve two mechanisms; the first is associated with the partition of the more nonpolar compounds in the hydrophobic interior of the membrane, while the second one includes the formation of hydrogen bonds between the polar head groups of lipids and the more hydrophilic flavonoids at the membrane interface. The membrane interactions and localization of flavonoids play a vital role in altering membrane-mediated cell signaling cascades [165].

The studies of Arora et al. [166] demonstrated that flavonoids and isoflavonoids preferentially enter into the hydrophobic core of membranes. In plant tissues, flavonoids occur mainly in the form of glycoside, and the presence of glycosidic residues in the flavonoid skeleton influences the hydrophobicity of flavonoids. Flavonoids with a greater hydrophobicity have been reported to influence the transmembrane potential to a greater extent than the less hydrophobic flavonols, probably because they can enter deeper into the lipid bilayer, thereby disrupting the compact packing of lipids [157]. Moreover, the spatial configuration is also important; a substantially higher affinity for artificial membranes was reported for flavonols (planar) than flavanones (tilted) [167]. Using the fluorescence anisotropy technique, it was reported that naringenin and naringin enhanced membrane fluidity, while membrane interaction with quercetin **(111)**, daidzein **(64)**, luteolin **(49)**, galangin **(109)**, kaempferol **(108)** and genistein **(65)** resulted in rigidified membranes [165]; however, the impact depended on the lipid composition of membranes.

Wu et al. [168] have shown the positive correlation between antibacterial capacity and membrane rigidification effect of the polyphenolic compounds. Authors have observed that flavonoids decreased the membrane fluidity with the potency being kaempferol **(108)** > chrysin **(47)** > baicalein **(45)** > quercetin **(111)** > luteolin **(49)**, whereas isoflavonoids increased the membrane fluidity with the potency being puerarin **(70)** > ononin **(75)** > daidzein **(64)** > genistin **(68)** [168]. Kaempferol **(108)**, located deeply in the hydrophobic core of the lipid bilayer, decreased the membrane fluidity most and exhibited the highest antibacterial capacity against *E. coli*. The number and the position of hydroxyl groups influenced the membrane interaction with polyphenols; the OH group at C-3 in the C ring was important for decreasing membrane fluidity. He et al. [169] suggested that for flavonoids to be effective antimicrobial agents, interaction with the polar head–group of the model membrane followed by penetration into the hydrophobic regions must occur. The antimicrobial efficacies of the flavonoids were consistent with liposome interaction activities and decreased in the order: kaempferol **(108)** > hesperetin **(83)** > (+)-catechin **(7)** > biochanin A **(76)**.

Sophoraflavanone B (8-prenylnaringenin) **(101)** caused cell wall weakening, membrane damage and intracellular constituents leaking from the cell of methicillin-resistant *S. aureus* [170]. In the study, the direct binding of sophoraflavanone B **(101)** to peptidoglycan was demonstrated. It also has been proposed that sophoraflavanone G **(98)** and (−)-epigallocatechin gallate **(18)** inhibited cytoplasmic membrane function [171].

4.5. Inhibition of Energy Metabolism

Many aspects of cellular metabolism revolve around ATP production and consumption, and ATP is regarded as the universal energy exchange factor that connects anabolism and catabolism but also enables processes such as motile contraction, phosphorylation, and active transport (uptake) of nutrients. Membrane-bound F1F0 ATP synthase from bacteria is an enzyme responsible for ATP production through oxidative phosphorylation or photophosphorylation. It has been demonstrated that morin **(110)**, baicalein **(45)**, silibinin **(138)**, and epicatechin caused complete inhibition of ATPase activity, while hesperidin **(86)**, chrysin **(47)**, kaempferol **(108)**, diosmin **(51)**, apigenin **(38)**, genistein **(65)**, or rutin **(121)** exert partial inhibition of about 40–60% [172].

Dadi et al. [173] demonstrated that resveratrol **(202)**, piceatannol **(203)**, quercetin **(111)**, quercitrin **(123)**, or quercetin-3-β-D glucoside **(126)** inhibited *E. coli* ATP synthase, but to different degrees. The most potent inhibitor was piceatannol **(203)** (~0 residual activity); inhibition by other compounds was partial and ranged from ~20% residual activity for quercetin **(111)** to ∼60% residual activity for quercitrin **(123)** or resveratrol **(202)**. Inhibition was identical in both F_1F_0 membrane preparations as well as in isolated, purified F_1, but in all cases, inhibition was reversible. Interestingly, resveratrol **(202)** and piceatannol **(203)** inhibited both ATPase and ATP synthesis, whereas quercetin **(111)**, quercitrin **(123)** or quercetin-3-β-D glucoside **(126)** inhibited only ATPase activity and not ATP synthesis. The membrane-bound ATPase activity of *E. coli* was also inhibited by eugenol **(185)** or carvacrol **(183)** [174]. Similar results were obtained for thymoquinone that completely inhibited both purified F_1 and membrane-bound F_1F_0 *E. coli* ATP synthase, and the process of inhibition was fully reversible [175].

It has been proposed that licochalcones A **(150)** and C **(151)** can inhibit energy metabolism [171]. Haraguchi et al. [176] have proved the antibacterial activity of licochalcones A **(150)** and C **(151)** against *S. aureus* and *Micrococcus luteus*, which resulted from inhibited oxidation of NADH in bacterial membranes. As licochalcones inhibited NADH-cytochrome c reductase, they exerted their antibacterial activity by inhibiting the bacterial respiratory electron transport chain.

EGCG **(18)** directly interacts with proteins and phospholipids in the plasma membrane and regulates signal transduction pathways, transcription factors, DNA methylation, as well as mitochondrial function and autophagy [177].

4.6. The Inhibition of Biofilm Formation and Interfering with Bacterial Quorum Sensing

Biofilm is an assemblage of microbial cells that are irreversibly linked to a surface with bacteria embedded in an extracellular matrix of self-produced biopolymers. The ability to form biofilm is an important property of various bacteria such as pathogenic species and is associated with quorum sensing (QS) or cell-to-cell communication. Therefore, bacterial cell-to-cell communication has received attention to manifest the role of quorum signals in the attachment and growth of pathogenic bacteria in foods. QS participates in the biofilm formation as well as controls the expression of various virulence factors, inter alia, the production of proteases that degrade connective tissue, the production of siderophores that facilitate iron uptake, the releasing of toxins that disrupt cellular processes, the formation of phenazines that favor the reactive oxygen species generation, and production of exopolysaccharides that are necessary for the phagocytosis-resistant capsules structure [178].

An anti-quorum sensing (anti-QS) agent curcumin **(216)** from *Curcuma longa* (turmeric) was shown to inhibit the biofilm formation of pathogens, such as *Escherichia coli*, *Pseudomonas aeruginosa* PAO1, *Proteus mirabilis* and *Serratia marcescens*, possibly by interfering with their QS systems [179], because the biofilm maturation was disturbed by a biomass reduction and by the interruption of swimming motility. Chlorogenic acid **(172)** was proved to significantly inhibit the formation of biofilm by *P. aeruginosa*, its ability to swarm, and virulence factors including protease and elastase activities and rhamnolipid and pyocyanin production. Moreover, the QS related genes were downregulated in *P. aeruginosa*, and the inhibitory rates were as follows: lasI 85.09%, lasR 48.63%, rhlI 27.98%, rhlR 34.7%, pqsA 73.08%, and pqsR 45.85%, respectively [180].

Quercetin **(111)** efficiently reduced the biofilm formation and other QS regulated phenotypes like violacein inhibition, exopolysaccharide production and alginate production in foodborne pathogens *K. pneumoniae*, *P. aeruginosa*, and *Y. enterocolitica* [181]. Furthermore, quercetin **(111)** significantly inhibited the swimming and swarming behavior of *P. aeruginosa* and *Y. enterocolitica*.

L. paracasei exposed to resveratrol **(202)** displayed changes in the physicochemical properties of their surface, especially with a global increase in negative charges, a more basic nature and an increase in their hydrophobicity. These changes may largely contribute to the enhanced adhesion, induced formation of bacterial aggregates and biofilm formation abilities of resveratrol-treated *L. paracasei* [182]. However, the majority of studied polyphenols have shown the opposite impact on the ability of biofilm formation. Apple flavonoid phloretin **(152)** was reported to control *E. coli* O157:H7 biofilm formation by a mechanism that implies repressing the curli genes (csgA and csgB), which are involved in fimbriae production [183]. Epigallocatechin-3-gallate **(18)** eliminates the biofilm matrix by directly interfering with the assembly of curli subunits into amyloid fibers and by triggering the σ^E cell envelope stress response and thereby reducing the expression of a crucial activator of curli and cellulose biosynthesis (CsgD) [184]. Recently, it has been shown that EGCG **(18)** act against biofilms by strongly interfering with the assembly of amyloid fibers and the production of phosphoethanolamine-modified cellulose fibrils [185].

EGCG **(18)** also inhibited the formation of *Streptococcus mutans* biofilms [186,187]. The growth of *Streptococcus mutans* decreased, and the biofilm formation was inhibited by pinocembrin **(95)**, apigenin **(38)**, quercetin **(111)**, while caffeic acid phenethyl ester decreased, probably due to changes in bacterial architecture [188].

According to Xu et al. [189], the ECGC **(18)** activity against *S. mutans* is due to disrupting at the transcriptional level the adherence of bacteria to surfaces and hence inhibiting the biofilm formation. Authors hypothesized that EGCG at sublethal concentrations directly suppressed the expression of gift genes encoding glucosyltransferases, enzymes that synthesize polysaccharides necessary for biofilm formation [189]. Moreover, EGCG **(18)** was found to inhibit the enzymatic activity of the F_1F_0-ATPase and lactate dehydrogenase [190].

Morin **(110)** at its sub-MICs demonstrated a significant dose-dependent inhibitory efficacy against *Listeria monocytogenes* biofilm formation [191]. Moreover, morin-treated *Listeria* showed a significant reduction in hemolysin secretion and a concentration-dependent decrease in the flagella directed swimming and swarming velocity. The biofilm formation and biofilm-related genes in *L. monocytogenes* also were inhibited by thymol **(184)**, carvacrol **(183)** and eugenol **(185)** [192].

Bap (biofilm-associated protein) is expressed by *Staphylococcus* sp. in order to adopt functional amyloid-like structures as scaffolds of the biofilm matrix. Quercetin **(111)**, myricetin **(114)** and scutellarein **(44)** specifically inhibited Bap-mediated biofilm formation of *S. aureus* and other staphylococcal species [97] by preventing the assembly of Bap-related amyloid-like structures.

The ability of bacteria to adhere was also inhibited by phenolic acids. Adhesion was less favorable when the bacteria were exposed to gallic acid **(159)** (*P. aeruginosa, S. aureus* and *L. monocytogenes*) and ferulic acid **(168)** (*P. aeruginosa* and *S. aureus*). Both phenolics were able to inhibit bacterial motility and prevented biofilm formation, as well as reducing the mass of biofilms formed by the Gram-negative bacteria [193]. Further studies proved that gallic **(159)** and ferulic **(168)** acids led to irreversible changes in membrane properties, such as its charge, intra and extracellular permeability, and physicochemical properties. Both acids caused changes in hydrophobicity and negative surface charge and induced the local rupture or pore formation in the cell membranes leading in consequence to essential intracellular constituent leakage [194].

4.7. Substrate Deprivation

When polyphenol forms complex with protein, the biological function might change. Depending on the function of a complexed protein, the influence on bacteria cells will differ. As mentioned above, a decreased or inhibited activity of enzymes can result in a lack of energy for bacteria and, in consequence, might lead to cell death. Lack of energy also means a disturbed transport of nutrients across the cell wall and cytoplasmic membranes, diminished bacteria proliferation and limited mobility, as well as inhibited ability to biofilm formation or even blocked sporulation [195].

The deprivation of the substrates required for microbial growth, especially essential mineral micronutrients such as iron and zinc (via proanthocyanidin chelation with the metals), together with the destabilization of the cytoplasmic membrane, the permeabilization of the cell membrane, the inhibition of extracellular microbial enzymes, direct actions on microbial metabolism, were supposed to be the mechanism of the antibacterial activity of the A-type proanthocyanidin **(23,24)** [196].

Scalbert [197] suggested that tannin toxicity for bacteria is due to the direct impact on bacterial metabolism by inhibiting the oxidative phosphorylation as well as by deprivation of the substrates required for microbial growth, especially an iron deprivation. Generally, tannins are reported to be strong inhibitors of many various hydrolytic enzymes such as α-amylase, pectinase, cellulase, xylanase, lactate dehydrogenase, malate dehydrogenase, peroxidase, β-glucosidase, so they can inhibit the activity, growth or proliferation of microorganisms [198]

The inhibitory effect of tannic acid **(215)** on the growth of intestinal bacteria may be due to its strong iron-binding capacity. The growth of *E. coli* was restored by the addition of iron to the medium after the precipitate caused by tannic acid **(215)** was removed [199]. In the study, neither *Bifidobacterium infantis* nor *Lactobacillus acidophilus* required iron for growth, which probably contributes to their resistance to tannic acid. It is known that only a few bacteria, including lactobacilli, do not require iron. It is an essential trace element for most gut bacteria, and many have active Fe transport systems and other mechanisms to scavenge Fe [200]. For example, *Bacteroides* spp. are highly dependent on heme and iron, whereas many members of the Enterobacteriaceae have developed mechanisms, including siderophores, to acquire Fe in competition with other bacteria and the host.

It is well known that catechol and gallol structure (Figure 3) and hence many polyphenolic compounds are effective metal chelators. After deprotonation, which is required for metal binding, catecholate and gallate groups may be complexed with metal ions that prefer octahedral geometry, such as Fe^{2+} and Fe^{3+} (Figure 3) [201].

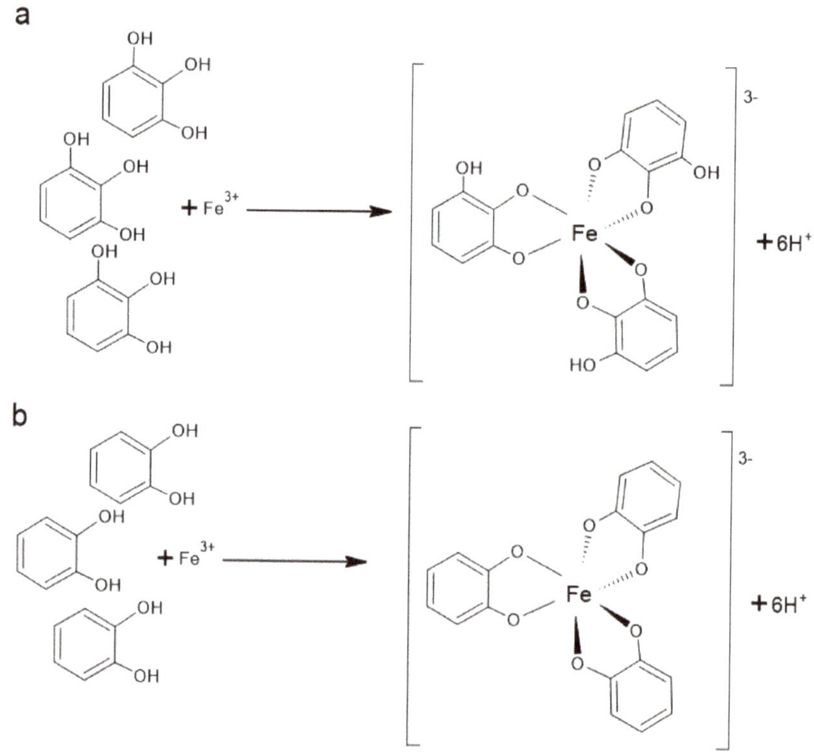

Figure 3. Expected octahedral coordination geometry of general iron-polyphenol complexes, (**a**) gallol, (**b**) catechols. Coordination requires deprotonation of the polyphenol ligands. Based on [201].

It was established for flavones and for the flavanone naringenin **(81)** that the binding metal sites are preferentially at the 5-hydroxyl and 4-oxo groups [202]. On the other hand, the study of Mladenka et al. [203] demonstrated that the most effective iron-binding site of flavonoids is 6,7-dihydroxy structure, present, for example, in baicalein **(45)**. The simultaneous presence of 3-hydroxy-4-keto conformation, 2,3-double bond and the catecholic B ring were associated with significant iron chelation; however, the catecholic B ring did not play an essential role in more acidic conditions. Quercetin **(111)** and myricetin **(114)** that contain all mentioned structural requirements, had activity similar to baicalein **(45)** at the neutral conditions but were clearly less active in lower pH. On the other hand, baicalein **(45)**, additionally possessing the 6,7-dihydroxyl groups, was very efficient even in the acidic condition. The 5-hydroxy-4-keto configuration has only moderate activity at all pH conditions. It was also proved that isolated keto, hydroxyl, methoxyl groups or ortho methoxy–hydroxy groups were not associated with iron chelation at all.

Polyphenols also have strong binding interactions with Cu^{2+}, and stability constants for Cu^{2+} catecholate complexes are even larger than for Fe^{2+} [201].

As bacteria contain various metalloenzymes, flavonoids by binding metal ions can inhibit their activity and lead to various metabolic disorders (enzyme inhibition, impairment of ion channel functions). Furthermore, the metabolic functions of the human gut microbiota that involve metalloenzymes may also be altered [204]. An enzyme methionine aminopeptidase (MetAP) carries out the removal of the initiator methionine residue from newly synthesized proteins, and this removal is critical for the activation, distribution and stability of many proteins. It was proved that the adjacent hydroxyl groups on the phenyl ring (catechol moiety) were essential for effective inhibition of the Fe (II)-a form of *E. coli* MetAP and growth inhibition of bacterial cells [205].

Polyphenols can also cause iron deficiency in the digestive tract, which will affect sensitive bacterial populations and change the composition of the intestinal microbiota. Oral bacterium Fusobacterium nucleatum is associated with colon cancer, causes erythrocytes lysis, and therefore releases hemoglobin, which provides an iron source to bacteria and other periodontopathogens, promoting their proliferation in periodontal pockets [206]. The tea polyphenols were proved to inhibit dose-dependently the hemolytic activity of *F. nucleatum*.

The virulence factors such as gelatinase, collagen-binding antigen, cytolysins, and proteases enhance colonization, survival and persistence of *E. faecalis* in the root canal. Treatment of *E. faecalis* with a sublethal concentration of EGCG **(18)** (2.5 mg/mL) significantly inhibited the expression of responsible genes (collagen adhesin (ace), cytolysins activator (cylA), gelatinase (gelE) and serine protease (sprE)) by >75% compared to the untreated control [207]. The elastase, protease and pyocyanin production in *P. aeruginosa* were inhibited by curcumin **(216)** in a dose-dependent manner [179]. EGCG **(18)** caused the inhibition of glucose uptake by *E. coli*, which can suggest that EGCG inhibits the major function of porin proteins, namely the passive transport of small hydrophilic molecules such as glucose, leading to growth inhibition of *E. coli* [114].

4.8. The Relationship between Polyphenols Structure and Antibacterial Activity

The mechanism of inhibition by polyphenols may differ depending both on the structure of the polyphenolic compound and bacteria species. The amphipathic character of flavonoids plays a very important role as hydrophilic and hydrophobic moieties must be present together and well-spaced in these compounds [208].

Flavans with prenyl group at the A ring were potent antibacterial compounds against *Staphylococcus aureus*, and the number and position of prenyl groups on this ring influenced the activity [91].

The number of hydroxyl groups in the B ring in flavonols and flavones is associated with the antimicrobial activity against lactic acid bacteria (LAB). Myricetin **(114)** is a flavonol possessing three hydroxyl groups in the B ring as pyrogallol structure, whereas quercetin **(111)** and kaempferol **(108)** have one and two hydroxyl groups less in the B ring than in myricetin, respectively. Myricetin, as a pure compound, significantly inhibited the growth of all tested LAB that originated from the human gastrointestinal tract, as well as the Gram-positive *Enterococcus faecalis* and *Bifidobacterium lactis*, while quercetin **(111)** and kaempferol **(108)**, with a more lipophilic nature, had no inhibitory impact on the above bacteria [47]. Flavone luteolin **(49)** has a structure similar to quercetin **(111)**, but it lacks the OH group at position 3 in ring C. Luteolin **(49)** was bacteriostatic against some of the tested LAB as well as against *E. faecalis* and *B. lactis*, while other flavone apigenin **(38)**, which has one hydroxyl group less in the B ring had no such effects.

Baicalein (flavon) **(45)** and myricetin (flavonol) **(114)** show the most significant antibacterial effects among the tested flavonoids. Both have a pyrogallol structure, but baicalein in ring A (5, 6, 7–OH) and myricetin in ring B. Results proved that the pyrogallol structure was an important element for the potent antibacterial activity for flavonoids [60]. Echeverria et al. [208] made the comparison between a flavone (planar) and flavanone (not planar) with similar lipophilicity and oxygenated substitution patterns in the A and B rings (e.g., pinocembrin **(95)** and 3-*O*-methylgalangin **(113)**) and showed that flavones have higher antibacterial activity. On the other hand, possessing at least one hydroxy group in the ring A (especially at position C-7) seems to be crucial for antibacterial activity of flavones, and an additional OH group in another position such as C-5 and C-6 can further increase the activity [91].

All the flavonols and flavanones with antibacterial activities had two hydroxyl substituents on C-5 and C-7 of ring A in common, such as quercetin **(111)**, rutin **(121)**, naringenin **(81)**, and hesperetin **(83)** [60]. Moreover, the authors suggest that flavanones were more active than the corresponding flavones. For example, naringenin **(81)** showed antibacterial effects on all the tested bacteria, whereas apigenin **(38)** showed almost no effect. Such results indicate that the saturation of the C2-C3 double bond increased the antibacterial activity.

On the other side, Wu et al. [168] demonstrated that flavonoids were more effective *E. coli* inhibitors than isoflavonoids with relative activity being as follows: kaempferol **(108)** > quercetin **(111)** > chrysin **(47)** > luteolin **(49)** > baicalein **(45)** > tangeretin **(57)** and daidzein **(64)** > genistin **(68)** > ononin **(75)** > puerarin **(70)**. The only structural difference between quercetin **(111)** and luteolin **(49)** is that quercetin has a hydroxyl group at position 3 in the C ring, while luteolin has none. It means that the 3-OH group is important to the activity of flavonoids against Gram-negative bacteria *E. coli*. Further analysis of structure–activity relationships revealed that the methylation of OH groups could decrease the antimicrobial activity of flavonols. It also has been shown a significant positive correlation between the antibacterial capacity of flavonoids and the membrane rigidification effect. A quantitative structure–activity relationship (QSAR) study revealed that the activity of the flavonoid compounds could be related to molecular hydrophobicity and charges on the C atom at position 3 [168].

The hydrophobic substituents such as prenyl groups, alkylamino chains, alkyl chains, and nitrogen or oxygen-containing heterocyclic moieties usually enhance the antibacterial activity for all the flavonoids [98]. It was concluded that hydroxyl groups on special sites are favorable for antibacterial activity, such as 5,7-dihydroxyl substitution for flavone and flavanone and 2′ or 4′ hydroxylation for chalcones. The hydroxyl group at position three on the C ring of flavone also increased the activity. However, the methylation of the hydroxyl groups generally decreased the activity. The lipophilicity of ring A is therefore of great importance for the activity of chalcones. In addition, hydroxy groups at 4′, 4, and 6 of A and B rings increase the activity of chalcones [91].

The substitution of the flavonoid ring system with prenyl groups increases the lipophilicity of the molecule and results in a strong affinity to biological membranes. Prenylated flavonoids, i.e., featuring C_5 isoprenoid substituents, have a relatively narrow distribution in the plant kingdom and are constitutively expressed in plants, as compared with prenylated isoflavonoids, which are produced in response to an attack or damage [209]. Xanthohumol **(145)** is the main component (80–90% of the total flavonoids) and is the most abundant prenylated chalcone in hops. It exerted high antimicrobial activity against *Bacteroides fragilis, Clostridium perfringens* and *Clostridium difficile* [210]. β-bitter acids (lupulones) were less effective, and the least effective against anaerobic pathogens were α-bitter acids (humulones). Xanthohumol **(145)**, naringenin **(81)**, chalconaringenin **(140)** and 4-hydroxy-4′-methoxychalcone inhibited the growth of *S. aureus* [211]. The presence of at least one hydroxyl group and especially at the C-4 position was crucial for the antibacterial activity against *S. aureus*. The lack of hydroxyl group or its replacement by a halogen atom (–Cl, –Br), nitro group ($-NO_2$), ethoxy group ($-O-CH_2CH_3$), or aliphatic groups ($-CH_2CH_3$), ($-CH_3$) led to inactivation of the compounds. Prenylated flavonoids, such as artocarpin **(62)** and isobavachalcone **(148)**, exhibited strong antibacterial activity towards *B. cereus*, *E. coli*, and *Pseudomonas putida* or only Gram-positive species, respectively [212]. It has been demonstrated that any isoflavonoid modification that results in the absence or cyclization of the prenyl group decreases the antibacterial activity of the compound.

Campos et al. [213] had demonstrated that hydroxycinnamic acids (p-coumaric **(166)**, caffeic **(167)** and ferulic **(168)** acids) induced greater potassium and phosphate leakage than hydroxybenzoic acids (protocatechuic **(157)**, gallic **(159)**, and vanillic **(161)** acids) across the membranes of *Oenococcus oeni* and *Lactobacillus hilgardii*.

Flavonoids can occur in two forms: free as "aglycons" or in the form of "glycosides", where an aglycon is combined with sugar moiety ("glycone"). Flavonoid glycosides occur in a diet generally in ring A or C as O-glycosides, and a corresponding substitution in ring A has a far greater impact on activity [180]. Aglycones of most flavonoids are more hydrophobic than their glycosides [91]. Both the number of glycosylation as well as the position and structure of saccharides are of great significance for the antioxidant, antibacterial, anticancer, anti-inflammatory and antidiabetic activity of a compound [180]. It has been postulated that glycosylation of flavonoids enhances antimicrobial activity, but their antioxidant, anti-inflammatory, anticancer and cardioprotective properties decreased [180]. However, it seems that the impact of glycosylation on antibacterial activity depends on the flavonoid class as well as the position at which sugar moiety is added. The results of Duda-Chodak [63] demonstrated that flavonoid aglycones, but not their glycosides, may inhibit the growth of some intestinal bacteria. In this study, rutin (quercetin 3-*O*-rutinoside) **(121)** had no inhibitory influence on the intestinal bacteria analyzed, and even slight stimulation of the growth of *Lactobacillus* spp. was observed. In contrast, its aglycone quercetin **(111)** exerted a dose-depended inhibitory effect on intestinal bacteria (except on *Bifidobacterium catenulatum*), and this was especially strong on *Ruminococcus gauvreauii*, *Bacteroides galacturonicus* and *Lactobacillus* spp. growth. The same was true for flavanones. Naringin **(85)** and hesperidin (flavanone 7-*O*-glycosides) **(86)** had no impact, but their aglycones (naringenin **(81)** and hesperetin **(83)**, respectively) inhibited the growth of almost all bacteria analyzed. A similar result, showing that 7-*O*-glycosylation of flavanones (naringenin and hesperetin) and flavones (baicalein **(45)**) decreased the antimicrobial activity against *E. coli*, *S. aureus*, *S. typhimurium*, *Enterobacter sakazakii* and *Vibrio parahemolyticus* were demonstrated by Xie et al. [60]. The opposite results were obtained by Adamczak et al. [67]; flavonol aglycones kaempferol **(108)** and quercetin **(111)** displayed a moderate activity only against *E. coli*, while quercetin 3-*O*-rutinoside **(121)** demonstrated inhibitory influence on all strains tested.

Docking results have revealed that the substitution of galloyl or glycosides at position 3 of heterocyclic pyrane ring in flavonoids enhanced the binding affinity to three targets, i.e., fumarate reductase flavoprotein subunit (FrdA), dihydroorotate dehydrogenase (PyrD) and NADH-dependent enoyl-ACP reductase (FabI). Such a phenomenon was observed for flavonoids and their glycosides; quercetin 3-rhamnoside **(123)** and myricetin 3-galactoside **(127)** were more potent inhibitors to PyrD, FabI, and DYR than quercetin **(111)** and myricetin **(114)**, respectively [121]. One of the most potent bacterial inhibitors among flavan-3-ol is EGCG **(18)**, possessing both pyrogallol and galloyl structures in a moiety. EGCG is a stronger inhibitor of pathogens than other flavan-3-ols having fewer or no galloyl groups and pyrogallols. Antibacterial activity of tea flavan-3-ols was in decreasing order EGCG **(18)** > ECG **(17)** > EC **(12)** $\geq$ theaflavins $\geq$ gallic acid **(159)** > EGC **(14)** against *S. aureus* and *P. aeruginosa* [214]. The importance of these free galloyl groups for antibacterial activity was also proved in the study of Puljula et al. [215]. Salicarinin A **(213)** and rugosin D **(211)** possess many free galloyl groups, inhibited the growth of *S. aureus* completely at a 0.5 mM concentration. Other ellagitannins, with lower numbers of galloyl or pyrogallol substituents, were less effective.

4.9. The Impact of Food Matrix on Polyphenol Activity

There are large discrepancies between the results, i.e., in some studies, it is shown that a given polyphenol class inhibits bacteria, and in others, it does not affect or even stimulates their growth. This may be due to the structure of the used polyphenol, including the molecule size, number and position of hydroxyl groups, their substitutions, the presence/absence and position of glycosylation, hydrophobicity and hydrophilicity of the moiety, and others. The observed discrepancies could also be attributed to the changes in the structure of polyphenols when dissolved in various solvents (water, ethanol, methanol, organic solvents and their mixtures) or after their addition to the medium with bacteria. It is because polyphenols do not dissolve in every solvent, and they can precipitate (affecting the actual concentration of the tested compound) after changing the solvent [216–218]. Polyphenols also have different rates of diffusion depending on the medium and environmental conditions.

Moreover, the type of microorganism (Gram-positive, Gram-negative, anaerobic or aerobic or microaerophile, etc.) also has a significant impact when the activity of polyphenols is assessed. It should be borne in mind that used assays, analytical methods, as well as conditions and incubation time, strains of microorganisms, inoculum size, and even concentrations of tested polyphenol may differ between scientific laboratories. There are also big differences between results when the impact of polyphenol on intestinal bacteria is assessed using pure polyphenols solution, plant extract containing polyphenols mixture of whole food in which polyphenols are bound to a food matrix. When in vitro studies are performed, usually pure cultures of bacteria are tested, and interactions with other members of gut microbiota, the impact of human digestive enzymes, the host health, or interactions with other components of a meal are not taken into account. However, all mentioned factors are important for the final results.

It is obvious that the results obtained from in vivo and in vitro studies should not be compared directly. When in vivo studies are conducted, the scientists introduce an ingredient into the diet and analyze changes in the abundance or composition of the gut microbiota, usually focusing on the effect on the entire bacterial population rather than on individual species. In such experiments, many factors contribute to the final results: the chemical composition of the food matrix, the bioaccessibility of polyphenols, their bioavailability, the interactions between particular bacterial strains present in the gut, the health of consumers and many more. Depending on the polyphenols present in the plant, different effects can be achieved because each polyphenol reacts differently with the components of plant tissues. Moreover, each plant differs in its composition. Tarko and Duda-Chodak [219] proved the differences between the bioaccessibility of polyphenolic compounds originating directly from fruits (black chokeberry, elderberry, hawthorn, Cornelian cherry, apple and Japanese quince) and that of those present in the fruit extracts during their digestion conducted in a simulated human gut. They proved significant differences in polyphenols bioavailability that resulted from their interactions with food matrixes. It was caused by polyphenols bounding to the matrix, which is known to modify the polyphenols extractability and susceptibility to digestive enzymes and bacterial metabolism [220]. The interaction with the food matrix also modulates the impact of polyphenols on bacteria inhabiting the colon.

During in vivo studies, it should also be considered that some polyphenols present in a diet are absorbed before they reach the colon, and hence, do not influence the microbiota. For example, quercetin glycosides can undergo partial hydrolysis by pepsin during their passage through the stomach [221], and the released aglycone quercetin **(111)** may be then absorbed in the stomach and secreted in the bile. Glycosides of other flavonoids can be hydrolyzed to aglycones in the small intestine due to the activity of human digestive enzymes, such as lactase phlorizin hydrolase and cytosolic β-glucosidase. It refers to the glycosides that contain glucose, xylose or galactose; as mentioned, humans enzymes e have an affinity for those sugars. It means that only polyphenols resistant to the action of human enzymes are not absorbed in the small intestine and pass to the colon, where they may exert their inhibitory or stimulatory activity towards microbiota, or they may be cleaved by bacterial enzymes to produce derivatives and metabolites of various activity.

Another important issue is the diversity of the chemical composition of plant tissues. For example, chokeberries and apples contain much higher amounts of pectin than the elderberry fruit, which resulted in small amounts of polyphenols in the sediment obtained after elderberry digestion [219]. Further, fruits of the Cornelian cherry are rich in pectin and also in low-molecular-weight phenolic acids that can firmly bind with pectin and so pass to the colon intact [219]. However, the differences between the food matrix could also be related to the cell wall composition of the fruit, resulting in an observed different bioaccessibility of polyphenols present in apples, chokeberries and Japanese fruit [219]. The flesh of Japanese quince fruits contains much pectin, whereas, in the cell walls, cellulose dominates [222]. On the other hand, apples are rich both in pectin and cellulose, but they also contain lignin [223]. The presence of lignins was believed to reduce the proanthocyanidin adsorption in skin cell walls when compared to that of the flesh cell walls [224], causing that unbound proanthocyanidins were more sensitive to enzymatic digestion and acidic pH in the stomach.

Proanthocyanidins are of neutral charge, so they are easily absorbed by the cell wall polysaccharides, while anthocyanins—which are positively charged molecules—could rather selectively bind to a negatively charged pectin [225]. The ratio of bound to free proanthocyanidins depends mainly on their concentration and degree of polymerization. The susceptibility of anthocyanins, anthocyanidins, and proanthocyanidins to digestion can also depend both on the structure of the cell wall polysaccharide network in fruits and the structure of pectin. Voragen et al. [226] have demonstrated that 47% of the structural elements of pectin in apples are neutral side chains, while in bilberry or black currant, more than 60% are homogalacturonan]. Yet another structure was reported for Japanese quince pectin, which consisted of four different populations, mainly arabinans and highly methylated homogalacturonans [227]. The simultaneous presence of pectin, cellulose and hemicellulose in food favors the bounding of procyanidins and anthocyanins and protects them against digestive enzyme activity. In consequence, they are not released from the food matrix at this digestion stage. Moreover, during proanthocyanidins degradation, free (+)-catechin **(7)** could be released, which can bind effectively to cellulose [228].

Tarko and Duda-Chodak [219] also revealed that procyanidin B1 in hawthorn was almost insensitive to digestive enzymes, and probably the saponins, which presence in the hawthorn fruit is characteristic, had such a protective impact. Saponins are poorly absorbed in the intestine mainly due to their unfavorable physicochemical traits, such as large molecular mass (>500 Da), high hydrogen binding capacity (>12), and high molecular flexibility (>10).

Concluding, the presence/absence of the food matrix, as well as its chemical composition, can affect the bioaccessibility, bioavailability and biological activity of polyphenols and their bidirectional interactions with the intestinal microbiota.

5. Polyphenols Biotransformation by Intestinal Bacteria

It is believed that only undigested and unabsorbed polyphenols can reach the large intestine and exert their impact on bacteria inhabiting there. As described above, many of the polyphenols can inhibit the growth of microbiota residing in the colon. However, some of the phenolic compounds act as prebiotics and stimulate the growth of particular species. Hence, polyphenols modulate the composition of human gut microbiota. On the other hand, only unabsorbed polyphenols can undergo biotransformation during the activity of bacterial enzymes. Products of bacterial metabolism can further be metabolized to various derivatives and absorbed into the human body [15].

Due to the great diversity of species forming the intestinal microbiota in different individuals, the profile of polyphenol metabolites that are generated and their final effect on the body are highly variable within the human population. The dietary polyphenols can be metabolized by various pathways leading to the formation of a number of different phenolic derivatives characterized by small and low molecular weight as well as a modified biological activity. For example, aglycones and oligomers are released by microbial glycosidases and esterases, which enhances their absorption [229]. On the other side, released aglycones can inhibit intestinal microbiota growth and activity, preventing the metabolism of other polyphenolic compounds from the diet. Some reactions of bacterial metabolism really improve the bioavailability and activity of polyphenolic compounds. In many situations, only the product of bacterial metabolism of a polyphenol can be absorbed and exert a beneficial impact in humans. However, other bacterial metabolites may be harmful to human cells or other members of the microbiota. Hence, apart from interindividual variation in a daily intake of polyphenols, interindividual differences in the composition of the human microbiota may lead to differences in bioavailability and bioefficiency of polyphenols and their metabolites and cause a different impact on host health [230–232].

The identified pathways of bacterial metabolism of the most important groups of polyphenolic compounds are presented below.

5.1. Isoflavonoids

One of the best examples of how significant the role is of the intestinal microbiota in polyphenols impact on human health are the nonsteroidal estrogens. A lack of particular species within the microbiota may cause that isoflavonoid cannot exert its expected effect even though it has been consumed. Isoflavonoids, including daidzein, genistein, and glycitein, are present in soybeans, but they are rather inactive. Only their metabolites, e.g., S-equol or O-desmethylangolensin (O-DMA), are able to exert their pro-healthy effects. Equol, because of its high binding affinity to the estrogen receptor (S-equol preferentially activates estrogen receptor ERβ), can alleviate the symptoms of menopause. Moreover, the antiandrogenic activity and inhibition of osteoclast formation, anticancer activities and anti-inflammatory effects have been observed [6]. It was demonstrated that equol has about 100 times higher estrogenic activity than the daidzein itself [232]. Although O-DMA did not exhibit agonistic or antagonistic activities toward the glucocorticoid receptor (TRa1, or TRb1) and has very weak agonistic activities against ERα and ERβ, it can influence the growth of cancer cells, osteoclast formation, scavenging superoxide radical or exert leptin secretion inhibitory activity [233]. Bacteria strains producing small to moderate amounts of dihydrodaidzein and/or O-DMA from daidzein and dihydrogenistein from genistein are recognized more often than equol producers [234]. O-DMA is found in 80–90% of the human population, whereas equol is found in only 30–50% of the population [235].

The possible metabolic pathways of daidzin and genistin degradation by bacteria are presented in Figures 4 and 5, respectively.

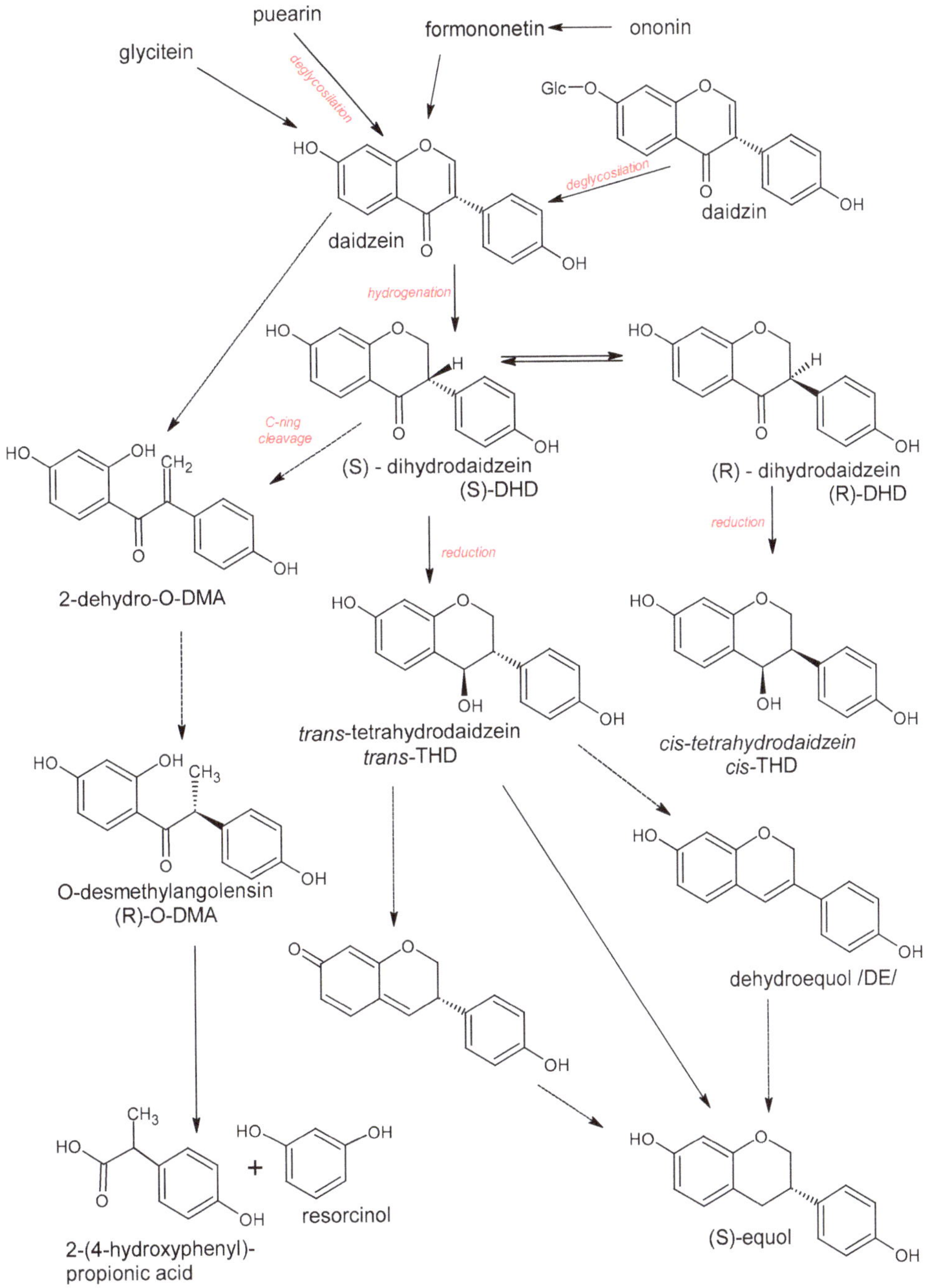

Figure 4. Possible pathways of microbial metabolism of daidzin and daidzein. Based on [236–243]. The dashed arrows indicate hypothesized reactions of microbiological degradation that were observed in vitro but were not reported in vivo.

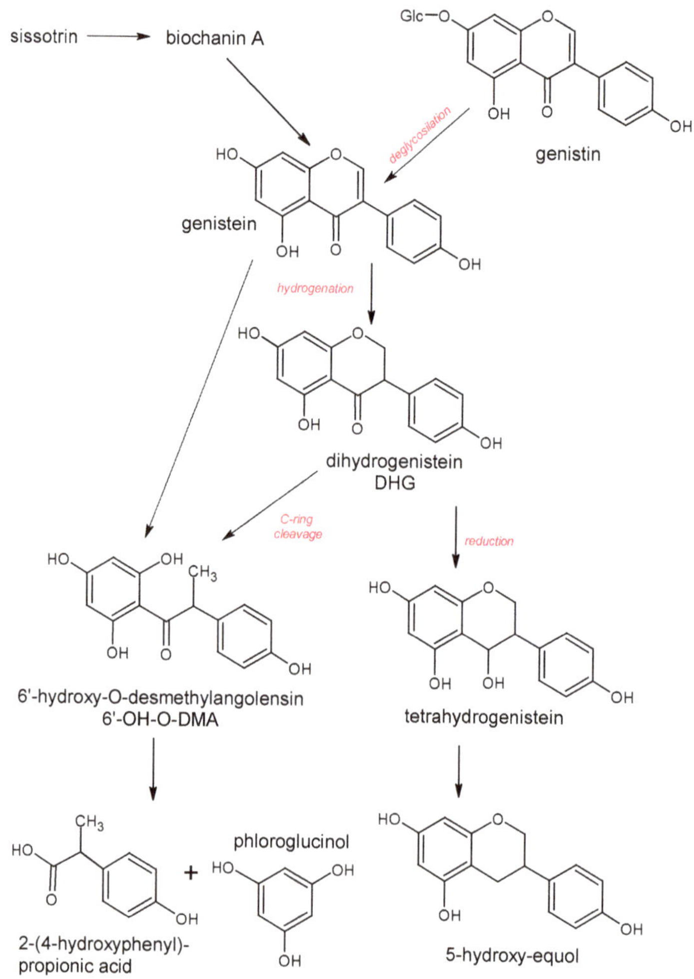

Figure 5. Possible pathways of microbial metabolism of genistin and genistein. Based on [236–243]. The dashed arrows indicate hypothesized pathways of microbiological degradation that were observed in vitro but were not reported in vivo.

The isoflavones biotransformation generally starts with glycoside hydrolysis to release the aglycon. For example, daidzin (daidzein 7-*O*-glycoside) can be hydrolyzed to daidzein by *Eubacterium ramulus* [244]. Then, the hydrogenation of the double bond between C2 and C3 in ring C of daidzein (DZN) and genistein (GN) generates dihydroisoflavones such as dihydrodaidzein (DHD) and dihydrogenistein (DHG), respectively [236]. Dihydroisoflavones are further subjected to bacterial metabolism and can undergo: (a) the reductive pathway leading to equol formation, (b) the cleavage of the C ring, followed by the fission of the molecules into two moieties. Equol is generally produced from daidzein through the reductive metabolism, through dihydrodaidzein (DHD), tetrahydrodaidzein (cis-THD and/or trans-THD) and dehydroequol (DE) as intermediates; however, some equol-producing bacteria have also been shown to convert the genistein into dihydrogenistein and finally to 5-hydroxy-equol [245,246]. *Clostridium* sp. strain HGH136 cleaved the C-ring of daidzein to produce O-desmethylangolensin, probably via 2-dehydro-*O*-demethylangolensin [241]. O-DMA may be further partially metabolized to resorcinol and 2-(4-hydroxyphenyl) propionic acid [233].

Daidzein was in part degraded by *E. ramulus* to *O*-desmethylangolensin, while genistein was completely degraded via 6′-hydroxy-*O*-desmethylangolensin to 2-(4-hydroxyphenyl)-propionic acid [243]. It means that the OH group in position 6′ of O-DMA was crucial for its further degradation. It is interesting that dihydrogenistein was neither observed as an intermediate in this transformation nor converted itself by growing cells of *E. ramulus*. Genistein-7-*O*-glucoside was partially transformed by way of genistein to the product 2-(4-hydroxyphenyl)-propionic acid.

E. ramulus, strain CG19-1 is capable of cleaving both 6′-hydroxy-*O*-desmethylangolensin and *O*-desmethylangolensin to phloroglucinol and resorcinol, respectively; and 2-(4-hydroxyphenyl) propionic acid was additionally formed from both O-DMA and 6′-OH-*O*-DMA [239].

A different metabolic pathway was revealed by Murota et al. [235]. They reported that the metabolites of genistein and glycitein that are primarily found in human urine were dihydrogenistein, 6′-OH-*O*-DMA, 2-(4-hydroxyphenyl)-propionic acid and phloroglucinol for genistein, while dihydroglycitein, 5′-methoxy-*O*-DMA and 6-methoxy-equol for glycitein. Moreover, strain CG19-1 cleaved both O-desmethylangolensin and 6′-hydroxy-*O*-desmethylangolensin to yield 2-(4-dihydroxyphenyl) propionic acid. The corresponding cleavage product, resorcinol, was only observed for O-desmethylangolensin.

According to Rossi et al. [242], the metabolites arising from glycitein include dihydroglycitein, which can be further O-demethylated to 6,7,4′-trihydroxyisoflavone (proved in vitro for *Eubacterium limosum*) and reduced to dihydro-6,7,4′-trihydroxyisoflavone, and further reduced to 6-hydroxyequol or cleaved to 5′-hydroxy-*O*-desmethylangolensin. The other pathway of dihydroglycitein degradation was through the C-ring cleavage producing 5′-*O*-methoxy-*O*-desmethylangolensin or reduction to 6-methoxy-equol (Figure 6).

Slackia isoflavoniconvertens is capable of contributing to the bioactivation of daidzein and genistein by the formation of equol and 5-hydroxy-equol, respectively [246].

It should be underlined that some bacteria can produce equol from either daidzein or its glycoside daidzin, but some cannot produce equol unless several other species of bacteria metabolize daidzin to aglycone daidzein and daidzein to DHD or other derivatives that are also present [235]. For example, *Clostridium* sp. strain HGH6 and *Lactobacillus* sp. Niu-O16 can reduce daidzein to dihydrodaidzein but did not convert dihydrodaidzein to equol [247]. On the other hand, *Eggerthella* sp. Julong 732 is capable of converting dihydrodaidzein, but not daidzein, to equol [238,247]. *Eggerthella* sp. Strain YY7918 converted substrates daidzein and dihydrodaidzein into S-equol but did not convert daidzin, glycitein, genistein, or formononetin into it [248]. Strain TM-40 (93% of homology with *Coprobacillus catenaformis*) isolated by Tamura et al. [249] produced dihydrodaidzein from both daidzein and daidzin. Decroos et al. [240] isolated from human feces a stable mixed microbial culture (*Enterococcus faecium* strain EPI1, *Lactobacillus mucosae* strain EPI2, *Finegoldia magna* strain EPI3 and an as yet undescribed species related to *Veillonella* sp.) that was able to covert daidzein into equol.

Among intestinal bacteria that were proved to metabolize the soya isoflavone daidzein and genistein to equol, DHD and/or O-DMA are *Slackia equolifaciens* (DZN to equol) [250], *Slackia isoflavoniconvertens*, *Adlercreutzia equolifaciens*, *Asaccharobacter celatus*, *Enterorhabdus mucosicola* (DZN to equol), *Peptoniphilus gorbachii* (DZN and GN to equol and O-DMA), *Gordonibacter urolithinfaciens* (DZN to O-DMA), some strains of *Eggerthella lenta* (DZN and GN to O-DMA), *Enterococcus lactis* (to O-DMA), some strains of *Bifidobacterium adolescentis* (DZN and GN to O-DMA), *B. animalis* (DZN to O-DMA) and *B. longum* (DZN and GN to O-DMA), some members of Coriobacteriaceae, e.g., *Collinsella massiliensis* (DZN and GN to O-DMA) and *C. aerofaciens* (DZN to O-DMA) [234], *Eggertella* strain Julong 732 (DHD via THD to equol) [238], *Lactococcus garvieae* strain 20–92 [251], *Eubacterium ramulus* Julong 601 (DZN to O-DMA, GN to 2-(4-hydroxyphenyl) propionic acid) [252], *Clostridium* sp. HGH 136 (DZN to DHD) [241] and HGH6 (DZN do DHD, GN to DHG), and *E. coli* HGH21 (DZN to DHD and GN to DHG) [253].

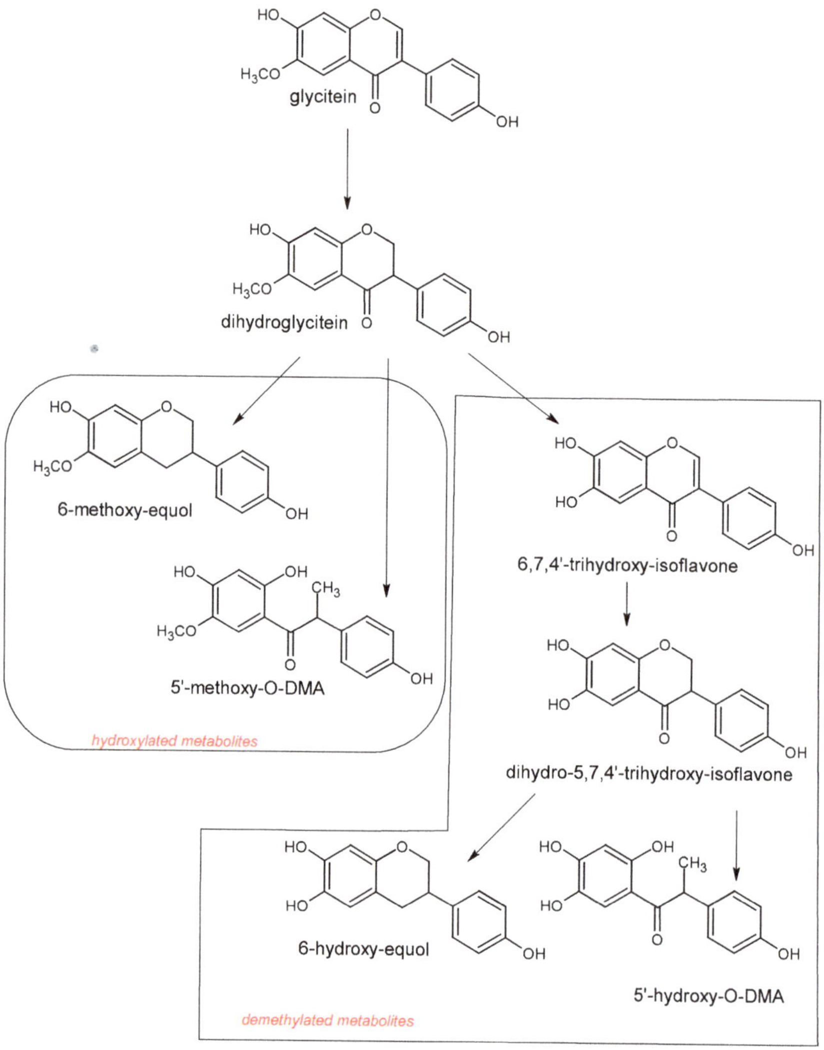

Figure 6. The pathways of bacterial metabolism of glycitin. Based on [237,242].

Puerarin is a daidzein 8-C-glucoside and was reported to be metabolized to daidzein by human intestinal flora such as *E. ramulus* CG 19-1 [239] or intestinal strain PUE, converting puerarin to daidzein by cleaving a C-glucosyl bond [254]. Formononetin and biochanin A are the principal isoflavones of red clover (and as a consequence, equol is present in cow milk) and can be consumed in the form of dietary supplements. Hur et al. [255] demonstrated that *Eubacterium limosum* is able to produce daidzein and genistein from formononetin and biochanin A, respectively. It means that due to bacterial metabolism, more potent phytoestrogens have been formed in the colon, as the estrogenic potencies of the mentioned compound for both estrogen receptors ERα and ERβ showed the affinities in the order of genistein > daidzein > biochanin A > formononetin. In the urine samples of volunteers consuming formononetin and biochanin A, other metabolites were also identified, such as dihydroformononetin and angolensin for formononetin and dihydrobiochanin A and 6′-hydroxyangolensin for biochanin A [256] (Figure 7).

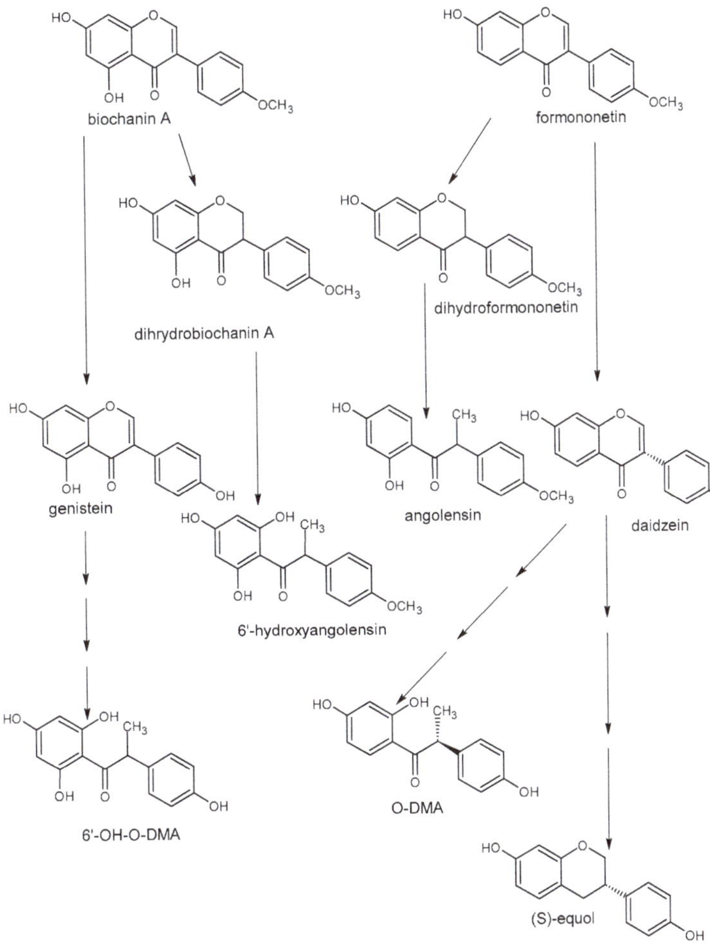

Figure 7. The human metabolism of formononetin and biochanin A. Based on [16,255,256].

5.2. Other Phytoestrogens

In addition to soy isoflavonoids, there are other ligands for estrogen receptors that are produced by intestinal microbiota, such as enterolactone, enterodiol, urolithins and 8-prenylnaringenin. Enterolactone and enterodiol are derivatives of plant lignans from sesame seed or flaxseed. It was proved at concentrations that can be achieved with high consumption of products rich in lignans, both, but enterolactone to a lesser extent can potently activate human estrogen receptors ERα and ERβ [257]. The bacterial transformation of lignans into phytoestrogens (Figure 8) was carried out mainly by *Peptostreptococcus* and *Eubacterium* species and included their demethylation and dihydroxylation, leading to enterolactone production [258]. Enterolactone can further be converted into enterodiol, and various studies proved that both the mentioned mammalian lignans are produced by human colonic microbiota from dietary precursors. Production of enterodiol is about 2000 times more efficient, meaning that the enterodiol-producing bacteria are dominant in the human gut.

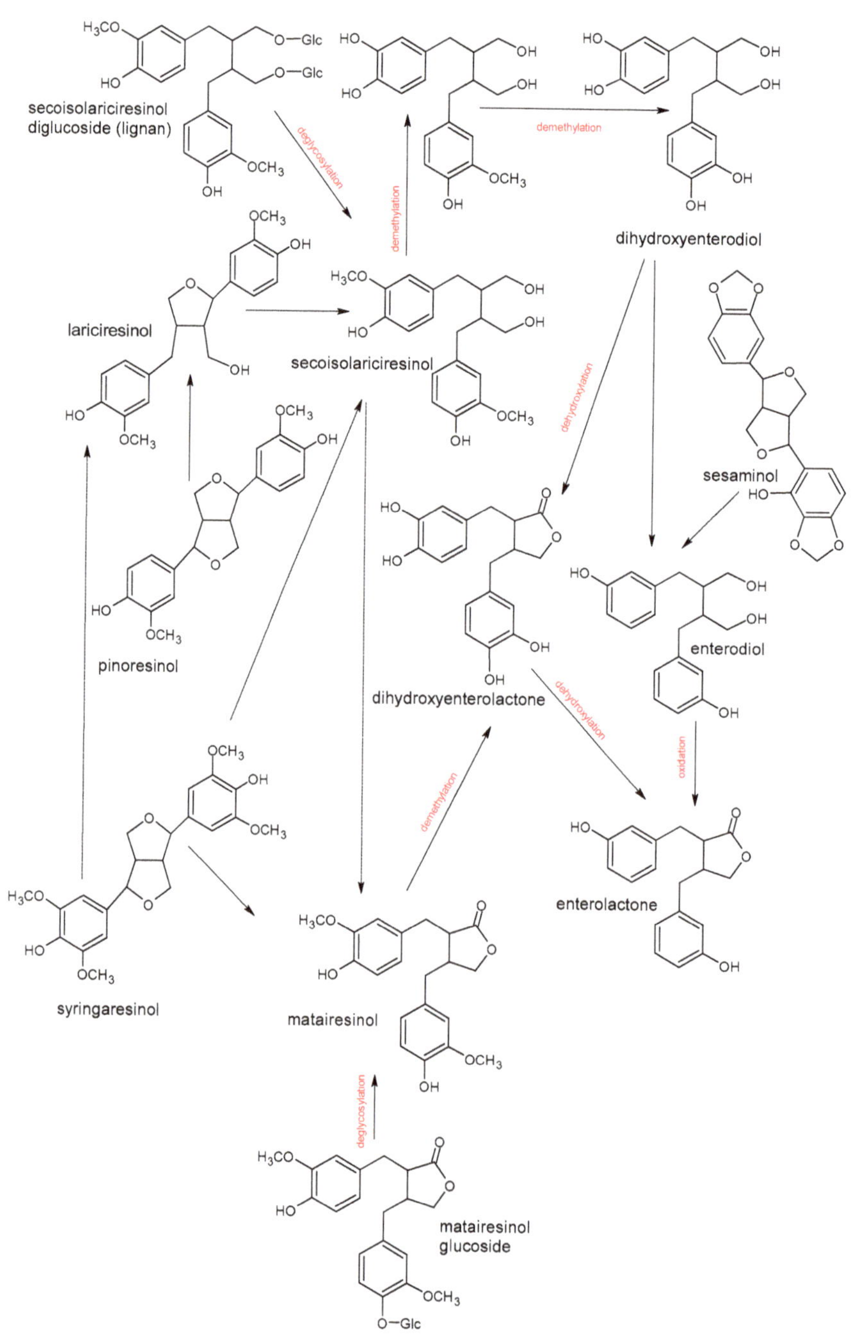

Figure 8. Lignans metabolism by gut microbiota. Based on [12,237,258–261]. The dashed arrows indicate hypothesized or multistep process.

The main bacteria converting lignans to enterolactone and enterodiol are *Peptostreptococcus* sp. SDG-1 and *Eubacterium* sp. SDG-2 [262], *Bacteroides distasonis, B. fragilis, B. ovatus, Eubacterium callanderi, Eubacterium limosum, Clostridium cocleatum, Clostridium scindens, Eggerthella lenta, Butyribacterium methylotrophicum, Butyribacterium pseudocatenulatum, Bifi-*

dobacterium longum, B > breve, B. catenulatum, B. pseudocateunaltum, Enterococcus faecalis, Ruminococcus sp. END-1 [261], *Clostridium saccharogumia* and *Lactonifactor longoviformis* [263]. Two organisms able to demethylate and dehydroxylate secoisolariciresinol were isolated from human feces. Based on 16S rRNA gene sequence analyses, they were named *Peptostreptococcus productus* SECO-Mt75m3 and *Eggerthella lenta* SECO-Mt75m2 [264]. It was demonstrated both in vivo and in vitro that the major metabolite of sesamin in humans is enterolactone [265]. The intestinal pathways of enterolactone and enterodiol production from lignan are presented in Figure 8.

Ellagitannins (ELT) are one of the main groups of hydrolyzable tannins that are characterized by high antioxidant activity. They are common in some fruits, such as pomegranates, black raspberries, raspberries and strawberries, as well as in walnuts and almonds. Chemically they are different esters of hexahydroxydiphenic acid (HHDP) and a polyol, usually glucose or quinic acid [266]. According to the number of HHDP groups linked to the sugar moiety, ellagitannins can be classified into monomeric, oligomeric, and polymeric ELT. The main ellagitannins identified in foods are punicalagin (Figure 9), sanguiin H-6 (dimer of casuarictin) **(212)**, lambertianin C (trimer of casuarictin) **(214)**, pedunculagin, castalagin **(209)**, casuarictin **(210)** and potentillin. Because of their size (634 Da for sanguiin H4 to up to 3740 Da for lambertianin D), these molecules are characterized by very low bioavailability and are not absorbed in the gastrointestinal tract until they are metabolized by gut bacteria. Intact ellagitannins and a product of their acidic or basic hydrolysis—ellagic acid (Figure 9), reach the distal part of the gastrointestinal tract where they are transformed by intestinal microbiota into dibenzopyran-6-one derivatives, known as urolithins, that are much better absorbed [267].

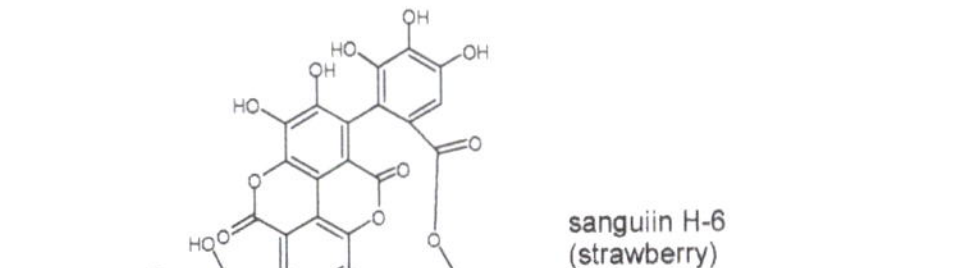

Figure 9. Degradation of ellagitannins to ellagic acid. Based on [268].

The generic name of urolithins includes different hydroxylated 6H-dibenzo[b,d] pyran-6-one derivatives. The bacterial transformation includes reduction of one of the two lactone groups followed by decarboxylation and sequential dehydroxylation involving a step-by-step reduction to tetrahydroxy (urolithin D), trihydroxy (urolithin C), dihydroxy (urolithin

A and isourolithin A), and monohydroxy dibenzopyranones (urolithin B). The pathway of bacterial metabolism is presented in Figure 10. Bacteria able to catalyze the biotransformation of ellagitannins to urolithins are *Gordonibacter urolithinfaciens* and *G. pamelaeae* that belong to the family Coriobacteriaceae [269,270] and *Ellagibacter isourolithinifaciens* from Eggerthellaceae [271].

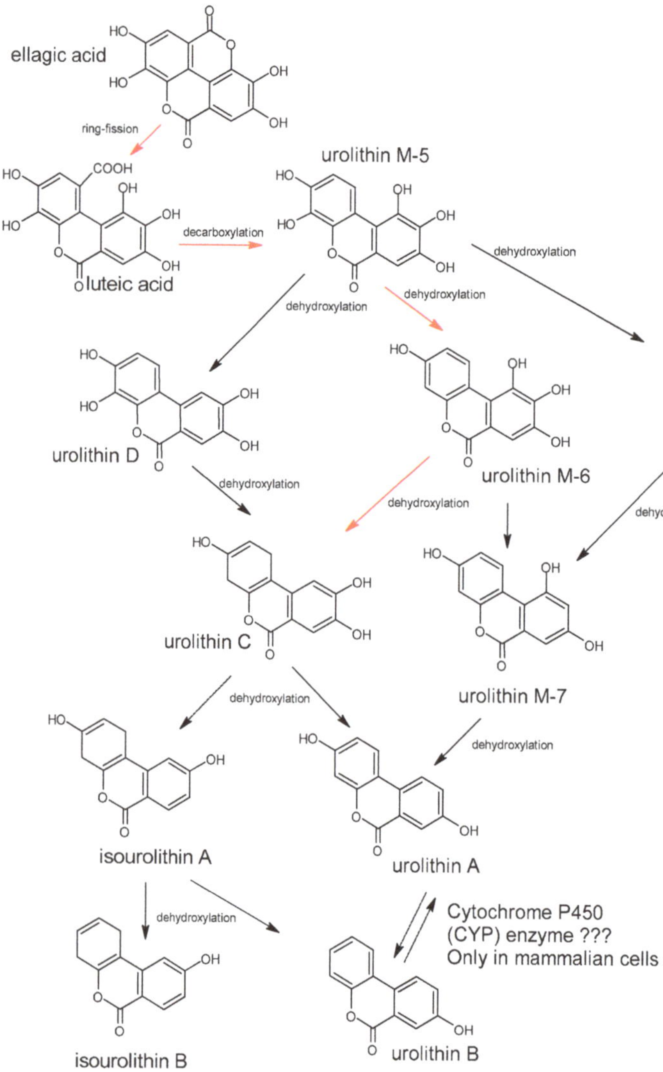

Figure 10. The bacterial metabolism of ellagic acid to urolithins and derivatives. The red arrows represent the pathways reported in *Gordonibacter urolithinfaciens* and *G. pamelaeae*. Based on [272–274].

Although urolithins are characterized by lower antioxidant activity than ellagitannins, they circulate in the plasma as glucuronide and sulfate conjugate and display benefit influence on human health due to their estrogenic and/or anti-estrogenic activity, as well as anticancer activities. It means that bacterial metabolism is crucial for the pro-healthy properties of various berries [266,272].

It has been shown that the production of the potent hop phytoestrogen 8-prenylnaringenin (8-PN) depends on the activity of human intestinal microbiota [275]. This compound is generated from xanthohumol and isoxanthohumol that unaltered reach the small intestine (Figure 11). Among bacteria that catalyze the demethylation of isoxanthohumol into 8-PN are *Eubacterium limosum* and *E. ramulus* [275–277]. In addition to a strong impact on the ERα receptor, 8-prenylnaringenin inhibits angiogenesis and metastasis, prevents bone loss in rats and exhibits antiandrogenic activity [275,278].

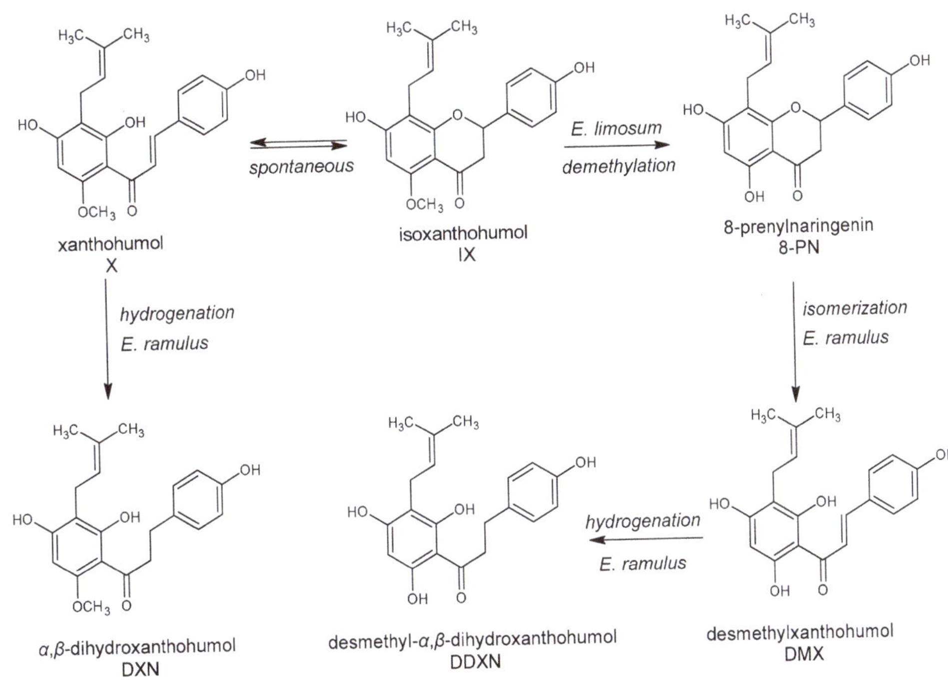

Figure 11. Bacterial biotransformation of prenylflavonoids xanthohumol and isoxanthohumol. Based on [275,276,279].

5.3. Bacterial Transformation of Anthocyanidins

Anthocyanidins (ACD) are plant pigments responsible for flower, fruit and vegetable color. Their structure and color depend on pH value and the presence of copigments. In plant tissues, ACD are generally present in the form of glycosides, called anthocyanins (ACN), that are susceptible to hydrolytic conversion into their corresponding anthocyanidins. Glucose, galactose, rhamnose and arabinose are the sugars most commonly encountered, usually as 3-*O*-glycosides or 3,5-*O*-diglycosides; however, rutinosides (6-*O*-L-rhamnosyl-D-glucosides), sophorosides (2-*O*-D-glucosyl-D-glucosides) and sambubiosides (2-*O*-D-xylosyl-D-glucosides) also occur, as do some 3,7-diglycosides and 3-triosides [280]. Moreover, some of the hydroxyl groups can be methylated, giving the big diversity of plant anthocyanidins. The anthocyanidins occur in the vacuole as an equilibrium of four molecular species that affects their color (Figure 12). However, after fruit and vegetable consumption, the form of the flavylium cation exists only in the stomach, while other forms are present in the lower parts of the gastrointestinal tract and in the tissues (if absorbed).

Bacterial metabolism of ACN involves the cleavage of glycosidic linkage and breakdown of the anthocyanidin heterocycle. Aura et al. [281] demonstrated that cyanidin-3-rutinoside was degraded through cyanidin-3-glucoside and cyanidin aglycone as intermediary metabolites. After hydrolysis of the protective 3-glycosidic linkage, the released aglycons are stable under acidic pH but unstable under neutral or slightly basic pH. It means that under physiological conditions in the small intestine, the cleavage of the heterocyclic flavylium ring occurs [274]. An attack of the flavylium carbon at position 2 produces an unstable hemiketal that rapidly forms a ketone (Figure 13). Through keto-enol tautomerism of the neighboring enol functionality, the resulting α-diketone is very reactive and is easily decomposed by gut microbiota to phenolic acids (mainly protocatechuic acid, syringic acid, vanillic acid) and aldehydes (mainly phloroglucinol aldehyde) [274].

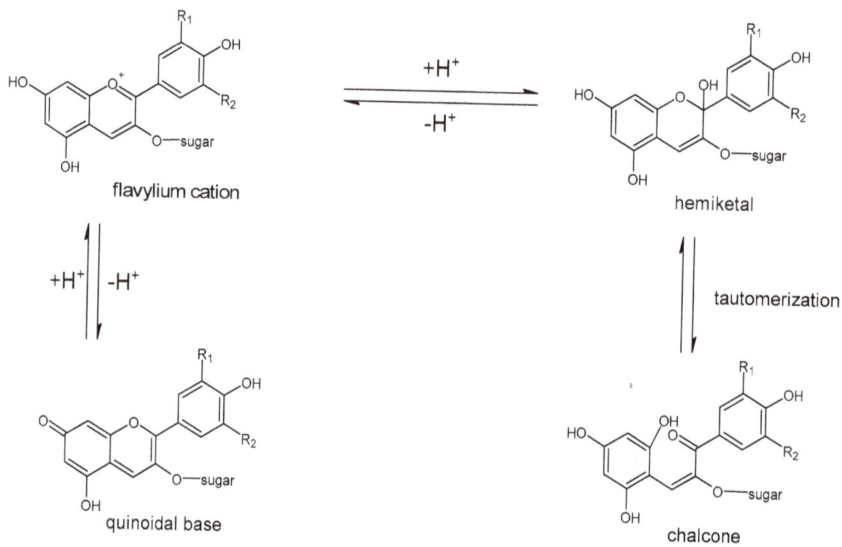

Figure 12. Anthocyanins equilibria [39].

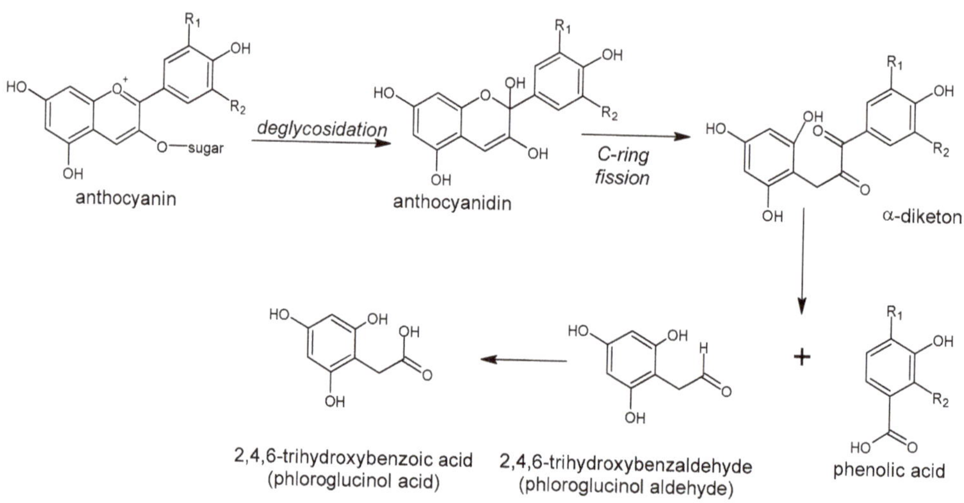

Figure 13. Main steps of anthocyanin degradation. Based on [274,282].

It was demonstrated for raspberry anthocyanins; when incubated with fecal suspensions under anaerobic conditions, that they underwent a transformation by the colonic microflora. After C-ring fission in cyanidin, aglycone phenolic acids were released, originating from both the A and B rings. It was proved that some of the colonic catabolites entered the circulation and were further metabolized before being excreted in urine (e.g., as hippuric acid) [283].

Phenolic acids may be utilized as a source of energy by the intestinal microflora. Keppler and Humpf [282] demonstrated that bacterial metabolism of the methoxyl derivatives as syringic acid and vanillic acid was accompanied by O-demethylation and resulted in the formation of gallic acid and protocatechuic acid (PCA), respectively. As phloroglucinol aldehyde (PHA) was degraded by the intestinal microflora very similar in comparison to the sterilized control samples, it was not possible to distinguish between the chemical or microbial transformation of the aldehyde. However, the phloroglucinol acid was detected as the oxidation product of PHA in very low amounts only in the non-sterilized inoculum filtrate [282], indicating the role of gut microbiota in the transformation.

Some in vitro studies revealed that the numbers of potentially beneficial bacteria (bifidobacteria and lactobacilli) increased after the consumption of purple sweet potato anthocyanins and grape seed extract [284]. As anthocyanins are hardly absorbed in the small intestine, they may be transformed into small molecular phenolic acids by colonic microbiota through ring cleavage, dihydroxylation and methylation reactions. Such metabolites generated from polyphenols may selectively stimulate the growth of beneficial bacterial, whereas the proliferation of harmful bacteria would be inhibited. Ávila et al. [285] analyzed various strains of *Lactobacillus plantarum* and *L. casei*, as well as probiotic strains *Lactobacillus acidophilus* LA-5 and *Bifidobacterium lactis* BB-12. They proved the enzymatic potential of selected strains for bioconversion of delphinidin and malvidin glycosides to their metabolites. Incubation of malvidin-3-glucoside with *B. lactis* BB-12, *L. plantarum* IFPL722, and *L. casei* LC-01 cell-free extracts led to different patterns of gallic, homogentisic and syringic acid formation.

It was also reported that gallic acid and free anthocyanins activated cell growth and the rate of malic acid degradation; vanillic acid showed a slight inhibiting effect, while protocatechuic acid had no effect. Finally, gallic acid and ACN were metabolized, especially by growing cells [88]. Incubation of malvidin-3-glucoside with fecal bacteria mainly resulted in the formation of syringic acid, while the mixture of anthocyanins resulted in the formation of gallic, syringic and *p*-coumaric acids [286].

The most abundant anthocyanins in fruit and vegetables are cyanidin, pelargonidin, petunidin, peonidin and delphinidin. The hypothesized pathways of their bacterial degradation are presented in Figures 14 and 15. Mayor ACN metabolites generated in the human colon by bacteria are protocatechuic acid, syringic acid, vanillic acid, gallic acid, phenylacetic acid, 3,4-dihydroxyphenylpropionic acid, 3,4-dihydroxyphenylacetic acid, 4-hydroxybenzoic acid, but also 4-hydroxyphenylethanol (tyrosol), catechol, benzoic acid [282,284,287–289].

Zhu et al. [284] reported that 2,4,6-trihydroxybenzoic acid, 4-hydroxybenzaldehyde, benzoic acid, phenylacetic acid, and phenylpropionic acid were found in the medium after bacterial metabolism od cyanidin-3-*O*-glucoside. The metabolism of cyanidin-3-*O*-glucoside and cyanidin-3-*O*-rutinoside mainly resulted in the formation of protocatechuic, vanillic, and *p*-coumaric acids, as well as 2,4,6-trihydroxybenzaldehyde, while the main metabolites of delphinidin-3-*O*-rutinoside were gallic acid, syringic acid and 2,4,6-trihydroxybenzaldehyde. Among minor metabolites identified after microbial metabolism of mentioned glycosides were: protocatechuic acid-glucoside, caffeic acid, tartaric acid, catechol, as well as pyrogallol, ferulic acid, 4-hydroxybenzoic acid [290]. This research indicated that the intake of ACNs might result in the appearance of specific metabolites that exert a protective effect in host physiology.

The main phenolic acid detected in fecal suspensions incubated with raspberry anthocyanins were: 3-phenylacetic acid, 3-(4′-hydroxyphenyl) lactic acid, tyrosol, 3-(4′-hydroxyphenyl) propionic acid, 3-(3′-hydroxyphenyl) propionic acid, 4-hydroxybenzoic acid and 3,4-dihydroxybenzoic acid, but lower amounts of catechol, resorcinol, pyrogallol and 3-(3′,4′-dihydroxyphenyl) propionic acid were also found [283]. Seven metabolites formed by human fecal bacteria were observed by LC/MS after incubation with cyanidin-

3-*O*-glucoside and cyanidin-3-*O*-galactoside, and protocatechuic acid (major metabolite), 2,4,5-trihydroxy-benzaldehyde, 5-hydroxy-2-(4′-hydroxyphenyl)-2H-chromen-3,7-dione, 3,5,7-trihydroxy-2-(3′,4′-dihydroxyphenyl)-2H-chromene, and 5-hydroxy-2-phenyl-2H-chromen-3,7-dione were identified [288]. The deglycosylation, decomposition, hydrogenation, and dihydroxylation reactions were involved in their generation.

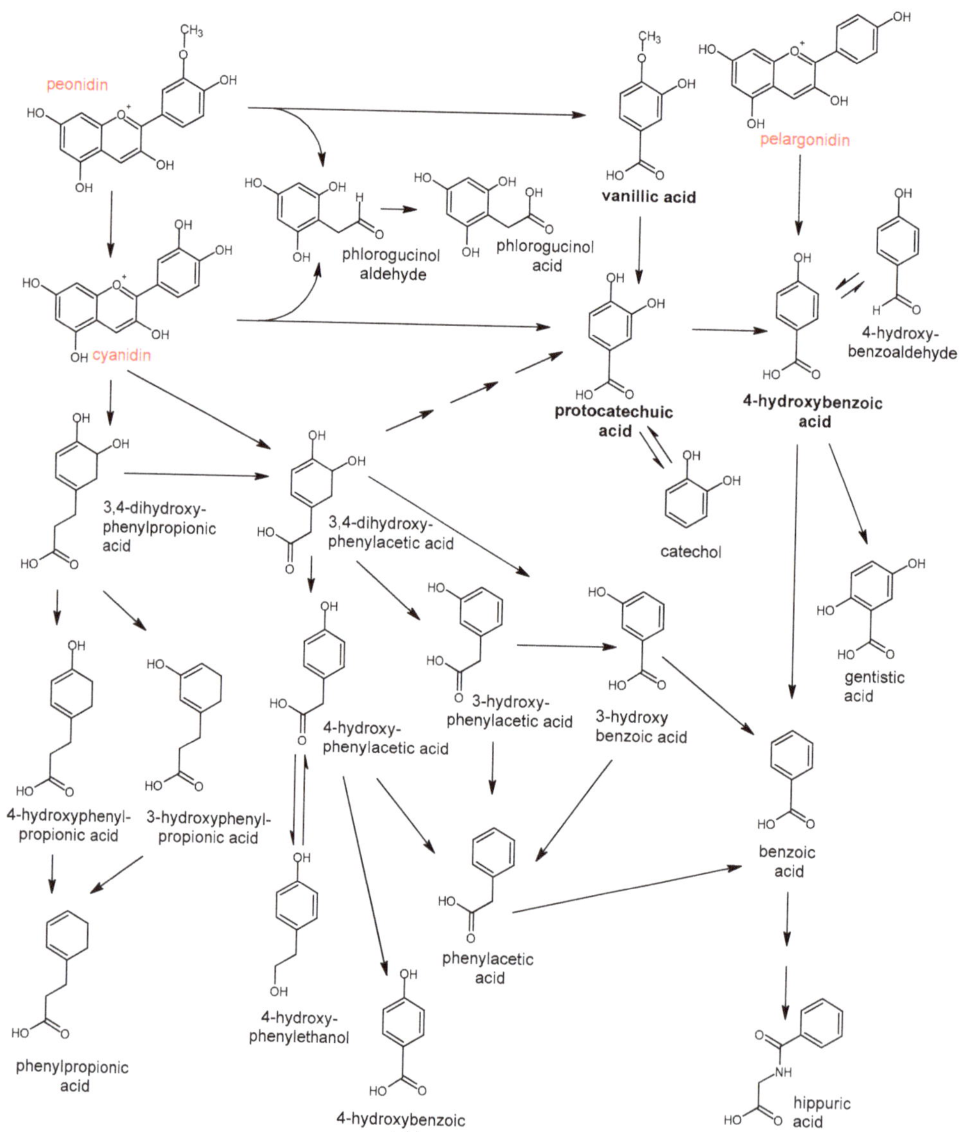

Figure 14. Biodegradation of anthocyanins and their main metabolites: cyanidin → 3,4-dihydroxybenzoic acid (protocatechuic acid), peonidin → 3-methoxy-4-hydroxybenzoic acid (vanillic acid) and pelargonidin → 4-hydroxybenzoic acid. Based on [282,284,287,288].

Malvidin-3-glucoside was completely degraded into syringic acid after incubation with a human fecal slurry for 24 h, whereas gallic acid, *p*-coumaric, and syringic acid were formed after a mixture of various anthocyanins were incubated with healthy human fecal bacteria [291].

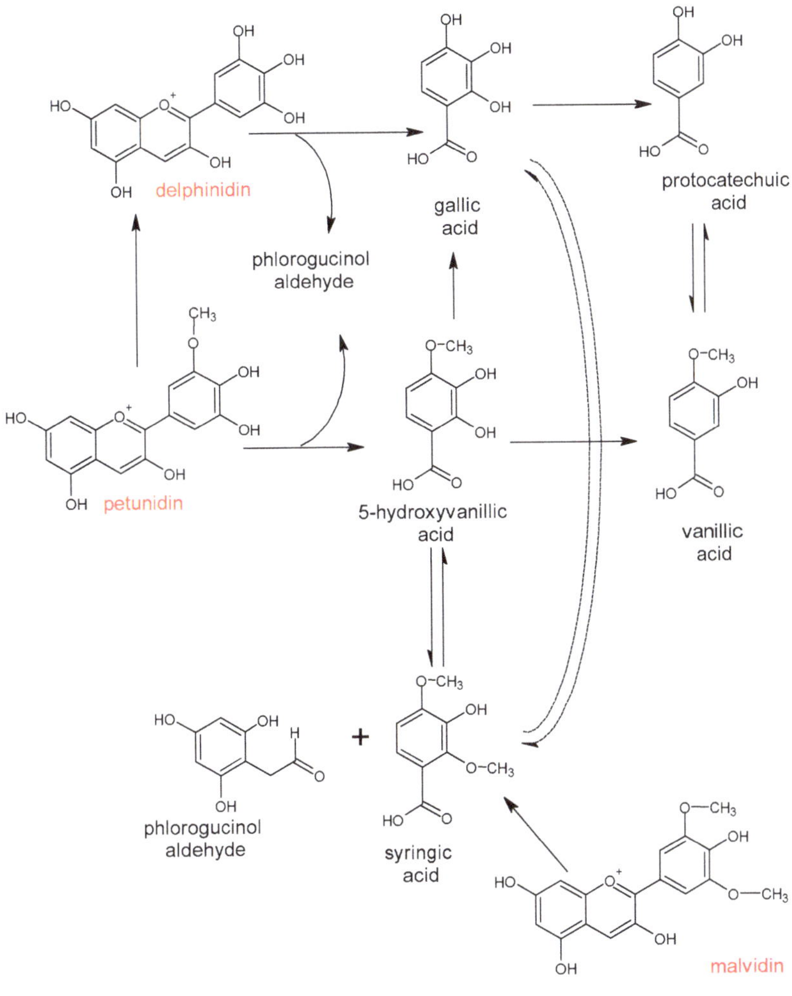

Figure 15. Biodegradation of malvidin, delphinidin and petunidin by intestinal bacteria. Based on [285,290,292].

5.4. Metabolism of Procyanidins and Catechins by Intestinal Bacteria

Condensed tannins (also called catechol-type tannins or non-hydrolyzable tannins) are an important class of polyphenols that do not contain sugar residues. They are also called proanthocyanidins as, under oxidative conditions, they depolymerize, yielding anthocyanidins. Therefore, different types of condensed tannins exist, such as the procyanidins, propelargonidins, prodelphinidins, profisetinidins, proteracacinidins, proguibourtinidins or prorobinetidins.

Condensed tannins are formed from flavan-3-ols or flavan-4-ols. These particular types of condensed tannins are procyanidins, which are not susceptible to cleavage by hydrolysis. Procyanidins are polymers of 2 to 50 (or more) catechin units (usually catechin and epicatechin molecules) joined by carbon–carbon bonds. The most ubiquitous are B-type procyanidins, abundant in apple, cocoa, pear, blueberries; these subunits are linked by single bond C4–C8 or C4–C6. In A-type procyanidins, present in cranberries, a double linkage exists; the C4–C8 or C4-C6 bond is accompanied by an additional C2–O–C7 or C2–O–C5 ether bond.

Some studies demonstrated that highly polymeric procyanidins (PPs) administration markedly decreased the Firmicutes/Bacteroidetes ratio and increased by eight times the proportion of *Akkermansia*, suggesting that PPs influence the gut microbiota and the intestinal metabolome to produce beneficial effects on metabolic homeostasis [76]. On the other hand, some species of intestinal bacteria are able to degrade oligomeric procyanidins. Spencer et al. [221] proved that procyanidin oligomers (trimer to hexamer) are hydrolyzed in simulated gastric juice to mixtures of epicatechin monomer and dimer, thus enhancing the potential for their absorption in the small intestine. Proanthocyanidins that undergo partial acid-catalyzed cleavage then decompose to monomeric flavan-3-ols, which are also metabolized by colonic bacteria. The bacterial degradation of flavan-3-ols and proanthocyanidins follow a similar pathway, and both lead to a generation of a unique compound 5-(3′,4′-dihydroxyphenyl)-γ-valerolactone, which further undergoes dihydroxylation and oxidation to produce phenolic acids [291,293] (Figure 16).

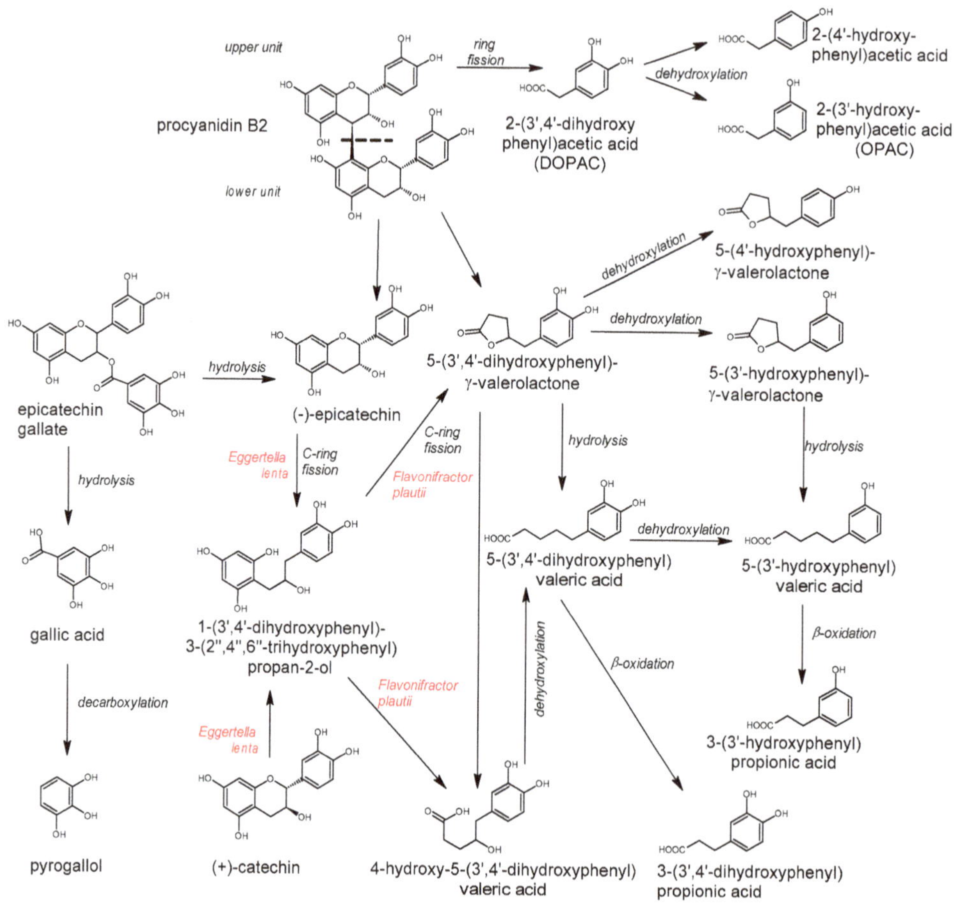

Figure 16. Microbial metabolism of procyanidin B2, (−)-epicatechin and (+)-catechin. Based on [274,291,294–297].

The incubation of purified (+)-catechin, (−)-epicatechin, procyanidins A2 and B2, as well as partially purified apple and cranberry procyanidins with human gut microbiota, resulted in their degradation. The common metabolites were benzoic acid, 2-phenylacetic acid, 3-phenylpropionic acid, 2-(3′-hydroxyphenyl) acetic acid (OPAC), 2-(4′-

hydroxyphenyl) acetic acid, 3-(3′-hydroxyphenyl) propionic acid, and hydroxyphenylvaleric acid. Interesting, that 5-(3′,4′-dihydroxyphenyl)-γ-valerolactone and 5-(3′-hydroxyphenyl)-γ-valerolactone were identified as the bacterial metabolites of epicatechin, catechin, procyanidin B2, and purified apple procyanidins, but not from the procyanidin A2 or cranberry procyanidin ferments, while 2-(3′,4′-dihydroxyphenyl) acetic acid was only found in the fermented broth of procyanidin B2, A2, apple, and cranberry procyanidins [298].

The monomeric flavan-3-ols, which are usually shared by condensed tannins, can be degraded by fecal microbiota into low molecular weight aromatic compounds, including phenylpropionic acid, 2-(3′-hydroxyphenyl) acetic acid, 3-(3′-hydroxyphenyl) propionic acid, 5-(3′-hydroxyphenyl) valeric acid, phenylacetic acid, and 2-(4′-hydroxyphenyl) acetic acid [294]. Similar results have been obtained Appeldoorn et al. [296]. Purified procyanidin dimers, when incubated with human microbiota, have been transformed and among major identified metabolites were 2-(3′,4′-dihydroxyphenyl) acetic acid (DOPAC) and 5-(3′,4′-dihydroxyphenyl)-γ-valerolactone. Other metabolites detected were: OPAC, 2-(4′-hydroxyphenyl) acetic acid, 3-(3′-hydroxyphenyl) propionic acid, phenylvaleric acids, monohydroxylated phenylvalerolactone, and 1-(3′,4′-dihydroxyphenyl)-3-(2″,4″,6″-trihydroxyphenyl) propan-2-ol. In studies of Deprez et al. [297], polymeric procyanidins were metabolized by colonic bacteria into low-molecular-weight phenolic acid, and the main metabolites were 3-phenylpropionic acid, 3-(4′-hydroxyphenyl) propionic acid, 3-(3′-hydroxyphenyl) propionic acid, 5-(3′-hydroxyphenyl) valeric acid, 2-(3′-hydroxyphenyl) acetic acid, and 2-(4′-hydroxyphenyl) acetic acid.

Among intestinal bacteria that are able to convert catechins are *Eggerthella lenta* and *Flavonifractor plautii* (formerly *Clostridium orbiscindens*) [295]. *Eggerthella lenta* rK3 reductively cleaved the heterocyclic C-ring of both (−)-epicatechin and (+)-catechin giving rise to 1-(3′,4′-dihydroxyphenyl)-3-(2″,4″,6″-trihydroxyphenyl) propan-2-ol (Figure 16). The conversion of catechin proceeded five times faster than that of epicatechin. *Flavonifractor plautii* aK2 and *Flavonifractor plautii* DSM 6740 further converted 1-(3′,4′-dihydroxyphenyl)-3-(2″,4″,6″-trihydroxyphenyl) propan-2-ol to 5-(3′,4′-dihydroxyphenyl)-γ-valerolactone and 4-hydroxy-5-(3′,4′-dihydroxyphenyl) valeric acid.

According to Tzounis et al. [59], the initial conversion of (+)-catechin to (+)-epicatechin is required to the generation of 5-(3′,4′-dihydroxyphenyl)-γ-valerolactone, 5-phenyl-γ-valerolactone and phenylpropionic acid as metabolites. The prebiotic effects of both (+)-catechin and (−)-epicatechin was observed, suggesting that the consumption of flavanol-rich foods may support gut health through their ability to exert prebiotic actions.

Monagas et al. [299] demonstrated that some phenolic acids, including 3-*O*-methyl gallic, gallic, caffeic, 3-(4′-hydroxyphenyl) propionic, phenylpropionic, and 2-(4′-hydroxyphenyl) acetic acids derived from the microbial degradation of tea catechins, were able to inhibit the growth of several pathogenic and non-beneficial intestinal bacteria without significantly affecting the growth of beneficial bacteria (*Lactobacillus* spp. and *Bifidobacterium* spp.). It is possible that *Bifidobacterium* sp. are resistant to flavan-3-ols, being the are important iron-chelating compounds, because these bacteria do not use heme-containing enzymes [237]. Growth of certain pathogenic bacteria such as *Clostridium perfringens*, *Clostridium difficile*, *Streptococcus pyogenes*, and *Str. pneumoniae* was significantly repressed by tea phenolics (catechin, epicatechin, gallic acid, caffeic acid), while commensal anaerobes like *Clostridium* spp., *Bifidobacterium* spp. and probiotics such as *Lactobacillus* sp. were less severely affected [70]. Similarly, the bacterial metabolites, such as 3-(4′-hydroxyphenyl) propionic acid, 3-phenylpropionic acid and 2-(4′-hydroxyphenyl) acetic acid, strongly inhibited the growth of *E. coli*, *S. aureus* and *Salmonella* sp. without influencing beneficial *L. casei* strain Shirota and *Bifidobacterium breve*.

Alakomi et al. [300] have shown that DOPAC, OPAC, 3-(3′,4′-dihydroxyphenyl) propionic acid, 3-(4′-hydroxyphenyl) propionic acid, 3-phenylpropionic acid, and 3-(3′-hydroxyphenyl) propionic acid efficiently destabilized the outer membrane of *Salmonella enterica* subsp. *enterica* serovar Typhimurium and *S. enterica* subsp. *enterica* serovar Infantis.

Moreover, DOPAC, OPAC and 3-(3′,4′-dihydroxyphenyl) propionic acid increased the susceptibility of *Salmonella* Typhimurium strains for novobiocin. It means that beneficial bacteria residing in the human gut can inhibit *Salmonella* growth by the transformation of food flavonoids to active antimicrobial metabolites.

5.5. The Bacterial Metabolism of Flavones and Flavonols

It is interesting that some compounds are common and can be generated by colonic microbiota during the metabolism of various polyphenols, although to a different extent. For example, flavan-3-ols, as well as flavonols and hydroxycinnamic acids, lead to the generation of 3-(3′,4′-dihydroxyphenyl)-propionic acid, 3-(3′-hydroxyphenyl) propionic acid, and 3-(4′-hydroxyphenyl) propionic acid. It means that some enzymes and metabolic pathways are quite common among bacteria. When quercetin is metabolized by bacteria, ring fission is done, leading to the generation of DOPAC, OPAC and protocatechuic acid (PCA) (Figure 17). Braune et al. [301] examined the degradation mechanism of the flavonol quercetin and the flavone luteolin and had demonstrated that *Eubacterium ramulus* converted quercetin through taxifolin and alphitonin, resulting in the formation of DOPAC and phloroglucinol. Flavonol luteolin was transformed by *E. ramulus* to 3-(3′,4′-dihydroxyphenyl) propionic acid via eriodictyol and derivatives. In both pathways, ring fission had taken place. Glycosides of quercetin, such as common in onions quercetin 4′-*O*-glucoside and quercetin 3-*O*-glucoside, are first hydrolyzed to aglycone and then are also catabolized, giving ring-fission products [235]. *Bifidobacterium animalis* subsp. *lactis* AD011, isolated from infant feces, has been shown to catalyze quercetin 3-*O*-glucoside and isorhamnetin 3-*O*-glucoside into quercetin and isorhamnetin, respectively [302].

The degradation of flavones and flavonols by *Clostridium orbiscindenss* was studied by Schoefer et al. [303]. They confirmed the quercetin degradation via taxifolin and alphitonin. Flavone apigenin and luteolin were converted to 3-(4′-hydroxyphenyl) propionic acid and 3-(3′,4′-dihydroxyphenyl) propionic acid, respectively, and phloroglucinol was released in both cases (Figure 17). The intermediate metabolites were naringenin and phloretin for apigenin and eriodictyol and dihydrochalcone for luteolin [303]. However, the isolated *C. orbiscindens* strain was unable to hydrolyze the glycosidic bonds of luteolin 3-*O*-glucoside, luteolin 5-*O*-glucoside, naringenin 7-*O*-neohesperidoside (naringin), quercetin 3-*O*-glucoside, quercetin 3-*O*-rutinoside (rutin), and phloretin 2′-*O*-glucoside, suggesting that other bacteria are required for the initial steps in the metabolism of flavonoid glycosides in the human intestine. Similar pathways were reported for degradation of myricetin, kaempferol as well as quercetin, apigenin and luteolin glycosides, and among bacteria involved in their metabolism were *Enterococcus casseliflavus*, *Eubacterium ramulus*, *Eubacterium oxidoreducens*, *Butyrivibrio* spp., *Clostridium orbiscidens*, *Eggerthella* sp., *Flavonifractor plautii*, *Bacteroides uniformis*, *Bacteroides ovatus*, *Bifidobacterium* spp., *Bacteroides distasonis*, and *Blautia* sp. [235,237,304–307].

Experiments using radiolabeled quercetin 4′-*O*-glucoside (Q4′G) revealed that Q4′G passes through the gastrointestinal tract of rats and that almost all of Q4′G is converted into phenolic acids, with DOPAC and OPAC being the most abundant, and a small amount of PCA was also generated [235]. Moreover, 69% of Q4′G radioactivity was recovered in the form of phenolic acid derivatives, such as OPAC and hippuric acid, in the urine. It means that the first ring-fission product is DOPAC, which is subsequently subjected to dehydroxylation to form OPAC, followed by further catabolism into hippuric or benzoic acids (Figure 17). DOPAC also has been identified as a major catabolite of quercetin glycosides, such as rutin, as well as procyanidins (Figure 16). It is important because DOPAC is known to be a metabolite of the neurotransmitter dopamine, suggesting the existence of a metabolic pathway for DOPAC in humans. It has been demonstrated that DOPAC exerts anticancer, anti-inflammatory, cardioprotective and neuroprotective impact. However, DOPAC may inhibit mitochondrial respiration in brain mitochondria (when NO radical is present) and thus lead to mitochondrial dysfunction, which is assumed to be an important mechanism involved in Parkinson's disease [308].

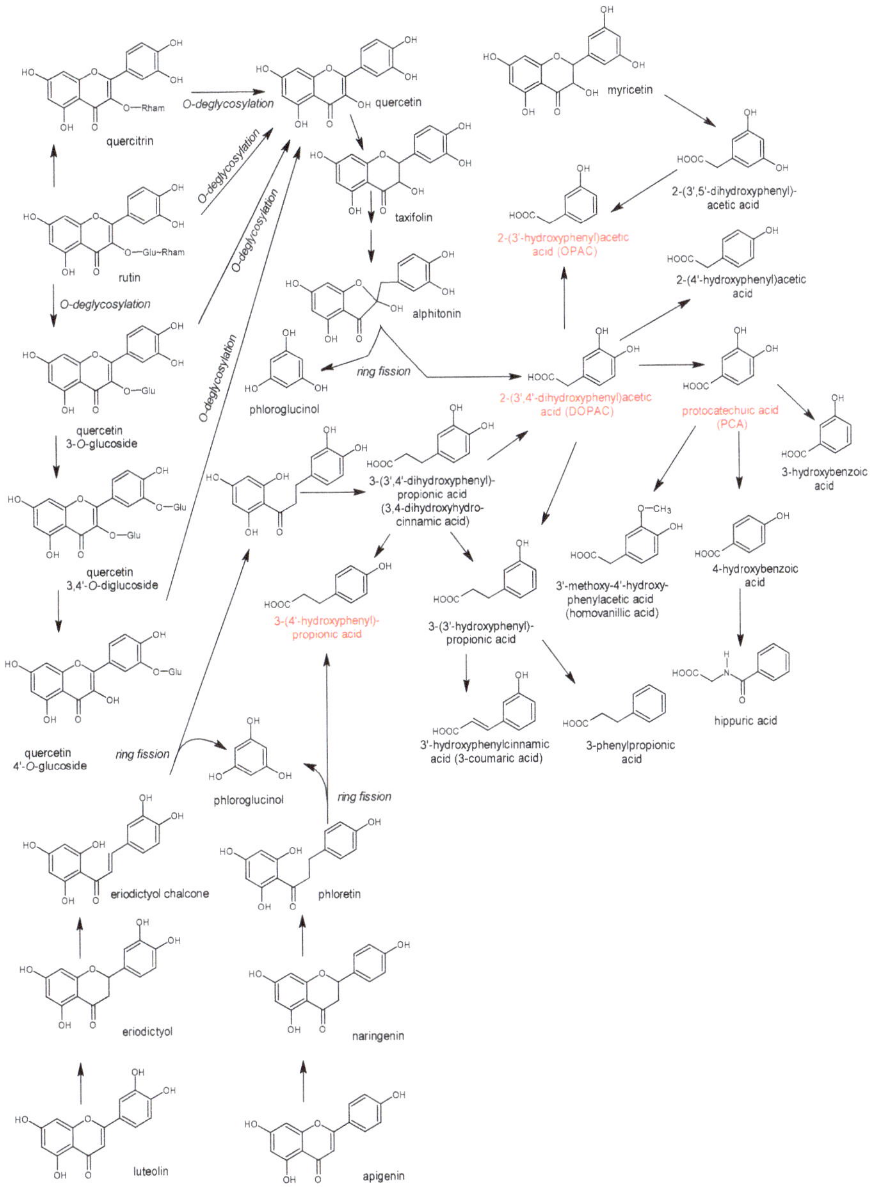

Figure 17. Possible pathways of the transformation of flavones and flavonols due to metabolism by intestinal bacteria. Based on [239,301,304,309,310].

5.6. Microbial Catabolism of Phenolic Acids

Phenolic acids can be delivered to the intestine with food, but they are also generated as the final metabolites during the degradation of various polyphenols. Phenolic acids play an important protective role in degenerative diseases as they exert antioxidant, antitumor, apoptotic, neuroprotective, hepatoprotective, anti-inflammatory and antimicrobial properties [35]. However, there has been some controversy about the bioactivity of polyphenols after metabolism. Once ingested, these molecules are metabolized and transformed into methylated, glucuronated and sulfated metabolites, and there is much evidence proving both the enhanced and decreased biological activity of phenolic acid metabolites [35]. Not all phenolic acids are absorbed, and some of them reach the colon and can be metabolized by bacteria. It is supposed that the presence of an ester moiety lowers hydroxycinnamic acids (HCAs) absorption. Actually, HCAs in a free form are rapidly absorbed throughout the gastrointestinal tract, while HCAs esters or HCAs attached to cell walls require to be hydrolyzed by bacterial esterases before absorption [311]. As a large interindividual variation of phenolic acid metabolites (Figure 18) was observed, it may suggest that the catabolic pathways of both chlorogenic acid and other phenolic acids depend mainly on the colon microbiota composition.

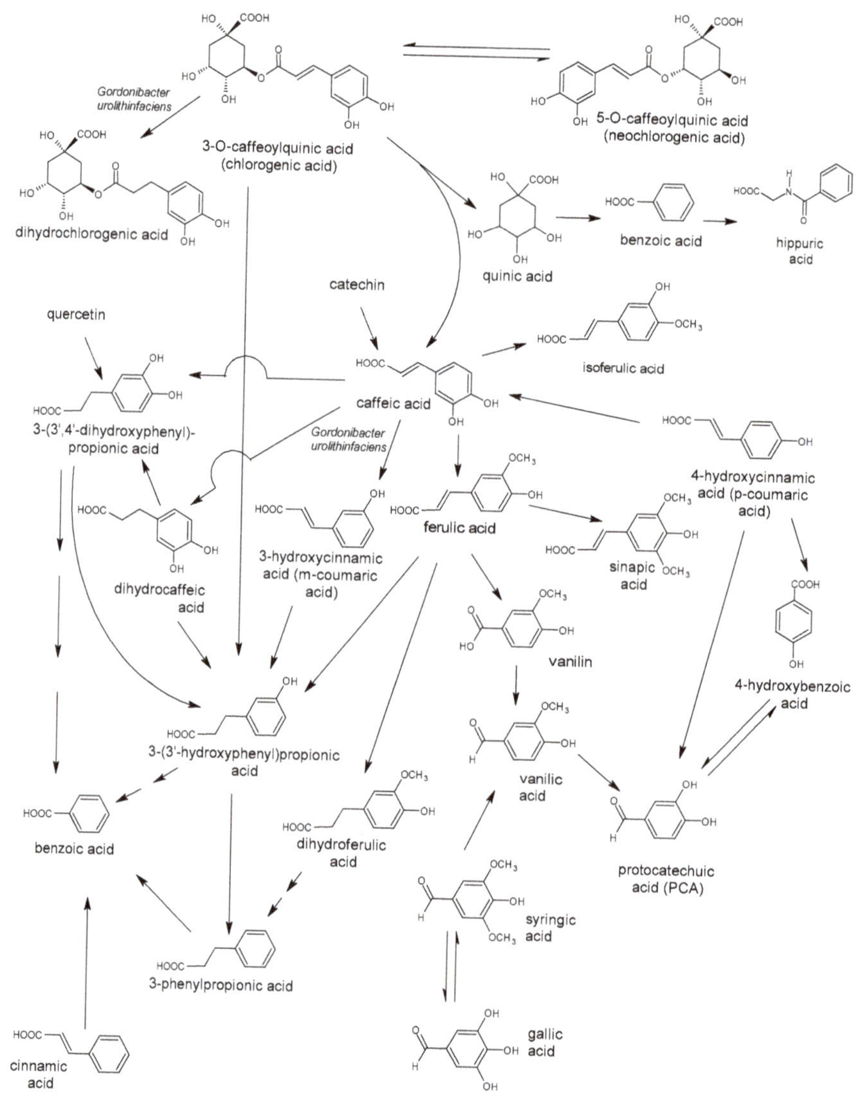

Figure 18. Pathways of some phenolic acid metabolism conducted by various bacteria. Based on [37,304,312–315].

Incubation of coffee samples (a rich source of phenolic acids) with the human fecal microbiota led to the rapid metabolism of chlorogenic acid and the production of dihydrocaffeic acid and dihydroferulic acid, while caffeine remained unmetabolized [316].

Caffeic acid esters can be rapidly transformed to 3-(3′-hydroxyphenyl) propionic acid by human fecal microbiota [304] by de-esterification followed by a reduction of a double bond and dehydroxylation at the C4 position (Figure 18). Monteiro et al. [317] revealed that the main chlorogenic acid metabolites identified in urine after coffee consumption were: dihydrocaffeic, gallic, isoferulic, ferulic, vanillic, caffeic, 5-*O*-caffeoylquinic, sinapic, 4-hydroxybenzoic, and *p*-coumaric acids, with gallic and dihydrocaffeic acids being the major ones. Similar results were reported by Clifford et al. [37]. One of the most abundant sources of caffeic acid in nature is 5-*O*-caffeoylquinic acid (neochlorogenic acid), which was also proved to be hydrolyzed to caffeic and quinic acids by esterases from colonic microflora and is not degraded and absorbed in the upper gastrointestinal tract [311]. The lack of colonic microbiota (e.g., in germfree rats) resulted in the inhibition of hippuric acid formation, indicating that esterase enzymes of the colonic microbiota are involved in this pathway [318].

5.7. Bacterial Metabolism of Resveratrol and Curcumin

Resveratrol is a natural polyphenol widely found in its *trans* isomer form in various fruits, especially grapes and berries, peanuts, and red wine. It was reported that purified resveratrol inhibited the growth of some pathogens, among other intestinal bacteria such as *Helicobacter pylori*, *Enterococcus faecalis*, *Pseudomonas aeruginosa*, *Vibrio* spp. [319]. However, some bacteria are able to metabolize *trans*-resveratrol. *Slackia equolifaciens* and *Adlercreutzia equolifaciens* [320] and *Eggerthella lenta* ATCC 4305 [319] converted resveratrol to dihydroresveratrol (Figure 19); while *Bacillus cereus* NCTR-466, *Achromobacter denitrificans* NCTR-774, and *E. coli* ATCC 47,004 metabolizes trans-resveratrol into resveratrol 3-*O*-glucoside (piceid) and resveratrol 4-*O*-glucoside (resveratroloside) [319]. Among other colonic metabolites of resveratrol, lunularin and 3,4′-dihydroxy-*trans*-stilbene [320] were identified. However, their bacterial producers are unknown (Figure 19). The 16S rRNA sequencing of fecal samples demonstrated the association of lunularin producers with a higher abundance of Bacteroidetes, actinobacteria, Verrucomicrobia, and Cyanobacteria and with a lower abundance of Firmicutes than either the dihydroresveratrol or mixed producers [321]. The bacterial metabolites of resveratrol can exert a beneficial impact on human health. Dihydroresveratrol reduced fatty acid-binding protein-4 expression, involved in fatty acid uptake in human macrophages treated with oxidized LDL and stimulates fatty acid oxidation in human fibroblasts, lunularin reduced the expression of proinflammatory mediators in endothelial cells [322], while 3,4′-dihydroxy-*trans*-stilbene increased glucose uptake and induced adenosine monophosphate kinase phosphorylation in C2C12 myotubes independently of insulin [323].

Jarosova et al. [324] examined the metabolism of six stilbenoids resveratrol, oxyresveratrol, piceatannol, thunalbene, batatasin III, and pinostilbene by colon microbiota from various donors. It was demonstrated that resveratrol, oxyresveratrol, piceatannol and thunalbene were subjected to metabolic transformation via double bond reduction, dihydroxylation, and demethylation (Figure 19), while batatasin III and pinostilbene were stable at simulated colon conditions. Authors reported strong interindividual differences in speed, intensity, and pathways of metabolism among the fecal samples obtained from the donors, suggesting that microbiota composition plays a crucial role in the influence of resveratrol on human health.

Curcumin is a lipophilic polyphenol characterized by quite poor bioavailability. It is supposed that curcumin passes through the stomach without any chemical modifications and reaches the large intestine, where it undergoes extensive phase I and II metabolism. The reductive pathways of metabolism by phase I enzymes lead to the formation of dihydrocurcumin, tetrahydrocurcumin, and hexahydrocurcumin (Figure 20) [325].

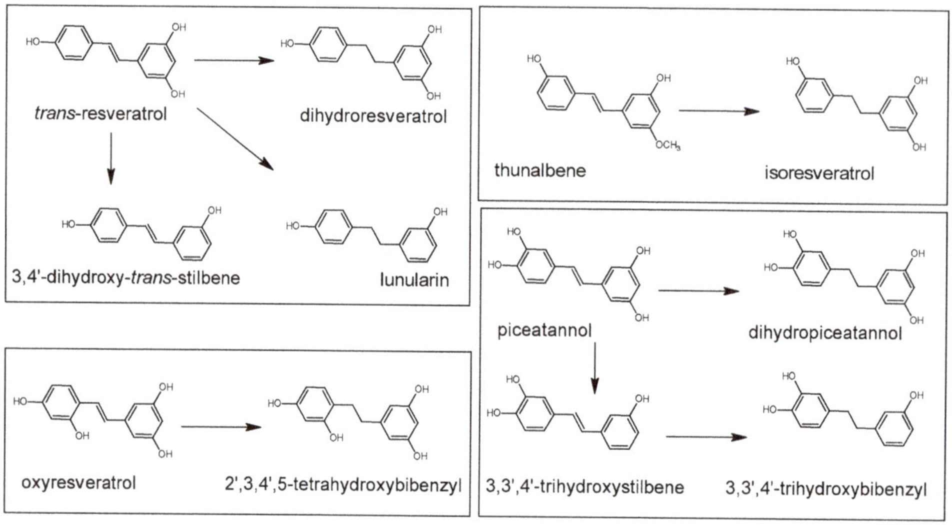

Figure 19. The effect of bacterial metabolism on resveratrol, oxyresveratrol, thunalbene and piceatannol. Based on [320,324].

However, consecutive reduction of the double bonds in the curcumin chain resulting in the formation of dihydrocurcumin, tetrahydrocurcumin, and hexahydrocurcumin can occur in the gut by a CurA reductase (NADPH-dependent curcumin/dihydrocurcumin reductase) that has been isolated from intestinal *E. coli* [326]. The 24-h fermentation of curcumin, demethoxycurcumin and bis-demethoxycurcumin by human fecal microbiota resulted in 24%, 61% and 87% degradation, respectively. Three main metabolites were identified: tetrahydrocurcumin, dihydroferulic acid and 1-(4-hydroxy-3-methoxyphenyl)-2-propanol [327]. A similar experiment was performed by Burapan et al. [328], but a mixture composed of curcumin, demethoxycurcumin, and bis-demethoxycurcumin was metabolized by the human intestinal bacterium *Blautia* sp. MRG-PMF1. New metabolites generated from curcumin and demethoxycurcumin by the methyl aryl ether cleavage reaction were identified. Demethylcurcumin and bisdemethylcurcumin were sequentially produced from curcumin, while demethyldemethoxycurcumin was produced from demethoxycurcumin [328]. Bis(demethyl)tetrahydrocurcumin and bis(demethyl)-hexahydrocurcumin were identified among colonic metabolites of curcumin, demethoxycurcumin and bis-demethoxycurcumin [329].

All these metabolites can undergo phase II metabolism by glucuronidases and sulfotransferases that are capable of conjugating glucuronic acid or sulfate molecule, respectively, to produce the corresponding glucuronide and sulfate O-conjugated metabolites. Furthermore, gut microbiota may deconjugate the phase II metabolites and convert them back to the corresponding phase I metabolites or to fission products such as ferulic acid and dihydroferulic acid in the colon [325].

It is interesting that CurA reductase, besides the ability to conversion of curcumin to tetrahydrocurcumin, is also able to metabolize resveratrol [326].

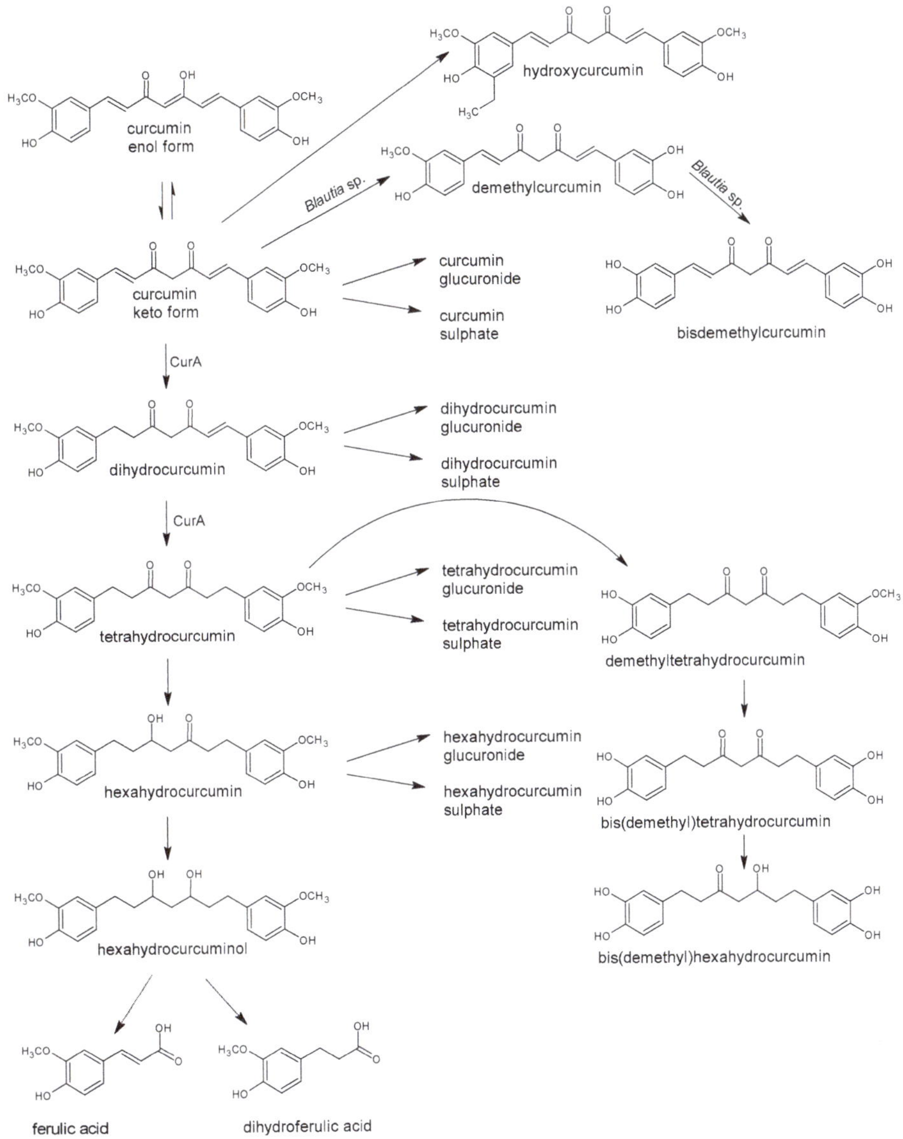

Figure 20. Metabolic pathway of curcumin. Reactions conducted by *E. coli* CurA and *Blautia* sp. are indicated. Based on [274,325,326,328–330].

6. Conclusions

This review describes a bidirectional relationship between polyphenols delivered with food and the human gut microbiota. The manuscript presents a compilation of the knowledge from two perspectives. The first part describes the impact of various polyphenols classes on bacteria, with particular emphasis on human intestinal microbiota representatives. The mechanism of inhibitory impact of polyphenols, including protein binding, inhibition of nucleic acid synthesis, interaction with the cell wall and bacterial membranes, substrate deprivation, inhibition of energy metabolism and changes in cell attachment and biofilm formation, are discussed in details. In the second part, the role and pathways of bacterial biotransformation of polyphenols are described, especially those reactions where bioactive metabolites with a significant impact on the human organism (both positive and negative) are produced. The role of interindividual variation in microbiota composition in the impact of food polyphenols on human health is explained. For example, the biotransformation of isoflavonoids and other phytoestrogens to bioactive O-DMA and S-equol, the generation of urolithins, the bacterial metabolites that can cross the blood–brain barrier, the degradation of complex condensed tannins and lignans as well as catabolic pathways of low-molecular-weight phenolic acids are elucidated.

The exact structures of all discussed phenolic compounds can be found in tables and figures, which will facilitate their comparison with the structures of other food ingredients and drawing one's own conclusions about their potential activity.

Author Contributions: All authors have read and agreed to the published version of the manuscript. Conceptualization, M.M. and A.D.-C.; Writing—Original Draft Preparation, M.M., I.D., T.T. and A.D.-C.; Writing—Review & Editing, M.M., I.D., T.T. and A.D.-C.; Visualization, M.M., I.D., T.T. and A.D.-C.; Supervision, A.D.-C.; Funding Acquisition, A.D.-C.

Funding: The APC was supported by the author's activation funds and through a research subsidy of the Department of Fermentation Technology and Microbiology, the University of Agriculture in Krakow.

Conflicts of Interest: The authors declare no conflict of interest.

References

1. Hooper, L.V.; Gordon, J.I. Commensal host-bacterial relationships in the gut. *Science* **2001**, *292*, 1115–1118. [CrossRef]
2. Zhu, B.; Wang, X.; Li, L. Human gut microbiome: The second genome of human body. *Protein Cell* **2010**, *1*, 718–725. [CrossRef]
3. Qin, J.; Li, R.; Raes, J.; Arumugam, M.; Burgdorf, K.S.; Manichanh, C.; Nielsen, T.; Pons, N.; Levenez, F.; Yamada, T.; et al. A human gut microbial gene catalogue established by metagenomic sequencing. *Nature* **2010**, *464*, 59–65. [CrossRef] [PubMed]
4. Eckburg, P.B.; Bik, E.M.; Bernstein, C.N.; Purdom, E.; Dethlefsen, L.; Sargent, M.; Gill, S.R.; Nelson, K.E.; Relman, D.A. Diversity of the human intestinal microbial flora. *Science* **2005**, *308*, 1635–1638. [CrossRef]
5. Lagier, J.-C.; Armougom, F.; Million, M.; Hugon, P.; Pagnier, I.; Robert, C.; Bittar, F.; Fournous, G.; Gimenez, G.; Maraninchi, M.; et al. Microbial culturomics: Paradigm shift in the human gut microbiome study. *Clin. Microbiol. Infect.* **2012**, *18*, 1185–1193. [CrossRef]
6. Duda-Chodak, A.; Tarko, T.; Satora, P.; Sroka, P. Interaction of dietary compounds, especially polyphenols, with the intestinal microbiota: A review. *Eur. J. Nutr.* **2015**, *54*, 325–341. [CrossRef]
7. Conlon, M.A.; Bird, A.R. The impact of diet and lifestyle on gut microbiota and human health. *Nutrients* **2015**, *7*, 17–44. [CrossRef]
8. Thursby, E.; Juge, N. Introduction to the human gut microbiota. *Biochem. J.* **2017**, *474*, 1823–1836. [CrossRef]
9. Hermon-Taylor, J. Gut pathogens: Invaders and turncoats in a complex cosmos. *Gut Pathog.* **2009**, *1*, 3. [CrossRef]
10. Acton, D.S.; Plat-Sinnige, M.J.; van Wamel, W.; de Groot, N.; van Belkum, A. Intestinal carriage of *Staphylococcus aureus*: How does its frequency compare with that of nasal carriage and what is its clinical impact? *Eur. J. Clin. Microbiol. Infect. Dis.* **2009**, *28*, 115–127. [CrossRef]
11. Kamada, N.; Chen, G.Y.; Inohara, N.; Núñez, G. Control of pathogens and pathobionts by the gut microbiota. *Nat. Immunol.* **2013**, *14*, 685–690. [CrossRef]
12. Rowland, I.; Gibson, G.; Heinken, A.; Scott, K.; Swann, J.; Thiele, I.; Tuohy, K. Gut microbiota functions: Metabolism of nutrients and other food components. *Eur. J. Nutr.* **2018**, *57*, 1–24. [CrossRef]
13. Possemiers, S.; Bolca, S.; Verstraete, W.; Heyerick, A. The intestinal microbiome: A separate organ inside the body with the metabolic potential to influence the bioactivity of botanicals. *Fitoterapia* **2011**, *82*, 53–66. [CrossRef]
14. Hervert-Hernández, D.; Goñi, I. Dietary polyphenols and human gut microbiota: A review. *Food Rev. Int.* **2011**, *27*, 154–169. [CrossRef]
15. Man, A.W.C.; Zhou, Y.; Xia, N.; Li, H. Involvement of gut microbiota, microbial metabolites and interaction with polyphenol in host immunometabolism. *Nutrients* **2020**, *12*, 3054. [CrossRef]
16. Del Rio, D.; Rodriguez-Mateos, A.; Spencer, J.P.E.; Tognolini, M.; Borges, G.; Crozier, A. Dietary (poly)phenolics in human health: Structures, bioavailability, and evidence of protective effects against chronic diseases. *Antioxid. Redox Signal.* **2013**, *18*, 1818–1892. [CrossRef]
17. Saarela, M.; Lähteenmäki, L.; Crittendenb, R.; Salminenc, S.; Mattila-Sandholm, T. Gut bacteria and health foods—The European perspective. *Int. J. Food Microbiol.* **2002**, *78*, 99–117. [CrossRef]
18. Marchesi, J.R.; Adams, D.H.; Fava, F.; Hermes, G.D.A.; Hirschfield, G.M.; Hold, G.; Quraishi, M.N.; Kinross, J.; Smidt, H.; Tuohy, K.M.; et al. The gut microbiota and host health: A new clinical frontier. *Gut* **2016**, *65*, 330–339. [CrossRef]

19. Cardona, F.; Andrés-Lacueva, C.; Tulipani, S.; Tinahones, F.J.; Queipo-Ortuño, M.I. Benefits of polyphenols on gut microbiota and implications in human health. *J. Nutr. Biochem.* **2013**, *24*, 1415–1422. [CrossRef]
20. Górniak, I.; Bartoszewski, R.; Króliczewski, J. Comprehensive review of antimicrobial activities of plant flavonoids. *Phytochem. Rev.* **2019**, *18*, 241–272. [CrossRef]
21. Piasecka, A.; Jedrzejczak-Rey, N.; Bednarek, P. Secondary metabolites in plant innate immunity: Conserved function of divergent chemicals. *New Phytol.* **2015**, *206*, 948–964. [CrossRef]
22. Mao, X.; Xiao, X.; Chen, D.; Yu, B.; He, J. Tea and its components prevent cancer: A review of the redox-related mechanism. *Int. J. Mol. Sci.* **2019**, *20*, 5249. [CrossRef]
23. Schaffer, S.; Asseburg, H.; Kuntz, S.; Muller, W.E.; Eckert, G.P. Effects of polyphenols on brain ageing and Alzheimer's disease: Focus on mitochondria. *Mol. Neurobiol.* **2012**, *46*, 161–178. [CrossRef]
24. Guasch-Ferré, M.; Merino, J.; Sun, Q.; Fitó, M.; Salas-Salvadó, J. Dietary polyphenols, mediterranean diet, prediabetes, and Type 2 diabetes: A narrative review of the evidence. *Oxid. Med. Cell. Longev.* **2017**, *2017*, 1–16. [CrossRef]
25. Kim, Y.; Keogh, J.B.; Clifton, P.M. Polyphenols and glycemic control. *Nutrients* **2016**, *8*, 17. [CrossRef]
26. Öztürk, E.; Arslan, A.K.K.; Yerer, M.B.; Bishayee, A. Resveratrol and diabetes: A critical review of clinical studies. *Biomed. Pharmacother.* **2017**, *95*, 230–234. [CrossRef]
27. Roleira, F.M.F.; Tavares-da-Silva, E.J.; Varela, C.L.; Costa, S.C.; Silva, T.; Garrido, J.; Borges, F. Plant derived and dietary phenolic antioxidants: Anticancer properties. *Food Chem.* **2015**, *183*, 235–258. [CrossRef] [PubMed]
28. Wang, S.; Moustaid-Moussa, N.; Chen, L.; Mo, H.; Shastri, A.; Su, R.; Bapat, P.; Kwun, I.; Shen, C.-L. Novel insights of dietary polyphenols and obesity. *J. Nutr. Biochem.* **2014**, *25*, 1–18. [CrossRef]
29. Fraga, C.G.; Croft, K.D.; Kennedy, D.O.; Tomás-Barberán, F.A. The effects of polyphenols and other bioactives on human health. *Food Funct.* **2019**, *10*, 514–528. [CrossRef] [PubMed]
30. Oppedisano, F.; Maiuolo, J.; Gliozzi, M.; Musolino, V.; Carresi, C.; Nucera, S.; Scicchitano, M.; Scarano, F.; Bosco, F.; Macrì, R.; et al. The potential for natural antioxidant supplementation in the early stages of neurodegenerative disorders. *Int. J. Mol. Sci.* **2020**, *21*, 2618. [CrossRef]
31. Silva, R.F.M.; Pogaĉnik, L. Polyphenols from food and natural products: Neuroprotection and safety. *Antioxidants* **2020**, *9*, 61. [CrossRef] [PubMed]
32. Singh, A.; Yau, Y.F.; Leung, K.S.; El-Nezami, H.; Lee, J.C.-Y. Interaction of polyphenols as antioxidant and anti-inflammatory compounds in brain–liver–gut axis. *Antioxidants* **2020**, *9*, 669. [CrossRef] [PubMed]
33. Stalmach, A.; Mullen, W.; Nagai, C.; Crozier, A. On-line HPLC analysis of the antioxidant activity of phenolic compounds in brewed, paper-filtered coffee. *Braz. J. Plant Physiol.* **2006**, *18*, 253–262. [CrossRef]
34. Ma, J.-N.; Ma, C.-M. Antifungal inhibitory activities of caffeic and quinic acid derivatives. In *Coffee in Health and Disease Prevention*; Preedy, V.R., Ed.; Elsevier Inc.; Academic Press: London, UK, 2015; pp. 635–641. [CrossRef]
35. Heleno, S.A.; Martins, A.; Queiroz, M.J.R.P.; Ferreira, I.C.F.R. Bioactivity of phenolic acids: Metabolites versus parent compounds: A review. *Food Chem.* **2015**, *173*, 501–513. [CrossRef] [PubMed]
36. He, F.; Pan, Q.H.; Shi, Y.; Duan, C.Q. Biosynthesis and genetic regulation of proanthocyanidins in plants. *Molecules* **2008**, *13*, 2674–2703. [CrossRef]
37. Clifford, M.N.; Jaganath, I.B.; Ludwig, I.A.; Crozier, A. Chlorogenic acids and the acyl-quinic acids: Discovery, biosynthesis, bioavailability and bioactivity. *Nat. Prod. Rep.* **2017**, *34*, 1391–1421. [CrossRef]
38. Tasdemir, D.; Lack, G.; Brun, R.; Ruedi, P.; Scapozza, L.; Perozzo, R. Inhibition of *Plasmodium falciparum* fatty acid biosynthesis: Evaluation of FabG, FabZ, and FabI as drug targets for flavonoids. *J. Med. Chem.* **2006**, *49*, 3345–3353. [CrossRef]
39. Routray, W.; Orsat, V. Blueberries and their anthocyanins: Factors affecting biosynthesis and properties. *Comprehen. Rev. Food Sci. Food Saf.* **2011**, *10*, 303–320. [CrossRef]
40. Naoi, M.; Wu, Y.; Shamoto-Nagai, M.; Maruyama, W. Mitochondria in neuroprotection by phytochemicals: Bioactive polyphenols modulate mitochondrial apoptosis system, function and structure. *Int. J. Mol. Sci.* **2019**, *20*, 2451. [CrossRef]
41. Qabaha, K.I. Antimicrobial and free radical scavenging activities of five Palestinian medicinal plants. *Afr. J. Tradit. Complement. Altern. Med.* **2013**, *10*, 101–108. [CrossRef]
42. Veličković, D.T.; Randjelović, N.V.; Ristić, M.S.; Veličković, A.S.; Šmelcerović, A.A. Chemical constituents and antimicrobial activity of the ethanol extracts obtained from the flower, leaf and stem of *Salvia officinalis* L. *J. Serb. Chem. Soc.* **2003**, *68*, 17–24. [CrossRef]
43. Bozyel, M.E.; Şenturan, M.; Benek, A.; Bozyel, E.M.; Canli, K.; Altuner, E.M. In vitro antimicrobial activity screening of *Heliotropium europaeum* against wide range of microorganisms and multi drug resistant (MDR) bacteria. *Eur. J. Biomed. Pharmac. Sci.* **2019**, *6*, 113–117.
44. Al-Juraifani, A.A. Antimicrobial activity of some medicinal plants used in Saudi Arabia. *CJPAS* **2011**, *5*, 509–1512.
45. Al-Bayati, F.A.; Sulaiman, K.D. In vitro antimicrobial activity of *Salvadora persica* L. extracts against some isolated oral pathogens in Iraq. *Turk. J. Biol.* **2008**, *32*, 57–62.
46. Rasooli, I.; Shayegh, S.; Taghizadeh, M.; Astaneh, S.D.A. Phytotherapeutic prevention of dental biofilm formation. *Phytother. Res.* **2008**, *22*, 1162–1167. [CrossRef] [PubMed]
47. Puupponen-Pimia, R.; Nohynek, L.; Meier, C.; Kahkonen, M.; Heinonen, M.; Hopia, A.; Oksman-Caldentey, K.-M. Antimicrobial properties of phenolic compounds from berries. *J. Appl. Microbiol.* **2001**, *90*, 494–507. [CrossRef] [PubMed]

48. Dall'Agnol, R.; Ferraz, A.; Bernardi, A.P.; Albring, D.; Nör, C.; Sarmento, L.; Lamb, L.; Hass, M.; von Poser, G.; Schapoval, E.E.S. Antimicrobial activity of some *Hypericum* species. *Phytomedicine* **2003**, *10*, 511–516. [CrossRef]
49. Rauha, J.-P.; Remes, S.; Heinonen, M.; Hopia, A.; Kahkonen, M.; Kujala, T.; Pihlaja, K.; Vuorela, H.; Vuorela, P. Antimicrobial effects of Finnish plant extracts containing flavonoids and other phenolic compounds. *Int. J. Food Microbiol.* **2000**, *56*, 3–12. [CrossRef]
50. Singh, R.; Shushni, M.A.M.; Belkheir, A. Antibacterial and antioxidant activities of *Mentha piperita* L. *Arab. J. Chem.* **2015**, *8*, 322–328. [CrossRef]
51. Osawa, K.; Saeki, T.; Yasuda, H.; Hamashima, H.; Sasatsu, M.; Arai, T. The antibacterial activities of peppermint oil and green tea polyphenols, alone and in combination, against enterohemorrhagic *Escherichia coli*. *Biocontrol Sci.* **1999**, *4*, 1–7. [CrossRef]
52. Vattem, D.A.; Lina, Y.-T.; Ghaedianb, R.; Shetty, K. Cranberry synergies for dietary management of *Helicobacter pylori* infections. *Process Biochem.* **2005**, *40*, 1583–1592. [CrossRef]
53. Si, W.; Gong, J.; Tsao, R.; Kalab, M.; Yang, R.; Yin, Y. Bioassay-guided purification and identification of antimicrobial components in Chinese green tea extract. *J. Chromatogr. A* **2006**, *1125*, 204–210. [CrossRef] [PubMed]
54. Shin, J.-S.; Chung, H.-S. Antibacterial activities of phenolic components from *Camellia sinensis* L. on pathogenic microorganisms. *J. Food Sci. Nutr.* **2007**, *12*, 135–140. [CrossRef]
55. Ankolekar, C.; Johnson, D.; da Silva Pinto, M.; Johnson, K.; Labbe, R.; Shetty, K. Inhibitory potential of tea polyphenolics and influence of extraction time against *Helicobacter pylori* and lack of inhibition of beneficial lactic acid bacteria. *J. Med. Food* **2011**, *14*, 1321–1329. [CrossRef] [PubMed]
56. Kohda, C.; Yanagawa, Y.; Shimamura, T. Epigallocatechin gallate inhibits intracellular survival of *Listeria monocytogenes* in macrophages. *Biochem. Biophys. Res. Commun.* **2008**, *365*, 310–315. [CrossRef]
57. Nakayama, M.; Shimatani, K.; Ozawa, T.; Shigemune, N.; Tomiyama, D.; Yui, K.; Katsuki, M.; Ikeda, K.; Nonaka, A.; Miyamoto, T. Mechanism for the antibacterial action of epigallocatechin gallate (EGCg) on *Bacillus subtilis*. *Biosci. Biotechnol. Biochem.* **2015**, *79*, 845–854. [CrossRef]
58. Coppo, E.; Marchese, A. Antbacterial activity of polyphenols. *Curr. Pharm. Biotechnol.* **2014**, *15*, 380–390. [CrossRef]
59. Tzounis, X.; Vulevic, J.; Kuhnle, G.G.C.; George, T.; Leonczak, J.; Gibson, G.R.; Kwik-Uribe, C.; Spencer, J.P.E. Flavanol monomer-induced changes to the human faecal microflora. *Br. J. Nutr.* **2008**, *99*, 782–792. [CrossRef]
60. Xie, Y.; Chen, J.; Xiao, A.; Liu, L. Antibacterial activity of polyphenols: Structure-activity relationship and influence of hyper-glycemic condition. *Molecules* **2017**, *22*, 1913. [CrossRef]
61. Bae, E.A.; Han, M.J.; Kim, D.H. *In vitro* anti-*Helicobacter pylori* activity of some flavonoids and their metabolites. *Planta Med.* **1999**, *65*, 442–443. [CrossRef]
62. Rodríguez Vaquero, M.J.; Alberto, M.R.; Manca de Nadra, M.C. Influence of phenolic compounds from wines on the growth of *Listeria monocytogenes*. *Food Control* **2007**, *18*, 587–593. [CrossRef]
63. Duda-Chodak, A. The inhibitory effect of polyphenols on human gut microbiota. *J. Physiol. Pharmacol.* **2012**, *63*, 497–503. [PubMed]
64. Cueva, C.; Moreno-Arribas, M.V.; Martín-Álvarez, P.J.; Bills, G.; Vicente, M.F.; Basilio, A.; Rivas, C.L.; Requena, T.; Rodríguez, J.M.; Bartolome, B. Antimicrobial activity of phenolic acids against commensal, probiotic and pathogenic bacteria. *Res. Microbiol.* **2010**, *161*, 372–382. [CrossRef] [PubMed]
65. Takahashi, T.; Kokubo, R.; Sakaino, M. Antimicrobial activities of eucalyptus leaf extracts and flavonoids from *Eucalyptus maculata*. *Lett. Appl. Microbiol.* **2004**, *39*, 60–64. [CrossRef]
66. Roldán, L.P.; Díaz, G.J.; Duringer, J.M. Composition and antibacterial activity of essential oils obtained from plants of the *Lamiaceae* family against pathogenic and beneficial bacteria. *Rev. Colomb. Cienc. Pec.* **2010**, *23*, 451–461.
67. Adamczak, A.; Ożarowski, M.; Karpiński, T.M. Antibacterial activity of some flavonoids and organic acids widely distributed in plants. *J. Clin. Med.* **2020**, *9*, 109. [CrossRef]
68. Tabasco, R.; Sánchez-Patán, F.; Monagas, M.; Bartolomé, B.; Moreno-Arribas, M.V.; Peláez, C.; Requena, T. Effect of grape polyphenols on lactic acid bacteria and bifidobacteria growth: Resistance and metabolism. *Food Microbiol.* **2011**, *28*, 1345–1352. [CrossRef]
69. Gwiazdowska, D.; Juś, K.; Jasnowska-Małecka, J.; Kluczyńska, K. The impact of polyphenols on *Bifidobacterium* growth. *Acta Biochim. Pol.* **2015**, *62*, 895–901. [CrossRef]
70. Lee, H.C.; Jenner, A.M.; Low, C.S.; Lee, Y.K. Effect of tea phenolics and their aromatic fecal bacterial metabolites on intestinal microbiota. *Res. Microbiol.* **2006**, *157*, 876–884. [CrossRef]
71. Dolara, P.; Luceri, C.; De Filippo, C.; Femia, A.P.; Giovannelli, L.; Caderni, G.; Cecchini, C.; Silvi, S.; Orpianesi, C.; Cresci, A. Red wine polyphenols influence carcinogenesis, intestinal microflora, oxidative damage and gene expression profiles of colonic mucosa in F344 rats. *Mutat. Res.* **2005**, *591*, 237–246. [CrossRef]
72. Smith, A.H.; Zoetendal, E.; Mackie, R.I. Bacterial mechanisms to overcome inhibitory effects of dietary tannins. *Microb. Ecol.* **2005**, *50*, 197–205. [CrossRef] [PubMed]
73. Henning, S.M.; Summanen, P.H.; Lee, R.-P.; Yang, J.; Finegold, S.M.; Heber, D.; Li, Z. Pomegranate ellagitannins stimulate the growth of *Akkermansia muciniphila* in vivo. *Anaerobe* **2017**, *43*, 56–60. [CrossRef] [PubMed]

74. Chen, M.-I.; Yi, L.; Zhang, Y.; Zhou, X.; Ran, L.; Yang, J.; Zhu, J.; Zhang, Q.; Mi, M. Resveratrol attenuates trimethylamine-N-oxide (TMAO)-induced atherosclerosis by regulating TMAO synthesis and bile acid metabolism via remodeling of the gut microbiota. *mBio* **2016**, *7*, 02210–02215. [CrossRef] [PubMed]
75. Qiao, Y.; Sun, J.; Xia, S.; Tang, X.; Le, Y.S.G. Effects of resveratrol on gut microbiota and fat storage in a mouse model with high-fat-inducedobesity. *Food Funct.* **2014**, *5*, 1241–1249. [CrossRef] [PubMed]
76. Masumoto, S.; Terao, A.; Yamamoto, Y.; Mukai, T.; Miura, T.; Shoji, T. Non-absorbable apple procyanidins prevent obesity associated with gut microbial and metabolomic changes. *Sci. Rep.* **2016**, *6*, 31208. [CrossRef] [PubMed]
77. Roopchand, D.E.; Carmody, R.N.; Kuhn, P.; Moskal, K.; Rojas-Silva, P.; Turnbaugh, P.J.; Raskin, I. Dietary polyphenols promote growth of the gut bacterium *Akkermansia muciniphila* and attenuate high-fat diet–induced metabolic syndrome. *Diabetes* **2015**, *64*, 2847–2858. [CrossRef] [PubMed]
78. Fogliano, V.; Corollaro, M.L.; Vitaglione, P.; Napolitano, A.; Ferracane, R.; Travaglia, F.; Arlorio, M.; Costabile, A.; Klinder, A.; Gibson, G. In vitro bioaccessibility and gut biotransformation of polyphenols present in the water-insoluble cocoa fraction. *Mol. Nutr. Food Res.* **2011**, *55*, S44–S55. [CrossRef] [PubMed]
79. Queipo-Ortuño, M.I.; Boto-Ordóñez, M.; Murri, M.; Gomez-Zumaquero, J.M.; Clemente-Postigo, M.; Estruch, R.; Diaz, F.C.; Andrés-Lacueva, C.; Tinahones, F.J. Influence of red wine polyphenols and ethanol on the gut microbiotaecology and biochemical biomarkers. *Am. J. Clin. Nutr.* **2012**, *95*, 1323–1334. [CrossRef] [PubMed]
80. Vendrame, S.; Guglielmetti, S.; Riso, P.; Arioli, S.; Klimis-Zacas, D.; Porrini, M. Six-week consumption of a wild blueberry powder drink increases bifidobacteria in the human gut. *J. Agric. Food Chem.* **2011**, *59*, 12815–12820. [CrossRef]
81. Moreno-Indias, I.; Sánchez-Alcoholado, L.; Pérez-Martínez, P.; Andrés-Lacueva, C.; Cardona, F.; Tinahones, F.; Queipo-Ortuño, M.I. Red wine polyphenols modulate fecal microbiota and reduce markers of the metabolic syndrome in obese patients. *Food Funct.* **2016**, *7*, 1775–1787. [CrossRef]
82. Bustos, I.; Garcia-Cayuela, T.; Hernández-Ledesma, B.; Peláez, C.; Requena, T.; Martinez-Cuesta, M.C. Effect of flavan-3-ols on the adhesion of potential probiotic Lactobacilli to intestinal cells. *J. Agric. Food Chem.* **2012**, *60*, 9082–9088. [CrossRef] [PubMed]
83. Volstatova, T.; Marsik, P.; Rada, V.; Geigerova, M.; Havlik, J. Effect of apple extracts and selective polyphenols on the adhesion of potential probiotic strains of *Lactobacillus gasseri* R and *Lactobacillus casei* FMP. *J. Funct. Foods* **2017**, *35*, 391–397. [CrossRef]
84. Howell, A.B. Bioactive compounds in cranberries and their role in prevention of urinary tract infections. *Mol. Nutr. Food Res.* **2007**, *51*, 732–737. [CrossRef] [PubMed]
85. Foo, L.Y.; Lu, Y.; Howell, A.B.; Vorsa, N. The structure of cranberry proanthocyanidins which inhibit adherence of uropathogenic P-fimbriated *Escherichia coli* in vitro. *Phytochemistry* **2000**, *54*, 173–181. [CrossRef]
86. Sánchez-Patán, F.; Cueva, C.; Monagas, M.; Walton, G.E.; Gibson, G.R.M.; Quintanilla-López, J.E.; Lebrón-Aguilar, R.; Martín-Álvarez, P.J.; Moreno-Arribas, M.V.; Bartolomé, B. In vitro fermentation of a red wine extract by human gut microbiota: Changes in microbial groups and formation of phenolic metabolites. *J. Agric. Food Chem.* **2012**, *60*, 2136–2147. [CrossRef]
87. Hervert-Hernández, D.; Pintado, C.; Rotger, R.; Goñi, I. Stimulatory role of grape pomace polyphenols on *Lactobacillus acidophilus* growth. *Int. J. Food Microbiol.* **2009**, *136*, 119–122. [CrossRef]
88. Vivas, N.; Lonvaud-Funel, A.; Glories, Y. Effect of phenolic acids and anthocyanins on growth, viability and malolactic activity of a lactic acid bacterium. *Food Microbiol.* **1997**, *14*, 291–300. [CrossRef]
89. Naito, Y.; Uchiyama, K.; Takagi, T. A next-generation beneficial microbe: *Akkermansia muciniphila*. *J. Clin. Biochem. Nutr.* **2018**, *63*, 33–35. [CrossRef]
90. Dehghanbanadaki, H.; Aazami, H.; Raftar, S.K.A.; Ashrafian, F.; Ejtahed, H.-S.; Hashemi, E.; Tavassol, H.Z.; Badi, S.A.; Siadat, S.D. Global scientific output trend for *Akkermansia muciniphila* research: A bibliometric and scientometric analysis. *BMC Med. Inform. Decis. Mak.* **2020**, *20*, 291. [CrossRef]
91. Farhadi, F.; Khameneh, B.; Iranshahi, M.; Iranshahy, M. Antibacterial activity of flavonoids and their structure–activity relationship: An update review. *Phytother. Res.* **2019**, *33*, 13–40. [CrossRef]
92. Mickymaray, S.; Alfaiz, F.A.; Paramasivam, A. Efficacy and mechanisms of flavonoids against the emerging opportunistic nontuberculous *Mycobacteria*. *Antibiotics* **2020**, *9*, 450. [CrossRef] [PubMed]
93. Pandey, A.K.; Kumar, S. Perspective on Plant Products as Antimicrobial Agents: A Review. *Pharmacologia* **2013**, *4*, 469–480. [CrossRef]
94. Papuc, C.; Goran, G.V.; Predescu, C.N.; Nicorescu, V.; Stefan, G. Plant polyphenols as antioxidant and antibacterial agents for shelf-life extension of meat and meat products: Classification, structures, sources, and action mechanisms. *Compr. Rev. Food Sci. Food Saf.* **2017**, *16*, 1243–1268. [CrossRef] [PubMed]
95. Cushnie, T.P.T.; Lamb, A.J. Antimicrobial activity of flavonoids. *Int. J. Antimicrob. Agents* **2005**, *26*, 343–356. [CrossRef] [PubMed]
96. Cushnie, T.P.T.; Lamb, A.J. Recent advances in understanding the antibacterial properties of flavonoids. *Int. J. Antimicrob. Agents* **2011**, *38*, 99–107. [CrossRef] [PubMed]
97. Matilla-Cuenca, L.; Gil, C.; Cuesta, S.; Rapún-Araiz, B.; Žiemytė, M.; Mira, A.; Lasa, I.; Valle, J. Antibiofilm activity of flavonoids on staphylococcal biofilms through targeting BAP amyloids. *Sci. Rep.* **2020**, *10*, 18968. [CrossRef]
98. Xie, Y.; Yang, W.; Tang, F.; Chen, X.; Ren, L. Antibacterial activities of flavonoids: Structure-activity relationship and mechanism. *Curr. Med. Chem.* **2015**, *22*, 132–149. [CrossRef]
99. Kumar, S.; Pandey, A.K. Chemistry and biological activities of flavonoids: An overview. *Sci. World J.* **2013**, 162750. [CrossRef]

100. Brudzynski, K.; Maldonado-Alvarez, L. Polyphenol-protein complexes and their consequences for the redox activity, structure and function of honey. A current view and new hypothesis—A review. *Pol. J. Food Nutr. Sci.* **2015**, *65*, 71–80. [CrossRef]
101. Seczyk, Ł.; Świeca, M.; Kapusta, I.; Gawlik-Dziki, U. Protein–phenolic interactions as a factor affecting the physicochemical properties of White Bean Proteins. *Molecules* **2019**, *24*, 408. [CrossRef]
102. Boath, A.S.; Grussu, D.; Stewart, D.; McDougall, G.J. Berry polyphenols inhibit digestive enzymes: A source of potential health benefits? *Food Dig.* **2012**, *3*, 1–7. [CrossRef]
103. Gonçalves, S.; Romano, A. Inhibitory properties of phenolic compounds against enzymes linked with human diseases. In *Phenolic Compounds—Biological Activity*; Soto-Hernandez, M., Palma-Tenango, M., Garcia-Mateos, M., Eds.; IntechOpen: Rijeka, Croatia, 2017; pp. 99–118. ISBN 978-953-51-5090-9. [CrossRef]
104. Martinez-Gonzalez, A.I.; Díaz-Sánchez, Á.G.; de la Rosa, L.A.; Vargas-Requena, C.L.; Bustos-Jaimes, I.; Alvarez-Parrilla, E. Polyphenolic compounds and digestive enzymes: In vitro non-covalent interactions. *Molecules* **2017**, *22*, 669. [CrossRef] [PubMed]
105. Sun, L.; Wang, Y.; Miao, M. Inhibition of α-amylase by polyphenolic compounds: Substrate digestion, binding interactions and nutritional intervention. *Trends Food Sci. Technol.* **2020**, *104*, 190–207. [CrossRef]
106. Tadera, K.; Minami, Y.; Takamatsu, K.; Matsuoka, T. Inhibition of alpha-glucosidase and alpha-amylase by flavonoids. *J. Nutr. Sci. Vitaminol.* **2006**, *52*, 149–153. [CrossRef]
107. Cowan, M.M. Plant products as antimicrobial agents. *Clin. Microbiol. Rev.* **1999**, *12*, 564–582. [CrossRef]
108. Yang, N.Y.; Hinner, M.J. Getting across the cell membrane: An overview for small molecules, peptides, and proteins. *Methods Mol. Biol.* **2015**, *1266*, 29–53. [CrossRef]
109. Hu, P.; Huang, P.; Chen, M.W. Curcumin reduces *Streptococcus mutans* biofilm formation by inhibiting sortase A activity. *Arch. Oral Biol.* **2013**, *58*, 1343–1348. [CrossRef]
110. Kang, S.S.; Kim, J.-G.; Lee, T.-H.; Oh, K.-B. Flavonols inhibit sortases and sortase-mediated *Staphylococcus aureus* clumping to fibrynogen. *Biol. Pharm. Bull.* **2006**, *29*, 1751–1755. [CrossRef]
111. Edziri, H.; Mastouri, M.; Mahjoub, M.A.; Mighri, Z.; Mahjoub, A.; Verschaeve, L. Antibacterial, antifungal and cytotoxic activities of two flavonoids from *Retama raetam* flowers. *Molecules* **2012**, *17*, 7284–7293. [CrossRef]
112. Chen, Y.; Liu, T.; Wang, K.; Hou, C.; Cai, S.; Huang, Y.; Du, Z.; Huang, H.; Kong, J.; Chen, Y. Baicalein inhibits *Staphylococcus aureus* biofilm formation and the quorum sensing system in vitro. *PLoS ONE* **2016**, *11*, e0153468. [CrossRef]
113. Yoda, Y.; Hu, Z.-Q.; Zhao, W.-H.; Shimamura, T. Different susceptibilities of *Staphylococcus* and Gram-negative rods to epigallocatechin gallate. *J. Infect. Chemother.* **2004**, *10*, 55–58. [CrossRef] [PubMed]
114. Nakayama, M.; Shimatani, K.; Ozawa, T.; Shigemune, N.; Tsugukuni, T.; Tomiyama, D.; Kurahachi, M.; Nonaka, A.; Miyamoto, T. A study of the antibacterial mechanism of catechins: Isolation and identification of *Escherichia coli* cell surface proteins that interact with epigallocatechin gallate. *Food Sci. Biotechnol.* **2013**, *33*, 433–439. [CrossRef]
115. Taylor, P.W.; Hamilton-Miller, J.M.T.; Stapleton, P.D. Antimicrobial properties of green tea catechins. *Food Sci. Technol. Bull.* **2005**, *2*, 71–81. [CrossRef] [PubMed]
116. Fowler, Z.L.; Shah, K.; Panepinto, J.C.; Jacobs, A.; Koffas, M.A.G. Development of non-natural flavanones as antimicrobial agents. *PLoS ONE* **2011**, *6*, e25681. [CrossRef] [PubMed]
117. Omosa, L.K.; Midiwo, J.O.; Mbaveng, A.T.; Tankeo, S.B.; Seukep, J.A.; Voukeng, I.K.; Dzotam, J.K.; Isemeki, J.; Derese, S.; Omolle, R.A.; et al. Antibacterial activities and structure—Activity relationships of a panel of 48 compounds from Kenyan plants against multidrug resistant phenotypes. *SpringerPlus* **2016**, *5*, 901. [CrossRef] [PubMed]
118. Randhawa, H.K.; Hundal, K.K.; Ahirrao, P.N.; Jachak, S.M.; Nandanwar, H.S. Efflux pump inhibitory activity of flavonoids isolated from *Alpinia calcarata* against methicillin-resistant *Staphylococcus aureus*. *Biologia* **2016**, *71*, 484–493. [CrossRef]
119. Brown, A.K.; Papaemmanouil, A.; Bhowruth, V.; Bhatt, A.; Dover, L.G.; Besra, G.S. Flavonoid inhibitors as novel antimycobacterial agents targeting Rv0636, a putative dehydratase enzyme involved in *Mycobacterium tuberculosis* fatty acid synthase II. *Microbiology* **2007**, *153*, 3314–3322. [CrossRef]
120. Zhang, Y.M.; Rock, C.O. Evaluation of epigallocatechin gallate and related plant polyphenols as inhibitors of the FabG and FabI reductases of bacterial type II fatty-acid synthase. *J. Biol. Chem.* **2004**, *279*, 30994–31001. [CrossRef]
121. Xiao, Z.-T.; Zhu, Q.; Zhang, H.-Y. Identifying antibacterial targets of flavonoids by comparative genomics and molecular modelling. *Open J. Genom.* **2014**, *3*, 1–8. [CrossRef]
122. Zhang, L.; Kong, Y.; Wu, D.; Zhang, H.; Wu, J.; Chen, J.; Ding, J.; Hu, L.; Jiang, H.; Shen, X. Three flavonoids targeting the β-hydroxyacyl-acyl carrier protein dehydratase from *Helicobacter pylori*: Crystal structure characterization with enzymatic inhibition assay. *Protein Sci.* **2008**, *17*, 1971–1978. [CrossRef]
123. Jeong, K.-W.; Lee, J.-Y.; Kang, D.-I.; Lee, J.-U.; Shin, S.Y.; Kim, Y. Screening of flavonoids as candidate antibiotics against *Enterococcus faecalis*. *J. Nat. Prod.* **2009**, *72*, 719–724. [CrossRef] [PubMed]
124. Hertel, W.; Peschel, G.; Ozegowski, J.-H.; Müller, P.-J. Inhibitory effects of triterpenes and flavonoids on the enzymatic activity of hyaluronic acid-splitting enzymes. *Arch. Pharm. Chem. Life Sci.* **2006**, *339*, 313–318. [CrossRef] [PubMed]
125. Brahmachari, G.; Mandal, N.C.; Jash, S.K.; Roy, R.; Mandal, L.C.; Mukhopadhyay, A.; Biswajit Behera, B.; Majhi, S.; Mondal, A.; Gangopadhyay, A. Evaluation of the antimicrobial potential of two flavonoids isolated from *Limnophila* plants. *Chem. Biodivers.* **2011**, *8*, 1139–1151. [CrossRef] [PubMed]
126. Xiao, Z.-P.; Shi, D.-H.; Li, H.-Q.; Zhang, L.N.; Xua, C.; Zhua, H.-L. Polyphenols based on isoflavones as inhibitors of *Helicobacter pylori* urease. *Bioorg. Med. Chem.* **2007**, *15*, 3703–3710. [CrossRef]

127. Ulrey, R.K.; Barksdale, S.M.; Zhou, W.; van Hoek, M.L. Cranberry proanthocyanidins have anti-biofilm properties against. *BMC Complement. Altern. Med.* **2014**, *14*, 499. [CrossRef]
128. Mori, A.; Nishino, C.; Enoki, N.; Tawata, S. Antibacterial activity and mode of action of plant flavonoids against *Proteus vulgaris* and *Staphylococcus aureus*. *Phytochem.* **1987**, *26*, 2231–2234. [CrossRef]
129. Lou, Z.; Wang, H.; Rao, S.; Sun, J.; Ma, C.; Li, J. p-Coumaric acid kills bacteria through dual damage mechanisms. *Food Control* **2012**, *25*, 550–554. [CrossRef]
130. Dzoyem, J.P.; Hamamoto, H.; Ngameni, B.; Ngadjui, B.T.; Sekimizu, K. Antimicrobial action mechanism of flavonoids from *Dorstenia* species. *Drug Discov. Ther.* **2013**, *7*, 66–72. [CrossRef]
131. Suriyanarayanan, B.; Shanmugam, K.; Santhosh, R.S. Synthetic quercetin inhibits mycobacterial growth possibly by interacting with DNA gyrase. *Rom. Biotechnol. Lett.* **2013**, *18*, 8587–8593.
132. Bandele, O.J.; Clawson, S.J.; Osheroff, N. Dietary polyphenols as topoisomerase II poisons: B ring and C ring substituents determine the mechanism of enzyme-mediated DNA cleavage enhancement. *Chem. Res. Toxicol.* **2008**, *21*, 1253–1260. [CrossRef]
133. Bandele, O.J.; Osheroff, N. (−)-Epigallocatechin gallate, a major constituent of green tea, poisons human type II topoisomerases. *Chem. Res. Toxicol.* **2008**, *21*, 936–943. [CrossRef] [PubMed]
134. Oblak, M.; Kotnik, M.; Solmajer, T. Discovery and Development of ATPase inhibitors of DNA gyrase as antibacterial agents. *Curr. Med. Chem.* **2007**, *14*, 2033–2047. [CrossRef] [PubMed]
135. Plaper, A.; Golob, M.; Hafner, I.; Oblak, M.; Solmajer, T.; Jerala, R. Characterization of quercetin binding site on DNA gyrase. *Biochem. Biophys. Res. Commun.* **2003**, *306*, 530–536. [CrossRef]
136. Gradišar, H.; Pristovšek, P.; Plaper, A.; Jerala, R. Green Tea catechins inhibit bacterial DNA gyrase by interaction with its ATP binding site. *J. Med. Chem.* **2007**, *50*, 264–271. [CrossRef] [PubMed]
137. Ohemeng, K.A.; Podlogar, B.L.; Nguyen, V.N.; Bernstein, J.I.; Krause, H.M.; Hilliard, J.J.; Barrett, J.F. DNA gyrase inhibitory and antimicrobial activities of some diphenic acid monohydroxamides. *J. Med. Chem.* **1997**, *40*, 3292–3296. [CrossRef] [PubMed]
138. Fang, Y.; Lu, Y.; Zang, X.; Wu, T.; Qi, X.J.; Pan, S.; Xu, X. 3D-QSAR and docking studies of flavonoids as potent *Escherichia coli* inhibitors. *Sci. Rep.* **2016**, *6*, 1–13. [CrossRef] [PubMed]
139. Bernard, F.-X.; Sable, S.; Cameron, B.; Provost, J.; Desnottes, J.-F.; Crouzet, J.; Blanche, F. Glycosylated flavones as selective inhibitors of topoisomerase IV. *Antimicrob. Agents Chemother.* **1997**, *41*, 992–998. [CrossRef] [PubMed]
140. Arima, H.; Hitoshi Ashida, H.; Danno, G. Rutin-enhanced antibacterial activities of flavonoids against *Bacillus cereus* and *Salmonella enteritidis*. *Biosci. Biotechnol. Biochem.* **2002**, *66*, 1009–1014. [CrossRef]
141. Khan, N.S.; Ahmad, A.; Hadi, S.M. Anti-oxidant, pro-oxidant properties of tannic acid and its binding to DNA. *Chem. Biol. Interact.* **2000**, *125*, 177–189. [CrossRef]
142. Cetin-Karaca, H.; Newman, M.C. Antimicrobial efficacy of plant phenolic compounds against *Salmonella* and *Escherichia coli*. *Food Biosci.* **2015**, *11*, 8–16. [CrossRef]
143. Stepanović, S.; Antić, N.; Dakić, I.; Švabić-Vlahović, M. In vitro antimicrobial activity of propolis and synergism between propolis and antimicrobial drugs. *Microbiol. Res.* **2003**, *158*, 353–357. [CrossRef] [PubMed]
144. Ikigai, H.; Nakae, T.; Hara, Y.; Shimamura, T. Bactericidal catechins damage the lipid bilayer. *Biochim. Biophys. Acta* **1993**, *1147*, 132–136. [CrossRef]
145. Vattem, D.A.; Lin, Y.-T.; Labbe, R.G.; Shetty, K. Phenolic antioxidant mobilization in cranberry pomace by solid-state bioprocessing using food grade fungus Lentinus edodes and effect on antimicrobial activity against select food borne pathogens. *Innov. Food Sci. Emerg. Technol.* **2004**, *5*, 81–91. [CrossRef]
146. Wu, D.; Kong, Y.; Han, C.; Chen, J.; Hu, L.; Jiang, H.; Shen, X. d-Alanine:d-alanine ligase as a new target for the flavonoids quercetin and apigenin. *Int. J. Antimicrob. Agents* **2008**, *32*, 421–426. [CrossRef] [PubMed]
147. Xiao, Z.-P.; Wei, W.; Liu, Q.; Wang, P.-F.; Luo, X.; Chen, F.-Y.; Cao, Y.; Huang, H.-X.; Liua, M.-M.; Zhu, H.L. C-7 modified flavonoids as novel tyrosyl-tRNA synthetase inhibitors. *RSC Adv.* **2017**, *7*, 6193–6201. [CrossRef]
148. Cai, W.; Fu, Y.; Zhang, W.; Chen, X.; Zhao, J.; Song, W.; Li, Y.; Huang, Y.; Wu, Z.; Sun, R.; et al. Synergistic effects of baicalein with cefotaxime against *Klebsiella pneumonia* through inhibiting CTX-M-1 gene expression. *BMC Microbiol.* **2016**, *16*, 181. [CrossRef]
149. Farooq, S.; Wahab, A.-T.; Fozing, C.D.A.; Rahman, A.-U.; Choudhary, M.I. Artonin I inhibits multidrug resistance in *Staphylococcus aureus* and potentiates the action of inactive antibiotics in vitro. *J. Appl. Microbiol.* **2014**, *117*, 996–1011. [CrossRef]
150. Zhao, W.-H.; Hu, Z.-Q.; Okubo, S.; Hara, Y.; Shimamura, T. Mechanism of synergy between epigallocatechin gallate and β-lactams against Methicillin-Resistant *Staphylococcus aureus*. *Antimicrob. Agents Chemother.* **2001**, *45*, 1737–1742. [CrossRef]
151. Delehanty, J.B.; Johnson, B.J.; Hickey, T.E.; Pons, T.; Ligler, F.S. Plant proanthocyanidins bind to and neutralize bacterial lipopolysaccharides. In *Naval Research Laboratory Reviews*; 2008; pp. 101–107.
152. Nohynek, L.J.; Alakomi, H.L.; Kähkönen, M.P.; Heinonen, M.; Helander, I.M.; Oksman-Caldentey, K.-M.; Puupponen-Pimiä, R.H. Berry phenolics: Antimicrobial properties and mechanisms of action against severe human pathogens. *Nutr. Cancer* **2006**, *54*, 18–32. [CrossRef]
153. Hartmann, M.; Berditsch, M.; Hawecker, J.; Ardakani, M.F.; Gerthsen, D.; Ulrich, A.S. Damage of the bacterial cell envelope by antimicrobial peptides gramicidin S and PGLa as revealed by transmission and scanning electron microscopy. *Antimicrob. Agents Chemother.* **2010**, *54*, 3132–3142. [CrossRef]
154. Reygaert, W.C. The antimicrobial possibilities of green tea. *Front. Microbiol.* **2014**, *5*, 434. [CrossRef] [PubMed]

155. Lee, H.; Woo, E.-R.; Lee, D.G. Apigenin induces cell shrinkage in *Candida albicans* by membrane perturbation. *FEMS Yeast Res.* **2018**, *18*, 1–9. [CrossRef] [PubMed]
156. Perumal, S.; Mahmud, R.; Ismail, S. Mechanism of action of isolated caffeic acid and epicatechin 3-gallate from *Euphorbia hirta* against *Pseudomonas aeruginosa*. *Pharmacogn. Mag.* **2017**, *13* (Suppl. 2), S311–S315. [CrossRef]
157. Matijašević, D.; Pantić, M.; Rašković, B.; Pavlović, V.; Duvnjak, D.; Sknepnek, A.; Nikšić, M. The antibacterial activity of *Coriolus versicolor* methanol extract and its effect on ultrastructural changes of *Staphylococcus aureus* and *Salmonella* Enteritidis. *Front. Microbiol.* **2016**, *7*, 1226. [CrossRef] [PubMed]
158. Fathima, A.; Rao, J.R. Selective toxicity of Catechin—A natural flavonoid towards bacteria. *Appl. Microbiol. Biotechnol.* **2016**, *100*, 6395–6402. [CrossRef]
159. Tagousop, C.N.; Tamokou, J.d.D.; Ekom, S.E.; Ngnokam, D.; Voutquenne-Nazabadioko, L. Antimicrobial activities of flavonoid glycosides from *Graptophyllum grandulosum* and their mechanism of antibacterial action. *BMC Complement. Altern. Med.* **2018**, *18*, 252. [CrossRef]
160. Ollila, F.; Halling, K.; Vuorela, P.; Vuorela, H.; Slotte, J.P. Characterization of flavonoid–biomembrane interactions. *Arch. Biochem. Biophys.* **2002**, *399*, 103–108. [CrossRef]
161. Chabot, S.; Bel-Rhlid, R.; Chenevert, R.; Piche, Y. Hyphal growth promotion in vitro of the VA mycorrhizal fungus, *Gigaspora margarita* Becker & Hall, by the activity of structurally specific flavonoid compounds under CO_2 -enriched conditions. *New Phytol.* **1992**, *122*, 461–467.
162. Tsuchiya, H. Stereospecificity in membrane effects of catechins. *Chem. Biol. Inter.* **2001**, *134*, 41–54. [CrossRef]
163. Carson, C.F.; Mee, B.J.; Riley, T.V. Mechanism of action of *Melaleuca alternifolia* (Tea Tree) oil on *Staphylococcus aureus* determined by time-kill, lysis, leakage, and salt tolerance assays and electron microscopy. *Antimicrob. Agents Chemother.* **2002**, *46*, 1914–1920. [CrossRef]
164. Tsuchiya, H. Membrane interactions of phytochemicals as their molecular mechanism applicable to the discovery of drug leads from plants. *Molecules* **2015**, *20*, 18923–18966. [CrossRef]
165. Selvaraj, S.; Krishnaswamy, S.; Devashya, V.; Sethuraman, S.; Krishnan, U.M. Influence of membrane lipid composition on flavonoid–membrane interactions: Implications on their biological activity. *Progr. Lipid Res.* **2015**, *58*, 1–13. [CrossRef] [PubMed]
166. Arora, A.; Byrem, T.M.; Nair, M.G.; Strasburg, G.M. Modulation of liposomal membrane fluidity by flavonoids and isoflavonoids. *Arch. Biochem. Biophys.* **2000**, *373*, 102–109. [CrossRef] [PubMed]
167. Van Dijk, C.; Driessen, A.J.M.; Recourt, K. The Uncoupling efficiency and affinity of flavonoids for vesicles. *Biochem. Pharmacol.* **2000**, *60*, 1593–1600. [CrossRef]
168. Wu, T.; He, M.; Zang, X.; Zhou, Y.; Qio, T.; Pan, S.; Xu, X. A structure-activity relationship study of flavonoids as inhibitors of *E. coli* by membrane interaction effect. *Biochim. Biophys. Acta* **2013**, *1828*, 2751–2756. [CrossRef]
169. He, M.; Wu, T.; Pan, S.; Xu, X. Antimicrobial mechanism of flavonoids against *Escherichia coli* ATCC25922 by model membrane study. *Appl. Surf. Sci.* **2014**, *305*, 515–521. [CrossRef]
170. Mun, S.-H.; Joung, D.-K.; Kim, S.-B.; Park, S.-J.; Seo, Y.-S.; Gong, R.; Choi, J.-G.; Shin, D.-W.; Rho, J.-R.; Kang, O.-H.; et al. The mechanism of antimicrobial activity of sophoraflavanone B against methicillin-resistant *Staphylococcus aureus*. *Foodborne Pathog. Dis.* **2014**, *11*, 234–239. [CrossRef]
171. Savoia, D. Plant-derived antimicrobial compounds: Alternatives to antibiotic. *Future Microbiol.* **2012**, *7*, 979–990. [CrossRef]
172. Chinnam, N.; Dadi, P.K.; Sabri, S.A.; Ahmad, M.; Kabir, M.A.; Ahmad, Z. Dietary bioflavonoids inhibit *Escherichia coli* ATP synthase in a differential manner. *Int. J. Biol. Macromol.* **2010**, *46*, 478–486. [CrossRef]
173. Dadi, P.K.; Ahmad, M.; Ahmad, Z. Inhibition of ATPase activity of *Escherichia coli* ATP synthase by polyphenols. *Int. J. Biol. Macromol.* **2009**, *45*, 72–79. [CrossRef]
174. Gill, A.O.; Holley, R.A. Inhibition of membrane bound ATPases of *Escherichia coli* and *Listeria monocytogenes* by plant oil aromatics. *Int. J. Food Microbiol.* **2006**, *111*, 170–174. [CrossRef] [PubMed]
175. Ahmad, Z.; Laughlin, T.F.; Kady, I.O. Thymoquinone inhibits *Escherichia coli* ATP synthase and cell growth. *PLoS ONE* **2015**, *10*, e0127802. [CrossRef] [PubMed]
176. Haraguchi, H.; Tanimoto, K.; Tamura, Y.; Mizutanit, K.; Kinoshita, T. Mode of antibacterial action of retrochalcones from *Glycyrrhiza inflata*. *Phytochemistry* **1998**, *48*, 125–129. [CrossRef]
177. Kim, H.-S.; Quon, M.J.; Kim, J. New insights into the mechanisms of polyphenols beyond antioxidant properties; lessons from the green tea polyphenol, epigallocatechin 3-gallate. *Redox Biol.* **2014**, *2*, 187–195. [CrossRef] [PubMed]
178. Castillo-Juárez, I.; Maeda, T.; Mandujano-Tinoco, E.A.; Tomás, M.; Pérez-Eretza, B.; García-Contreras, S.J.; Wood, T.K.; García-Contreras, R. Role of quorum sensing in bacterial infections. *World J. Clin. Cases* **2015**, *3*, 575–598. [CrossRef] [PubMed]
179. Packiavathy, I.A.S.V.; Priya, S.; Pandian, S.K.; Ravi, A.V. Inhibition of biofilm development of uropathogens by curcumin—An anti-quorum sensing agent from *Curcuma longa*. *Food Chem.* **2014**, *148*, 453–460. [CrossRef]
180. Wang, T.-Y.; Li, Q.; Bi, K.-S. Bioactive flavonoids in medicinal plants: Structure, activity and biological fate. *Asian J. Pharm. Sci.* **2018**, *13*, 12–23. [CrossRef]
181. Gopu, V.; Meena, C.K.; Shetty, P.H. Quercetin influences quorum sensing in food borne bacteria: In-vitro and in-silico evidence. *PLoS ONE* **2015**, *10*, e0134684. [CrossRef]

182. Al Azzaz, J.; Al Tarraf, A.; Heumann, A.; Da Silva Barreira, D.; Laurent, J.; Assifaoui, A.; Rieu, A.; Guzzo, J.; Lapaquette, P. Resveratrol favors adhesion and biofilm formation of *Lacticaseibacillus paracasei* subsp. *paracasei* Strain ATCC334. *Int. J. Mol. Sci.* **2020**, *21*, 5423. [CrossRef]
183. Lee, J.-H.; Regmi, S.C.; Kim, J.-A.; Cho, M.H.; Yun, H.; Lee, C.-S.; Lee, J. Apple flavonoid phloretin inhibits *Escherichia coli* O157:H7 biofilm formation and ameliorates colon inflammation in rats. *Infect. Immun.* **2011**, *79*, 4819–4827. [CrossRef]
184. Serra, D.O.; Mika, F.; Richter, A.M.; Hengge, R. The green tea polyphenol EGCG inhibits *E. coli* biofilm formation by impairing amyloid curli fibre assembly and downregulating the biofilm regulator CsgD via the r E -dependent sRNA RybB. *Mol. Microbiol.* **2016**, *101*, 136–151. [CrossRef] [PubMed]
185. Hengge, R. Targeting bacterial biofilms by the Green Tea polyphenol EGCG. *Molecules* **2019**, *24*, 2403. [CrossRef] [PubMed]
186. Bai, L.; Takagi, S.; Ando, T.; Yoneyama, H.; Ito, K.; Mizugai, H.; Isogai, E. Antimicrobial activity of tea catechin against canine oral bacteria and the functional mechanisms. *J. Vet. Med. Sci.* **2016**, *78*, 1439–1445. [CrossRef] [PubMed]
187. Wu, C.-Y.; Su, T.Y.; Wang, M.Y.; Yang, S.-F.; Mar, K.; Hung, S.-L. Inhibitory effects of tea catechin epigallocatechin-3-gallate against biofilms formed from *Streptococcus mutans* and a probiotic *Lactobacillus* strain. *Arch. Oral Biol.* **2018**, *94*, 69–77. [CrossRef]
188. Veloz, J.J.; Alvear, M.; Salazar, L.A. Antimicrobial and antibiofilm activity against *Streptococcus mutans* of individual and mixtures of the main polyphenolic compounds found in chilean propolis. *BioMed Res. Int.* **2019**, *2019*, 7602343. [CrossRef]
189. Xu, X.; Zhou, X.D.; Wu, C.D. Tea catechin epigallocatechin gallate inhibits *Streptococcus mutans* biofilm formation by suppressing gtf genes. *Arch. Oral Biol.* **2012**, *57*, 678–683. [CrossRef]
190. Xu, X.; Zhou, X.D.; Wu, C.D. The tea catechin epigallocatechin gallate suppresses cariogenic virulence factors of *Streptococcus mutans*. *Antimicrob. Agents Chemother.* **2011**, *55*, 1229–1236. [CrossRef]
191. Sivaranjani, M.; Gowrishankar, S.; Kamaladevi, A.; Pandian, S.K.; Balamurugan, K.; Ravi, A.V. Morin inhibits biofilm production and reduces the virulence of *Listeria monocytogenes*—An in vitro and in vivo approach. *Int. J. Food Microbiol.* **2016**, *237*, 73–82. [CrossRef]
192. Upadhyay, A.; Upadhyaya, I.; Kollanoor-Johny, A.; Venkitanarayanan, K. Antibiofilm effect of plant derived antimicrobials on *Listeria monocytogenes*. *Food Microbiol.* **2013**, *36*, 79–89. [CrossRef]
193. Borges, A.; Saavedra, M.J.; Simões, M. The activity of ferulic and gallic acids in biofilm prevention and control of pathogenic bacteria. *J. Bioadh. Biofilm Res.* **2012**, *28*, 755–767. [CrossRef]
194. Borges, A.; Ferreira, C.; Saavedra, M.J.; Simões, M. Antibacterial activity and mode of action of ferulic and gallic acids against pathogenic bacteria. *Microb. Drug Resist.* **2013**, *19*, 256–265. [CrossRef] [PubMed]
195. Muñoz-Cazares, N.; García-Contreras, R.; Pérez-López, M.; Castillo-Juárez, I. Phenolic compounds with anti-virulence properties. In *Phenolic Compounds—Biological Activity*; Soto-Hernandez, M., Palma-Tenango, M., del Rosario Garcia-Mateos, M., Eds.; IntechOpen: Rijeka, Croatia, 2017; pp. 139–167. [CrossRef]
196. Daglia, M. Polyphenols as antimicrobial agents. *Curr. Opin. Biotechnol.* **2012**, *23*, 174–181. [CrossRef] [PubMed]
197. Scalbert, A. Antimicrobial properties of tannins. *Phytochemistry* **1991**, *30*, 3875–3883. [CrossRef]
198. Field, S.A.; Lettinga, G. Toxicity of tannic compounds to microorganisms. In *Plant Polyphenols*, 1st ed.; Hemingway, R.W., Laks, P.E., Eds.; Plenum Press: New York, NY, USA, 1992; pp. 673–692.
199. Chung, K.T.; Lu, Z.; Chou, M.W. Mechanism of inhibition of tannic acid and related compounds on the growth of intestinal bacteria. *Food Chem. Toxicol.* **1998**, *36*, 1053–1060. [CrossRef]
200. Andrews, S.C.; Robinson, A.K.; Rodriguez-Quiñones, F. Bacterial iron homeostasis. *FEMS Microbiol. Rev.* **2003**, *27*, 215–237. [CrossRef]
201. Perron, N.R.; Brumaghim, J.L. A Review of the antioxidant mechanisms of polyphenol compounds related to iron binding. *Cell Biochem. Biophys.* **2009**, *53*, 75–100. [CrossRef]
202. Fernandez, M.T.; Mira, M.L.; Florencio, M.H.; Jennings, K.R. Iron and copper chelation by flavonoids: An electrospray mass spectrometry study. *J. Inorg. Biochem.* **2002**, *92*, 105–111. [CrossRef]
203. Mladěnka, P.; Macáková, K.; Filipský, T.; Zatloukalová, L.; Jahodář, L.; Bovicelli, P.; Silvestri, I.P.; Hrdina, R.; Saso, L. In vitro analysis of iron chelating activity of flavonoids. *J. Inorg. Biochem.* **2011**, *105*, 693–701. [CrossRef]
204. Rajakovich, L.J.; Balskus, E.P. Metabolic functions of the human gut microbiota: The role of metalloenzymes. *Nat. Prod. Rep.* **2019**, *36*, 593. [CrossRef]
205. Wang, W.-L.; Chai, S.C.; Huang, M.; He, H.-Z.; Hurley, T.D.; Ye, Q.-Z. Discovery of inhibitors of *Escherichia coli* methionine aminopeptidase with the Fe(II)-form selectivity and antibacterial activity. *J. Med. Chem.* **2008**, *51*, 6110–6120. [CrossRef]
206. Lagha, A.B.; Haas, B.; Grenier, D. Tea polyphenols inhibit the growth and virulence properties of *Fusobacterium nucleatum*. *Sci. Rep.* **2017**, *7*, 44815. [CrossRef] [PubMed]
207. Lee, P.; Tan, K.S. Effects of epigallocatechin gallate against *Enterococcus faecalis* biofilm and virulence. *Arch. Oral Biol.* **2015**, *60*, 393–399. [CrossRef] [PubMed]
208. Echeverría, J.; Opazo, J.; Mendoza, L.; Urzúa, A.; Wilkens, M. Structure-activity and lipophilicity relationships of selected antibacterial natural flavones and flavanones of Chilean flora. *Molecules* **2017**, *22*, 608. [CrossRef]
209. Botta, B.; Vitali, A.; Menendez, P.; Misiti, D.; Delle Monache, G. Prenylated flavonoids: Pharmacology and biotechnology. *Curr. Med. Chem.* **2005**, *12*, 713–739. [CrossRef] [PubMed]

210. Cermak, P.; Olsovska, J.; Mikyska, A.; Dusek, M.; Kadleckova, Z.; Vanicek, J.; Nyc, O.; Sigler, K.; Bostikova, V.; Bostik, P. Strong antimicrobial activity of xanthohumol and other derivatives from hops (*Humulus lupulus* L.) on gut anaerobic bacteria. *APMIS* **2017**, *125*, 1033–1038. [CrossRef] [PubMed]
211. Stompor, M.; Zarowska, B. Antimicrobial activity of xanthohumol and its selected structural analogues. *Molecules* **2016**, *21*, 608. [CrossRef] [PubMed]
212. Jamil, S.; Lathiff, S.M.A.; Abdullah, S.A.; Jemaon, N.; Sirat, H.M. Antimicrobial flavonoids from *Artocarpus anisophyllus* Miq. and *Artocarpus lowii* King. *J. Tekchnologi* **2014**, *71*, 95–99. [CrossRef]
213. Campos, F.M.; Couto, J.A.; Figueiredo, A.R.; Tóth, I.V.; Rangel, A.O.S.S.; Hogg, T.A. Cell membrane damage induced by phenolic acids on wine lactic acid bacteria. *Int. J. Food Microbiol.* **2009**, *135*, 144–151. [CrossRef]
214. Thakur, D.; Das, S.C.; Sabhapondit, S.; Tamuly, P.; Deka, D.K. Antimicrobial activities of Tocklai vegetative tea clones. *Indian J. Microbiol.* **2011**, *51*, 450–455. [CrossRef]
215. Puljula, E.; Walton, G.; Woodward, M.J.; Karonen, M. Antimicrobial activities of ellagitannins against *Clostridiales perfringens*, *Escherichia coli*, *Lactobacillus plantarum* and *Staphylococcus aureus*. *Molecules* **2020**, *25*, 3714. [CrossRef]
216. Tarko, T.; Duda-Chodak, A.; Wajda, Ł.; Satora, P.; Sroka, P.; Semik-Szczurak, D. Application of principal component analysis for optimization of polyphenol extraction from alternative plant sources. *J. Food Nutr. Res.* **2017**, *56*, 61–72.
217. Boeing, J.S.; Barizão, E.O.; Costa e Silva, B.; Montanher, F.P.; de Cinque Almeida, V.; Visentainer, J.V. Evaluation of solvent effect on the extraction of phenolic compounds and antioxidant capacities from the berries: Application of principal component analysis. *Chem. Cent. J.* **2014**, *8*, 48. [CrossRef] [PubMed]
218. Borges, A.; José, H.; Homem, V.; Simões, M. Comparison of techniques and solvents on the antimicrobial and antioxidant potential of extracts from *Acacia dealbata* and *Olea europaea*. *Antibiotics* **2020**, *9*, 48. [CrossRef] [PubMed]
219. Tarko, T.; Duda-Chodak, A.; Soszka, A. Changes in phenolic compounds and antioxidant activity of fruit musts and fruit wines during simulated digestion. *Molecules* **2020**, *25*, 5574. [CrossRef] [PubMed]
220. Renard, C.M.G.C.; Watrelot, A.A.; Le Bourvellec, C. Interactions between polyphenols and polysaccharides: Mechanisms and consequences in food processing and digestion. *Trends Food Sci. Technol.* **2017**, *60*, 43–51. [CrossRef]
221. Spencer, J.P.E. Metabolism of tea flavonoids in the gastrointestinal tract. *J. Nutr.* **2003**, *133*, 3255S–3261S. [CrossRef]
222. Thomas, M.; Thibault, J.-F. Cell-wall polysaccharides in the fruits of Japanese quince (*Chaenomeles japonica*): Extraction and preliminary characterisation. *Carbohydr. Polym.* **2002**, *49*, 345–355. [CrossRef]
223. Ma, Y.; Luo, J.; Xu, Y. Co-preparation of pectin and cellulose from apple pomace by a sequential process. *J. Food. Sci. Technol.* **2019**, *56*, 4091–4100. [CrossRef]
224. Bindon, K.A.; Bacic, A.; Kennedy, J.A. Tissue-specific and developmental modifications of grape cell walls influence the adsorption of proanthocyanidins. *J. Agric. Food Chem.* **2012**, *60*, 9249–9260. [CrossRef]
225. Phan, A.D.T.; Flanagan, B.M.; D'Arcy, B.R.; Gidley, M.J. Binding selectivity of dietary polyphenols to different plant cell wall components: Quantification and mechanism. *Food Chem.* **2017**, *233*, 216–227. [CrossRef]
226. Voragen, A.G.J.; Coenen, G.-J.; Verhoef, R.P.; Schols, H.A. Pectin, a versatile polysaccharide present in plant cell walls. *Struct. Chem.* **2009**, *20*, 263–275. [CrossRef]
227. Thomas, M.; Guillemin, F.; Guillon, F.; Thibault, J.-F. Pectins in the fruits of Japanese quince (*Chaenomeles japonica*). *Carbohydr. Polym.* **2003**, *53*, 361–372. [CrossRef]
228. Liu, Y.; Ying, D.; Sanguansri, L.; Cai, Y.; Le, X. Adsorption of catechin onto cellulose and its mechanism study: Kinetic models, characterization and molecular simulation. *Food Res. Int.* **2018**, *112*, 225–232. [CrossRef] [PubMed]
229. Selma, M.V.; Espin, J.C.; Tomás-Barberán, F.A. Interaction between phenolics and gut microbiota: Role in human health. *J. Agric. Food Chem.* **2009**, *57*, 6485–6501. [CrossRef] [PubMed]
230. Healey, G.R.; Murphy, R.; Brough, L.; Butts, C.A.; Coad, J. Interindividual variability in gut microbiota and host response to dietary interventions. *Nutr Rev.* **2017**, *75*, 1059–1080. [CrossRef]
231. Gross, G.; Jacobs, D.M.; Peters, S.; Possemiers, S.; van Duynhoven, J.; Vaughan, E.E.; van de Wiele, T. In vitro bioconversion of polyphenols from black tea and red wine/grape juice by human intestinal microbiota displays strong interindividual variability. *J. Agric. Food Chem.* **2010**, *58*, 10236–10246. [CrossRef]
232. Hwang, C.S.; Kwak, H.S.; Lim, H.J.; Lee, S.H.; Kang, Y.S.; Choe, T.B.; Hur, H.G.; Han, K.O. Isoflavone metabolites and their in vitro dual functions: They can act as an estrogenic agonist or antagonist depending on the estrogen concentration. *J. Steroid. Biochem. Mol. Biol.* **2006**, *101*, 246–253. [CrossRef]
233. Frankenfeld, C.L. O-desmethylangolensin: The importance of equol's lesser known cousin to human health. *Adv. Nutr.* **2011**, *2*, 317–324. [CrossRef]
234. Vázquez, L.; Llórez, A.B.; Redruello, B.; Mayo, B. Metabolism of soy isoflavones by intestinal bacteria: Genome analysis of an *Adlercreutzia equolifaciens* strain that does not produce equol. *Biomolecules* **2020**, *10*, 950. [CrossRef]
235. Murota, K.; Nakamura, Y.; Uehara, M. Flavonoid metabolism: The interaction of metabolites and gut microbiota. *Biosci. Biotechnol. Biochem.* **2018**, *82*, 600–610. [CrossRef]
236. Mace, T.A.; Ware, M.B.; King, S.A.; Loftus, S.; Farren, M.R.; McMichael, E.; Scoville, S.; Geraghty, C.; Young, G.; Carson, W.E., III; et al. Soy isoflavones and their metabolites modulate cytokine-induced natural killer cell function. *Sci. Rep.* **2019**, *9*, 5068. [CrossRef] [PubMed]

237. Marín, L.; Miguélez, E.M.; Villar, C.J.; Lombó, F. Bioavailability of dietary polyphenols and gut microbiota metabolism: Antimicrobial properties. *Biomed Res. Int.* **2015**, *2015*, 905215. [CrossRef] [PubMed]
238. Kim, M.; Kim, S.-I.; Han, J.; Wang, X.-L.; Song, D.-G.; Kim, S.-U. Stereospecific biotransformation of dihydrodaidzein into (3S)-equol by the human intestinal bacterium *Eggerthella* strain Julong 732. *Appl. Environ. Microbiol.* **2009**, *75*, 3062–3068. [CrossRef] [PubMed]
239. Braune, A.; Blaut, M. Bacterial species involved in the conversion of dietary flavonoids in the human gut. *Gut Microbes* **2016**, *7*, 216–234. [CrossRef] [PubMed]
240. Decroos, K.; Vanhemmens, S.; Cattoir, S.; Boon, N.; Verstraete, W. Isolation and characterisation of an equol-producing mixed microbial culture from a human faecal sample and its activity under gastrointestinal conditions. *Arch. Microbiol.* **2005**, *183*, 45–55. [CrossRef]
241. Hur, H.-G.; Beger, R.D.; Heinze, T.M.; Lay, J.O., Jr.; Freeman, J.P.; Dore, J.; Rafii, F. Isolation of an anaerobic intestinal bacterium capable of cleaving the C-ring of the isoflavonoid daidzein. *Arch. Microbiol.* **2002**, *178*, 8–12. [CrossRef]
242. Rossi, M.; Amaretti, A.; Roncaglia, L.; Leonardi, A.; Raimondi, S. Dietary isoflavones and intestinal microbiota: Metabolism and transformation into bioactive compounds. In *Isoflavones Biosynthesis, Occurence and Health Effects*; Thompson, M.J., Ed.; Nova Science Publishers Inc.: New York, NY, USA, 2010; pp. 137–216. ISBN 9781617281136.
243. Schoefer, L.; Mohan, R.; Braune, A.; Birringer, M.; Blaut, M. Anaerobic C-ring cleavage of genistein and daidzein by *Eubacterium ramulus*. *FEMS Microbiol. Lett.* **2002**, *208*, 197–202. [CrossRef]
244. Kim, M.; Han, J.; Kim, S.-U. Isoflavone daidzein: Chemistry and bacterial metabolism. *J. Appl. Biol. Chem.* **2008**, *51*, 253–261. [CrossRef]
245. Matthies, A.; Clavel, T.; Gütschow, M.; Engst, W.; Haller, D.; Blaut, M.; Braune, A. Conversion of daidzein and genistein by an anaerobic bacterium newly isolated from the mouse intestine. *Appl. Environ. Microbiol.* **2008**, *74*, 4847–4852. [CrossRef]
246. Matthies, A.; Loh, G.; Blaut, M.; Braune, A. Daidzein and genistein are converted to equol and 5-hydroxy-equol by human intestinal *Slackia isoflavoniconvertens* in gnotobiotic rats. *J. Nutr.* **2012**, *142*, 40–46. [CrossRef]
247. Wang, X.-L.; Kim, H.-K.; Kang, S.-I.; Kim, S.-U.; Hur, H.-G. Production of phytoestrogen S-equol from daidzein in mixed culture of two anaerobic bacteria. *Arch. Microbiol.* **2007**, *187*, 155–160. [CrossRef] [PubMed]
248. Yokoyama, S.-i.; Suzuki, T. Isolation and characterization of a novel equol-producing bacterium from human feces. *Biosci. Biotechnol. Biochem.* **2008**, *72*, 2660–2666. [CrossRef] [PubMed]
249. Tamura, M.; Tsushida, T.; Shinohara, K. Isolation of an isoflavone-metabolizing, Clostridium-like bacterium, strainTM-40, from human faeces. *Anaerobe* **2007**, *13*, 32–35. [CrossRef]
250. Jin, J.-S.; Kitahara, M.; Sakamoto, M.; Hattori, M.; Benno, Y. *Slackia equolifaciens* sp. nov., a human intestinal bacterium capable of producing equol. *Int. J. Syst. Evol. Microbiol.* **2010**, *60*, 1721–1724. [CrossRef] [PubMed]
251. Shimada, Y.; Yasuda, S.; Takahashi, M.; Hayashi, T.; Miyazawa, N.; Sato, I.; Abiru, Y.; Uchiyama, S.; Hishigaki, H. Cloning and expression of a novel NADP(H)-dependent daidzein reductase, an enzyme involved in the metabolism of daidzein, from equol-producing *Lactococcus* Strain 20-92. *Appl. Environ. Microbiol.* **2010**, *76*, 5892–5901. [CrossRef]
252. Wang, X.L.; Kim, K.T.; Lee, J.H.; Hur, H.G.; Kim, S.I. C-Ring cleavage of isoflavones daidzein and genistein by a newly-isolated human intestinal bacterium *Eubacterium ramulus* Julong 601. *J. Microbiol. Biotechnol.* **2004**, *14*, 766–771.
253. Hur, H.-G.; Lay, J.O., Jr.; Beger, R.D.; Freeman, J.P.; Rafii, F. Isolation of human intestinal bacteria metabolizing the natural isoflavone glycosides daidzin and genistin. *Arch. Microbiol.* **2000**, *174*, 422–428. [CrossRef]
254. Jin, J.-S.; Nishihata, T.; Kakiuchi, N.; Hattori, M. Biotransformation of C-glucosylisoflavone puerarin to estrogenic (3S)-equol in co-culture of two human intestinal bacteria. *Biol. Pharm. Bull.* **2008**, *31*, 1621–1625. [CrossRef]
255. Hur, H.-G.; Raffi, F. Biotransformation of the isoflavonoids biochanin A, formononetin, and glycitein by *Eubacterium limosum*. *FEMS Microbiol. Lett.* **2000**, *192*, 21–25. [CrossRef]
256. Heinonen, S.-M.; Wähälä, K.; Adlercreutz, H. Identification of urinary metabolites of the red clover isoflavones formononetin and biochanin A in human subjects. *J. Agric. Food Chem.* **2004**, *52*, 6802–6809. [CrossRef]
257. Carreau, C.; Flouriot, G.; Bennetau-Pelissero, C.; Potier, M. Enterodiol and enterolactone, two major diet-derived polyphenol metabolites have different impact on ER_ transcriptional activation in human breast cancer cells. *J. Steroid Biochem. Mol.* **2008**, *110*, 176–185. [CrossRef] [PubMed]
258. Wang, L.-Q. Mammalian phytoestrogens: Enterodiol and enterolactone. *J. Chromatogr. B* **2002**, *777*, 289–309. [CrossRef]
259. Dinkova-Kostova, A.T.; Gang, D.R.; Davin, L.B.; Bedgar, D.L.; Chu, A.; Lewis, N.G. (1)-pinoresinol/(1)-lariciresinol reductase from *Forsythia intermedia*. *J. Biol. Chem.* **1996**, *271*, 29473–29482. [CrossRef] [PubMed]
260. Landete, J.M. Plant and mammalian lignans: A review of source, intake, metabolism, intestinal bacteria and health. *Food Res. Int.* **2012**, *46*, 410–424. [CrossRef]
261. Yoder, S.C.; Lancaster, S.M.; Hullar, M.A.J.; Lampe, J.W. Gut microbial metabolism of plant lignans: Influence on human health. In *Diet-Microbe Interactions in the Gut. Effects on Human Health and Disease*; Tuohy, K., Del Rio, D., Eds.; Elsevier Inc.: Oxford, UK, 2015; pp. 103–117. [CrossRef]
262. Wang, L.-Q.; Meselhy, M.R.; Li, Y.; Qin, G.-W.; Hattori, M. Human intestinal bacteria capable of transforming secoisolariciresinol diglucoside to mammalian lignans, enterodiol and enterolactone. *Chem. Pharm. Bull.* **2000**, *48*, 1606–1610. [CrossRef] [PubMed]

263. Clavel, T.; Lippman, R.; Gavini, F.; Dore´, J.; Blaut, M. *Clostridium saccharogumia* sp. nov. and *Lactonifactor longoviformis* gen. nov., sp. nov., two novel human faecal bacteria involved in the conversion of the dietary phytoestrogen secoisolariciresinol diglucoside. *Syst. Appl. Microbiol.* **2007**, *30*, 16–26. [CrossRef]
264. Clavel, T.; Henderson, G.; Alpert, C.-A.; Philippe, C.; Rigottier-Gois, L.; Doré, J.; Blaut, M. Intestinal bacterial communities that produce active estrogen-like compounds enterodiol and enterolactone in humans. *Appl. Environ. Microbiol.* **2005**, *71*, 6077–6085. [CrossRef]
265. Peñalvo, J.L.; Heinonen, S.-M.; Aura, A.-M.; Adlercreutz, H. Dietary sesamin is converted to enterolactone in humans. *J. Nutr.* **2005**, *135*, 1056–1062. [CrossRef]
266. Villalba, K.J.O.; Barka, F.V.; Pasos, C.V.; Rodríguez, P.E. Food ellagitannins: Structure, metabolomic fate, and biological properties. In *Tannins—Structural Properties, Biological Properties and Current Knowledge Food Ellagitannins: Structure, Metabolomic Fate, and Biological Properties*; Aires, A., Ed.; IntechOpen: London, UK, 2019. [CrossRef]
267. Wu, S.; Tian, L. Diverse phytochemicals and bioactivities in the ancient fruit and modern functional food pomegranate (*Punica granatum*). *Molecules* **2017**, *22*, 1606. [CrossRef]
268. Larrosa, M.; Tomás-Barberán, F.A.; Espín, J.C. The dietary hydrolysable tannin punicalagin releases ellagic acid that induces apoptosis in human colon adenocarcinoma Caco-2 cells by using the mitochondrial pathway. *J. Nutr. Biochem.* **2006**, *17*, 611–625. [CrossRef]
269. Selma, M.V.; Beltrán, D.; García-Villalba, R.; Espín, J.C.; Tomás-Barberán, F.A. Description of urolithin production capacity from ellagic acid of two human intestinal *Gordonibacter* species. *Food Funct.* **2014**, *5*, 1779–1784. [CrossRef] [PubMed]
270. Selma, M.V.; Tomás-Barberán, F.A.; Beltran, D.; Garcıa-Villalba, R.; Espın, J.C. *Gordonibacter urolithinfaciens* sp. nov., a urolithin-producing bacterium isolated from the human gut. *Int. J. Syst. Evol. Microbiol.* **2014**, *64*, 2346–2352. [CrossRef] [PubMed]
271. Beltrán, D.; Romo-Vaquero, M.; Espín, J.C.; Tomás-Barberán, F.A.; Selma, M.V. *Ellagibacter isourolithinifaciens* gen. nov., sp. nov., a new member of the family Eggerthellaceae, isolated from human gut. *Int. J. Syst. Evol. Microbiol.* **2018**, *68*, 1707–1712. [CrossRef] [PubMed]
272. Espín, J.C.; Larrosa, M.; García-Conesa, M.T.; Tomás-Barberán, F. Biological significance of urolithins, the gut microbial ellagic acid-derived metabolites: The evidence so far. *Evid.-Based Complementary Altern. Med.* **2013**, *2013*, 1–16. [CrossRef]
273. Tomás-Barberán, F.A.; González-Sarrías, A.; García-Villalba, R.; Núñez-Sánchez, M.A.; Selma, M.V.; García-Conesa, M.T.; Espín, J.C. Urolithins, the rescue of "old" metabolites to understand a "new" concept: Metabotypes as a nexus among phenolic metabolism, microbiota dysbiosis, and host health status. *Mol. Nutr. Food Res.* **2017**, *61*, 1500901. [CrossRef]
274. Kawabata, K.; Yoshioka, Y.; Terao, J. Role of intestinal microbiota in the bioavailability and physiological functions of dietary polyphenols. *Molecules* **2019**, *24*, 370. [CrossRef]
275. Possemiers, S.; Heyerick, A.; Robbens, V.; Keukeleire, D.D.; Verstraete, W. Activation of proestrogens from hops (*Humulus lupulus* L.) by intestinal microbiota; conversion of isoxanthohumol into 8-prenylnaringenin. *J. Agric. Food Chem.* **2005**, *53*, 6281–6288. [CrossRef]
276. Paraiso, I.L.; Plagmann, L.S.; Yang, L.; Zielke, R.; Gombart, A.F.; Maier, C.S.; Sikora, A.E.; Blakemore, P.R.; Stevens, J.F. Reductive metabolism of xanthohumol and 8-prenylnaringenin by the intestinal bacterium *Eubacterium ramulus*. *Mol. Nutr. Food Res.* **2019**, *63*, 1800923. [CrossRef]
277. Nikolic, D.; Li, Y.; Chadwick, L.R.; Pauli, G.F.; van Breemen, R.B. Metabolism of xanthohumol and isoxanthohumol, prenylated flavonoids from hops (*Humulus lupulus* L.), by human liver microsomes. *J. Mass Spectrom.* **2005**, *40*, 289–299. [CrossRef]
278. Pepper, M.S.; Hazel, S.J.; Hümpel, M.; Schleuning, W.-D. 8-prenylnaringenin, a novel phytoestrogen, inhibits angiogenesis in vitro and in vivo. *J. Cell. Physiol.* **2004**, *199*, 98–107. [CrossRef]
279. Hameed, A.S.S.; Rawat, P.S.; Meng, X.; Liu, W. Biotransformation of dietary phytoestrogens by gut microbes: A review on bidirectional interaction between phytoestrogen metabolism and gut microbiota. *Biotechnol. Adv.* **2020**, *43*, 107576. [CrossRef] [PubMed]
280. Clifford, M.N. Review. Anthocyanins—Nature, occurrence and dietary burden. *J. Sci. Food Agric.* **2000**, *80*, 1063–1072. [CrossRef]
281. Aura, A.-M.; Martin-Lopez, P.; O'Leary, K.A.; Williamson, G.; Oksman-Caldentey, K.-M.; Poutanen, K.; Santos-Buelga, C. In vitro metabolism of anthocyanins by human gut microflora. *Eur. J. Nutr.* **2005**, *44*, 133–142. [CrossRef] [PubMed]
282. Keppler, K.; Humpf, H.-U. Metabolism of anthocyanins and their phenolic degradation products by the intestinal microflora. *Bioorg. Med. Chem.* **2005**, *13*, 5195–5205. [CrossRef] [PubMed]
283. Gonzalez-Barrio, R.; Edwards, C.A.; Crozier, A. Colonic catabolism of ellagitannins, ellagic acid, and raspberry anthocyanins: In vivo and in vitro studies. *Drug Metab. Dispos.* **2011**, *39*, 1680–1688. [CrossRef]
284. Zhu, Y.; Sun, H.; He, S.; Lou, Q.; Yu, M.; Tang, M.; Tu, L. Metabolism and prebiotics activity of anthocyanins from black rice (*Oryza sativa* L.) in vitro. *PLoS ONE* **2018**, *13*, e0195754. [CrossRef]
285. Ávila, M.; Hidalgo, M.; Sánchez-Moreno, C.; Pelaez, C.; Requena, T.; de Pascual-Teresa, S. Bioconversion of anthocyanin glycosides by *Bifidobacteria* and *Lactobacillus*. *Food Res. Int.* **2009**, *42*, 1453–1461. [CrossRef]
286. Hidalgo, M.; Oruna-Concha, J.; Kolida, S.; Walton, G.E.; Kallithraka, S.; Spencer, J.P.E.; Gibson, G.; de Pascual-Teresa, S. Metabolism of anthocyanins by human gut microflora and their influence on gut bacterial growth. *J. Agric. Food Chem.* **2012**, *60*, 3882–3890. [CrossRef]
287. Fleschhut, J.; Kratzer, F.; Rechkemmer, G.; Kulling, S.E. Stability and biotransformation of various dietary anthocyanins in vitro. *Eur. J. Nutr.* **2006**, *45*, 7–18. [CrossRef]

288. Racova, Z.; Anzenbacherova, E.; Papouskova, B.; Poschner, S.; Kucova, P.; Gausterer, J.C.; Gabor, F.; Kolar, M.; Anzenbacher, P. Metabolite profiling of natural substances in human: *in vitro* study from fecal bacteria to colon carcinoma cells (Caco-2). *J. Nutr. Biochem.* **2020**, *85*, 108482. [CrossRef]
289. Xie, L.; Lee, S.G.; Vance, T.M.; Wang, Y.; Kim, B.; Lee, J.-Y.; Chun, O.K.; Bolling, B.W. Bioavailability of anthocyanins and colonic polyphenol metabolites following consumption of aronia berry extract. *Food Chem.* **2016**, *211*, 860–868. [CrossRef]
290. Chen, Y.; Li, Q.; Zhao, T.; Zhang, Z.; Mao, G.; Feng, W.; Wu, X.; Yang, L. Biotransformation and metabolism of three mulberry anthocyanin monomers by rat gut microflora. *Food Chem.* **2017**, *237*, 887–894. [CrossRef] [PubMed]
291. Li, H.; Christman, L.M.; Li, R.; Gu, L. Synergic interactions between polyphenols and gut microbiota in mitigating inflammatory bowel diseases. *Food Funct.* **2020**, *11*, 4878–4891. [CrossRef] [PubMed]
292. Barbosa, P.; Araújo, P.; Oliveira, J.; Fraga, I.; Pissarra, J.; Amaral, C. Metabolic pathways of degradation of malvidin-3-O-monoglucoside by *Candida oleophila*. *Int. Biodeterior. Biodegrad.* **2019**, *144*, 104768. [CrossRef]
293. Sánchez-Patán, F.; Cueva, C.; Monagas, M.; Walton, G.E.; Gibson, G.R.; Martín-Álvarez, P.J.; Moreno-Arribas, M.V.; Bartolomé, B. Gut microbial catabolism of grape seed flavan-3-ols by human faecal microbiota. Targetted analysis of precursor compounds, intermediate metabolites and end-products. *Food Chem.* **2012**, *131*, 337–347. [CrossRef]
294. Mena, P.; Calani, L.; Bruni, R.; Del Rio, D. Bioactivation of high-molecular-weight polyphenols by the gut microbiome. In *Diet-Microbe Interactions in the Gut. Effects on Human Health and Disease*; Tuohy, K., Del Rio, D., Eds.; Elsevier Inc.: Oxford, UK, 2015; pp. 73–101. [CrossRef]
295. Kutschera, M.; Engst, W.; Blaut, M.; Braune, A. Isolation of catechin-converting human intestinal bacteria. *J. Appl. Microbiol.* **2011**, *111*, 165–175. [CrossRef] [PubMed]
296. Appeldoorn, M.M.; Vincken, J.-P.; Gruppen, H.; Hollman, P.C.H. Procyanidin dimers A1, A2, and B2 are absorbed without conjugation or methylation from the small intestine of rats. *J. Nutr.* **2009**, *139*, 1469–1473. [CrossRef] [PubMed]
297. Déprez, S.; Brezillon, C.; Rabot, S.; Philippe, C.; Mila, I.; Lapierre, C.; Scalbert, A. Polymeric proanthocyanidins are catabolized by human colonic microflora into low-molecular-weight phenolic acids. *J. Nutr.* **2000**, *130*, 2733–2738. [CrossRef]
298. Ou, K.; Sarnoski, P.; Schneider, K.R.; Song, K.; Khoo, C.; Gu, L. Microbial catabolism of procyanidins by human gut microbiota. *Mol. Nutr. Food Res.* **2014**, *58*, 2196–2205. [CrossRef]
299. Monagas, M.; Urpi-Sarda, M.; Sánchez-Patán, F.; Llorach, R.; Garrido, I.; Gómez-Cordovés, C.; Andres-Lacueva, C.; Bartolomé, B. Insights into the metabolism and microbial biotransformation of dietary flavan-3-ols and the bioactivity of their metabolites. *Food Funct.* **2010**, *1*, 233–253. [CrossRef]
300. Alakomi, H.L.; Puupponen-Pimiä, R.; Aura, A.M.; Helander, I.M.; Nohynek, L.; Oksman-Caldentey, K.M.; Saarela, M. Weakening of *Salmonella* with selected microbial metabolites of berry-derived phenolic compounds and organic acids. *J. Agric. Food Chem.* **2007**, *55*, 3905–3912. [CrossRef] [PubMed]
301. Braune, A.; Blaut, M. Deglycosylation of puerarin and other aromatic C-glucosides by a newly isolated human intestinal bacterium. *Environ. Microbiol.* **2011**, *13*, 482–494. [CrossRef] [PubMed]
302. Ahn, H.J.; You, H.L.; Park, M.S.; Li, Z.; Choe, D.; Johnston, T.V.; Ku, S.; Ji, G.E. Microbial biocatalysis of quercetin-3-glucoside and isorhamnetin-3-glucoside in *Salicornia herbacea* and their contribution to improved anti-inflammatory activity. *RSC Adv.* **2020**, *10*, 5339–5350. [CrossRef]
303. Schoefer, L.; Mohan, R.; Schwiertz, A.; Braune, A.; Blaut, M. Anaerobic degradation of flavonoids by *Clostridium orbiscindens*. *Appl. Environ. Microbiol.* **2003**, *69*, 5849–5854. [CrossRef]
304. Aura, A.-M. Microbial metabolism of dietary phenolic compounds in the colon. *Phytochem. Rev.* **2008**, *7*, 407–429. [CrossRef]
305. Corrêa, T.A.F.; Rogero, M.M.; Hassimotto, N.M.A.; Lajolo, F.M. The two-way polyphenols-microbiota interactions and their effects on obesity and related metabolic diseases. *Front. Nutr.* **2019**, *6*, 188. [CrossRef]
306. Hanske, L.; Loh, G.; Sczesny, S.; Blaut, M.; Braune, A. The bioavailability of apigenin-7-glucoside is influenced by human intestinal microbiota in rats. *J. Nutr.* **2009**, *139*, 1095–1102. [CrossRef]
307. Schneider, H.; Blaut, M. Anaerobic degradation of flavonoids by *Eubacterium ramulus*. *Arch. Microbiol.* **2000**, *173*, 71–75. [CrossRef]
308. Nunes, C.; Almeida, L.; Laranjinha, J. Synergistic inhibition of respiration in brain mitochondria by nitric oxide and dihydroxyphenylacetic acid (DOPAC): Implications for Parkinson's disease. *Neurochem. Int.* **2005**, *47*, 173–182. [CrossRef] [PubMed]
309. Najmanová, I.; Pourová, J.; Vopršalová, M.; Pilařová, V.; Semecký, V.; Nováková, L.; Mladěnka, P. Flavonoid metabolite 3-(3-hydroxyphenyl)propionic acid formed by human microflora decreases arterial blood pressure in rats. *Mol. Nutr. Food Res.* **2016**, *60*, 981–991. [CrossRef] [PubMed]
310. Rechner, A.R.; Smith, M.A.; Kuhnle, G.; Gibson, G.R.; Debnam, E.S.; Srai, S.K.S.; Moore, K.P.; Rice-Evans, C.A. Colonic metabolism of dietary polyphenols: Influence of structure on microbial fermentation products. *Free Rad. Biol. Med.* **2004**, *36*, 212–225. [CrossRef] [PubMed]
311. Sova, M.; Saso, L. Natural sources, pharmacokinetics, biological activities and health benefits of hydroxycinnamic acids and their metabolites. *Nutrients* **2020**, *12*, 2190. [CrossRef] [PubMed]
312. Zhang, L.; Gao, W.; Chen, X.; Wang, H. The effect of bioprocessing on the phenolic acid composition and antioxidant activity of wheat bran. *Cereal Chem.* **2014**, *91*, 255–261. [CrossRef]
313. Farah, A.; Monteiro, M.; Donangelo, C.M.; Lafay, S. Chlorogenic acids from green coffee extract are highly bioavailable in humans. *J. Nutr.* **2008**, *138*, 2309–2315. [CrossRef] [PubMed]

314. Gonthier, M.P.; Verny, M.A.; Besson, C.; Rémésy, C.; Scalbert, A. Chlorogenic acid bioavailability largely depends on its metabolisr by the gut microflora in rats. *J. Nutr.* **2003**, *133*, 1853–1859. [CrossRef] [PubMed]
315. García-Villalba, R.; Beltrán, D.; Frutos, M.D.; Selma, M.V.; Espín, J.C.; Tomás-Barberán, F.A. Metabolism of different dietary ph nolic compounds by the urolithin-producing human-gut bacteria *Gordonibacter urolithinfaciens* and *Ellagibacter isourolithinifacien Food Funct.* **2020**, *11*, 7012–7022. [CrossRef]
316. Mills, C.E.; Tzounis, X.; Oruna-Concha, M.J.; Mottram, D.S.; Gibson, G.R.; Spencer, J.P. *In vitro* colonic metabolism of coffee an chlorogenic acid results in selective changes in human faecal microbiota growth. *Br. J. Nutr.* **2015**, *113*, 1220–1227. [CrossRef]
317. Monteiro, M.; Farah, A.; Perrone, D.; Trugo, L.C.; Donangelo, C. Chlorogenic acid compounds from coffee are differentiall absorbed and metabolized in humans. *J. Nutr.* **2007**, *137*, 2196–2201. [CrossRef]
318. Ogawa, M. Coffee and hippuric acid. In *Coffee in Health and Disease Prevention*; Preedy, V.R., Ed.; Academic Press: Cambridge, MA USA, 2014; pp. 209–216.
319. Jung, C.M.; Heinze, T.M.; Schnackenberg, L.K.; Mullis, L.B.; Elkins, S.A.; Elkins, C.A.; Steele, R.S.; Sutherland, J.B. Interaction o dietary resveratrol with animal-associated bacteria. *FEMS Microbiol. Lett.* **2009**, *297*, 266–273. [CrossRef]
320. Bode, L.M.; Bunzel, D.; Huch, M.; Cho, G.S.; Ruhland, D.; Bunzel, M.; Bub, A.; Franz, C.M.; Kulling, S.E. In vivo and in vitr metabolism of trans-resveratrol by human gut microbiota. *Am. J. Clin. Nutr.* **2013**, *97*, 295–309. [CrossRef]
321. Springer, M.; Moco, S. Resveratrol and its human metabolites-effects on metabolic health and obesity. *Nutrients* **2019**, *11*, 143 [CrossRef] [PubMed]
322. Chaplin, A.; Carpéné, C.; Mercader, J. Resveratrol, metabolic syndrome, and gut microbiota. *Nutrients* **2018**, *10*, 1651. [CrossRef] [PubMed]
323. Ito-Nagahata, T.; Kurihara, C.; Hasebe, M.; Ishii, A.; Yamashita, K.; Iwabuchi, M.; Sonoda, M.; Fukuhara, K.; Sawada, R. Matsuoka, A.; et al. Stilbene analogs of resveratrol improve insulin resistance through activation of AMPK. *Biosci. Biotechnol. Biochem.* **2013**, *77*, 1229–1235. [CrossRef] [PubMed]
324. Jarosova, V.; Vesely, O.; Marsik, P.; Jaimes, J.D.; Smejkal, K.; Kloucek, P.; Havlik, J. Metabolism of stilbenoids by human faecal microbiota. *Molecules* **2019**, *24*, 1155. [CrossRef]
325. Scazzocchio, B.; Minghetti, L.; D'Archivio, M. Interaction between gut microbiota and curcumin: A new key of understanding for the health effects of curcumin. *Nutrients* **2020**, *12*, 2499. [CrossRef]
326. Tan, S.; Calani, L.; Bresciani, L.; Dall'asta, M.; Faccini, A.; Augustin, M.A.; Gras, S.L.; Del Rio, D. The degradation of curcuminoids in a human faecal fermentation model. *Int. J. Food Sci. Nutr.* **2015**, *66*, 790–796. [CrossRef]
327. Hassaninasab, A.; Hashimoto, Y.; Tomita-Yokotani, K.; Kobayashi, M. Discovery of the curcumin metabolic pathway involving a unique enzyme in an intestinal microorganism. *Proc. Natl. Acad. Sci. USA* **2011**, *108*, 6615–6620. [CrossRef]
328. Burapan, S.; Kim, M.; Han, J. Curcuminoid demethylation as an alternative metabolism by human intestinal microbiota. *J. Agric. Food Chem.* **2017**, *65*, 3305–3310. [CrossRef]
329. Bresciani, L.; Favari, C.; Calani, L.; Francinelli, V.; Riva, A.; Petrangolini, G.; Allegrini, P.; Mena, P.; Del Rio, D. The effect of formulation of curcuminoids on their metabolism by human colonic microbiota. *Molecules* **2020**, *25*, 940. [CrossRef]
330. Anand, P.; Kunnumakkara, A.J.; Newman, R.A.; Aggarwal, B.B. Bioavailability of curcumin: Problems and promises. *Mol. Pharmac.* **2007**, *4*, 807–818. [CrossRef]

antioxidants

MDPI

Review

Translational Approaches with Antioxidant Phytochemicals against Alcohol-Mediated Oxidative Stress, Gut Dysbiosis, Intestinal Barrier Dysfunction, and Fatty Liver Disease

Jacob W. Ballway * and Byoung-Joon Song *

Section of Molecular Pharmacology and Toxicology, National Institute on Alcohol Abuse and Alcoholism, 9000 Rockville Pike, Bethesda, MD 20892, USA

* Correspondence: jake.ballway@nih.gov (J.W.B.); bj.song@nih.gov (B.-J.S.)

Abstract: Emerging data demonstrate the important roles of altered gut microbiomes (dysbiosis) in many disease states in the peripheral tissues and the central nervous system. Gut dysbiosis with decreased ratios of Bacteroidetes/Firmicutes and other changes are reported to be caused by many disease states and various environmental factors, such as ethanol (e.g., alcohol drinking), Western-style high-fat diets, high fructose, etc. It is also caused by genetic factors, including genetic polymorphisms and epigenetic changes in different individuals. Gut dysbiosis, impaired intestinal barrier function, and elevated serum endotoxin levels can be observed in human patients and/or experimental rodent models exposed to these factors or with certain disease states. However, gut dysbiosis and leaky gut can be normalized through lifestyle alterations such as increased consumption of healthy diets with various fruits and vegetables containing many different kinds of antioxidant phytochemicals. In this review, we describe the mechanisms of gut dysbiosis, leaky gut, endotoxemia, and fatty liver disease with a specific focus on the alcohol-associated pathways. We also mention translational approaches by discussing the benefits of many antioxidant phytochemicals and/or their metabolites against alcohol-mediated oxidative stress, gut dysbiosis, intestinal barrier dysfunction, and fatty liver disease.

Keywords: gut microbiome; dysbiosis; leaky gut; endotoxemia; fatty liver disease; ethanol; oxidative stress; inflammation; phytochemicals; antioxidant

Citation: Ballway, J.W.; Song, B.-J. Translational Approaches with Antioxidant Phytochemicals against Alcohol-Mediated Oxidative Stress, Gut Dysbiosis, Intestinal Barrier Dysfunction, and Fatty Liver Disease. *Antioxidants* **2021**, *10*, 384. https://doi.org/10.3390/antiox10030384

Academic Editor: Baojun Xu

Received: 15 January 2021
Accepted: 25 February 2021
Published: 4 March 2021

1. Introduction

Ample research conducted over the past decade has revealed the expansive and critical role of the human microbiota in numerous physiological processes and pathological consequences. Before elaborating further on these microbial communities and their abundance, the terms microbiota and microbiome should first be distinguished, considering they are sometimes used interchangeably. A microbiota specifically refers to the "assemblage of microorganisms present in a defined environment" [1], while the microbiome includes "the microorganisms (i.e., bacteria, archaea, lower and higher eukaryotes, and viruses), their genomes (i.e., genes), and the surrounding environmental conditions" [1].

Various microbiomes within the human body have been well defined, including the gastrointestinal (GI) tract, lung, skin, urinary, and oral microbiomes, and collectively amount to trillions of bacterial cells that work in concert with (or sometimes in opposition to) human cells and experimental rodent models [2,3]. Current estimates suggest that 1.3 bacterial cells are present for every 1 human cell, contrary to previous suggestions of a 10:1 bacterial to human cell ratio [4]. However, these reduced estimates should not understate the breadth of microbial influence on bodily habitats. Despite differences in microbial composition among bodily habitats, the various microbiomes often perform similar functions, such as immune regulation in the GI tract, respiratory, oral, skin, and urinary microbiomes and promoting nutrient availability in the skin and oral microbiomes [2].

Yet, while bacterial–human symbiosis can occur, so too can antagonistic interactions arise, resulting in diseases, such as periodontal disease (oral microbiome), urinary tract infection (UTI) (urinary tract microbiome), and pharyngitis and pneumonia (respiratory tract microbiome), to name several [2,3]. This review briefly addresses the nature of one of the most significant microbiomes: the gut microbiome. We also describe the mechanisms of the alcohol-mediated changes in gut microbiota, leaky gut, endotoxemia, and fatty liver disease. Based on the mechanistic insights, we finally propose the translational applications by describing the benefits of many antioxidant phytochemicals and/or their metabolites in preventing alcohol-mediated oxidative stress, gut leakiness, and fatty liver disease via changing gut microbiota.

2. The Gut Microbiome

2.1. Microbiome Present in Many Tissues

It is known that different parts of the GI tract may vary in bacterial and fungal composition and abundance, with the vast majority of gut microbes colonizing the colon [5–8]. The gut microbiota encompasses all the microbes found within the human GI tract, which stretches from the mouth, stomach, duodenum, jejunum, and ileum to the end of the large intestine (colon) at the rectum and anal canal. Among all the bacterial phylum present in the gut microbiota, the majority of bacteria fall into three phyla, Bacteroidetes, Firmicutes, or Actinobacteria, with a heavy slant toward Bacteroidetes and Firmicutes [9,10]. The major genera of these phyla found in gut microbial samples include *Bacteroides*, *Alistipes*, and *Prevotella* from the Bacteroidetes phylum, *Faecalibacterium*, and *Ruminococcus* from the Firmicutes phylum, and *Bifidobacterium* and *Collinsella* from the Actinobacteria phylum [9]. Three main enterotypes/clusters, derived from three of these specific genera, were developed to classify gut microbiotas based on their bacterial composition and their energy metabolism tendencies and capacities: Bacteroides (Enterotype 1), Prevotella (Enterotype 2) and Ruminococcus (Enterotype 3) [9,10]. However, as suggested in the study, usage of human fecal samples alone does not necessarily provide a comprehensive catalog of the gut microbial abundance and composition [9,11].

Indeed, a variety of tissues/organs lie between the mouth and anal canal, including the esophagus, stomach, and small and large intestines. Unsurprisingly, the physiological environments of the various organs/tissues of the GI tract vary considerably. Thus, although the gut microbiota is considered less diverse than other bodily microbiomes [12], microbial diversity does exist among the subsections of the GI tract. Indeed, sequencing of bacterial 16S ribosomal RNA from various regions of the GI tracts in healthy fasting adults revealed distinct microbial communities in the saliva, upper GI tract, lower GI tract, and feces [13]. For example, though *Bacteroides uniformis* levels in the salivary and upper GI regions were minimal, a significant increase was detected in the lower GI tract, which further increased in feces [13]. Additionally, diversity and heterogeneity of microbial communities are noted to decrease in regions farther down the digestive tract, owing to the increase in selective pressures and environmental changes present between the stomach and intestinal regions [13]. For instance, *Prevotella melaninogenica* abundance was high in the salivary region; yet, an insignificant decrease was noted toward the end of the upper GI tract, followed by a significant reduction in the lower GI tract, where it was nearly absent [13].

Taking a closer look at the lower GI tract, a marked environmental contrast exists between the acidic, oxygen- and antimicrobial-rich small intestine, and the more basic, oxygen- and antimicrobial-depleted large intestine (colon) [11]. Facultative anaerobes from the *Lactobacillaceae* (*Firmicutes* phylum) and *Enterobacteriaceae* (*Proteobacteria* phylum) families colonize the harsh small intestinal conditions, while anaerobes from families, such as *Bacteroidaceae (Bacteriodetes* phylum) and *Lachnospiraceae (Firmicutes* phylum), inhabit the large intestine in great numbers, owing to the tolerant conditions of the colon [11,14]. Additional factors determining microbial composition of the small intestine and colon include differences in number, amount, and composition of mucus layers and nutrient availability (reviewed extensively in [11]).

Naturally, these differences in the bacterial composition suggest distinct functions for microbes at these particular regions. Indeed, in the intestines, the gut microbiota serves an important role in the metabolism of various endogenous and exogenous compounds, the catabolism of complex carbohydrates, amino acids, and fatty acids into smaller molecules, the synthesis of vitamins or short chain fatty acids (SCFAs), and the degradation of bile acids for the benefit of the host [2,11,12,15]. By the same token, it is also possible that a few microorganisms can metabolize some substrates and produce potentially harmful compounds, such as ethanol [16–18] and trimethylamine (TMA), or disturb the balance of certain secondary bile acids [19–21]. However, many metabolites of this microbial-mediated metabolism, especially SCFAs, are essential for host energy and signaling pathways as well as epigenetic regulation [2,11,12,15,19]. Additionally, the gut microbiota plays a role in immune regulation and host defense [15]. Specifically, the gut microbiota and its metabolites were shown to regulate T-lymphocytes and macrophages, interact with the enteric nervous system (e.g., by producing and/or increasing production of serotonin and melatonin), and mitigates pathogenic bacterial colonization [15,22–25]. Importantly, these functions may be altered or even compromised when changes to the composition of the gut microbiota take place. Indeed, the gut microbiota composition can vary not only between individuals and throughout the different stages of an individual's life, especially during infancy (thoroughly reviewed in [10]), but also in response to numerous environmental factors or other pathophysiological states.

2.2. Gut Microbiota Altered by Aging, Disease States, and Environmental Factors

Changes in the gut microbiota due to normal aging or aging-related pathological states have been well described, and a significant portion of age-related changes occur during early human life [10,26]. Beginning with the transfer of maternal microbes during fetal development, numerous factors will subsequently alter and reshape the human gut microbiota, including the type of birth delivery, the method of milk administration, the weaning of the infant from milk onto solid food, and the interactions with family and the geographical environment during early development [26] (see [10] for a comprehensive list of microbial changes during development). Eventually, the fluctuations experienced by early gut microbiota as a result of these factors will give way to the adult gut microbiota, which face an array of environmental pressures [27]. Even so, the adult microbiota continues to fluctuate with continued aging, especially considering the results of one study, where notable differences in microbial composition, such as a significant increase in *Bacteroidetes* phylum members in elderly individuals compared to young adults, were observed [28]. Additionally, not only does the composition of the gut microbiota change in aging, but also the effectiveness of the microbiota in performing important functions. For instance, although the diversity of microbes actively synthesizing proteins was increased in elderly individuals, the levels of proteins involved in tryptophan and indole metabolism (TnaA and TrpB) were decreased in elderly people compared to infants. In fact, a predicted ~90% drop-off in tryptophan and indole production was observed for individuals 34 years old or above, which may alter immunological and neurological functioning [29].

In addition to normal aging and related physiological changes, pathological states can also play a role in altering the gut microbiota composition. Traumatic brain injury (TBI) represents one such example, and a recent study examining the fecal microbiome of individuals who have suffered chronic TBI (examined years after the acute incident) found significant decreases in *Prevotella* and *Bacteroides* species in addition to increases in the *Ruminococcaceae* family of bacteria, when compared to healthy individuals [30]. Additionally, the researchers suspect that the increases in *Ruminococcaceae* members, coupled with decreases in *Prevotella* species, specifically *Prevotella copri*, may explain the changes in amino acid metabolism and inflammation experienced by these individuals [30]. However, another study examining repetitive, mild TBI found no major alterations in the gut microbiota composition. Interestingly, an increase in the percent relative abundance of members from the *Desulfovibrionaceae* family was observed, which, as mentioned by the researchers, has been connected to cognitive impairments in previous studies, though no direct effect was evaluated or confirmed in this study [30,31].

Moreover, gut dysbiosis can also be affected by environmental factors such as various diets, including Western-style high-fat diets (HFDs) [32] containing fructose [33,34] or predominantly fructose alone [35]. Western-style HFDs containing n-6 polyunsaturated fatty acids can drive the incidence of obesity [36], which may progress to worse complications (e.g., insulin resistance, hypertension, etc.) characteristic of metabolic syndrome [37,38] or even type 2 diabetes. Western-style HFDs impact the resident gut microbiota, contributing to gut leakiness, lipopolysaccharide (LPS) translocation, and inflammatory damage in host intestinal tissue [39,40]. Specifically, a Western-style HFD appears to reduce the presence of carbohydrate-metabolizing proteins, in favor of proteins implicated in amino acid metabolism, with decreased amounts of the Ruminococcaceae family members, since they are involved in carbohydrate degradation [41]. Importantly, the increased abundance of members of the Proteobacteria phylum following HFD exposure to mice [42] is also physiologically important, considering that members of this phylum produce endotoxin LPS, which can enter host circulation due to HFD-mediated damage to host intestinal barrier and stimulate inflammatory damage in the GI tract and other tissues [39]. Interestingly, both Western-style high-fat and fructose diets can increase the abundance of members of the Proteobacteria phylum with a corresponding decrease in Bacteroidetes members, which, as previously stated, likely contributes to the increased leaky gut and LPS translocation [39,43]. Furthermore, chronic and binge alcohol (ethanol) consumption or exposure also affects the amount and composition of gut microbiomes [44,45], resulting in increased oxidative stress, intestinal permeability, endotoxemia, and damage in many tissues. In addition, Chen et al. [46] recently showed that high mobility group box-1 (HMGB1) contained in extracellular vesicles (exosomes) derived from gut dysbiosis can also promote non-alcoholic fatty liver disease (NAFLD) in adapter protein ASC-null mice after exposure to an HFD.

All these conditions clearly indicate the important role of changes in the amounts and composition of the gut microbiota, resulting in impairment or damage in various tissues or organs through the gut–liver–brain axis. In this review, we specifically focus on the role of alcohol-mediated oxidative stress in promoting intestinal barrier dysfunction, and fatty liver disease. In addition, we explain the mechanisms of alcohol-mediated gut dysbiosis, resulting in increased leaky gut and fatty liver (steatosis) and/or inflammation (steatohepatitis) via increased oxidative and nitrosative/nitrative stress. Finally, we briefly describe the beneficial effects of various antioxidant phytochemicals and their mechanisms of action against gut dysbiosis, intestinal barrier dysfunction, and fatty liver disease.

3. Oxidative Alcohol Metabolism and Progression to Alcoholic Liver Disease

Following consumption, most alcohol (ethanol) molecules can be oxidatively metabolized to acetaldehyde and then irreversibly converted to acetate in numerous organs/tissues, including the liver, stomach, and possibly brain at very low levels. However, the primary site of oxidative alcohol metabolism occurs in the hepatocytes of the liver [47,48], although the stomach is also known to be involved in alcohol metabolism [49]. In addition, alcohol can be metabolized by the non-oxidative metabolic pathway such as fatty acid ethyl esters by cholesterol esterase, etc. [50,51]. Furthermore, unmetabolized ethanol can be excreted from the body through the breath, skin, sweat, and urine [52].

Owing to its low Km and high expression in the liver, cytoplasmic alcohol dehydrogenase (ADH) oxidatively metabolizes the majority of alcohol entering the liver into acetaldehyde, a highly reactive intermediate, while simultaneously reducing the cofactor NAD^+ to NADH [47,48]. The reactive (and potentially harmful) intermediate acetaldehyde is then oxidized in the mitochondria by the low Km aldehyde dehydrogenase 2 (ALDH2) with the reduction of a cofactor NAD^+ to NADH, to generate acetate, which is converted into acetyl-CoA for metabolic use or subsequently exported to other organs, including the brain [47,48]. The ADH- and ALDH-mediated oxidative ethanol metabolism pathway provides a reliable and efficient means of metabolizing the majority of consumed alcohol. Aside from the physiological consequences of alcohol overconsumption, social problems arise when alcohol consumption increases in frequency and amount. In fact, more than ~75% of alcohol-associated with sociomedical consequences are ascribed to the consumption of large amounts of alcohol in short periods of time [53,54].

Naturally, increased amount and frequency of alcohol consumption begins to deplete the cellular levels of the NAD^+ cofactor by the ADH- and ALDH2-dependent reactions. A decrease in NAD^+ levels, coupled with an increase of NADH levels, will cause redox changes and alter numerous cellular functions, resulting in elevated fat synthesis, decreased fat oxidation, and increased cell death processes, among others, with characteristic hallmarks of early alcoholic liver disease (ALD), specifically, steatosis and liver inflammation [47,55]. Furthermore, two additional enzymes are involved in oxidizing ethanol in the liver: peroxisome-resident catalase and endoplasmic reticulum (ER)- and mitochondria-localized ethanol-inducible cytochrome P450-2E1 (CYP2E1), which represents a major component of the microsomal ethanol oxidizing system (MEOS) [48,56]. However, the role of catalase in hepatic ethanol metabolism is minor in comparison possibly due to the limited availability of hydrogen peroxide [56]. Thus, the CYP2E1 enzyme, constitutively expressed under normal physiological states and then induced by ethanol, at least via protein stabilization [57,58], becomes functionally important in ethanol metabolism, as well as in alcohol-mediated oxidative liver damage, particularly because CYP2E1-mediated alcohol metabolism results in the production of both highly reactive acetaldehyde and reactive oxygen species (ROS), such as the superoxide anion (O_2^-) and hydrogen peroxide (H_2O_2) [47,59,60]. In fact, *Cyp2e1*-null mice are resistant to alcohol-mediated liver injury [61], while transgenic mice with overexpressed CYP2E1 [62] and *Cyp2e1* knock-in mice were more sensitive to liver injury by alcohol [63] or non-alcoholic substances, including a diet with 20% fat-derived calories [64]. Overwhelming increases in ROS levels strain the cellular antioxidant defense mechanisms, resulting in elevated oxidative stress, various post-translational protein modifications (PTMs), and apoptotic cellular damage through increased lipid peroxidation, ER stress, mitochondrial dysfunction, and DNA damage with genomic instability [47,65–69]. If alcohol consumption persists, hepatic damage will continue to increase and prime the liver for progression into more severe stages of ALD such as liver fibrosis, cirrhosis, and hepatocarcinoma [70,71].

ALD pathogenesis has been well reviewed elsewhere [72–75]; however, here we will briefly address the main phases of ALD and the pathological hallmarks of each stage that will be relevant when discussing the contributing role of the gut microbiota in ALD pathogenesis. The first major stage of ALD is the development of fatty liver, also termed steatosis, which involves the accumulation of lipid droplets within hepatocytes. The mechanisms underlying the development of alcohol-induced fat accumulation are numerous. Important mechanisms include a redox change with the decreased NAD^+/NADH ratio during the oxidative ethanol metabolism and increased fat synthesis in the cytoplasm, through activated transcription factors, such as sterol regulatory element binding protein (SREBP-1c). In addition, ethanol intake can cause fat accumulation via decreased fat degradation, resulting from the suppressed mitochondrial enzymes for the fat oxidation pathway and acetaldehyde-mediated decreased transcription of peroxisome proliferator activator receptor α (PPARα) needed for fatty acid export and degradation, and increased import/transport of free fatty acids from adipose tissues after lipolysis. Persistent fat accumulation and oxidative stress can severely damage hepatocytes, resulting in apoptosis, which promotes the activation of liver-resident macrophages (Kupffer cells) and attracts infiltrating neutrophils, leading to inflammation and steatohepatitis. During steatohepatitis, Kupffer cells and other liver cells respond to both damage-associated molecular pattern (DAMP) molecules from apoptotic hepatocytes, in addition to other molecules, such as gut-derived pathogen-associated molecular pattern (PAMP) molecules such as

LPS, leading to the further secretion of proinflammatory cytokines and the persistence of oxidative stress, thus exacerbating liver damage [47,76]. Eventually, hepatic stellate cells can be activated by transforming growth factor-β (TGF-β) secreted by Kupffer cells attempting to resolve inflammation, which may propel the liver toward fibrosis [47]. Liver fibrosis arises as structural proteins, such as collagen and α-smooth muscle actin (α-SMA) derived from transformed hepatic stellate cells, assemble into the extracellular matrix to form a network of rigid, fibrotic scar tissue to surround damaged portions of the liver [77]. Acetaldehyde, produced during the oxidative ethanol metabolism and potentially elevated due to inactivation of ALDH2 under oxidative stress [65,78–80], is recognized to modulate important aspects of fibrosis, such as inhibiting PPARγ or increasing the transcriptional activity of C/EBPβ to stimulate collagen α1(I) expression [67]. Persistent fibrosis leads to liver failure during cirrhosis, as unresolved fibrotic tissue continues to damage liver architecture and hinders liver recovery and function [47,74]. Development of hepatocellular carcinoma can arise not only from the formation of various adducts between DNA or proteins and acetaldehyde, malondialdehyde (MDA), or 4-hydroxynonenal (4-HNE) (likely resulting at least partially from CYP2E1-mediated oxidative stress) [81], but also from the release of PAMP and DAMP molecules, that can activate immune cells and indirectly contribute to the abundance and increased activity of tumor-initiating stem-cell-like cells (TICs) [74].

With this basic overview of ALD in mind, we can now analyze the direct effect of alcohol on the gut function and the gut microbiota. Following this analysis, we will systematically address the effects of alcohol-induced changes to the gut microbiota during ALD. The contribution of specific bacterial groups and/or species to liver and gut damage following alcohol exposure will be discussed. Additionally, this review will highlight some of the numerous therapeutic options that may mitigate alcohol-induced oxidative stress, gut dysbiosis, leaky gut, and fatty liver by various dietary supplements, such as antioxidant phytochemicals, probiotics, small molecule metabolites, and traditional/ancient medications.

4. The Mechanisms of Alcohol-Mediated Gut Dysbiosis, Intestinal Barrier Dysfunction, and Consequences

Over the past decade, several well-published articles have reviewed the role of alcohol-induced gut dysbiosis on ALD pathogenesis [20,82–85]. In fact, it has been reported that people with ALD, including liver cirrhosis, have elevated levels of serum endotoxin compared to control subjects [44], indicating increased intestinal permeability or gut leakiness (leaky gut). This seminal observation with AUD people was replicated by many other laboratories [86–88]. Furthermore, the elevated gut dysbiosis and leaky gut following long-term and/or binge ethanol exposure was also observed in experimental models with mice [89] and rats [86,90], indicating a common phenomenon conserved among different species. The following section will highlight alcohol-induced gut dysbiosis, the specific changes in their amounts and composition during ALD, and potential translational approaches designed to remedy these alterations.

4.1. The Effect of Alcohol on the Amounts and Composition of Gut Microbiota

Many microbes are affected by the presence of alcohol in the various parts of the GI tract, and the changes in the abundances and composition of the gut microbiota have been extensively studied. Conveniently, a recent study [91] has compiled data from many publications describing changes in the gut microbiota in humans, such as individuals with AUD [92], those who have a history of chronic overconsumption of alcohol [93], and those in different stages of ALD [94]. Examining the data from a subset of these publications (in addition to several very recent publications) reveals important trends to consider for ALD and gut dysbiosis prevention.

Alcohol intake is known to increase the degree of small intestine bacterial overgrowth (SIBO), as originally reported [44,45,95]. In addition, several important phyla, including the major *Proteobacteria*, *Bacteroidetes*, *Firmicutes*, and *Actinobacteria*, are all impacted by the presence of alcohol in the GI tract. Numerous studies examining human colonic biopsies [96] and human [93] and mouse [97] feces indicate a higher abundance of members of the *Proteobacteria* phylum in response to alcohol [91]. As suggested elsewhere, this change can conceivably result from the ability of microbes in this phylum to persist in the high ROS environment generated by increased inducible nitric oxide synthase (iNOS) following alcohol exposure [98], since they are predominantly facultative anaerobes, which can withstand these conditions [92]. Specifically, at the family level, *Enterobacteriaceae* abundance was increased [91,92,96], in addition to elevated levels of certain genera from this family, such as *Escherichia* [91,92,96]. Owing to their Gram-negative status and endotoxin (e.g., LPS) producing capabilities, *Proteobacteria*, such as those from the genera *Escherichia*, are unsurprisingly seen as potential instigators of gut barrier dysfunction during alcohol consumption [93,97]. In particular, the ability of species in the *Escherichia* genus to produce harmful PAMP molecules, such as LPS [93], and to metabolize alcohol (in some strains) [99], producing the toxic metabolite acetaldehyde, supports the hypothesized harmful role of *Proteobacteria* in alcohol-induced gut dysbiosis. Additionally, although the *Proteobacteria* genera *Sutterella* displayed a decreased relative abundance in the colonic biopsies of heavy drinkers [96], another study examining the feces of individuals with a history of chronic alcohol overconsumption found a higher relative abundance of this genera and views the *Sutterella* increase in light of previous studies reporting its role in promoting inflammation [93]. This same study provides evidence for an inverse relationship between *Proteobacteria* and the presence of the anti-inflammatory SCFA butyric acid, which was decreased in the feces of individuals with a history of chronic AUD, although as noted by the authors, confirming this correlation is hindered by the nature of the study [93].

Alcohol exposure has been shown to decrease the abundance of members of the phylum *Bacteroidetes* in the feces of mice exposed to chronic alcohol [97] and from colonic biopsies [87] and feces [92] of AUD individuals. However, several other reports using chronic alcohol mouse models have described an increase in *Bacteroidetes* presence following the sequencing of colonic and cecal contents [20,100,101]. More conflicting results emerge when examining the genus *Bacteroides* in this phylum. Although one recent study did not detect any difference in the abundance of *Bacteroides* members in colonic biopsies from heavy drinkers [96], other studies examining stool from AUD patients found an increase in *Bacteroides* members [102], while the sequencing of feces from alcoholic individuals [92] found decreased abundance of this genus. As mentioned by the authors, unlike facultative anaerobic *Proteobacteria*, *Bacteroides* members are obligate anaerobes and may, therefore, struggle to survive in the presence of significant ROS during prolonged alcohol exposure, which may explain the decrease in this study [92]. Although *Bacteroides* are not active ethanol metabolizers [92], they are known to play a role in the metabolism of bile acid, which could interfere with farnesoid X receptor (FXR) signaling (due to its regulation by bile acids) [20]; however, elevated Bacteroides levels could lead to increased production of the gamma-aminobutyric acid (GABA) neurotransmitter [102], suggesting a potential interplay between members of the *Bacteroides* genus and the brain during ALD. Interestingly, one study showed that members of another genus, *Prevotella*, have the capacity to metabolize ethanol and generate acetaldehyde in vitro, suggesting a possible role for members of this genus in contributing to acetaldehyde production in vivo [92]. However, like the *Bacteroides* data, conflicting reports have emerged regarding the relative abundance of *Prevotella* during alcohol exposure, where *Prevotella* members displayed increased abundance in the stool of AUD patients [102]; yet, numerous studies have described a decrease in the relative abundance of members from this genus [91].

In a mouse model, the presence of members of the *Actinobacteria* phylum was noted to increase in the feces of alcohol-exposed mice, and this elevation, in conjunction with the amplified presence of *Proteobacteria*, has been suggested to play a role in the intestinal manifestations of ALD [97]. Indeed, several interesting genera from this Gram-positive phylum are altered in the gut following alcohol exposure, including *Corynebacterium*, *Bifidobacterium*, and *Collinsella* [91,97]. *Corynebacterium* was found to be increased in the feces of mice chronically exposed to alcohol and, although the relevance of this elevation in ALD pathogenesis remains unknown, the authors posit that this amplification could be noteworthy, since other studies have found *Corynebacterium* infection in ALD [97,103,104]. The other genera, *Bifidobacterium* and *Collinsella*, display conflicting alterations among studies. While *Bifidobacterium* was decreased in human feces of individuals who habitually drink alcohol [92], this genus was found in increased abundance in the feces of active alcoholic patients with cirrhosis and severe alcoholic hepatitis, compared to individuals with cirrhosis and lacking severe alcoholic hepatitis [94]. Similarly, while *Collinsella* was found in increased abundance in the stool of AUD individuals [102], analysis of the relative abundance of *Collinsella* showed no significant change in the abundance of these genera in the feces of individuals who habitually consume alcohol [92]. Interestingly, both *Bifidobacterium* and *Collinsella* were characterized as potential acetaldehyde accumulators in in vitro aerobic conditions [105], and, in particular, the absence of *Bifidobacterium* was hypothesized to contribute to the observed decrement in alcohol metabolism in the feces of alcoholic individuals [92]. Some have postulated that the ability for *Collinsella* to metabolize ethanol may permit their observed overgrowth in the stool of AUD patients [102], or could potentially allow them to persist longer than other microbes.

Although the phylum *Firmicutes* was demonstrated to decrease in abundance in mouse fecal samples following chronic alcohol exposure [97], this diverse phylum has several genera that were both increased, including *Streptococcus, Coprobacillus, Holdemania,* and *Clostridium*, and decreased, including *Ruminococcus, Faecalibacterium, Subdoligranulum, Roseburia*, and *Lactobacillus*, among others, following alcohol exposure [91–93,96,101,102]. *Roseburia* and *Lactobacillus*, two therapeutically relevant bacterial genera, showed decreased abundance in colonic biopsies of heavy drinkers [96] and colonic contents of rats chronically exposed to alcohol [101], respectively. Similarly, *Faecalibacterium* and anti-inflammatory members of the genus, such as *F. prausnitzii*, are decreased in the feces of heavy drinkers [96] and individuals who have a history of AUD [93]. Expectedly, these decreased levels of *F. prausnitzii* during alcohol exposure likely impact the levels of beneficial SCFAs in the intestines [91], such as butyrate, and, unsurprisingly, a positive correlation was observed between *Faecalibacterium* and butyric acid levels in the feces of AUD individuals [93]. Overall, the Firmicutes phylum is quite diverse. While obligate anaerobes from the Ruminococcus genus are observed to decrease [92], likely the result of the increased oxidative stress in the gut following alcohol consumption, as previously reported [98], facultative anaerobes from the Streptococcus genus have been shown to elevate in both the stool of patients with AUD [102] and in the feces of alcoholic individuals [92]. Indeed, as others have postulated, the observed decrease in obligate anaerobes may give members of the Streptococcus genus (and other facultative anaerobes) an opportunity to proliferate, which may be of concern, considering that infections from Streptococcus members have been noted during cirrhosis [91].

Lastly, from the phylum *Verrucomicrobia*, the *Akkermansia* genus was noted to decrease in both the stool of people with AUD [102] and in colonic biopsies from heavy alcohol consumers [96]. The *Akkermansia* genus and, specifically, *Akkermansia muciniphila*, have numerous beneficial roles in the intestines, including protecting the gut barrier and aiding in the production of epithelial cell-protective mucus [91,96,102], and supplementation was shown to prevent manifestations of ALD in binge and chronic mouse models of alcohol exposure [106]. Importantly, in patients with AUD, low *Akkermansia* levels negatively correlated with the inflammatory marker MCP-1, indicating a potentially significant role for these microbes in inflammation regulation during alcohol consumption [102].

Importantly, when analyzing the alcohol-induced changes in the composition of the gut microbiota, one needs to take special consideration for the specific tissue/sample being examined. For instance, examining microbial changes in the feces of mice following alcohol exposure [97] or humans who have consumed alcohol at some time [92,93] does not necessarily provide a comprehensive assessment of the gut-wide microbial status. As suggested by others, analysis of microbial changes in samples other than feces is needed to pinpoint changes in the diverse environments of the gut [91]. For example, one study examining the microbial changes in jejunal and colonic contents of rats chronically exposed to alcohol found significant changes in colonic microbiota, but hardly any impact on composition in the jejunal microbiota [101]. Thus, future research may target specific areas of the intestines (duodenal, jejunal, ileal, colonic, rectal, etc.), which should provide a more complete profile of the gut microbial changes following alcohol exposure.

4.2. Mechanisms of Alcohol-Induced Damage to the Intestines, Resulting in Leaky Gut and Endotoxemia

Before ethanol molecules reach bodily destinations, such as the liver and brain, they must first pass through the tissues of the GI tract. Alcohol can be absorbed in the mouth and esophagus (albeit in limited quantities), and absorption rates will begin to increase further down the GI tract, especially at the stomach, duodenum, and jejunum and, to a lesser extent, at the ileum [107,108]. Alcohol absorption at the proximal regions of the small intestine (duodenum and jejunum) [109] will result in the passage of these molecules into the capillaries and blood vessels, including the portal vein, where they will be delivered to the liver [110]. Although alcohol passes from the intestinal lumen into the bloodstream through simple diffusion [107,111], ethanol molecules first must pass through the layers of the intestinal barrier and can trigger changes at these various regions.

Considering the structure of the small and large intestinal barrier, both consist of a monolayer of intestinal epithelial cells with a layer of mucus on the luminal side and the immune cell-rich lamina propria on the non-luminal side. Aside from the chief intestinal epithelial cell, the enterocyte, the monolayer can contain numerous other cell types. These include intestinal epithelial stem cells (IESCs), which can differentiate to replace cells in the monolayer, Paneth cells, which produce antimicrobial peptides and regulate IESCs, and goblet cells, which produce and secrete mucin glycoproteins for incorporation into the luminal mucus layer [112,113]. Several differences exist between the small and large intestinal barriers. Specifically, unlike the small intestine, the colon contains two mucus layers, a loose and a thick layer [11], a proportionally greater number of goblet cells, and an absence of Paneth cells [112].

In the intestinal lumen, ethanol will encounter the mucus layer, which contains glycosylated mucin proteins and lipids, among other components, and serves as both a protective barrier and a regulator of the microbial environment, whereby species, such as *Akkermansia muciniphila*, will use barrier-generated mucus resources [11,112]. Interestingly, mRNA levels of mucin proteins 1-4 do not increase in the small or large intestines in response to binge alcohol [114]. However, examining the glycosylation status of mucin glycoproteins, studies using chronic models of alcohol exposure found altered patterns of mucin glycosylation in the intestinal mucosa, especially increased galactosylation [115], which may influence bacterial binding to the mucus layer or may prevent mucus adherence to the intestinal barrier, as demonstrated in a gastric mucosa study [116]. Importantly, one study has suggested that ethanol may also decrease the hydrophobicity of the mucus layer, specifically, through the ethanol-mediated dissolution of mucus-layer free fatty acids, which function as an absorptive barrier, thus likely promoting ethanol-induced gut permeability dysfunction [117].

Eventually, from the mucus layer, ethanol will pass into the monolayer through simple diffusion, where it will either continue to diffuse into the circulation for delivery to various bodily sites or be metabolized in the barrier [108]. Indeed, the presence of ADH and ALDH isozymes and their activities have been detected in the small and large intestines, although their contribution to alcohol metabolism is not as significant as the liver [118]. Although differences in ADH and ALDH expression vary across the intestinal landscape, importantly, ADH expression and activity appears greater in both the small and large intestines compared to those of ALDH, suggesting a greater buildup of reactive acetaldehyde over acetate in the monolayer following alcohol metabolism [118]. In addition, the presence of CYP2E1 and its elevated levels in the GI tract after alcohol exposure [119–121] are likely to produce elevated levels of ROS and acetaldehyde, especially due to low levels of ALDH2 expression in the gut [122,123]. As reviewed thoroughly elsewhere [108], the metabolism of ingested alcohol can also occur through non-oxidative pathways. Ingested alcohol may also interact with and be metabolized by gut microbes. Indeed, some species are capable of metabolizing ingested alcohol, such as certain strains of *Escherichia coli* [99], while other gut microbes display limited alcohol metabolism ability, such as some members of the lactobacillus genus [124]. Additionally, a recent study examining the feces of Japanese alcoholic individuals categorized several bacterial groups, such as the *Ruminococcus* and *Collinsella* genera, as potential acetaldehyde accumulators for their ability to metabolize ethanol in aerobic in vitro conditions [105,125].

With these factors in mind, along with the previous discussion of the alcohol-mediated alterations to the gut microbial composition, we can now examine the mechanisms and impact of gut dysbiosis and alcohol exposure on the gut barrier, specifically, on the increased intestinal permeability observed following chronic and/or binge alcohol exposure. Normally, the intestinal monolayer forms a tightly linked barrier with various proteins forming the intestinal tight junction (TJ), adherent junction (AJ), and desmosomes that keep microbes in the intestinal lumen out of the bloodstream [126], while also permitting the passage of luminal nutrients into the bloodstream, thus ensuring a useful, non-toxic blood supply for recipient organs, such as the liver via the portal vein [127]. However, alcohol exposure can alter the permeability of this intestinal barrier, resulting in an influx of harmful luminal molecules, such as bacterial endotoxin LPS and peptidoglycan, which can induce inflammation and oxidative damage both in the gut and in other organs [127]. Based on this information, we will concisely address the role of ethanol, acetaldehyde, ROS, and other factors in triggering this intestinal permeability dysfunction and will subsequently address the impact of this leaky barrier on the various stages of ALD pathogenesis, as illustrated in Figure 1. This review will place a particular emphasis on the role of oxidative and nitrosative/nitrative stress in gut dysfunction and its impact on ALD pathogenesis, in addition to the mechanistic role of enterocyte apoptosis and PTMs of paracellular junctional complex proteins in initiating gut barrier dysfunction (leaky gut), leading to inflammatory liver injury (Figure 1).

Early studies determined that exposure of ethanol to the human colonic Caco-2 cell line triggered apoptosis in these epithelial cells [128]. Although a mouse model of chronic ethanol exposure alone displayed no significant changes in apoptotic protein markers or intestinal permeability in the jejunum [129], binge alcohol exposure elevated endotoxin levels in rodent models, pointing to the development of a leaky gut in these animals [130,131]. The effects of binge alcohol exposure on time- and dose-dependent gut permeability change, elevated endotoxemia, and fatty liver injury were confirmed by other laboratories [121,132,133]. Mouse and rat models of binge alcohol exposure showed increased levels of apoptotic protein markers, such as BAX and cleaved caspase-3, and histological evidence for the apoptosis of intestinal epithelial cells, consequently indicating increased gut permeability [121]. In rodent models, treatment with antioxidants or a specific inhibitor of CYP2E1 significantly prevented binge alcohol-mediated leaky gut and fatty liver disease, while *Cyp2e1*-knockout mice were also quite resistant to these changes [120,121,132]. These results clearly indicate the important roles of CYP2E1 and consequent oxidative stress in promoting intestinal barrier dysfunction and ALD. Furthermore, a recent study using a mouse model of chronic plus binge alcohol exposure found increased apoptotic markers in the proximal small intestine, likely mediated by ER stress, and this corresponded to the observed increase in bacterial product translocation at this region [134]. Importantly, however, the effect of ethanol on the intestinal barrier is not limited to the induction of apoptosis in the epithelial cells within the gut monolayer.

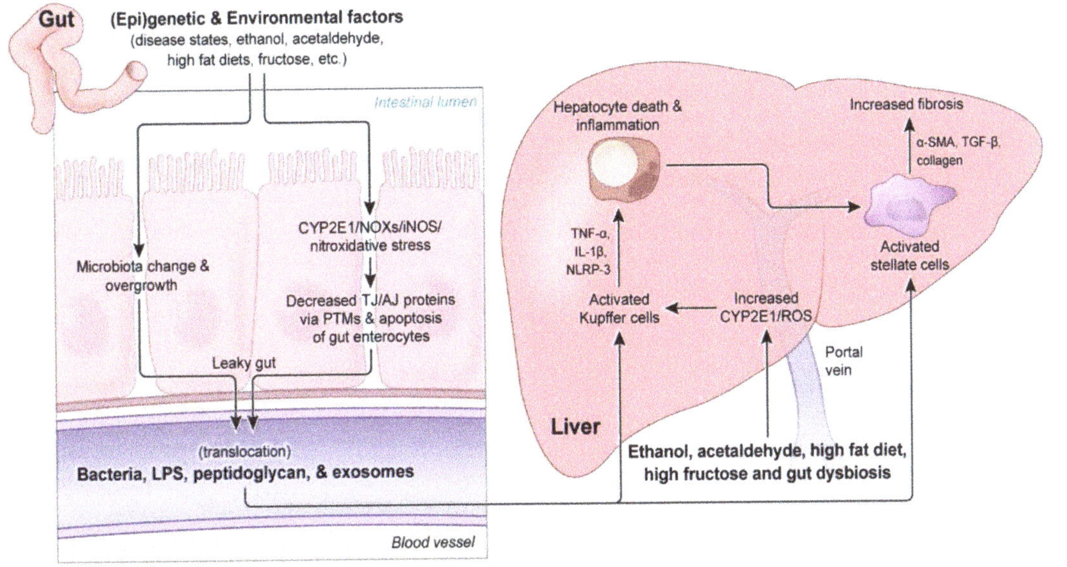

Figure 1. Schematic overview of gut–liver communication and damage prompted by intestinal disorders or consumption of exogenous agents. Numerous exogenous agents (e.g., alcohol, high-fat diet (HFD), fructose, etc.) or underlying intestinal disorders (e.g., Crohn's disease, ulcerative colitis, etc.) can elicit changes to the abundance and/or composition of the gut microbiota. Elevated gut CYP2E1 and NADPH oxidases (NOXs) can increase oxidative stress. The resulting gut dysbiosis alters gut metabolism and damages the intestinal barrier through various mechanisms, including the oxidative stress-mediated post-translational modifications (PTMs), leading to decreases in paracellular junction complex proteins. Sustained damage to the barrier causes gut leakiness and, subsequently, a gut-localized immune response and increased levels of harmful gut-derived compounds (e.g., lipopolysaccharide (LPS), peptidoglycan, exosomes, etc.) into the circulation. LPS (and other gut-derived metabolites) and alcohol will reach the liver and drive alcoholic liver disease (ALD) pathogenesis and progression. Kupffer cell activation, mediated by LPS and oxidative stress driven by metabolism of the ethanol by hepatic CYP2E1 and from activated NOXs, increases inflammatory cytokine levels (e.g., TNF-α, etc.) and instigates hepatocyte apoptosis. Eventually, sustained oxidative stress, LPS infiltration, and hepatocyte damage will lead to the activation of hepatic stellate cells, driving liver fibrosis and continued liver damage.

Several important proteins that normally connect cells of the monolayer and enhance the impermeability of the intestinal barrier are affected by ethanol. To limit the travel of molecules between adjacent monolayer cells (the paracellular pathway), TJ complexes composed of proteins, such as claudins, occludin, and zonula occludens-1 (ZO-1), along with AJ complexes, composed of proteins, such as E-cadherin and β-catenin [135,136], form between monolayer cells to prevent the mass influx of particles into the bloodstream. Exposure of ethanol to Caco-2 cells resulted in a time-dependent increase in epithelial cell permeability, which correlated with the time- and dose-dependent decrease in ZO-1 and increase in claudin-1 protein levels [137]. Additionally, ZO-1 and claudin-1 were noted to be irregularly distributed upon localization. A similar observation was noted in Caco-2 cells treated with ethanol (10, 20, or 40 mM) for 3 h, whereby ZO-1 and occludin showed decreased localization to the membranes for intercellular interactions, which correlated with increased barrier permeability, despite no observed change in ZO-1 (and other tight junction proteins) mRNA levels [138]. Thus, the presence and proper localization of junctional complex proteins involved in paracellular transport are crucial for maintaining intestinal integrity, especially considering that occludin knockout mice exposed to ethanol displayed both increased permeability at the colon and increased triglyceride accumulation in the liver compared to wild-type (WT) mice exposed to ethanol [139]. In addition, ethanol and consequently CYP2E1-mediated oxidative and nitrosative/nitrative stress appear to induce changes to the PTM landscape of intestinal proteins on a global level, with regard to increased acetylation, nitration, and ubiquitination [120] as well as phosphorylation [140–143].

4.3. Mechanisms of Gut Leakiness via Oxidative Stress and PTMs of Paracellular Proteins

Interestingly, ethanol appears to induce changes to the PTM landscape of intestinal proteins on a global level, with regard to acetylation, oxidation, nitration, and ubiquitination [120,121]. One recent study found that the removal of intestinal NAD^+-dependent class III deacetylase SIRT1 blunts alcohol-induced liver damage in an acute on chronic alcohol mouse model, likely through prevention of ferroptosis, which may indicate a potential role for global acetylation in intestinal barrier function in a model of ALD pathogenesis [144]. However, in particular, various PTMs of proteins involved in the blockage of paracellular transport, such as those comprising TJ, AJ, and cytoskeletal proteins, have received considerable interest for their possible role in stimulating gut leakiness following alcohol exposure. Indeed, many paracellular transport proteins, especially TJ proteins, are known to be post-translationally modified, which play a role in their functional capabilities [145,146].

In the context of alcohol exposure, acetylation of α-tubulin appears to interfere with ZO-1 recruitment to the membrane, which likely contributes to increased permeability observed in the in vitro Caco-2 cell model following exposure to ethanol or acetaldehyde [138]. Additionally, increased nitration and ubiquitin-conjugation of α-tubulin in the intestines of mice exposed to binge alcohol correlated with decreased protein levels of α-tubulin and endotoxemia in these mice [121]. Modification of specific TJ and AJ proteins has also been well documented. In this same study, increased nitration and ubiquitin-conjugation of β-catenin (AJ), plakoglobin (adherens junction/desmosomes), claudin-1 (TJ), and claudin-4 (TJ) was observed in the intestines of binge alcohol-exposed mice, which, once again, correlated with endotoxemia and, thus, gut leakiness and fatty liver disease [121]. The significantly decreased gut TJ and AJ proteins in binge alcohol-exposed rats compared to controls were further confirmed by quantitative mass-spectral analysis. Additionally, the role of paracellular protein phosphorylation in both TJ and AJ formation has been well characterized and is altered by acetaldehyde [140,147]. Specifically, the presence of acetaldehyde leads to the persistent phosphorylation of the TJ protein occludin and AJ proteins E-cadherin and β-catenin, which are hypothesized to prevent the binding of these proteins to actin filaments, evidenced by the decreased amounts of these proteins in actin-rich Triton-insoluble fraction, which, as indicated by the authors, is an accurate predicator of TJ and AJ complex integrity [148]. Further studies have pinpointed the specific enzymes involved in regulating paracellular protein phosphorylation status following alcohol exposure. The protein phosphatase 2A (PP2A)-mediated dephosphorylation of threonine residues on occludin [149] and the altered activity of protein tyrosine phosphatase 1B-mediated dephosphorylation of AJ proteins [147] are two such examples of alcohol-induced changes to the paracellular protein landscape. Further supporting the role of phosphorylation in paracellular protein regulation, the results of a study examining brain endothelial cells exposed to ethanol suggest that the increased permeability of bovine brain microvascular endothelial cells (BBMEC) observed following ethanol exposure is linked to the phosphorylation of claudin-5 and occludin, likely mediated by myosin light chain kinase (MLCK) [150].

Acetaldehyde has also received considerable attention for its ability to alter the intestinal barrier function following alcohol consumption, and this subject has been thoroughly reviewed elsewhere [151]. Like ethanol, acetaldehyde was also shown to increase membrane permeability of an in vitro 3-D Caco-2 spheroid model exposed to acetaldehyde concentrations as low as 0.025 mM; yet, this exposure did not affect the mRNA levels of TJ proteins when this same model was exposed to 0.2 mM acetaldehyde [138]. Additionally, acetaldehyde also affects the localization of ZO-1 and causes an increase in global protein acetylation, including α-tubulin [138]. Indeed, considerable research has been devoted to determining the role of acetaldehyde in altering the localization of ZO-1 and other paracellular barrier proteins (for a comprehensive list, see [151]). As previously mentioned, various PTMs appear to play a significant role in determining the integrity of the barrier, whereby the presence of acetaldehyde leads to the persistent phosphorylation of several paracellular proteins, thus contributing to gut leakiness [140]. Furthermore, in vivo studies using WT and ALDH2(+/−) mice demonstrated increased intestinal permeability at the distal and proximal colon, jejunum, and ileum in ALDH2(+/−) mice following ethanol exposure, compared to the corresponding WT, which only displayed increased permeability at the distal colon [152]. This suggests an important role for acetaldehyde in mediating gut leakiness and for ALDH2 in moderating elevated acetaldehyde levels and thus affecting the re-distribution of TJ and AJ proteins in the mouse ileum and colon [152]. Interestingly, acetaldehyde was also shown to negatively impact an in vitro model of mucin-producing goblet cells, particularly through another catalyst of intestinal barrier dysfunction: oxidative stress [153].

Indeed, many adverse cellular manifestations, such as lipid peroxidation, and DNA damage arise as a result of oxidative stress, which ultimately leads to inflammation and apoptosis [47,74]. Oxidative stress was observed in the intestines of rats exposed to chronic alcohol conditions, evidenced by the increase in iNOS protein levels, nitrate and nitrite levels, and jejunal, ileal, and colonic protein nitration and oxidation, which correlated with increased intestinal permeability [98,154,155]. Specifically, the iNOS-dependent increase in miR-212 levels in Caco-2 cells exposed to ethanol appears to contribute to perturbations in the cell permeability [156]. In addition, an iNOS-mediated decrease in ZO-1 expression was observed in the ethanol-exposed Caco-2 cells [156]. The role of iNOS in instigating in-vitro increases in Caco-2 cell permeability was confirmed in-vivo using iNOS-KO mice, which displayed significantly less intestinal barrier dysfunction compared to the corresponding WT mice following exposure to alcohol in both mouse strains [156]. Additionally, elevated and activated iNOS appears to contribute to the activation of the Snail transcription factor, which was demonstrated to play a role in increased gut permeability [157]. However, iNOS is not the only enzyme responsible for the alcohol-induced development of oxidative stress. In a binge alcohol rat model, both iNOS and CYP2E1 protein levels were elevated in the intestines 1 or 2 h following the last binge dose [132]. CYP2E1 is able to oxidize ethanol into acetaldehyde, but in doing so, generates ROS molecules, such as the 1-hydroxyethyl radical [60,158]. Indeed, the role of intestinal CYP2E1 on gut leakiness observed following alcohol exposure has been well reviewed and involves important circadian proteins, such as CLOCK and PER2 [155]. Importantly, the observed increase in plasma endotoxin levels, oxidative stress (determined via increased 3-nitrotyrosine (3-NT) levels), and intestinal permeability was dependent on intestinal CYP2E1, since *Cyp2e1*-null mice were quite resistant to binge alcohol-mediated gut leakiness and subsequent fatty liver [120,132,159]. Additionally, the levels of nitrated and ubiquitin conjugated TJ and AJ proteins were markedly diminished in the same *Cyp2e1*-null mice exposed to binge alcohol, suggesting a role of CYP2E1 and/or CYP2E1-generated oxidative and nitrosative/nitrative stress in modulating the intestinal PTM landscape [120,132,159].

Furthermore, it is also possible that ROS can be provided through activated NADPH-oxidase isozyme(s) (NOXs) present in the colon and in immune cells in the lamina propria of the GI tract [32,160,161]. However, treatment of Caco-2 cells with acetaldehyde, ethanol, and the NADPH oxidase inhibitor diphenyleneiodonium did not ameliorate evidence of increased barrier permeability and dysfunction as *N*-acetyl cysteine was demonstrated to do [162]. The oxidative stress-induced mitogen-activated protein kinases (MAPKs) [e.g., p38 protein kinase (p38k), c-Jun *N*-terminal protein kinase (JNK), and extracellular signal regulated protein kinase (ERK)] are also implicated in gut barrier damage, since they are observed to be phosphorylated (activated) in Caco-2 cells and played a role in TJ disruption and gut barrier dysfunction, particularly through increasing MLCK mRNA [163], whose activity has been implicated in regulating intestinal permeability [164]. The oxidative stress-mediated decrease in hepatocyte nuclear factor-4α (HNF-4α) was also implicated in tight junction disruption following alcohol exposure [165]. Additionally, studies have implicated the fatty acid ethyl esters ethyl oleate and ethyl palmitate in increased ROS production in Caco-2 cells and these esters also caused redistribution of the paracellular proteins ZO-1 and occludin and caused increased ROS-mediated permeability of these cultured cells [166].

4.4. Crosstalk among Gut Dysbiosis, Intestinal Barrier Dysfunction, and ALD

Bidirectional gut–liver communication occurs as gut-derived molecules pass through the intestinal barrier and enter the portal vein to reach the liver, while liver-derived molecules pass through the biliary tract to interact with the intestines [20,167]. Typically, in a resting, non-disease state, SCFAs, secondary bile acids and other diet-derived metabolites will pass into the liver to perform various functions, such as driving fatty acid oxidation in the case of SCFAs [168], or, in the case of secondary bile acids, which have been modified in the gut, these metabolites can be recycled for future use [20]. Concurrently, the liver will secrete primary bile acids, among other liver-derived molecules, to the intestines to increase lipid absorption [169] and potentially alter the gut microbial landscape through their antimicrobial properties [170]. Heavy alcohol consumption will alter communication between the gut and liver, not only by increasing the presence of circulating alcohol [108] and subsequently produced acetaldehyde [171], which can damage the liver at high concentrations, but also by enabling harmful gut microbiota-derived molecules to leak into the circulation and inflict damage on both the intestinal barrier and the liver [167].

The non-luminal side of the intestinal barrier is home to the lamina propria containing a wide variety of immune cells, which can respond to changes in gut barrier dysfunction. The response of immune cells (specifically leukocytes) to alcohol consumption does not appear to be uniform across the lower GI tract [172], possibly due to the differences in microbial composition and abundance found throughout the GI tract, gut environmental factors, and/or differing ethanol concentrations at these regions. Predictably, following ethanol consumption, these cells and others will interact with and be activated by luminal-derived molecules (LPS/PAMPs) that traversed the damaged, leaky intestinal barrier [127]. LPS is a significant luminal-derived molecule that will enter circulation and directly alter local (gut) regions, triggering inflammation and injury to downstream organs, such as the liver, inflicting further damage [127,172] and enterocyte barrier dysfunction [173]. Although passage of LPS through the intestinal barrier has long been speculated to occur via paracellular transport, a recent study examining LPS transport across the intestinal barrier during lipid absorption (non-alcohol conditions) observed CD63- and lipid raft-mediated transcellular transport of LPS across the barrier and absorption into the portal vein [174]. Regardless of the transport method across the barrier, the manifestation of endotoxemia is a common occurrence in individuals with ALD [175,176] and serum LPS levels appear to correlate with the amount of alcohol being consumed [177]. However, before LPS and other gut-derived harmful molecules (ethanol, acetaldehyde, cytolysin, candidalysin [178], DAMPs, exosomes, etc.) can travel through the portal vein to the liver, the gut-localized immune response will be affected [174]. For example, an early study examining mice chronically exposed to alcohol demonstrated that the mRNA levels of certain cytokines (IL-1β and TNFα) are upregulated in the ileum by ethanol alone, while others (IL-6 and IL-11) are upregulated in the presence of both ethanol and LPS [179]. A more recent study using a binge on chronic mouse model of alcohol exposure found increased production of IL-17 from Paneth cells, which contributed to inflammasome activation in the small intestine, evidenced by increases in activated caspase-1 and IL-18 [134]. Furthermore, alcohol exposure has been shown to alter populations of immune cells in the gut, such as T-lymphocyte populations in the small intestine of Rhesus macaques chronically exposed to alcohol [180]. Importantly, considering that NADPH oxidase levels are increased during exposure of mice to the Western-style HFD [32,161] and play a role in inflammation of the intestines during dextran sodium sulfate (DSS)-mediated colitis [160], it is possible that ROS generated by NADPH-oxidase isozyme(s) present in immune cells in the GI tract may also contribute to gut-localized inflammation in response to alcohol.

LPS and other luminal-derived molecules will travel through the portal vein and reach the liver, where they will interact with various cell types and influence the different stages of ALD pathogenesis [181–183]. Briefly, depending on the specific DAMP and/or PAMP, gut-derived molecules will interact with specific toll-like receptor (TLR) complexes present in resident liver macrophages, Kupffer cells (and other liver cells) to initiate a specific inflammatory cascade [181–183]. In particular, the pathway by which LPS induces inflammatory cascades in the liver has been investigated thoroughly [184]. Briefly, after the binding of LPS to LPS-binding protein (LBP) and CD14, LPS will bind MD2 to interact with TLR4 on the membrane, thus activating the complex [181,182,184]. The resulting TLR4 activation can elicit intracellular signaling cascades which, depending on the specific proteins involved [182,184], can increase the expression of proinflammatory cytokines, such as TNFα, which will both increase apoptotic liver damage and hepatic inflammation [185,186]. Indeed, LPS has been shown to drive steatosis and inflammation through mechanisms such as decreased autophagic response in the liver [187] and increased pro-inflammatory cytokine production [188], respectively. Additionally, the LPS-mediated induction of the liver damage has also been implicated in increased hepatic stellate cell (HSC) response to TGF-β and fibrosis onset, although LPS may also exhibit anti-fibrotic properties through by targeting HSC proliferation, as recently reviewed [189]. Importantly, a key mediator of gut-induced damage of the liver is oxidative stress and, specifically, ROS, generated not only by LPS-activated Kupffer cells but via the oxidative metabolism of ethanol by CYP2E1 as high concentrations of ethanol enter the liver from the circulation [67].

A recent study demonstrated that increased ROS levels, likely generated by CYP2E1, occur alongside ethanol-induced decreases in autophagy in alcohol-exposed mice [187]. Furthermore, studies have confirmed the role of CYP2E1 in LPS-induced liver damage via the production of ROS and activation of oxidative stress-sensitive downstream kinases, such as the MAP kinases, and mitochondrial damage [190,191]. MAP kinases are not only activated by CYP2E1-generated ROS [68,79], but also by the NADPH oxidase-mediated production of ROS, which activate ERK in a CYP2E1-independent manner, leading to TNF-α production [192]. Another study found that arachidonic acid supplementation with alcohol activated ERK via ROS and, subsequently, increased TNF-α levels, supporting the notion that other factors, such as dietary n-6 fatty acids, can act alongside LPS in inducing oxidative stress-mediated liver damage following alcohol exposure [193] or during other liver pathologies, such as palmitic acid-mediated NAFLD [194]. In a binge alcohol model, NADPH oxidase-mediated ROS production is also important for inflammatory signaling by increasing interleukin-1 receptor-associated kinase (IRAK) levels in Kupffer cells 21 h post binge alcohol exposure, a change that was dependent on NF-κB activity and which correlated with increased TNF-α levels in these Kupffer cells [195]. Interestingly, at the PTM level of regulation, LPS and acetate (and/or acetaldehyde) were demonstrated to decrease hepatic SIRT1 levels, with, expectedly, increased hyperacetylation of a subunit of nuclear transcription factor kappa B (NF-κB), leading to increased inflammatory response in in vitro rat Kupffer cells [196]. Though not explicitly evaluated, the authors hypothesized that ROS may regulate this mechanism.

Kupffer cell-localized NADPH oxidase (NOX) was hypothesized early on to generate ROS during infiltration of LPS and/or neutrophils following alcohol exposure [197] and the role of specific members of the NOX family of NADPH oxidases, such as NOX4, has been described. Indeed, recent in vitro and in vivo models of alcohol-induced liver damage confirmed the role of NOX4 in increasing mitochondrial ROS and mitochondrial-mediated apoptosis, in addition to a partial role in steatosis development following alcohol exposure [198]. Translational approaches to combat macrophage-mediated ROS production during ALD have also revealed the mechanistic regulation of macrophage-localized NADPH oxidase, as globular adiponectin was shown to inhibit ROS and NOX2 expression through activation of liver kinase B1 (LKB1) and AMP-dependent protein kinase (AMPK) [199]. Besides immune cells, HSCs can also be targeted by LPS and, mechanistically, oxidative stress also appears to drive LPS-mediated increases in MCP-1 and IL-6 in HSCs [200]. Although gut-mediated hepatic oxidative stress represents an important contributing factor to ALD pathogenesis and progression, not all aspects of ALD pathogenesis appear to rely on oxidative stress, since other factors, such as insulin resistance, may also play a role [201].

5. The Antioxidant Properties, Metabolisms, and Health Benefits of Various Phytochemicals against Gut Dysbiosis, Intestinal Barrier Dysfunction, and Fatty Liver Disease

As described in the previous sections, gut dysbiosis with decreased ratios of Bacteroidetes/Firmicutes and changes in the levels of various endogenous metabolites, including ethanol and SCFAs, are associated with many disease states. Thus, there have been many efforts to normalize or restore the gut microbiome by using different diets, such as microbiota-targeted vegan (vegetarian) diets and/or consuming microbiota-accessible carbohydrates (oligocarbohydrates) or dietary supplements with various phytochemicals [19]. In addition, many phytochemicals represent diverse classes of antioxidants contained in a variety of fruits and vegetables as well as medicinal plants [202]. Most of these phytochemicals are known to have very low bioavailabilities due to their water insolubilities [202,203]. These phytochemicals include different chemical classes, such as various polyphenols, lignans such as phytoestrogens, carotenoids, phytosterols/phytostanols, alkaloids and glucosinolates, sulfur-containing compounds, etc., as recently reviewed [202]. In one class of phytochemical, there are many chemical subgroups. For instance, polyphenols represent antioxidant chemicals of many different subgroups. Common subgroups include flavonoids (e.g., apigenin, quercetin) polyphenolic amides (e.g., capsaicin) and polyphenolic acids [204], which include important polyphenols such as resveratrol, curcumin, capsaicin, quercetin, rutin, genistein, daidzein, ellagic acid, and proanthocyanidins (tannoids), such as (-)-epicatechin (EC), (-)-epigallocatechin (EGC), (-)-epicatechin-3-gallate (ECG), and (-)-epigallate catechin-3-gallate (EGCG), among others, as recently reviewed [205]. Despite the different chemical structures of these phytochemicals, they have exhibited their beneficial effects in some human studies, many experimental rodent models, and in vitro cell culture models.

One of the characteristic properties of these phytochemicals is their very low water solubility, despite the presence of many hydroxyl groups in polyphenols [206,207]. It is known that some phytochemicals can be metabolized or modified by intestinal bacteria (specifically hydrolyzed, reduced, deglycosylated, degraded, conjugated, etc.) and some of these alterations may enhance absorption of these chemicals in the intestine [208–210]. For instance, resveratrol, contained in grapes, berries, and red wine, can be metabolized to dihydroxyresveratrol, 3,4′dihydroxybibenzyl (lunularin), and 3,4′-dihydroxy-trans-stilbene by gut bacteria or converted to the more bioavailable metabolite piceid [205,211]. Likewise, 3,4-dihydroxyphenylacetic acid, a gut-derived metabolite of the flavonoid quercetin, found that various fruits, vegetables, and beverages (mostly as quercetin glycosides) [212] can both act as a free radical scavenger and reduce markers of advanced glycation end products [194]. A multitude of polyphenols (e.g., flavonoids, thearubigins, chlorogenic acids) are known to be metabolized by gut microbiota [213] as well as conjugated, glucuronated, deglycosylated, etc., in the intestinal epithelial cells and liver, although each compound is supposed to produce its own unique structural derivatives. As reviewed elsewhere, these chemicals will undergo gut bacteria-mediated metabolic transformations or conjugation, which may increase absorption, as in the case of anthocyanins [214] and phytoestrogens [215]. In addition, bacterial transformations may yield new metabolites with beneficial effects, as in the case of equol derived from daidzein [215], where equol is known to possess anti-inflammatory and anticancer properties [214]; however, not all microbial-derived metabolites exhibit beneficial effects or increased absorption [214,215]. In addition, many phytochemicals, despite their low bioavailabilities, are known to exert their biological activities in the gut independent of tissue absorption. In fact, these phytochemicals are likely to be present in the highest abundance in the gut after oral consumption, mainly due to very low absorption rates through the gut membrane [208–210]. These phytochemicals exhibit their beneficial effects by altering the rates of transcription of certain genes and the absorption or levels of many essential nutrients or compounds, such as cholesterol, triglycerides, and bile acids [208–210]. Furthermore, they are known to exert their functional activities by their unique antioxidant activities by suppressing the activities of pro-oxidant enzymes and transcription factors, including NF-kB, which transcriptionally regulates many proinflammatory downstream targets such as TNFα and iNOS. These antioxidant compounds also show anti-inflammatory effects with decreased levels of pro-inflammatory cytokines and chemokines against gut-related disorders such as inflammatory bowel disease [216], in addition to combatting other metabolic syndrome-related symptoms (e.g., obesity-related fatty liver and type 2 diabetes associated cardiovascular disorders) [210] and possibly preventing neurodegenerative abnormalities, including Alzheimer's disease, through the gut(–liver)–brain axis [217].

As illustrated in Figure 2, oxidative and nitrosative/nitrative stress, inflammation and leaky gut, endotoxemia and inflammatory tissue injury arise following exposure to many environmental compounds such as ethanol (alcohol drinking), Western-style high-fat diets, and high fructose or genetic risk factors and/or certain disease states. However, certain phytochemicals and/or their metabolites are capable of ameliorating these pathological manifestations caused by some of these insults, particularly through changing (usually normalizing) the amounts and components of the gut microbiota, resulting in improved composition (i.e., gut eubiosis) [202,218–220] from gut dysbiosis. Additionally, these beneficial agents can alter the metabolism and production of many endogenous compounds such as ethanol, acetaldehyde, SCFAs, bile acids, and trimethylamine (TMA) by altering the compositions of the gut microbiome [202,221] or perhaps even ethanol and acetaldehyde, which have been shown to be produced by some gut bacterial strains [202,218–220,222]. These antioxidant phytochemicals and/or their gut metabolites can also suppress the increased oxidative and nitrosative/nitrative stress by inhibiting the enzymes, such as CYP2E1, NOXs, iNOS, and NF-κB [223–225], as listed in the Figure. Furthermore, these beneficial antioxidant compounds from dietary supplements can affect key enzymes/proteins in the cellular signaling pathways, such as AMPK [226] and hepatic Sirt-1 [227], or transcription factors, including NF-κB and PGC-1α [228,229], to show their biological effects, as demonstrated by polyphenols in green and black teas [230]. In addition, these antioxidant phytochemicals may improve the disease states by modifying the genetic and epigenetic regulations in the gut and other tissues possibly through the NAD^+-dependent non-histone deacetylase Sirt-1 [227] and its isoforms. Finally, gut dysbiosis-mediated altered levels of SCFAs, especially propionate and butyrate, which are inhibitors of histone deacetylase [19], may result in different epigenetic profiles, which further regulate the gene transcription rates compared to those during eubiosis and/or in phytochemical-treated cases. All these changes are likely to contribute to the beneficial effects of phytochemicals and improvement of various disease states in experimental models, as well as in a few human studies.

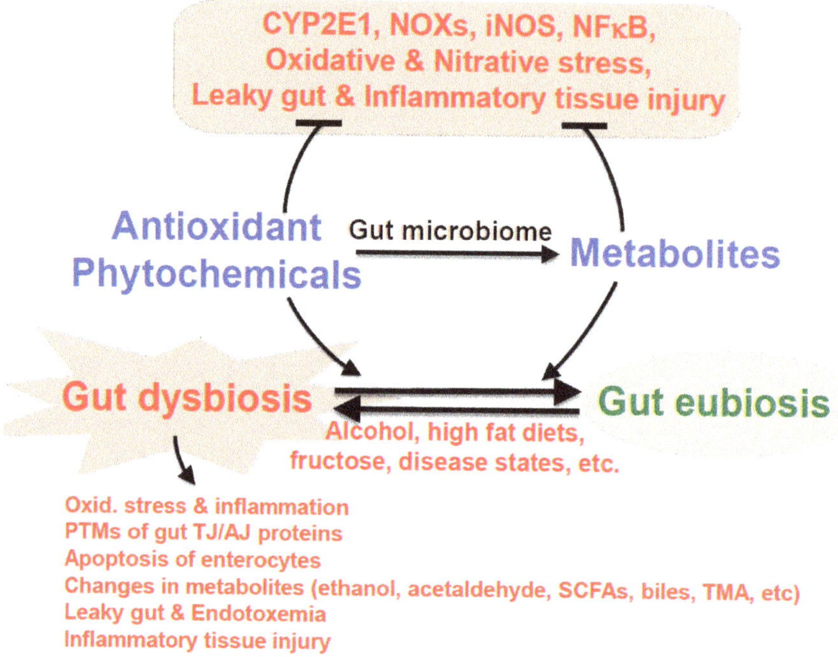

Figure 2. Proposed mechanisms of the beneficial effects of antioxidant phytochemicals on gut dysbiosis and oxidative stress-mediated intestinal barrier dysfunction and inflammatory liver injury. As described in the text, many phytochemicals contained in various fruits, vegetables and dietary supplements can be metabolized by gut microbiota for improved absorption, leading to greater bioavailability. By improving the oxidative stress and gut dysbiosis, these antioxidant phytochemicals and/or their metabolites prevent leaky gut, endotoxemia, inflammation, and alcoholic and/or non-alcoholic fatty liver disease.

6. Translational Approaches and Therapeutics against Alcohol-Mediated Oxidative Stress, Gut Dysbiosis, Intestinal Barrier Dysfunction and Fatty Liver Disease

The presence of pathogenic microbes and their metabolites in the gut may potentially promote ALD and its progression, since pretreatment with non-absorbable antibiotics such as neomycin and polymyxin B [231] or rifaximin [232] significantly prevented the severity of alcohol-mediated liver disease in different animal models. Yet, although neomycin and polymyxin B helped ameliorate fructose-mediated steatosis development [233], rifaximin only showed limited success in combatting certain manifestations of non-alcoholic steatosis and steatohepatitis (NAFLD/NASH) [234] and thioacetamide-induced liver injury [235]. Furthermore, gut microbiome analyses of control and age-matched cirrhosis patients revealed that altered gut microbiota is also positively associated with cirrhosis and progression [236], suggesting an important role of gut dysbiosis in various liver diseases. The opposite case may also be true, where the absence of all bacteria is also liable to drive liver damage, since one study showed that germ-free mice exposed to binge alcohol conditions displayed greater hepatic fat accumulation and inflammation compared to WT mice, in addition to increased CYP2E1 mRNA in the proximal small intestine, which may trend toward elevated CYP2E1 protein levels and, likely, a stimulation of ROS production and intestinal and liver damage [237]. This germ-free study and reports characterizing gut dysbiosis following alcohol exposure demonstrate that while the gut dysbiosis has the potential to escalate ALD, the complete absence of gut microbes does not necessarily attenuate manifestations of ALD. Thus, maintenance of the normal balance of microbes should help prevent the initiation of the 'second hit' in the two-hit hypothesis of ALD progression, whereby early, mild manifestations of ALD (fatty liver) can progress to more severe insults (inflammation or fibrosis) through a second hit, such as certain gut-derived molecules, such as LPS and reactive oxygen species, as proposed in NAFLD/NASH [238–241]. Indeed, numerous therapeutic studies have sought approaches to moderate oxidative and nitrosative/nitrative stress, gut dysbiosis, and intestinal barrier dysfunction following alcohol exposure. These approaches include supplementation with commensal microbes, gut-protective factors, and a wide range of dietary options from both synthetic and natural origins.

Numerous studies have reported the beneficial effects of commensal microbial (probiotics) supplementation on ALD manifestations and gut dysbiosis [242]. Recent exciting results with clinical studies revealed that fecal microbiota transplantation (FMT) from healthy donor people seems to be relatively safe and effective in attenuating the severity with improvement of health outcome in mouse models of alcoholic liver injury [243] and patients with severe alcoholic hepatitis or cirrhosis [244–246], as reviewed [85]. In addition to whole FMT, supplementation of a specific microbial strain(s) such as lactobacillus [247] can also improve systemic endotoxemia and liver conditions following alcohol consumption. Additional evidence for the beneficial role of microbial strain supplementation includes the decrease in high blood ALT and AST levels following ethanol exposure to rats supplemented with yogurt or cream cheese made with *Lactococcus chungangensis* CAU 28 [248], the decline in hepatic iNOS and global nitration levels following *Lactobacillus fermentum* administration, depending on time point [249], and the *Roseburia intestinalis*-mediated reduction in gut permeability through increased occludin mRNA and protein levels [250]. Additionally, other members of the Lactobacillus genus such as *Lactobacillus plantarum*, have been shown to reduce inflammatory markers, triglyceride levels and gut leakiness associated with ALD by way of epidermal growth factor receptor (EGFR) activity [251]. Inflammation, steatosis, and gut leakiness were also mitigated in a chronic alcohol mouse model supplemented with *Akkermansia muciniphila* gut population [106], which, as previously stated, is found to be decreased in individuals with steatohepatitis [106] and alcohol use disorder [20,102].

Alongside probiotics, certain dietary supplements have been shown to attenuate ALD pathogenesis and limit gut dysbiosis and leakiness. For example, administration of the comestible cricket *Gryllus bimaculatus* prior to acute alcohol exposure was demonstrated to mitigate ethanol-induced increases in intestinal oxidative stress (8-OHdG levels), hepatic apoptotic markers, and hepatic triglyceride accumulation [252]. Pomegranate extracts and indole-3-carbinol (I3C), derived from Brassica vegetables, both mitigated inflammation of the liver and hepatocyte and enterocyte apoptosis in binge alcohol [120] and chronic plus binge alcohol exposure models, respectively [159]; although, unlike pomegranate [120], I3C supplementation did not attenuate fatty liver [159]. Additionally, ellagic acid (EA) and urolithin A (UA), two polyphenols derived from pomegranate extracts, mitigated ethanol-induced increases in gut permeability in T84 colon cells [120]. Furthermore, EA, UA, and I3C were all capable of reducing ethanol-induced increases in hepatic CYP2E1 levels, in vivo (I3C) [120,159] and in vitro (EA, UA) [120], which should reduce hepatic oxidative stress and attenuate liver injury. Interestingly, the PTM landscape is also affected by these supplements, whereby both I3C and pomegranate supplementation decrease ethanol-induced hyperacetylation in the liver and intestines, respectively [120,159]. Pomegranate extract, in particular, reduced ethanol-induced global nitration and ubiquitination of intestinal proteins and, specifically, ethanol-induced claudin-1 nitration and ubiquitination, thus preventing its degradation, and all of these PTM changes correlated with the prevention of intestinal barrier damage, and endotoxemia [120]. Additionally, rats chronically administered alcohol and fed oats exhibited decreases in ethanol-induced protein nitration and oxidation in all small intestine subregions and the colon, which was suggested to be due to a decrease in oxidative stress as a result of oat-mediated decreases in iNOS, nitrite, and nitrate levels [253]. Additionally, β-glucans from various foods, such as oats and barley [224], were effective against gut dysbiosis in non-alcohol-related diseases, thus supporting human health [225]. When co-administered with ethanol at various percentages, fish oil (high in n-3 polyunsaturated fatty acids) was shown to attenuate liver manifestations of ALD (steatosis and inflammation) [90], plasma endotoxin levels [90], alcohol-induced intestinal permeability dysfunction [254] and even impacted gut microbial composition, through a recovery of fecal Bifidobacterium members [254] and an increase in Bacteroidetes members (especially with supplementation of 25% fish oil), thus decreasing the ratio of Firmicutes to Bacteroidetes following alcohol exposure [90]. However, studies have reported that oxidation of fatty acids in fish oil (prior to administration) actually worsens liver outcomes following alcohol exposure and even increases the abundance of members of the Gram-negative, LPS-producing Proteobacteria phylum [255]. Thus, proper maintenance of fish oil in non-oxidized states appears to be very important to ensure its beneficial effects against alcohol-mediated fatty liver injury [256], possibly through preventing leaky gut [254].

Supplementation of zinc [130] may also attenuate liver and gut dysfunction following alcohol exposure. Zinc levels were observed to decrease in the ileum of mice chronically exposed to alcohol, which correlated with plasma endotoxemia, ileal oxidative stress and permeability, and decreased tight junction protein levels [257]. Mechanistically, by hampering the activity of enzymatic regulators of hepatic apoptosis (e.g., caspase-3), zinc supplementation can prevent hepatocyte death following alcohol exposure [258] and may benefit from, but does not require, the zinc-binding ability of metallothionein to exert its therapeutic effect [259].

Some alternative and complimentary remedies, such as traditional herbal medicines in China [260–262], Korea [263], and India [264,265], have recently proved effective in both combatting inflammation and in altering the gut microbial composition in models of ALD or models of intestinal disease. For example, co-administration of ethanol with the fungi *Wolfporia cocos* (or, more specifically, the water-insoluble polysaccharides from their fruiting bodies) decreased the hepatic triglyceride levels and MCP-1 levels, indicating decreased steatosis and inflammation of the liver [261]. Interestingly, these polysaccharides increased Firmicutes abundance and decreased the abundance of the Gram-negative, LPS-producing Proteobacteria phylum following chronic alcohol exposure [261]. Additionally, leaf extract from the plant *Dendropanax morbifera* leaf extract also altered microbial composition in rats acutely exposed to alcohol, whereby, for example, members of the Bacteroides operational taxonomic unit (OTU) increased upon co-administration of leaf extract with ethanol, compared to the ethanol group [263]. The *Dendropanax morbifera* leaf extract co-administered with ethanol contained several phytochemicals (e.g., rutin, caffeic acid, etc.) [263] and it is highly likely that the list of beneficial antioxidant phytochemicals capable of protecting against alcohol-mediated oxidative stress, gut dysbiosis, epithelial barrier dysfunction, and ALD (as well as NAFLD) will be increased in the future.

Microbiota-derived molecules may also help combat alcohol-induced tissue damage and ALD pathogenesis. Indeed, while microbial products derived from the colon and feces of alcohol-exposed mice were demonstrated to increase intestinal permeability in vitro, in addition to instigating in vitro T-cell activation [266], certain beneficial microbial products may reverse these adverse outcomes. SCFAs, in particular, butyric acid supplemented as tributyrin, may improve intestinal barrier functioning through the preservation of ZO-1 and occludin protein levels and proper localization at the barrier [267]. Additionally, butyrate can restrain cytokine (TNFα and MCP-1) production following acute alcohol exposure, although it does not appear to ameliorate alcohol-induced steatosis [267]. Indirect benefits of SCFA supplementation might also be achieved through administration of *Pediococcus pentosaceus*, which increased the levels of the SCFAs possibly due to the partial recovery of bacterial genera, such as Clostridium, whose levels were decreased in the mouse model of chronic plus binge alcohol exposure and whose recovery could increase SCFA production, as suggested by the positive correlation found between Clostridium and butyric acid levels in this study [268]. Administration of other SCFAs or other beneficial microbial products may yield similar results to butyrate supplementation, such as administration of indole-3-propionic acid (IPA), an indole metabolite generated by the metabolism of tryptophan, which elevated gut tight junction proteins, thus preventing intestinal barrier dysfunction and liver damage in rats exposed to a Western-style HFD, possibly by the gut–liver axis [269]. However, another study exposing mice to HFD did not find a reduction in liver triglyceride levels with co-administration of IPA (in addition to other parameters of bodily damage or inflammation), which could be attributed to minor differences in the HFD or murine model used [270]. Nevertheless, these differing results highlight the need for continued studies on the role of microbial products in potentially ameliorating liver and tissue damage induced by a wide range of stimuli.

Certain synthetic drugs, such as sennoside A [271] and metformin [272,273], can exhibit their beneficial effects on leaky gut and endotoxemia, in addition to preventing NAFLD and/or learning and memory impairment. However, it is also known that some antibiotics and chemotherapeutic agents such as 5-fluouracil and vancomycin are known to cause dysbiosis of gut microbiota [274]. In the latter cases, a probiotic with digestive enzymes was formulated to protect against cancer-drug-related dysbiosis, suggesting a reminder of careful interpretation of the results in using some drugs to regulate gut dysbiosis. As mentioned in the previous sections, usage of some antibiotics such as neomycin and polymyxin B or rifaximin could be considered for treating alcohol-mediated gut endotoxemia and steatotic and inflammatory liver damage through repurposing their applications, since these beneficial changes were observed in rodent models [231,232]. Additionally, the composition of the gut microbiota was changed in several orders (e.g., Erysipelotrichales) following ethanol and ethanol with rifaximin administration; however, the physiological role of these compositional changes and others should be further studied [232]. These antibiotics may represent possible options for treating gut dysbiosis, but future studies will have to confirm this proposition. However, aside from already-approved antibiotics, the usage of other synthetic compounds may not be practical, since large-scale randomized clinical studies are likely to require considerable time, cost, and effort in evaluating the safety and efficacy tests needed for FDA approval.

The benefits of naturally occurring compounds found in many edible plants, fruits, vegetables, and dietary supplements should also be considered as alternative approaches. For instance, various phytochemicals (berberine, curcumin, resveratrol, and the numerous components of silymarin) from different foods and plants can prevent NAFLD and, in some cases, can normalize gut dysbiosis and intestinal permeability changes, especially in rodent models [262]. The antioxidant *N*-acetylcysteine (NAC) was shown to mitigate LPS-induced intestinal permeability in the IPEC-J2 intestinal porcine enterocyte cell line [275], and experiments using Caco-2 cells showing that NAC supplementation with alcohol can prevent an EtOH-induced increase in CLOCK and PER2 expression helped formulate the hypothesis that ROS generated by CYP2E1 in response to ethanol could increase the expression of these circadian rhythm proteins, leading to intestinal dysfunction and permeability, as recently reviewed [276]. Some of these phytochemicals, such as resveratrol, curcumin, silymarin, genistein, quercetin, rutin, anthocyanidins, ellagic acid, etc., may have very limited bioavailability, although some may be better absorbed than others [277]. In theory, intakes of extremely large amounts of these natural compounds are needed to observe their beneficial effects due to their very low bioavailability, though experiments testing this hypothesis using curcumin or ellagic acid suggest this may is not the case [205,208–210,226,278]. These phytochemical compounds may have paradoxically high functional activity despite the setback of their low bioavailability [205]. For instance, a subset of French people, who usually consume a diet with relatively high saturated fatty acids and cholesterol, exhibit low risks of major diseases, like coronary heart disease, and deaths known as the French paradox [279,280] possibly due to their habits of daily drinking of wine, which contains resveratrol and other antioxidants [281]. Part of the reason could be that these polyphenol compounds may have direct antioxidant effects that counteract potentially oxidative enzymes such as NADPH oxidases [282] and CYP2E1 [223,283]. Alternatively, these chemicals may possibly exhibit their benefits by preserving the activities and/or levels of certain antioxidizing enzymes and proteins, as suggested by the importance of the presence of SOD2 for the resveratrol-mediated prevention of cytotoxicity in mouse hippocampal neurons pretreated with this polyphenol [284]. Additionally, other antioxidant-related proteins, such as ALDH2 and glutathione peroxidase (Gpx), which were demonstrated to be inactivated in the presence of acetaminophen [285], alcohol or non-alcoholic substances [68,79], may also be restored by the administration of antioxidant phytochemicals. Furthermore, these phytochemicals, including various flavonoid compounds, demonstrate their benefits by preventing gut dysbiosis and normalizing the amount and composition of beneficial bacteria such as *Lactobacillus*, *Akkermansia*, and *Bifidobacteria*, while decreasing the population of potentially harmful microbes, as suggested elsewhere [219,262,286]. In addition to their known beneficial interaction with the gut microbiota [202], certain phytochemicals may also affect bacterial strains known to produce ethanol and acetaldehyde, both of which are known to cause leaky gut and fatty liver disease [120,121], although this hypothesis has yet to be tested. In general, the usage of these phytochemicals and/or other prebiotics or probiotics as dietary supplements could be important in treating patients with alcoholic steatohepatitis with leaky gut and endotoxemia, especially considering the increasing incidence of ALD and the proposed high mortality rate (30~40%) at 1 month for individuals with severe alcoholic hepatitis [287–289]. Unfortunately, there is no clinically proven drug approved for effectively treating alcoholic hepatitis patients, although a few clinical studies are being conducted [290,291]. Early clinical trials using FMT from healthy donors into patients with alcoholic hepatitis, cirrhosis, AUD, etc., offer preliminary data suggesting that FMT may be a safe and promising therapeutic option [85,244–246], although additional large-scale and long-term multi-center studies are needed to confirm the safety/toxicity profile and efficacy of the newly emerging FMT therapy against various liver diseases, including ALD. Requests have been submitted for several phytochemicals listed in this review, such as trans-resveratrol, urolithin A, and curcumin, to be recognized as generally regarded as safe (GRAS by the FDA definition), and, if given FDA approval, these could be recommended as dietary supplements for ALD patients in addition to their clinical treatment protocols. However, these newly emerging areas need to be further studied to completely understand the benefits of various antioxidants contained in fruits, vegetables and dietary supplements.

7. Conclusions

In this review, we have briefly described the role of gut microbiota alteration during normal development as well as in pathological conditions. We also mentioned the role of gut dysbiosis caused by various environmental factors as well as epigenetic and/or genetic risks. We also describe the patterns of gut microbiome changes in various pathophysiological conditions and provide evidence for the benefits of ameliorating gut dysbiosis by many different agents in human populations and experimental rodents as well as cell culture models. Moreover, we describe the causal role of increased oxidative stress in promoting intestinal barrier dysfunction, subsequently leading to elevated endotoxemia and fatty liver disease. We also described potentially safe methods of translational approaches against alcohol-mediated oxidative stress, gut dysbiosis, leaky gut, and ALD. These beneficial methods and agents include healthy lifestyle changes with proper intake of healthy diets, containing beneficial chemicals contained in many fruits and vegetables, or n-3 polyunsaturated fatty acids. Indeed, these beneficial agents have demonstrated their effectiveness against gut dysbiosis, intestinal barrier dysfunction, endotoxemia, and ALD and NAFLD in many experimental models. In particular, supplementation with certain phytochemicals represents an exciting prospective therapeutic option for mitigating alcohol-mediated tissue damage owing to the breadth of known phytochemicals and the observed benefits of these chemicals and/or their metabolites on the gut microbiome and redox regulation, despite many compounds having low bioavailabilities. Importantly, future certification of many phytochemicals in various fruits, vegetables, and plants as GRAS could permit usage of these chemicals as dietary supplements in treating ALD. However, randomized large-scale clinical studies need to be conducted in the future to accurately demonstrate the efficacy of some of these antioxidant phytochemicals.

Author Contributions: Conceptualization by J.W.B. and B.-J.S.; writing—original by J.W.B. and B.-J.S.; review and revisions by J.W.B. and B.-J.S. All authors have read and agreed to the published version of the manuscript.

Funding: This work was supported by the Intramural Research Fund of the National Institute of Alcohol Abuse and Alcoholism, National Institutes of Health.

Acknowledgments: Both authors are grateful to Wiramon Rungratanawanich for her excellent help in designing the Figures and significant editorial assistance.

Conflicts of Interest: The authors declare no conflict of interest.

References

1. Marchesi, J.R.; Ravel, J. The vocabulary of microbiome research: A proposal. *Microbiome* **2015**, *3*, 31. [CrossRef] [PubMed]
2. Dekaboruah, E.; Suryavanshi, M.V.; Chettri, D.; Verma, A.K. Human microbiome: An academic update on human body site specific surveillance and its possible role. *Arch. Microbiol.* **2020**, *202*, 1–21. [CrossRef] [PubMed]
3. Kostic, A.D.; Howitt, M.R.; Garrett, W.S. Exploring host-microbiota interactions in animal models and humans. *Genes Dev.* **2013**, *27*, 701–718. [CrossRef] [PubMed]
4. Gilbert, J.A.; Blaser, M.J.; Caporaso, J.G.; Jansson, J.K.; Lynch, S.V.; Knight, R. Current understanding of the human microbiome. *Nat. Med.* **2018**, *24*, 392–400. [CrossRef]
5. Shanahan, F. The Colonic Microbiota and Colonic Disease. *Curr. Gastroenterol. Rep.* **2012**, *14*, 446–452. [CrossRef]
6. Hoffmann, C.; Dollive, S.; Grunberg, S.; Chen, J.; Li, H.; Wu, G.D.; Lewis, J.D.; Bushman, F.D. Archaea and Fungi of the Human Gut Microbiome: Correlations with Diet and Bacterial Residents. *PLoS ONE* **2013**, *8*, e66019. [CrossRef]
7. Zhang, Y.-J.; Li, S.; Gan, R.-Y.; Zhou, T.; Xu, D.-P.; Li, H.-B. Impacts of Gut Bacteria on Human Health and Diseases. *Int. J. Mol. Sci.* **2015**, *16*, 7493–7519. [CrossRef]
8. Sam, Q.H.; Chang, M.W.; Chai, L.Y.A. The Fungal Mycobiome and Its Interaction with Gut Bacteria in the Host. *Int. J. Mol. Sci.* **2017**, *18*, 330. [CrossRef]
9. Arumugam, M.; Raes, J.; Pelletier, E.; Le Paslier, D.; Yamada, T.; Mende, D.R.; Fernandes, G.R.; Tap, J.; Bruls, T.; Batto, J.M.; et al. Enterotypes of the human gut microbiome. *Nature* **2011**, *473*, 174–180. [CrossRef] [PubMed]
10. Rinninella, E.; Raoul, P.; Cintoni, M.; Franceschi, F.; Miggiano, G.A.D.; Gasbarrini, A.; Mele, M.C. What is the Healthy Gut Microbiota Composition? A Changing Ecosystem across Age, Environment, Diet, and Diseases. *Microorganisms* **2019**, *7*, 14. [CrossRef] [PubMed]
11. Donaldson, G.P.; Lee, S.M.; Mazmanian, S.K. Gut biogeography of the bacterial microbiota. *Nat. Rev. Genet.* **2016**, *14*, 20–32. [CrossRef]
12. Thursby, E.; Juge, N. Introduction to the human gut microbiota. *Biochem. J.* **2017**, *474*, 1823–1836. [CrossRef]
13. Vasapolli, R.; Schütte, K.; Schulz, C.; Vital, M.; Schomburg, D.; Pieper, D.H.; Vilchez-Vargas, R.; Malfertheiner, P. Analysis of Transcriptionally Active Bacteria throughout the Gastrointestinal Tract of Healthy Individuals. *Gastroenterology* **2019**, *157*, 1081–1092.e3. [CrossRef]
14. Zoetendal, E.G.; Raes, J.; Bogert, B.V.D.; Arumugam, M.; Booijink, C.C.G.M.; Troost, F.J.; Bork, P.; Wels, M.; De Vos, W.M.; Kleerebezem, M. The human small intestinal microbiota is driven by rapid uptake and conversion of simple carbohydrates. *ISME J.* **2012**, *6*, 1415–1426. [CrossRef] [PubMed]
15. Kho, Z.Y.; Lal, S.K. The Human Gut Microbiome—A Potential Controller of Wellness and Disease. *Front. Microbiol.* **2018**, *9*, 1835. [CrossRef] [PubMed]
16. Cope, K.; Risby, T.; Diehl, A.M. Increased gastrointestinal ethanol production in obese mice: Implications for fatty liver disease pathogenesis. *Gastroenterology* **2000**, *119*, 1340–1347. [CrossRef]
17. Engstler, A.J.; Aumiller, T.; Degen, C.; Dürr, M.; Weiss, E.; Maier, I.B.; Schattenberg, J.M.; Jin, C.J.; Sellmann, C.; Bergheim, I. Insulin resistance alters hepatic ethanol metabolism: Studies in mice and children with non-alcoholic fatty liver disease. *Gut* **2016**, *65*, 1564–1571. [CrossRef]
18. Elshaghabee, F.M.F.; Bockelmann, W.; Emeske, D.; Vrese, M.E.; Ewalte, H.-G.; Eschrezenmeir, J.; Heller, K.J. Ethanol Production by Selected Intestinal Microorganisms and Lactic Acid Bacteria Growing under Different Nutritional Conditions. *Front. Microbiol.* **2016**, *7*, 47. [CrossRef] [PubMed]
19. Gentile, C.L.; Weir, T.L. The gut microbiota at the intersection of diet and human health. *Science* **2018**, *362*, 776–780. [CrossRef]
20. Tripathi, A.; Debelius, J.; Brenner, D.A.; Karin, M.; Loomba, R.; Schnabl, B.; Knight, R. The gut–liver axis and the intersection with the microbiome. *Nat. Rev. Gastroenterol. Hepatol.* **2018**, *15*, 397–411. [CrossRef]

21. Kang, D.J.; Hylemon, P.B.; Gillevet, P.M.; Sartor, R.B.; Betrapally, N.S.; Kakiyama, G.; Sikaroodi, M.; Takei, H.; Nittono, H.; Zhou, H.; et al. Gut microbial composition can differentially regulate bile acid synthesis in humanized mice. *Hepatol. Commun.* **2017**, *1*, 61–70. [CrossRef] [PubMed]
22. Bäumler, A.J.; Sperandio, V. Interactions between the microbiota and pathogenic bacteria in the gut. *Nat. Cell Biol.* **2016**, *535*, 85–93. [CrossRef] [PubMed]
23. Ubeda, C.; Djukovic, A.; Isaac, S. Roles of the intestinal microbiota in pathogen protection. *Clin. Transl. Immunol.* **2017**, *6*, e128. [CrossRef]
24. Petra, A.I.; Panagiotidou, S.; Hatziagelaki, E.; Stewart, J.M.; Conti, P.; Theoharides, T.C. Gut-Microbiota-Brain Axis and Its Effect on Neuropsychiatric Disorders With Suspected Immune Dysregulation. *Clin. Ther.* **2015**, *37*, 984–995. [CrossRef]
25. Hyland, N.P.; Cryan, J.F. Microbe-host interactions: Influence of the gut microbiota on the enteric nervous system. *Dev. Biol.* **2016**, *417*, 182–187. [CrossRef]
26. Rodríguez, J.M.; Murphy, K.; Stanton, C.; Ross, R.P.; Kober, O.I.; Juge, N.; Avershina, E.; Rudi, K.; Narbad, A.; Jenmalm, M.C.; et al. The composition of the gut microbiota throughout life, with an emphasis on early life. *Microb. Ecol. Health Dis.* **2015**, *26*, 26050. [CrossRef]
27. Wang, M.; Monaco, M.H.; Donovan, S.M. Impact of early gut microbiota on immune and metabolic development and function. *Semin. Fetal Neonatal Med.* **2016**, *21*, 380–387. [CrossRef]
28. Claesson, M.J.; Cusack, S.; O'Sullivan, O.; Greene-Diniz, R.; De Weerd, H.; Flannery, E.; Marchesi, J.R.; Falush, D.; Dinan, T.G.; Fitzgerald, G.F.; et al. Composition, variability, and temporal stability of the intestinal microbiota of the elderly. *Proc. Natl. Acad. Sci. USA* **2011**, *108* (Suppl. 1), 4586–4591. [CrossRef]
29. Ruiz-Ruiz, S.; Sanchez-Carrillo, S.; Ciordia, S.; Mena, M.C.; Méndez-García, C.; Rojo, D.; Bargiela, R.; Zubeldia-Varela, E.; Martínez-Martínez, M.; Barbas, C.; et al. Functional microbiome deficits associated with ageing: Chronological age threshold. *Aging Cell* **2019**, *19*, e13063. [CrossRef]
30. Urban, R.J.; Pyles, R.B.; Stewart, C.J.; Ajami, N.; Randolph, M.K.; Durham, W.J.; Danesi, C.P.; Dillon, E.L.; Summons, M.J.R.; Singh, C.K.; et al. Altered Fecal Microbiome Years after Traumatic Brain Injury. *J. Neurotrauma* **2020**, *37*, 1037–1051. [CrossRef] [PubMed]
31. Angoa-Pérez, M.; Zagorac, B.; Anneken, J.H.; Briggs, D.I.; Winters, A.D.; Greenberg, J.M.; Ahmad, M.; Theis, K.R.; Kuhn, D.M. Repetitive, mild traumatic brain injury results in a progressive white matter pathology, cognitive deterioration, and a transient gut microbiota dysbiosis. *Sci. Rep.* **2020**, *10*, 1–11. [CrossRef] [PubMed]
32. Cremonini, E.; Wang, Z.; Bettaieb, A.; Adamo, A.M.; Daveri, E.; Mills, D.A.; Kalanetra, K.M.; Haj, F.G.; Karakas, S.; Oteiza, P.I. (-)-Epicatechin protects the intestinal barrier from high fat diet-induced permeabilization: Implications for steatosis and insulin resistance. *Redox Biol.* **2018**, *14*, 588–599. [CrossRef]
33. Charlton, M.; Krishnan, A.; Viker, K.; Sanderson, S.; Cazanave, S.; McConico, A.; Masuoko, H.; Gores, G. Fast food diet mouse: Novel small animal model of NASH with ballooning, progressive fibrosis, and high physiological fidelity to the human condition. *Am. J. Physiol. Liver Physiol.* **2011**, *301*, G825–G834. [CrossRef] [PubMed]
34. Rahman, K.; Desai, C.; Iyer, S.S.; Thorn, N.E.; Kumar, P.; Liu, Y.; Smith, T.; Neish, A.S.; Li, H.; Tan, S.; et al. Loss of Junctional Adhesion Molecule A Promotes Severe Steatohepatitis in Mice on a Diet High in Saturated Fat, Fructose, and Cholesterol. *Gastroenterology* **2016**, *151*, 733–746.e12. [CrossRef] [PubMed]
35. Mastrocola, R.; Ferrocino, I.; Liberto, E.; Chiazza, F.; Cento, A.S.; Collotta, D.; Querio, G.; Nigro, D.; Bitonto, V.; Cutrin, J.C.; et al. Fructose liquid and solid formulations differently affect gut integrity, microbiota composition and related liver toxicity: A comparative in vivo study. *J. Nutr. Biochem.* **2018**, *55*, 185–199. [CrossRef] [PubMed]
36. D'Angelo, S.; Motti, M.L.; Meccariello, R. ω-3 and ω-6 Polyunsaturated Fatty Acids, Obesity and Cancer. *Nutrients* **2020**, *12*, 2751. [CrossRef]
37. Dabke, K.; Hendrick, G.; Devkota, S. The gut microbiome and metabolic syndrome. *J. Clin. Investig.* **2019**, *129*, 4050–4057. [CrossRef]
38. Gurung, M.; Li, Z.; You, H.; Rodrigues, R.; Jump, D.B.; Morgun, A.; Shulzhenko, N. Role of gut microbiota in type 2 diabetes pathophysiology. *EBioMedicine* **2020**, *51*, 102590. [CrossRef]
39. Murphy, E.A.; Velazquez, K.T.; Herbert, K.M. Influence of high-fat diet on gut microbiota: A driving force for chronic disease risk. *Curr. Opin. Clin. Nutr. Metab. Care* **2015**, *18*, 515–520. [CrossRef]
40. Araújo, J.R.; Tomas, J.; Brenner, C.; Sansonetti, P.J. Impact of high-fat diet on the intestinal microbiota and small intestinal physiology before and after the onset of obesity. *Biochimie* **2017**, *141*, 97–106. [CrossRef]
41. Daniel, H.; Gholami, A.M.; Berry, D.; Desmarchelier, C.; Hahne, H.; Loh, G.; Mondot, S.; Lepage, P.; Rothballer, M.; Walker, A.; et al. High-fat diet alters gut microbiota physiology in mice. *ISME J.* **2014**, *8*, 295–308. [CrossRef] [PubMed]
42. Singh, R.P.; Abu Halaka, D.; Hayouka, Z.; Tirosh, O. High-Fat Diet Induced Alteration of Mice Microbiota and the Functional Ability to Utilize Fructooligosaccharide for Ethanol Production. *Front. Cell. Infect. Microbiol.* **2020**, *10*, 376. [CrossRef]
43. Do, M.H.; Lee, E.; Oh, M.-J.; Kim, Y.; Park, H.-Y. High-Glucose or -Fructose Diet Cause Changes of the Gut Microbiota and Metabolic Disorders in Mice without Body Weight Change. *Nutrients* **2018**, *10*, 761. [CrossRef] [PubMed]
44. Bode, J.C.; Bode, C.; Heidelbach, R.; Dürr, H.K.; Martini, G.A. Jejunal microflora in patients with chronic alcohol abuse. *Hepatogastroenterology* **1984**, *31*, 30–34.
45. Baraona, E.; Julkunen, R.; Tannenbaum, L.; Lieber, C.S. Role of intestinal bacterial overgrowth in ethanol production and metabolism in rats. *Gastroenterology* **1986**, *90*, 103–110. [CrossRef]

46. Chen, Y.; Sun, H.; Bai, Y.; Zhi, F. Gut dysbiosis-derived exosomes trigger hepatic steatosis by transiting HMGB1 from intestinal to liver in mice. *Biochem. Biophys. Res. Commun.* **2019**, *509*, 767–772. [CrossRef] [PubMed]
47. Osna, N.A.; Donohue, T.M., Jr.; Kharbanda, K.K. Alcoholic Liver Disease: Pathogenesis and Current Management. *Alcohol Res. Curr. Rev.* **2017**, *38*, 147–161.
48. Wilson, D.F.; Matschinsky, F.M. Ethanol metabolism: The good, the bad, and the ugly. *Med. Hypotheses* **2020**, *140*, 109638. [CrossRef]
49. Lieber, C.S.; Leo, M.A. Metabolism of Ethanol and Some Associated Adverse Effects on the Liver and the Stomach. *Recent Dev. Alcohol.* **1998**, *14*, 7–40. [CrossRef] [PubMed]
50. Lange, L.G. Nonoxidative ethanol metabolism: Formation of fatty acid ethyl esters by cholesterol esterase. *Proc. Natl. Acad. Sci. USA* **1982**, *79*, 3954–3957. [CrossRef]
51. Heier, C.; Xie, H.; Zimmermann, R. Nonoxidative ethanol metabolism in humans—from biomarkers to bioactive lipids. *IUBMB Life* **2016**, *68*, 916–923. [CrossRef] [PubMed]
52. Holford, N.H.G. Clinical Pharmacokinetics of Ethanol. *Clin. Pharmacokinet.* **1987**, *13*, 273–292. [CrossRef]
53. CDC. Alcohol Use and Your Health. Centers of Disease Control and Prevention. 2020. Available online: https://www.cdc.gov/alcohol/fact-sheets/alcohol-use.htm#:~{}:text=Over%20time%2C%20excessive%20alcohol%20use,liver%20disease%2C%20and%20digestive%20problems.&text=Cancer%20of%20the%20breast%2C%20mouth,esophagus%2C%20liver%2C%20and%20colon (accessed on 17 December 2020).
54. NIAAA. Alcohol Facts and Statistics. The National Institute on Alcohol Abuse and Alcoholism (NIAAA). 2020. Available online: https://www.niaaa.nih.gov/sites/default/files/AlcoholFactsAndStats.pdf (accessed on 17 December 2020).
55. Zakhari, S. Alcohol Metabolism and Epigenetics Changes. *Alcohol Res. Curr. Rev.* **2013**, *35*, 6–16.
56. Lieber, C.S. Microsomal Ethanol-Oxidizing System (MEOS): The First 30 Years (1968–1998)—A Review. *Alcohol. Clin. Exp. Res.* **1999**, *23*, 991–1007. [CrossRef]
57. Song, B.J.; Gelboin, H.V.; Park, S.S.; Yang, C.S.; Gonzalez, F.J. Complementary DNA and protein sequences of ethanol-inducible rat and human cytochrome P-450s. Transcriptional and post-transcriptional regulation of the rat enzyme. *J. Biol. Chem.* **1986**, *261*, 16689–16697. [CrossRef]
58. Roberts, B.J.; Song, B.J.; Soh, Y.; Park, S.S.; Shoaf, S.E. Ethanol induces CYP2E1 by protein stabilization. Role of ubiquitin conjugation in the rapid degradation of CYP2E1. *J. Biol. Chem.* **1995**, *270*, 29632–29635. [CrossRef]
59. Song, B.J. Ethanol-inducible cytochrome P450 (CYP2E1): Biochemistry, molecular biology and clinical relevance: 1996 update. *Alcohol. Clin. Exp. Res.* **1996**, *20*, 138–146. [CrossRef]
60. Caro, A.A.; Cederbaum, A.I. Oxidative stress, toxicology, and pharmacology of CYP2E1. *Annu. Rev. Pharmacol. Toxicol.* **2004**, *44*, 27–42. [CrossRef]
61. Lu, Y.; Zhuge, J.; Wang, X.; Bai, J.; Cederbaum, A.I. Cytochrome P450 2E1 contributes to ethanol-induced fatty liver in mice. *Hepatology* **2008**, *47*, 1483–1494. [CrossRef] [PubMed]
62. Morgan, K.; French, S.W.; Morgan, T.R. Production of a cytochrome P450 2E1 transgenic mouse and initial evaluation of alcoholic liver damage. *Hepatology* **2002**, *36*, 122–134. [CrossRef]
63. Cederbaum, A.I. CYP2E1 potentiates toxicity in obesity and after chronic ethanol treatment. *Drug Metab. Drug Interact.* **2012**, *27*, 125–144. [CrossRef] [PubMed]
64. Kathirvel, E.; Morgan, K.; French, S.W.; Morgan, T.R. Overexpression of liver-specific cytochrome P4502E1 impairs hepatic insulin signaling in a transgenic mouse model of nonalcoholic fatty liver disease. *Eur. J. Gastroenterol. Hepatol.* **2009**, *21*, 973–983. [CrossRef] [PubMed]
65. Moon, K.-H.; Hood, B.L.; Kim, B.-J.; Hardwick, J.P.; Conrads, T.P.; Veenstra, T.D.; Song, B.J. Inactivation of oxidized andS-nitrosylated mitochondrial proteins in alcoholic fatty liver of rats. *Hepatology* **2006**, *44*, 1218–1230. [CrossRef] [PubMed]
66. Kim, B.-J.; Hood, B.L.; Aragon, R.A.; Hardwick, J.P.; Conrads, T.P.; Veenstra, T.D.; Song, B.J. Increased oxidation and degradation of cytosolic proteins in alcohol-exposed mouse liver and hepatoma cells. *Proteomics* **2006**, *6*, 1250–1260. [CrossRef]
67. Ceni, E.; Mello, T.; Galli, A. Pathogenesis of alcoholic liver disease: Role of oxidative metabolism. *World J. Gastroenterol.* **2014**, *20*, 17756–17772. [CrossRef] [PubMed]
68. Song, B.-J.; Akbar, M.; Abdelmegeed, M.A.; Byun, K.; Lee, B.; Yoon, S.K.; Hardwick, J.P. Mitochondrial dysfunction and tissue injury by alcohol, high fat, nonalcoholic substances and pathological conditions through post-translational protein modifications. *Redox Biol.* **2014**, *3*, 109–123. [CrossRef] [PubMed]
69. Galligan, J.J.; Smathers, R.L.; Shearn, C.T.; Fritz, K.S.; Backos, D.S.; Jiang, H.; Franklin, C.C.; Orlicky, D.J.; MacLean, K.N.; Petersen, D.R. Oxidative Stress and the ER Stress Response in a Murine Model for Early-Stage Alcoholic Liver Disease. *J. Toxicol.* **2012**, *2012*, 1–12. [CrossRef]
70. Song, B.-J.; Abdelmegeed, M.A.; Cho, Y.-E.; Akbar, M.; Rhim, J.S.; Song, M.-K.; Hardwick, J.P. Contributing Roles of CYP2E1 and Other Cytochrome P450 Isoforms in Alcohol-Related Tissue Injury and Carcinogenesis. *Adv. Exp. Med. Biol.* **2019**, *1164*, 73–87. [CrossRef]
71. Seitz, H.K. The role of cytochrome P4502E1 in the pathogenesis of alcoholic liver disease and carcinogenesis. *Chem. Interact.* **2020**, *316*, 108918. [CrossRef]
72. Szabo, G. Gut–Liver Axis in Alcoholic Liver Disease. *Gastroenterology* **2015**, *148*, 30–36. [CrossRef]
73. Gao, B.; Bataller, R. Liver Fibrosis in Alcoholic Liver Disease. *Semin. Liver Dis.* **2015**, *35*, 146–156. [CrossRef] [PubMed]

74. Seitz, H.K.; Bataller, R.; Cortez-Pinto, H.; Gao, B.; Gual, A.; Lackner, C.; Mathurin, P.; Mueller, S.; Szabo, G.; Tsukamoto, H. Alcoholic liver disease. *Nat. Rev. Dis. Primers* **2018**, *4*, 16. [CrossRef] [PubMed]
75. Diehl, A.M. Developmental Morphogens & Recovery from Alcoholic Liver Disease. *Results Probl. Cell Differ.* **2018**, *1032*, 145–151. [CrossRef]
76. Wang, H.; Mehal, W.; Nagy, L.E.; Rotman, Y. Immunological mechanisms and therapeutic targets of fatty liver diseases. *Cell. Mol. Immunol.* **2021**, *18*, 73–91. [CrossRef]
77. Hernandez-Gea, V.; Friedman, S.L. Pathogenesis of Liver Fibrosis. *Annu. Rev. Pathol. Mech. Dis.* **2011**, *6*, 425–456. [CrossRef]
78. Doorn, J.A.; Hurley, T.D.; Petersen, D.R. Inhibition of Human Mitochondrial Aldehyde Dehydrogenase by 4-Hydroxynon-2-enal and 4-Oxonon-2-enal†. *Chem. Res. Toxicol.* **2006**, *19*, 102–110. [CrossRef] [PubMed]
79. Song, B.-J.; Abdelmegeed, M.A.; Yoo, S.-H.; Kim, B.-J.; Jo, S.A.; Jo, I.; Moon, K.-H. Post-translational modifications of mitochondrial aldehyde dehydrogenase and biomedical implications. *J. Proteom.* **2011**, *74*, 2691–2702. [CrossRef]
80. Fritz, K.S.; Galligan, J.J.; Hirschey, M.D.; Verdin, E.; Petersen, D.R. Mitochondrial Acetylome Analysis in a Mouse Model of Alcohol-Induced Liver Injury Utilizing SIRT3 Knockout Mice. *J. Proteome Res.* **2012**, *11*, 1633–1643. [CrossRef]
81. Rungratanawanich, W.; Qu, Y.; Wang, X.; Essa, M.M.; Song, B.-J. Advanced glycation end products (AGEs) and other adducts in aging-related diseases and alcohol-mediated tissue injury. *Exp. Mol. Med.* **2021**, 1–21. [CrossRef]
82. Hartmann, P.; Seebauer, C.T.; Schnabl, B. Alcoholic Liver Disease: The Gut Microbiome and Liver Cross Talk. *Alcohol. Clin. Exp. Res.* **2015**, *39*, 763–775. [CrossRef]
83. Engen, P.A.; Green, S.J.; Voigt, R.M.; Forsyth, C.B.; Keshavarzian, A. The Gastrointestinal Microbiome: Alcohol Effects on the Composition of Intestinal Microbiota. *Alcohol Res.* **2015**, *37*, 223–236.
84. Dubinkina, V.B.; Tyakht, A.V.; Odintsova, V.Y.; Yarygin, K.S.; Kovarsky, B.A.; Pavlenko, A.V.; Ischenko, D.S.; Popenko, A.S.; Alexeev, D.G.; Taraskina, A.Y.; et al. Links of gut microbiota composition with alcohol dependence syndrome and alcoholic liver disease. *Microbiome* **2017**, *5*, 1–14. [CrossRef]
85. Bajaj, J.S. Alcohol, liver disease and the gut microbiota. *Nat. Rev. Gastroenterol. Hepatol.* **2019**, *16*, 235–246. [CrossRef]
86. Mutlu, E.; Keshavarzian, A.; Engen, P.; Forsyth, C.B.; Sikaroodi, M.; Gillevet, P. Intestinal Dysbiosis: A Possible Mechanism of Alcohol-Induced Endotoxemia and Alcoholic Steatohepatitis in Rats. *Alcohol. Clin. Exp. Res.* **2009**, *33*, 1836–1846. [CrossRef]
87. Mutlu, E.A.; Gillevet, P.M.; Rangwala, H.; Sikaroodi, M.; Naqvi, A.; Engen, P.A.; Kwasny, M.; Lau, C.K.; Keshavarzian, A. Colonic microbiome is altered in alcoholism. *Am. J. Physiol. Liver Physiol.* **2012**, *302*, G966–G978. [CrossRef] [PubMed]
88. Queipo-Ortuño, M.I.; Boto-Ordóñez, M.; Murri, M.; Gomez-Zumaquero, J.M.; Clemente-Postigo, M.; Estruch, R.; Diaz, F.C.; Andrés-Lacueva, C.; Tinahones, F.J. Influence of red wine polyphenols and ethanol on the gut microbiota ecology and biochemical biomarkers. *Am. J. Clin. Nutr.* **2012**, *95*, 1323–1334. [CrossRef] [PubMed]
89. Xiao, J.; Zhang, R.; Zhou, Q.; Liu, L.; Huang, F.; Deng, Y.; Ma, Y.; Wei, Z.; Tang, X.; Zhang, M. Lychee (*Litchi chinensis* Sonn.) Pulp Phenolic Extract Provides Protection against Alcoholic Liver Injury in Mice by Alleviating Intestinal Microbiota Dysbiosis, Intestinal Barrier Dysfunction, and Liver Inflammation. *J. Agric. Food Chem.* **2017**, *65*, 9675–9684. [CrossRef]
90. Chen, Y.-L.; Shirakawa, H.; Lu, N.-S.; Peng, H.-C.; Xiao, Q.; Yang, S.-C. Impacts of fish oil on the gut microbiota of rats with alcoholic liver damage. *J. Nutr. Biochem.* **2020**, *86*, 108491. [CrossRef] [PubMed]
91. Litwinowicz, K.; Choroszy, M.; Waszczuk, E. Changes in the composition of the human intestinal microbiome in alcohol use disorder: A systematic review. *Am. J. Drug Alcohol Abus.* **2019**, *46*, 4–12. [CrossRef] [PubMed]
92. Tsuruya, A.; Kuwahara, A.; Saito, Y.; Yamaguchi, H.; Tsubo, T.; Suga, S.; Inai, M.; Aoki, Y.; Takahashi, S.; Tsutsumi, E.; et al. Ecophysiological consequences of alcoholism on human gut microbiota: Implications for ethanol-related pathogenesis of colon cancer. *Sci. Rep.* **2016**, *6*, 27923. [CrossRef]
93. Bjørkhaug, S.T.; Aanes, H.; Neupane, S.P.; Bramness, J.G.; Malvik, S.; Henriksen, C.; Skar, V.; Medhus, A.W.; Valeur, J. Characterization of gut microbiota composition and functions in patients with chronic alcohol overconsumption. *Gut Microbes* **2019**, *10*, 663–675. [CrossRef] [PubMed]
94. Ciocan, D.; Voican, C.S.; Wrzosek, L.; Hugot, C.; Rainteau, D.; Humbert, L.; Cassard, A.-M.; Perlemuter, G. Bile acid homeostasis and intestinal dysbiosis in alcoholic hepatitis. *Aliment. Pharmacol. Ther.* **2018**, *48*, 961–974. [CrossRef] [PubMed]
95. Ghosh, G.; Jesudian, A.B. Small Intestinal Bacterial Overgrowth in Patients with Cirrhosis. *J. Clin. Exp. Hepatol.* **2019**, *9*, 257–267. [CrossRef] [PubMed]
96. Gurwara, S.; Dai, A.; Ajami, N.J.; Graham, D.Y.; White, D.L.; Chen, L.; Jang, A.; Chen, E.; El-Serag, H.B.; Petrosino, J.F.; et al. Alcohol use alters the colonic mucosa–associated gut microbiota in humans. *Nutr. Res.* **2020**, *83*, 119–128. [CrossRef]
97. Bull-Otterson, L.; Feng, W.; Kirpich, I.; Wang, Y.; Qin, X.; Liu, Y.; Gobejishvili, L.; Joshi-Barve, S.; Ayvaz, T.; Petrosino, J.; et al. Metagenomic Analyses of Alcohol Induced Pathogenic Alterations in the Intestinal Microbiome and the Effect of Lactobacillus rhamnosus GG Treatment. *PLoS ONE* **2013**, *8*, e53028. [CrossRef]
98. Keshavarzian, A.; Farhadi, A.; Forsyth, C.B.; Rangan, J.; Jakate, S.; Shaikh, M.; Banan, A.; Fields, J.Z. Evidence that chronic alcohol exposure promotes intestinal oxidative stress, intestinal hyperpermeability and endotoxemia prior to development of alcoholic steatohepatitis in rats. *J. Hepatol.* **2009**, *50*, 538–547. [CrossRef] [PubMed]
99. Nyfors, S.; Salaspuro, V.; Jousimies-Somer, H.; Siitonen, A.; Salaspuro, M.; Heine, R. Ethanol oxidation and acetaldehyde production in vitro by human intestinal strains of Escherichia coli under aerobic, microaerobic, and anaerobic conditions. *Scand. J. Gastroenterol.* **1999**, *34*, 967–973.

100. Yan, A.W.; Fouts, D.E.; Brandl, J.; Stärkel, P.; Torralba, M.; Schott, E.; Tsukamoto, H.; Nelson, K.E.; Brenner, D.A.; Schnabl, B. Enteric dysbiosis associated with a mouse model of alcoholic liver disease. *Hepatology* **2010**, *53*, 96–105. [CrossRef]
101. Fan, Y.; Ya-E, Z.; Ji-Dong, W.; Yu-Fan, L.; Ying, Z.; Ya-Lun, S.; Meng-Yu, M.; Rui-Ling, Z. Comparison of Microbial Diversity and Composition in Jejunum and Colon of the Alcohol-dependent Rats. *J. Microbiol. Biotechnol.* **2018**, *28*, 1883–1895. [CrossRef]
102. Addolorato, G.; Ponziani, F.R.; Dionisi, T.; Mosoni, C.; Vassallo, G.A.; Sestito, L.; Petito, V.; Picca, A.; Marzetti, E.; Tarli, C.; et al. Gut microbiota compositional and functional fingerprint in patients with alcohol use disorder and alcohol-associated liver disease. *Liver Int.* **2020**, *40*, 878–888. [CrossRef]
103. Harnisch, J.P.; Tronca, E.; Nolan, C.M.; Turck, M.; Holmes, K.K. Diphtheria among Alcoholic Urban Adults. *Ann. Intern. Med.* **1989**, *111*, 71–82. [CrossRef]
104. Cericco, M.; Iglicki, F.; Guillaumont, M.P.; Schmitt, J.L.; Dupas, J.L.; Capron, J.P. [Corynebacterium xerosis endocarditis associated with alcoholic cirrhosis]. *Gastroentérologie Clin. Biol.* **1996**, *20*, 514.
105. Tsuruya, A.; Kuwahara, A.; Saito, Y.; Yamaguchi, H.; Tenma, N.; Inai, M.; Takahashi, S.; Tsutsumi, E.; Suwa, Y.; Totsuka, Y.; et al. Major Anaerobic Bacteria Responsible for the Production of Carcinogenic Acetaldehyde from Ethanol in the Colon and Rectum. *Alcohol Alcohol.* **2016**, *51*, 395–401. [CrossRef] [PubMed]
106. Grander, C.; Adolph, T.E.; Wieser, V.; Lowe, P.; Wrzosek, L.; Gyongyosi, B.; Ward, D.V.; Grabherr, F.; Gerner, R.R.; Pfister, A.; et al. Recovery of ethanol-induced *Akkermansia muciniphila* depletion ameliorates alcoholic liver disease. *Gut* **2017**, *67*, 891–901. [CrossRef]
107. Halsted, C.H.; Robles, E.A.; Mezey, E. Distribution of ethanol in the human gastrointestinal tract. *Am. J. Clin. Nutr.* **1973**, *26*, 831–834. [CrossRef] [PubMed]
108. Elamin, E.E.; Masclee, A.; Dekker, J.; Jonkers, D. Ethanol metabolism and its effects on the intestinal epithelial barrier. *Nutr. Rev.* **2013**, *71*, 483–499. [CrossRef] [PubMed]
109. Cederbaum, A.I. Alcohol metabolism. *Clin. Liver Dis.* **2012**, *16*, 667–685. [CrossRef] [PubMed]
110. Zakhari, S. Overview: How Is Alcohol Metabolized by the Body? *Alcohol Res. Health* **2006**, *29*, 245–254. [PubMed]
111. Bode, C.; Bode, J.C. Alcohol's role in gastrointestinal tract disorders. *Alcohol Health Res. World* **1997**, *21*, 76–83.
112. Okumura, R.; Takeda, K. Roles of intestinal epithelial cells in the maintenance of gut homeostasis. *Exp. Mol. Med.* **2017**, *49*, e338. [CrossRef]
113. Thoo, L.; Noti, M.; Krebs, P. Keep calm: The intestinal barrier at the interface of peace and war. *Cell Death Dis.* **2019**, *10*, 1–13. [CrossRef]
114. Hammer, A.M.; Khan, O.M.; Morris, N.L.; Li, X.; Movtchan, N.V.; Cannon, A.R.; Choudhry, M.A. The Effects of Alcohol Intoxication and Burn Injury on the Expression of Claudins and Mucins in the Small and Large Intestines. *Shock* **2016**, *45*, 73–81. [CrossRef]
115. Kaur, J. Chronic ethanol feeding affects intestinal mucus lipid composition and glycosylation in rats. *Ann. Nutr. Metab.* **2002**, *46*, 38–44. [CrossRef]
116. Slomiany, A.; Piotrowski, E.; Piotrowski, J.; Slomiany, B.L. Impact of ethanol on innate protection of gastric mucosal epithelial surfaces and the risk of injury. *J. Physiol. Pharmacol.* **2000**, *51*, 433–447. [PubMed]
117. Qin, X.; Deitch, E.A. Dissolution of lipids from mucus: A possible mechanism for prompt disruption of gut barrier function by alcohol. *Toxicol. Lett.* **2015**, *232*, 356–362. [CrossRef]
118. Chiang, C.-P.; Wu, C.-W.; Lee, S.-P.; Ho, J.-L.; Lee, S.-L.; Nieh, S.; Yin, S.-J. Expression Pattern, Ethanol-Metabolizing Activities, and Cellular Localization of Alcohol and Aldehyde Dehydrogenases in Human Small Intestine. *Alcohol. Clin. Exp. Res.* **2012**, *36*, 2047–2058. [CrossRef] [PubMed]
119. Roberts, B.J.; Shoaf, S.; Jeong, K.; Song, B.-J. Induction of CYP2E1 in Liver, Kidney, Brain and Intestine During Chronic Ethanol Administration and Withdrawal: Evidence That CYP2E1 Possesses a Rapid Phase Half-Life of 6 Hours or Less. *Biochem. Biophys. Res. Commun.* **1994**, *205*, 1064–1071. [CrossRef] [PubMed]
120. Cho, Y.-E.; Song, B.-J. Pomegranate prevents binge alcohol-induced gut leakiness and hepatic inflammation by suppressing oxidative and nitrative stress. *Redox Biol.* **2018**, *18*, 266–278. [CrossRef]
121. Cho, Y.-E.; Yu, L.-R.; Abdelmegeed, M.A.; Yoo, S.-H.; Song, B.-J. Apoptosis of enterocytes and nitration of junctional complex proteins promote alcohol-induced gut leakiness and liver injury. *J. Hepatol.* **2018**, *69*, 142–153. [CrossRef]
122. Salaspuro, M. Bacteriocolonic Pathway for Ethanol Oxidation: Characteristics and Implications. *Ann. Med.* **1996**, *28*, 195–200. [CrossRef]
123. Koivisto, T.; Salaspuro, M. Aldehyde Dehydrogenases of the Rat Colon: Comparison with Other Tissues of the Alimentary Tract and the Liver. *Alcohol. Clin. Exp. Res.* **1996**, *20*, 551–555. [CrossRef]
124. Nosova, T.; Jousimies-Somer, H.; Jokelainen, K.; Heine, R.; Salaspuro, M. Acetaldehyde production and metabolism by human indigenous and probiotic lactobacillus and bifidobacterium strains. *Alcohol Alcohol.* **2000**, *35*, 561–568. [CrossRef]
125. Bagchi, D.; Carryl, O.R.; Tran, M.X.; Krohn, R.L.; Bagchi, D.J.; Garg, A.; Bagchi, M.; Mitra, S.; Stohs, S.J. Stress, diet and alcohol-induced oxidative gastrointestinal mucosal injury in rats and protection by bismuth subsalicylate. *J. Appl. Toxicol.* **1998**, *18*, 3–13. [CrossRef]
126. Odenwald, M.A.; Turner, J.R. The intestinal epithelial barrier: A therapeutic target? *Nat. Rev. Gastroenterol. Hepatol.* **2017**, *14*, 9–21. [CrossRef] [PubMed]

127. Bishehsari, F.; Magno, E.; Swanson, G.; Desai, V.; Voigt, R.M.; Forsyth, C.B.; Keshavarzian, A. Alcohol and Gut-Derived Inflammation. *Alcohol Res.* **2017**, *38*, 163–171.
128. Asai, K.; Buurman, W.A.; Reutelingsperger, C.P.M.; Schutte, B.; Kaminishi, M. Low concentrations of ethanol induce apoptosis in human intestinal cells. *Scand. J. Gastroenterol.* **2003**, *38*, 1154–1161. [CrossRef] [PubMed]
129. Klingensmith, N.J.; Fay, K.T.; Lyons, J.D.; Chen, C.-W.; Otani, S.; Liang, Z.; Chihade, D.B.; Burd, E.M.; Ford, M.L.; Coopersmith, C.M. Chronic Alcohol Ingestion Worsens Survival and Alters Gut Epithelial Apoptosis and CD8+ T Cell Function After Pseudomonas Aeruginosa Pneumonia-Induced Sepsis. *Shock* **2019**, *51*, 453–463. [CrossRef]
130. Lambert, J.C.; Zhou, Z.; Wang, L.; Song, Z.; McClain, C.J.; Kang, Y.J. Prevention of Alterations in Intestinal Permeability Is Involved in Zinc Inhibition of Acute Ethanol-Induced Liver Damage in Mice. *J. Pharmacol. Exp. Ther.* **2003**, *305*, 880–886. [CrossRef]
131. Lambert, J.C.; Zhou, Z.; Wang, L.; Song, Z.; McClain, C.J.; Kang, Y.J. Preservation of Intestinal Structural Integrity by Zinc Is Independent of Metallothionein in Alcohol-Intoxicated Mice. *Am. J. Pathol.* **2004**, *164*, 1959–1966. [CrossRef]
132. Abdelmegeed, M.A.; Banerjee, A.; Jang, S.; Yoo, S.-H.; Yun, J.-W.; Gonzalez, F.J.; Keshavarzian, A.; Song, B.-J. CYP2E1 potentiates binge alcohol-induced gut leakiness, steatohepatitis, and apoptosis. *Free. Radic. Biol. Med.* **2013**, *65*, 1238–1245. [CrossRef] [PubMed]
133. Cresci, G.A. The Gut Microbiome: A New Frontier for Alcohol Investigation. *Alcohol. Clin. Exp. Res.* **2015**, *39*, 947–949. [CrossRef]
134. Gyongyosi, B.; Cho, Y.; Lowe, P.; Calenda, C.D.; Iracheta-Vellve, A.; Satishchandran, A.; Ambade, A.; Szabo, G. Alcohol-induced IL-17A production in Paneth cells amplifies endoplasmic reticulum stress, apoptosis, and inflammasome-IL-18 activation in the proximal small intestine in mice. *Mucosal Immunol.* **2019**, *12*, 930–944. [CrossRef]
135. Turner, J.R. Intestinal mucosal barrier function in health and disease. *Nat. Rev. Immunol.* **2009**, *9*, 799–809. [CrossRef]
136. Neunlist, M.; Van Landeghem, L.; Mahé, M.M.; Derkinderen, P.; Varannes, S.B.D.; Rolli-Derkinderen, M. The digestive neuronal-glial-epithelial unit: A new actor in gut health and disease. *Nat. Rev. Gastroenterol. Hepatol.* **2012**, *10*, 90–100. [CrossRef]
137. Wang, Y.; Tong, J.; Chang, B.; Wang, B.; Zhang, D.; Wang, B. Effects of alcohol on intestinal epithelial barrier permeability and expression of tight junction-associated proteins. *Mol. Med. Rep.* **2014**, *9*, 2352–2356. [CrossRef] [PubMed]
138. Elamin, E.; Jonkers, D.; Juuti-Uusitalo, K.; Van Ijzendoorn, S.; Troost, F.; Duimel, H.; Broers, J.; Verheyen, F.; Dekker, J.; Masclee, A. Effects of Ethanol and Acetaldehyde on Tight Junction Integrity: In Vitro Study in a Three Dimensional Intestinal Epithelial Cell Culture Model. *PLoS ONE* **2012**, *7*, e35008. [CrossRef]
139. Mir, H.; Meena, A.S.; Chaudhry, K.K.; Shukla, P.K.; Gangwar, R.; Manda, B.; Padala, M.K.; Shen, L.; Turner, J.R.; Dietrich, P.; et al. Occludin deficiency promotes ethanol-induced disruption of colonic epithelial junctions, gut barrier dysfunction and liver damage in mice. *Biochim. Biophys. Acta (BBA)—Gen. Subj.* **2016**, *1860*, 765–774. [CrossRef] [PubMed]
140. Atkinson, K.J.; Rao, R.K. Role of protein tyrosine phosphorylation in acetaldehyde-induced disruption of epithelial tight junctions. *Am. J. Physiol. Liver Physiol.* **2001**, *280*, G1280–G1288. [CrossRef] [PubMed]
141. Rao, R.K.; Basuroy, S.; Rao, V.U.; Karnaky, K.J., Jr.; Gupta, A. Tyrosine phosphorylation and dissociation of occludin-ZO-1 and E-cadherin-beta-catenin complexes from the cytoskeleton by oxidative stress. *Biochem. J.* **2002**, *368*(Pt. 2), 471–481. [CrossRef]
142. Kale, G.; Naren, A.P.; Sheth, P.; Rao, R.K. Tyrosine phosphorylation of occludin attenuates its interactions with ZO-1, ZO-2, and ZO-3. *Biochem. Biophys. Res. Commun.* **2003**, *302*, 324–329. [CrossRef]
143. Samak, G.; Aggarwal, S.; Rao, R.K. ERK is involved in EGF-mediated protection of tight junctions, but not adherens junctions, in acetaldehyde-treated Caco-2 cell monolayers. *Am. J. Physiol. Liver Physiol.* **2011**, *301*, G50–G59. [CrossRef] [PubMed]
144. Zhou, Z.; Ye, T.J.; DeCaro, E.; Buehler, B.; Stahl, Z.; Bonavita, G.; Daniels, M.; You, M. Intestinal SIRT1 Deficiency Protects Mice from Ethanol-Induced Liver Injury by Mitigating Ferroptosis. *Am. J. Pathol.* **2020**, *190*, 82–92. [CrossRef] [PubMed]
145. Dörfel, M.J.; Huber, O. Modulation of Tight Junction Structure and Function by Kinases and Phosphatases Targeting Occludin. *J. Biomed. Biotechnol.* **2012**, *2012*, 1–14. [CrossRef] [PubMed]
146. Reiche, J.; Huber, O. Post-translational modifications of tight junction transmembrane proteins and their direct effect on barrier function. *Biochim. Biophys. Acta (BBA)—Biomembr.* **2020**, *1862*, 183330. [CrossRef] [PubMed]
147. Sheth, P.; Seth, A.; Atkinson, K.J.; Gheyi, T.; Kale, G.; Giorgianni, F.; Desiderio, D.M.; Li, C.; Naren, A.; Rao, R. Acetaldehyde dissociates the PTP1B-E-cadherin-beta-catenin complex in Caco-2 cell monolayers by a phosphorylation-dependent mechanism. *Biochem. J.* **2007**, *402*, 291–300. [CrossRef] [PubMed]
148. Basuroy, S.; Sheth, P.; Mansbach, C.M.; Rao, R.K. Acetaldehyde disrupts tight junctions and adherens junctions in human colonic mucosa: Protection by EGF and l-glutamine. *Am. J. Physiol. Liver Physiol.* **2005**, *289*, G367–G375. [CrossRef] [PubMed]
149. Dunagan, M.; Chaudhry, K.; Samak, G.; Rao, R.K. Acetaldehyde disrupts tight junctions in Caco-2 cell monolayers by a protein phosphatase 2A-dependent mechanism. *Am. J. Physiol. Liver Physiol.* **2012**, *303*, G1356–G1364. [CrossRef]
150. Haorah, J.; Heilman, D.; Knipe, B.; Chrastil, J.; Leibhart, J.; Ghorpade, A.; Miller, D.W.; Persidsky, Y. Ethanol-Induced Activation of Myosin Light Chain Kinase Leads to Dysfunction of Tight Junctions and Blood-Brain Barrier Compromise. *Alcohol. Clin. Exp. Res.* **2005**, *29*, 999–1009. [CrossRef]
151. Elamin, E.E.; Masclee, A.A.; Jonkers, D.M. Chapter 14—Effects of Acetaldehyde on Intestinal Barrier Function. In *Molecular Aspects of Alcohol and Nutrition*; Patel, V.B., Ed.; Academic Press: San Diego, CA, USA, 2016; pp. 171–186.
152. Chaudhry, K.K.; Chaudhry, K.K.; Samak, G.; Shukla, P.K.; Mir, H.; Gangwar, R.; Manda, B.; Isse, T.; Kawamoto, T.; Salaspuro, M.; et al. ALDH2 Deficiency Promotes Ethanol-Induced Gut Barrier Dysfunction and Fatty Liver in Mice. *Alcohol. Clin. Exp. Res.* **2015**, *39*, 1465–1475. [CrossRef]

153. Elamin, E.; Masclee, A.; Troost, F.; Dekker, J.; Jonkers, D. Cytotoxicity and metabolic stress induced by acetaldehyde in human intestinal LS174T goblet-like cells. *Am. J. Physiol. Liver Physiol.* **2014**, *307*, G286–G294. [CrossRef]
154. Tang, Y.; Forsyth, C.B.; Farhadi, A.; Rangan, J.; Jakate, S.; Shaikh, M.; Banan, A.; Fields, J.Z.; Keshavarzian, A. Nitric Oxide-Mediated Intestinal Injury Is Required for Alcohol-Induced Gut Leakiness and Liver Damage. *Alcohol. Clin. Exp. Res.* **2009**, *33*, 1220–1230. [CrossRef]
155. Forsyth, C.B.; Voigt, R.M.; Keshavarzian, A. Intestinal CYP2E1: A mediator of alcohol-induced gut leakiness. *Redox Biol.* **2014**, *3*, 40–46. [CrossRef] [PubMed]
156. Tang, Y.; Zhang, L.; Forsyth, C.B.; Shaikh, M.; Song, S.; Keshavarzian, A. The Role of miR-212 and iNOS in Alcohol-Induced Intestinal Barrier Dysfunction and Steatohepatitis. *Alcohol. Clin. Exp. Res.* **2015**, *39*, 1632–1641. [CrossRef]
157. Forsyth, C.B.; Tang, Y.; Shaikh, M.; Zhang, L.; Keshavarzian, A. Role of snail activation in alcohol-induced iNOS-mediated disruption of intestinal epithelial cell permeability. *Alcohol. Clin. Exp. Res.* **2011**, *35*, 1635–1643. [CrossRef] [PubMed]
158. Leung, T.-M.; Nieto, N. CYP2E1 and oxidant stress in alcoholic and non-alcoholic fatty liver disease. *J. Hepatol.* **2013**, *58*, 395–398. [CrossRef]
159. Choi, Y.; Abdelmegeed, M.A.; Song, B.-J. Preventive effects of indole-3-carbinol against alcohol-induced liver injury in mice via antioxidant, anti-inflammatory, and anti-apoptotic mechanisms: Role of gut-liver-adipose tissue axis. *J. Nutr. Biochem.* **2018**, *55*, 12–25. [CrossRef] [PubMed]
160. Yu, T.; Wan, P.; Zhu, X.-D.; Ren, Y.-P.; Wang, C.; Yan, R.-W.; Guo, Y.; Bai, A.-P. Inhibition of NADPH oxidase activities ameliorates DSS-induced colitis. *Biochem Pharmacol* **2018**, *158*, 126–133. [CrossRef] [PubMed]
161. Li, X.; Wei, X.; Sun, Y.; Du, J.; Li, X.; Xun, Z.; Li, Y.C. High-fat diet promotes experimental colitis by inducing oxidative stress in the colon. *Am. J. Physiol. Gastrointest. Liver Physiol.* **2019**, *317*, G453–G462. [CrossRef] [PubMed]
162. Samak, G.; Gangwar, R.; Meena, A.S.; Rao, R.G.; Shukla, P.K.; Manda, B.; Narayanan, D.; Jaggar, J.H.; Rao, R. Calcium Channels and Oxidative Stress Mediate a Synergistic Disruption of Tight Junctions by Ethanol and Acetaldehyde in Caco-2 Cell Monolayers. *Sci. Rep.* **2016**, *6*, 38899. [CrossRef] [PubMed]
163. Elamin, E.; Masclee, A.; Troost, F.; Pieters, H.-J.; Keszthelyi, D.; Aleksa, K.; Dekker, J.; Jonkers, D. Ethanol Impairs Intestinal Barrier Function in Humans through Mitogen Activated Protein Kinase Signaling: A Combined In Vivo and In Vitro Approach. *PLoS ONE* **2014**, *9*, e107421. [CrossRef] [PubMed]
164. Cunningham, K.E.; Turner, J.R. Myosin light chain kinase: Pulling the strings of epithelial tight junction function. *Ann. N. Y. Acad. Sci.* **2012**, *1258*, 34–42. [CrossRef]
165. Zhong, W.; Zhao, Y.; McClain, C.J.; Kang, Y.J.; Zhou, Z. Inactivation of hepatocyte nuclear factor-4α mediates alcohol-induced downregulation of intestinal tight junction proteins. *Am. J. Physiol. Gastrointest. Liver Physiol.* **2010**, *299*, G643–G651. [CrossRef]
166. Elamin, E.; Masclee, A.; Juuti-Uusitalo, K.; Van Ijzendoorn, S.; Troost, F.; Pieters, H.-J.; Dekker, J.; Jonkers, D. Fatty Acid Ethyl Esters Induce Intestinal Epithelial Barrier Dysfunction via a Reactive Oxygen Species-Dependent Mechanism in a Three-Dimensional Cell Culture Model. *PLoS ONE* **2013**, *8*, e58561. [CrossRef]
167. Stärkel, P.; Schnabl, B. Bidirectional Communication between Liver and Gut during Alcoholic Liver Disease. *Semin. Liver Dis.* **2016**, *36*, 331–339. [CrossRef]
168. Besten, G.D.; van Eunen, K.; Groen, A.K.; Venema, K.; Reijngoud, D.-J.; Bakker, B.M. The role of short-chain fatty acids in the interplay between diet, gut microbiota, and host energy metabolism. *J. Lipid Res.* **2013**, *54*, 2325–2340. [CrossRef]
169. Hofmann, A.F. The Continuing Importance of Bile Acids in Liver and Intestinal Disease. *Arch. Intern. Med.* **1999**, *159*, 2647–2658. [CrossRef]
170. Urdaneta, V.; Casadesús, J. Interactions between Bacteria and Bile Salts in the Gastrointestinal and Hepatobiliary Tracts. *Front. Med.* **2017**, *4*, 163. [CrossRef] [PubMed]
171. Korsten, M.A.; Matsuzaki, S.; Feinman, L.; Lieber, C.S. High Blood Acetaldehyde Levels after Ethanol Administration. *N. Engl. J. Med.* **1975**, *292*, 386–389. [CrossRef] [PubMed]
172. Barr, T.; Lewis, S.A.; Sureshchandra, S.; Doratt, B.; Grant, K.A.; Messaoudi, I. Chronic ethanol consumption alters lamina propria leukocyte response to stimulation in a region-dependent manner. *FASEB J.* **2019**, *33*, 7767–7777. [CrossRef]
173. Ling, X.; Linglong, P.; Weixia, D.; Hong, W. Protective Effects of Bifidobacterium on Intestinal Barrier Function in LPS-Induced Enterocyte Barrier Injury of Caco-2 Monolayers and in a Rat NEC Model. *PLoS ONE* **2016**, *11*, e0161635. [CrossRef] [PubMed]
174. Akiba, Y.; Maruta, K.; Takajo, T.; Narimatsu, K.; Said, H.; Kato, I.; Kuwahara, A.; Kaunitz, J.D. Lipopolysaccharides transport during fat absorption in rodent small intestine. *Am. J. Physiol. Liver Physiol.* **2020**, *318*, G1070–G1087. [CrossRef]
175. Parlesak, A.; Schäfer, C.; Schütz, T.; Bode, J. Increased intestinal permeability to macromolecules and endotoxemia in patients with chronic alcohol abuse in different stages of alcohol-induced liver disease. *J. Hepatol.* **2000**, *32*, 742–747. [CrossRef]
176. Schäfer, C.; Parlesak, A.; Schütt, C.; Bode, J.C.; Bode, C. Concentrations of lipopolysaccharide-binding protein, bactericidal/permeability-increasing protein, soluble cd14 and plasma lipids in relation to endotoxaemia in patients with alcoholic liver disease. *Alcohol Alcohol.* **2002**, *37*, 81–86. [CrossRef]
177. Liangpunsakul, S.; Toh, E.; Ross, R.A.; Heathers, L.E.; Chandler, K.; Oshodi, A.; McGee, B.; Modlik, E.; Linton, T.; Mangiacarne, D.; et al. Quantity of alcohol drinking positively correlates with serum levels of endotoxin and markers of monocyte activation. *Sci. Rep.* **2017**, *7*, 4462. [CrossRef]
178. Gao, B.; Emami, A.; Nath, S.; Schnabl, B. Microbial Products and Metabolites Contributing to Alcohol-Related Liver Disease. *Mol. Nutr. Food Res.* **2020**, e2000023. [CrossRef]

179. Fleming, S.; Toratani, S.; Shea-Donohue, T.; Kashiwabara, Y.; Vogel, S.N.; Metcalf, E.S. Pro- and anti-inflammatory gene expression in the murine small intestine and liver after chronic exposure to alcohol. *Alcohol Clin. Exp. Res.* **2001**, *25*, 579–589. [CrossRef]
180. Veazey, R.S.; Amedee, A.; Wang, X.; Kaack, M.B.; Porretta, C.; Dufour, J.; Welsh, D.; Happel, K.; Pahar, B.; Molina, P.E.; et al. Chronic Binge Alcohol Administration Increases Intestinal T-Cell Proliferation and Turnover in Rhesus Macaques. *Alcohol. Clin. Exp. Res.* **2015**, *39*, 1373–1379. [CrossRef]
181. Rao, R. Endotoxemia and gut barrier dysfunction in alcoholic liver disease. *Hepatology* **2009**, *50*, 638–644. [CrossRef] [PubMed]
182. Mencin, A.; Kluwe, J.; Schwabe, R.F. Toll-like receptors as targets in chronic liver diseases. *Gut* **2009**, *58*, 704–720. [CrossRef] [PubMed]
183. Gao, B.; Ahmad, M.F.; Nagy, L.E.; Tsukamoto, H. Inflammatory pathways in alcoholic steatohepatitis. *J. Hepatol.* **2019**, *70*, 249–259. [CrossRef] [PubMed]
184. Soares, J.-B.; Pimentel-Nunes, P.; Roncon-Albuquerque, R.; Leitemoreira, A.F. The role of lipopolysaccharide/toll-like receptor 4 signaling in chronic liver diseases. *Hepatol. Int.* **2010**, *4*, 659–672. [CrossRef] [PubMed]
185. An, L.; Wang, X.; Cederbaum, A.I. Cytokines in alcoholic liver disease. *Arch. Toxicol.* **2012**, *86*, 1337–1348. [CrossRef]
186. Kawaratani, H.; Tsujimoto, T.; Douhara, A.; Takaya, H.; Moriya, K.; Namisaki, T.; Noguchi, R.; Yoshiji, H.; Fujimoto, M.; Fukui, H. The Effect of Inflammatory Cytokines in Alcoholic Liver Disease. *Mediat. Inflamm.* **2013**, *2013*, 1–10. [CrossRef] [PubMed]
187. Kong, X.; Yang, Y.; Ren, L.; Shao, T.; Li, F.; Zhao, C.; Liu, L.; Zhang, H.; McClain, C.J.; Feng, W. Activation of autophagy attenuates EtOH-LPS-induced hepatic steatosis and injury through MD2 associated TLR4 signaling. *Sci. Rep.* **2017**, *7*, 9292. [CrossRef] [PubMed]
188. Ceccarelli, S.; Nobili, V.; Alisi, A. Toll-like receptor-mediated signaling cascade as a regulator of the inflammation network during alcoholic liver disease. *World J. Gastroenterol.* **2014**, *20*, 16443–16451. [CrossRef]
189. Gandhi, C.R. Pro- and Anti-fibrogenic Functions of Gram-Negative Bacterial Lipopolysaccharide in the Liver. *Front. Med.* **2020**, *7*, 130. [CrossRef]
190. Cederbaum, A.I.; Yang, L.; Wang, X.; Wu, D. CYP2E1 Sensitizes the Liver to LPS- and TNF α-Induced Toxicity via Elevated Oxidative and Nitrosative Stress and Activation of ASK-1 and JNK Mitogen-Activated Kinases. *Int. J. Hepatol.* **2012**, *2012*, 582790. [CrossRef]
191. Cederbaum, A.I.; Lu, Y.; Wang, X.; Wu, D. Synergistic Toxic Interactions Between CYP2E1, LPS/TNF? and JNK/p38 MAP Kinase and Their Implications in Alcohol-Induced Liver Injury. *Results Probl. Cell Differ.* **2014**, *815*, 145–172. [CrossRef]
192. Thakur, V.; Pritchard, M.T.; McMullen, M.R.; Wang, Q.; Nagy, L.E. Chronic ethanol feeding increases activation of NADPH oxidase by lipopolysaccharide in rat Kupffer cells: Role of increased reactive oxygen in LPS-stimulated ERK1/2 activation and TNF-alpha production. *J. Leukoc. Biol.* **2006**, *79*, 1348–1356. [CrossRef]
193. Cubero, F.J.; Nieto, N. Arachidonic acid stimulates TNFα production in Kupffer cells via a reactive oxygen species-pERK1/2-Egr1-dependent mechanism. *Am. J. Physiol. Gastrointest. Liver Physiol.* **2012**, *303*, G228–G239. [CrossRef]
194. Kim, S.Y.; Jeong, J.-M.; Kim, S.J.; Seo, W.H.; Kim, M.-H.; Choi, W.-M.; Yoo, W.; Lee, J.-H.; Shim, Y.-R.; Yi, H.-S.; et al. Pro-inflammatory hepatic macrophages generate ROS through NADPH oxidase 2 via endocytosis of monomeric TLR4-MD2 complex. *Nat. Commun.* **2017**, *8*, 2247. [CrossRef]
195. Yamashina, S.; Takei, Y.; Ikejima, K.; Enomoto, N.; Kitamura, T.; Sato, N. Ethanol-Induced Sensitization to Endotoxin in Kupffer Cells Is Dependent upon Oxidative Stress. *Alcohol. Clin. Exp. Res.* **2005**, *29*, 246S–250S. [CrossRef]
196. Shen, Z.; Ajmo, J.M.; Rogers, C.Q.; Liang, X.; Le, L.; Murr, M.M.; Peng, Y.; You, M. Role of SIRT1 in regulation of LPS- or two ethanol metabolites-induced TNF-α production in cultured macrophage cell lines. *Am. J. Physiol. Liver Physiol.* **2009**, *296*, G1047–G1053. [CrossRef]
197. Wheeler, M.D.; Kono, H.; Yin, M.; Nakagami, M.; Uesugi, T.; Arteel, G.E.; Gabele, E.; Rusyn, I.; Yamashina, S.; Froh, M. The role of kupffer cell oxidant production in early ethanol-induced liver disease1, 2 1Guest Editor: Arthur Cederbaum 2This article is part of a series of reviews on "Alcohol, Oxidative Stress and Cell Injury". The full list of papers may be found on the homepage of the journal. *Free Radic. Biol. Med.* **2001**, *31*, 1544–1549.
198. Sun, Q.; Zhang, W.; Zhong, W.; Sun, X.; Zhou, Z. Pharmacological inhibition of NOX4 ameliorates alcohol-induced liver injury in mice through improving oxidative stress and mitochondrial function. *Biochim. Biophys. Acta (BBA)—Gen. Subj.* **2017**, *1861*, 2912–2921. [CrossRef] [PubMed]
199. Kim, M.J.; Nagy, L.E.; Park, P.-H. Globular Adiponectin Inhibits Ethanol-Induced Reactive Oxygen Species Production through Modulation of NADPH Oxidase in Macrophages: Involvement of Liver Kinase B1/AMP-Activated Protein Kinase Pathway. *Mol. Pharmacol.* **2014**, *86*, 284–296. [CrossRef] [PubMed]
200. Shi, H.; Dong, L.; Dang, X.; Liu, Y.; Jiang, J.; Wang, Y.; Lu, X.; Guo, X. Effect of chlorogenic acid on LPS-induced proinflammatory signaling in hepatic stellate cells. *Inflamm. Res.* **2013**, *62*, 581–587. [CrossRef] [PubMed]
201. Setshedi, M.; Longato, L.; Petersen, D.R.; Ronis, M.; Chen, W.C.; Wands, J.R.; De La Monte, S.M. Limited Therapeutic Effect of N-Acetylcysteine on Hepatic Insulin Resistance in an Experimental Model of Alcohol-Induced Steatohepatitis. *Alcohol. Clin. Exp. Res.* **2011**, *35*, 2139–2151. [CrossRef] [PubMed]
202. Dingeo, G.; Brito, A.; Samouda, H.; Iddir, M.; La Frano, M.R.; Bohn, T. Phytochemicals as modifiers of gut microbial communities. *Food Funct.* **2020**, *11*, 8444–8471. [CrossRef] [PubMed]
203. Holst, B.; Williamson, G. Nutrients and phytochemicals: From bioavailability to bioefficacy beyond antioxidants. *Curr. Opin. Biotechnol.* **2008**, *19*, 73–82. [CrossRef] [PubMed]

204. Tsao, R. Chemistry and Biochemistry of Dietary Polyphenols. *Nutrients* **2010**, *2*, 1231–1246. [CrossRef]
205. Luca, S.V.; Macovei, I.; Bujor, A.; Miron, A.; Skalicka-Woźniak, K.; Aprotosoaie, A.C.; Trifan, A. Bioactivity of dietary polyphenols: The role of metabolites. *Crit. Rev. Food Sci. Nutr.* **2020**, *60*, 626–659. [CrossRef] [PubMed]
206. Summerlin, N.; Soo, E.; Thakur, S.; Qu, Z.; Jambhrunkar, S.; Popat, A. Resveratrol nanoformulations: Challenges and opportunities. *Int. J. Pharm.* **2015**, *479*, 282–290. [CrossRef] [PubMed]
207. Mouhid, L.; Corzo-Martínez, M.; Torres, C.; Vázquez, L.; Reglero, G.; Fornari, T.; De Molina, A.R. ImprovingIn VivoEfficacy of Bioactive Molecules: An Overview of Potentially Antitumor Phytochemicals and Currently Available Lipid-Based Delivery Systems. *J. Oncol.* **2017**, *2017*, 1–34. [CrossRef] [PubMed]
208. Marín, L.; Miguélez, E.M.; Villar, C.J.; Lombó, F. Bioavailability of Dietary Polyphenols and Gut Microbiota Metabolism: Antimicrobial Properties. *BioMed Res. Int.* **2015**, *2015*, 1–18. [CrossRef]
209. Murota, K.; Nakamura, Y.; Uehara, M. Flavonoid metabolism: The interaction of metabolites and gut microbiota. *Biosci. Biotechnol. Biochem.* **2018**, *82*, 600–610. [CrossRef] [PubMed]
210. Koudoufio, M.; Desjardins, Y.; Feldman, F.; Spahis, S.; Delvin, E.; Levy, E. Insight into Polyphenol and Gut Microbiota Crosstalk: Are Their Metabolites the Key to Understand Protective Effects against Metabolic Disorders? *Antioxidants* **2020**, *9*, 982. [CrossRef]
211. Chaplin, A.; Carpéné, C.; Mercader, J. Resveratrol, Metabolic Syndrome, and Gut Microbiota. *Nutrients* **2018**, *10*, 1651. [CrossRef]
212. Salehi, B.; Machin, L.; Monzote, L.; Sharifi-Rad, J.; Ezzat, S.M.; Salem, M.A.; Merghany, R.M.; El Mahdy, N.M.; Kılıç, C.S.; Sytar, O.; et al. Therapeutic Potential of Quercetin: New Insights and Perspectives for Human Health. *ACS Omega* **2020**, *5*, 11849–11872. [CrossRef]
213. Williamson, G.; Clifford, M.N. Role of the small intestine, colon and microbiota in determining the metabolic fate of polyphenols. *Biochem. Pharmacol.* **2017**, *139*, 24–39. [CrossRef]
214. Duda-Chodak, A.; Tarko, T.; Satora, P.; Sroka, P. Interaction of dietary compounds, especially polyphenols, with the intestinal microbiota: A review. *Eur. J. Nutr.* **2015**, *54*, 325–341. [CrossRef] [PubMed]
215. Possemiers, S.; Bolca, S.; Verstraete, W.; Heyerick, A. The intestinal microbiome: A separate organ inside the body with the metabolic potential to influence the bioactivity of botanicals. *Fitoterapia* **2011**, *82*, 53–66. [CrossRef] [PubMed]
216. Rahimi, R.; Abdollahi, M. The Role of Dietary Polyphenols in the Management of Inflammatory Bowel Disease. *Curr. Pharm. Biotechnol.* **2015**, *16*, 196–210. [CrossRef]
217. Godos, J.; Currenti, W.; Angelino, D.; Mena, P.; Castellano, S.; Caraci, F.; Galvano, F.; Del Rio, D.; Ferri, R.; Grosso, G. Diet and Mental Health: Review of the Recent Updates on Molecular Mechanisms. *Antioxidants* **2020**, *9*, 346. [CrossRef] [PubMed]
218. Tomás-Barberán, F.A.; Selma, M.V.; Espín, J.C. Interactions of gut microbiota with dietary polyphenols and consequences to human health. *Curr. Opin. Clin. Nutr. Metab. Care* **2016**, *19*, 471–476. [CrossRef] [PubMed]
219. Carrera-Quintanar, L.; Roa, R.I.L.; Quintero-Fabián, S.; Sánchez-Sánchez, M.A.; Vizmanos, B.; Ortuño-Sahagún, D. Phytochemicals That Influence Gut Microbiota as Prophylactics and for the Treatment of Obesity and Inflammatory Diseases. *Mediat. Inflamm.* **2018**, *2018*, 1–18. [CrossRef]
220. Liu, H.; Liu, M.; Fu, X.; Zhang, Z.; Zhu, L.; Zheng, X.; Liu, J. Astaxanthin Prevents Alcoholic Fatty Liver Disease by Modulating Mouse Gut Microbiota. *Nutrients* **2018**, *10*, 1298. [CrossRef]
221. Iglesias-Carres, L.; Hughes, M.D.; Steele, C.N.; Ponder, M.A.; Davy, K.P.; Neilson, A.P. Use of dietary phytochemicals for inhibition of trimethylamine N-oxide formation. *J. Nutr. Biochem.* **2021**, 108600. [CrossRef] [PubMed]
222. Yuan, J.; Chen, C.; Cui, J.; Lu, J.; Yan, C.; Wei, X.; Zhao, X.; Li, N.; Li, S.; Xue, G.; et al. Fatty Liver Disease Caused by High-Alcohol-Producing Klebsiella pneumoniae. *Cell Metab.* **2019**, *30*, 675–688.e7. [CrossRef]
223. Peiyuan, H.; Zhiping, H.; Chengjun, S.; Chunqing, W.; Bingqing, L.; Imam, M.U. Resveratrol Ameliorates Experimental Alcoholic Liver Disease by Modulating Oxidative Stress. *Evid.-Based Complement. Altern. Med.* **2017**, *2017*, 1–10. [CrossRef]
224. Du, B.; Meenu, M.; Liu, H.; Xu, B. A Concise Review on the Molecular Structure and Function Relationship of β-Glucan. *Int. J. Mol. Sci.* **2019**, *20*, 4032. [CrossRef] [PubMed]
225. Jayachandran, M.; Chen, J.; Chung, S.S.M.; Xu, B. A critical review on the impacts of β-glucans on gut microbiota and human health. *J. Nutr. Biochem.* **2018**, *61*, 101–110. [CrossRef] [PubMed]
226. Li, Y.; Gao, X.; Lou, Y. Interactions of tea polyphenols with intestinal microbiota and their implication for cellular signal conditioning mechanism. *J. Food Biochem.* **2019**, *43*, e12953. [CrossRef] [PubMed]
227. Sorrenti, V.; Fortinguerra, S.; Caudullo, G.; Buriani, A. Deciphering the Role of Polyphenols in Sports Performance: From Nutritional Genomics to the Gut Microbiota toward Phytonutritional Epigenomics. *Nutrients* **2020**, *12*, 1265. [CrossRef]
228. Mopuri, R.; Islam, M.S. Medicinal plants and phytochemicals with anti-obesogenic potentials: A review. *Biomed Pharmacother.* **2017**, *89*, 1442–1452. [CrossRef] [PubMed]
229. Davinelli, S.; De Stefani, D.; De Vivo, I.; Scapagnini, G. Polyphenols as Caloric Restriction Mimetics Regulating Mitochondrial Biogenesis and Mitophagy. *Trends Endocrinol. Metab.* **2020**, *31*, 536–550. [CrossRef]
230. Rothenberg, D.O.; Zhou, C.; Zhang, L. A Review on the Weight-Loss Effects of Oxidized Tea Polyphenols. *Molecules* **2018**, *23*, 1176. [CrossRef]
231. Adachi, Y.; Moore, L.E.; Bradford, B.U.; Gao, W.; Thurman, R.G. Antibiotics prevent liver injury in rats following long-term exposure to ethanol. *Gastroenterology* **1995**, *108*, 218–224. [CrossRef]

232. Kitagawa, R.; Kon, K.; Uchiyama, A.; Arai, K.; Yamashina, S.; Kuwahara-Arai, K.; Kirikae, T.; Ueno, T.; Ikejima, K. Rifaximin prevents ethanol-induced liver injury in obese KK-A(y) mice through modulation of small intestinal microbiota signature. *Am. J. Physiol. Gastrointest. Liver Physiol.* **2019**, *317*, G707–G715. [CrossRef]
233. Bergheim, I.; Weber, S.; Vos, M.; Krämer, S.; Volynets, V.; Kaserouni, S.; McClain, C.J.; Bischoff, S.C. Antibiotics protect against fructose-induced hepatic lipid accumulation in mice: Role of endotoxin. *J. Hepatol.* **2008**, *48*, 983–992. [CrossRef]
234. Gangarapu, V.; Ince, A.T.; Baysal, B.; Kayar, Y.; Kılıç, U.; Gök, Ö.; Uysal, Ö.; Şenturk, H. Efficacy of rifaximin on circulating endotoxins and cytokines in patients with nonalcoholic fatty liver disease. *Eur. J. Gastroenterol. Hepatol.* **2015**, *27*, 840–845. [CrossRef]
235. Harputluoglu, M.M.M.; Demirel, U.; Gul, M.; Temel, I.; Gursoy, S.; Selcuk, E.B.; Aladag, M.; Bilgic, Y.; Gunduz, E.; Seckin, Y. Effects of Rifaximin on Bacterial Translocation in Thioacetamide-Induced Liver Injury in Rats. *Inflammation* **2012**, *35*, 1512–1517. [CrossRef]
236. Bajaj, J.S.; Heuman, D.M.; Hylemon, P.B.; Sanyal, A.J.; White, M.B.; Monteith, P.; Noble, N.A.; Unser, A.B.; Daita, K.; Fisher, A.R.; et al. Altered profile of human gut microbiome is associated with cirrhosis and its complications. *J. Hepatol.* **2014**, *60*, 940–947. [CrossRef] [PubMed]
237. Chen, P.; Miyamoto, Y.; Mazagova, M.; Lee, K.-C.; Eckmann, L.; Schnabl, B. Microbiota Protects Mice against Acute Alcohol-Induced Liver Injury. *Alcohol. Clin. Exp. Res.* **2015**, *39*, 2313–2323. [CrossRef] [PubMed]
238. Day, C.P.; James, O.F. Steatohepatitis: A tale of two "hits"? *Gastroenterology* **1998**, *114*, 842–845. [CrossRef]
239. Abdelmegeed, M.A.; Yoo, S.-H.; Henderson, L.E.; Gonzalez, F.J.; Woodcroft, K.J.; Song, B.-J. PPARalpha expression protects male mice from high fat-induced nonalcoholic fatty liver. *J. Nutr.* **2011**, *141*, 603–610. [CrossRef] [PubMed]
240. Abdelmegeed, M.A.; Banerjee, A.; Yoo, S.-H.; Jang, S.; Gonzalez, F.J.; Song, B.-J. Critical role of cytochrome P450 2E1 (CYP2E1) in the development of high fat-induced non-alcoholic steatohepatitis. *J. Hepatol.* **2012**, *57*, 860–866. [CrossRef]
241. Takaki, A.; Kawai, D.; Yamamoto, K. Multiple Hits, Including Oxidative Stress, as Pathogenesis and Treatment Target in Non-Alcoholic Steatohepatitis (NASH). *Int. J. Mol. Sci.* **2013**, *14*, 20704–20728. [CrossRef]
242. Gu, Z.; Liu, Y.; Hu, S.; You, Y.; Wen, J.; Li, W.; Wang, Y. Probiotics for Alleviating Alcoholic Liver Injury. *Gastroenterol. Res. Pract.* **2019**, *2019*, 1–8. [CrossRef]
243. Ferrere, G.; Wrzosek, L.; Cailleux, F.; Turpin, W.; Puchois, V.; Spatz, M.; Ciocan, D.; Rainteau, D.; Humbert, L.; Hugot, C.; et al. Fecal microbiota manipulation prevents dysbiosis and alcohol-induced liver injury in mice. *J. Hepatol.* **2017**, *66*, 806–815. [CrossRef]
244. Bajaj, J.S.; Fagan, A.; Gavis, E.A.; Kassam, Z.; Sikaroodi, M.; Gillevet, P.M. Long-term Outcomes of Fecal Microbiota Transplantation in Patients With Cirrhosis. *Gastroenterology* **2019**, *156*, 1921–1923.e3. [CrossRef]
245. Bajaj, J.S.; Salzman, N.H.; Acharya, C.; Sterling, R.K.; White, M.B.; Gavis, E.A.; Fagan, A.; Hayward, M.; Holtz, M.L.; Matherly, S.; et al. Fecal Microbial Transplant Capsules Are Safe in Hepatic Encephalopathy: A Phase 1, Randomized, Placebo-Controlled Trial. *Hepatology* **2019**, *70*, 1690–1703. [CrossRef]
246. Bajaj, J.S.; Gavis, E.A.; Fagan, A.; Wade, J.B.; Thacker, L.R.; Fuchs, M.; Patel, S.; Davis, B.; Meador, J.; Puri, P.; et al. A Randomized Clinical Trial of Fecal Microbiota Transplant for Alcohol Use Disorder. *Hepatology* **2020**. [CrossRef]
247. Nanji, A.A.; Khettry, U.; Sadrzadeh, S.M.H. Lactobacillus Feeding Reduces Endotoxemia and Severity of Experimental Alcoholic Liver (Disease). *Exp. Biol. Med.* **1994**, *205*, 243–247. [CrossRef]
248. Konkit, M.; Kim, K.; Kim, J.-H.; Kim, W. Protective effects of *Lactococcus chungangensis* CAU 28 on alcohol-metabolizing enzyme activity in rats. *J. Dairy Sci.* **2018**, *101*, 5713–5723. [CrossRef] [PubMed]
249. Barone, R.; Rappa, F.; Macaluso, F.; Bavisotto, C.C.; Sangiorgi, C.; Di Paola, G.; Tomasello, G.; Di Felice, V.; Marcianò, V.; Farina, F.; et al. Alcoholic Liver Disease: A Mouse Model Reveals Protection by *Lactobacillus fermentum*. *Clin. Transl. Gastroenterol.* **2016**, *7*, e138. [CrossRef] [PubMed]
250. Seo, B.; Jeon, K.; Moon, S.; Lee, K.; Kim, W.-K.; Jeong, H.; Cha, K.H.; Lim, M.Y.; Kang, W.; Kweon, M.-N.; et al. Roseburia spp. Abundance Associates with Alcohol Consumption in Humans and Its Administration Ameliorates Alcoholic Fatty Liver in Mice. *Cell Host Microbe* **2020**, *27*, 25–40.e6. [CrossRef] [PubMed]
251. Shukla, P.K.; Meena, A.S.; Manda, B.; Gomes-Solecki, M.; Dietrich, P.; Dragatsis, I.; Rao, R. *Lactobacillus plantarum* prevents and mitigates alcohol-induced disruption of colonic epithelial tight junctions, endotoxemia, and liver damage by an EGF receptor-dependent mechanism. *FASEB J.* **2018**, *32*, 6274–6292. [CrossRef] [PubMed]
252. Hwang, B.B.; Chang, M.H.; Lee, J.H.; Heo, W.; Kim, J.K.; Pan, J.H.; Kim, Y.J.; Kim, J.H. The Edible Insect *Gryllus bimaculatus* Protects against Gut-Derived Inflammatory Responses and Liver Damage in Mice after Acute Alcohol Exposure. *Nutrients* **2019**, *11*, 857. [CrossRef] [PubMed]
253. Tang, Y.; Forsyth, C.B.; Banan, A.; Fields, J.Z.; Keshavarzian, A. Oats Supplementation Prevents Alcohol-Induced Gut Leakiness in Rats by Preventing Alcohol-Induced Oxidative Tissue Damage. *J. Pharmacol. Exp. Ther.* **2009**, *329*, 952–958. [CrossRef]
254. Chen, J.-R.; Chen, Y.-L.; Peng, H.-C.; Lu, Y.-A.; Chuang, H.-L.; Chang, H.-Y.; Wang, H.-Y.; Su, Y.-J.; Yang, S.-C. Fish Oil Reduces Hepatic Injury by Maintaining Normal Intestinal Permeability and Microbiota in Chronic Ethanol-Fed Rats. *Gastroenterol. Res. Pract.* **2016**, *2016*, 1–10. [CrossRef] [PubMed]
255. Feng, R.; Ma, L.-J.; Wang, M.; Liu, C.; Yang, R.; Su, H.; Yang, Y.; Wan, J.-B. Oxidation of fish oil exacerbates alcoholic liver disease by enhancing intestinal dysbiosis in mice. *Commun. Biol.* **2020**, *3*, 1–13. [CrossRef] [PubMed]

256. Song, B.-J.; Moon, K.-H.; Olsson, N.U.; Salem, N. Prevention of alcoholic fatty liver and mitochondrial dysfunction in the rat by long-chain polyunsaturated fatty acids. *J. Hepatol.* **2008**, *49*, 262–273. [CrossRef]
257. Zhong, W.; McClain, C.J.; Cave, M.; Kang, Y.J.; Zhou, Z. The role of zinc deficiency in alcohol-induced intestinal barrier dysfunction. *Am. J. Physiol. Liver Physiol.* **2010**, *298*, G625–G633. [CrossRef]
258. Lambert, J.C.; Zhou, Z.; Kang, Y.J. Suppression of Fas-Mediated Signaling Pathway is Involved in Zinc Inhibition of Ethanol-Induced Liver Apoptosis. *Exp. Biol. Med.* **2003**, *228*, 406–412. [CrossRef]
259. Zhou, Z.; Sun, X.; Lambert, J.C.; Saari, J.T.; Kang, Y.J. Metallothionein-Independent Zinc Protection from Alcoholic Liver Injury. *Am. J. Pathol.* **2002**, *160*, 2267–2274. [CrossRef]
260. Lou, Z.; Wang, J.; Chen, Y.; Xu, C.; Chen, X.; Shao, T.; Zhang, K.; Pan, H. Linderae radix ethanol extract attenuates alcoholic liver injury via attenuating inflammation and regulating gut microbiota in rats. *Braz. J. Med Biol. Res.* **2019**, *52*, e7628. [CrossRef] [PubMed]
261. Sun, S.; Wang, K.; Sun, L.; Cheng, B.; Qiao, S.; Dai, H.; Shi, W.; Ma, J.; Liu, H. Therapeutic manipulation of gut microbiota by polysaccharides of *Wolfiporia cocos* reveals the contribution of the gut fungi-induced PGE2 to alcoholic hepatic steatosis. *Gut Microbes* **2020**, *12*, 1830693. [CrossRef] [PubMed]
262. Yan, T.; Yan, N.; Wang, P.; Xia, Y.; Hao, H.; Wang, G.; Gonzalez, F.J. Herbal drug discovery for the treatment of nonalcoholic fatty liver disease. *Acta Pharm. Sin. B* **2020**, *10*, 3–18. [CrossRef]
263. Eom, T.; Ko, G.; Kim, K.C.; Kim, J.-S.; Unno, T. *Dendropanax morbifera* Leaf Extracts Improved Alcohol Liver Injury in Association with Changes in the Gut Microbiota of Rats. *Antioxidants* **2020**, *9*, 911. [CrossRef]
264. Peterson, C.T.; Denniston, K.; Chopra, D. Therapeutic Uses of Triphala in Ayurvedic Medicine. *J. Altern. Complement. Med.* **2017**, *23*, 607–614. [CrossRef]
265. Cota, D.; Mishra, S.; Shengule, S. Arjunarishta alleviates experimental colitis via suppressing proinflammatory cytokine expression, modulating gut microbiota and enhancing antioxidant effect. *Mol. Biol. Rep.* **2020**, *47*, 1–11. [CrossRef]
266. Samuelson, D.R.; Gu, M.; Shellito, J.E.; Molina, P.E.; Taylor, C.M.; Luo, M.; Welsh, D.A. Intestinal Microbial Products From Alcohol-Fed Mice Contribute to Intestinal Permeability and Peripheral Immune Activation. *Alcohol. Clin. Exp. Res.* **2019**, *43*, 2122–2133. [CrossRef]
267. Cresci, G.A.; Bush, K.; Nagy, L.E. Tributyrin Supplementation Protects Mice from Acute Ethanol-Induced Gut Injury. *Alcohol. Clin. Exp. Res.* **2014**, *38*, 1489–1501. [CrossRef]
268. Jiang, X.-W.; Li, Y.-T.; Ye, J.-Z.; Lv, L.-X.; Yang, L.-Y.; Bian, X.-Y.; Wu, W.-R.; Wu, J.-J.; Shi, D.; Wang, Q.; et al. New strain of *Pediococcus pentosaceus* alleviates ethanol-induced liver injury by modulating the gut microbiota and short-chain fatty acid metabolism. *World J. Gastroenterol.* **2020**, *26*, 6224–6240. [CrossRef] [PubMed]
269. Zhao, Z.-H.; Xin, F.-Z.; Xue, Y.; Hu, Z.; Han, Y.; Ma, F.; Zhou, D.; Liu, X.-L.; Cui, A.; Liu, Z.; et al. Indole-3-propionic acid inhibits gut dysbiosis and endotoxin leakage to attenuate steatohepatitis in rats. *Exp. Mol. Med.* **2019**, *51*, 1–14. [CrossRef]
270. Lee, D.M.; Ecton, K.E.; Trikha, S.R.J.; Wrigley, S.D.; Thomas, K.N.; Battson, M.L.; Wei, Y.; Johnson, S.A.; Weir, T.L.; Gentile, C.L. Microbial metabolite indole-3-propionic acid supplementation does not protect mice from the cardiometabolic consequences of a Western diet. *Am. J. Physiol. Liver Physiol.* **2020**, *319*, G51–G62. [CrossRef]
271. Wei, Z.; Shen, P.; Cheng, P.; Lu, Y.; Wang, A.; Sun, Z. Gut Bacteria Selectively Altered by Sennoside A Alleviate Type 2 Diabetes and Obesity Traits. *Oxidative Med. Cell. Longev.* **2020**, *2020*, 1–16. [CrossRef] [PubMed]
272. Brandt, A.; Hernández-Arriaga, A.; Kehm, R.; Sánchez, V.; Jin, C.J.; Nier, A.; Baumann, A.; Camarinha-Silva, A.; Bergheim, I. Metformin attenuates the onset of non-alcoholic fatty liver disease and affects intestinal microbiota and barrier in small intestine. *Sci. Rep.* **2019**, *9*, 1–14. [CrossRef] [PubMed]
273. Ahmadi, S.; Razazan, A.; Nagpal, R.; Jain, S.; Wang, B.; Mishra, S.P.; Wang, S.; Justice, J.; Ding, J.; McClain, D.A.; et al. Metformin Reduces Aging-Related Leaky Gut and Improves Cognitive Function by Beneficially Modulating Gut Microbiome/Goblet Cell/Mucin Axis. *J. Gerontol. Ser. A Boil. Sci. Med. Sci.* **2020**, *75*, e9–e21. [CrossRef] [PubMed]
274. Ichim, T.E.; Kesari, S.; Shafer, K. Protection from chemotherapy- and antibiotic-mediated dysbiosis of the gut microbiota by a probiotic with digestive enzymes supplement. *Oncotarget* **2018**, *9*, 30919–30935. [CrossRef]
275. Lee, S.I.; Kang, K.S. N-acetylcysteine modulates lipopolysaccharide-induced intestinal dysfunction. *Sci. Rep.* **2019**, *9*, 1–10. [CrossRef] [PubMed]
276. Voigt, R.M.; Forsyth, C.B.; Keshavarzian, A. Circadian rhythms: A regulator of gastrointestinal health and dysfunction. *Expert Rev. Gastroenterol. Hepatol.* **2019**, *13*, 411–424. [CrossRef]
277. Manach, C.; Williamson, G.; Morand, C.; Scalbert, A.; Rémésy, C. Bioavailability and bioefficacy of polyphenols in humans. I. Review of 97 bioavailability studies. *Am. J. Clin. Nutr.* **2005**, *81*, 230S–242S. [CrossRef]
278. Potì, F.; Santi, D.; Spaggiari, G.; Zimetti, F.; Zanotti, I. Polyphenol Health Effects on Cardiovascular and Neurodegenerative Disorders: A Review and Meta-Analysis. *Int. J. Mol. Sci.* **2019**, *20*, 351. [CrossRef]
279. Simini, B. Serge Renaud: From French paradox to Cretan miracle. *Lancet* **2000**, *355*, 48. [CrossRef]
280. Ferrières, J. The French paradox: Lessons for other countries. *Heart* **2004**, *90*, 107–111. [CrossRef] [PubMed]
281. Yang, J.; Xiao, Y.-Y. Grape Phytochemicals and Associated Health Benefits. *Crit. Rev. Food Sci. Nutr.* **2013**, *53*, 1202–1225. [CrossRef]
282. Cremonini, E.; Fraga, C.G.; Oteiza, P.I. (–)-Epicatechin in the control of glucose homeostasis: Involvement of redox-regulated mechanisms. *Free. Radic. Biol. Med.* **2019**, *130*, 478–488. [CrossRef]

283. Wu, X.; Li, C.; Xing, G.; Qi, X.; Ren, J. Resveratrol Downregulates Cyp2e1 and Attenuates Chemically Induced Hepatocarcinogenesis in SD Rats. *J. Toxicol. Pathol.* **2013**, *26*, 385–392. [CrossRef] [PubMed]
284. Fukui, M.; Choi, H.J.; Zhu, B.T. Mechanism for the protective effect of resveratrol against oxidative stress-induced neuronal death. *Free. Radic. Biol. Med.* **2010**, *49*, 800–813. [CrossRef]
285. Abdelmegeed, M.A.; Jang, S.; Banerjee, A.; Hardwick, J.P.; Song, B.-J. Robust protein nitration contributes to acetaminophen-induced mitochondrial dysfunction and acute liver injury. *Free. Radic. Biol. Med.* **2013**, *60*, 211–222. [CrossRef]
286. Anhê, F.F.; Pilon, G.; Roy, D.; Desjardins, Y.; Levy, E.; Marette, A. Triggering Akkermansia with dietary polyphenols: A new weapon to combat the metabolic syndrome? *Gut Microbes* **2016**, *7*, 146–153. [CrossRef]
287. Singal, A.K.; Kamath, P.S.; Gores, G.J.; Shah, V.H. Alcoholic Hepatitis: Current Challenges and Future Directions. *Clin. Gastroenterol. Hepatol.* **2014**, *12*, 555–564. [CrossRef] [PubMed]
288. Younossi, Z.; Henry, L. Contribution of Alcoholic and Nonalcoholic Fatty Liver Disease to the Burden of Liver-Related Morbidity and Mortality. *Gastroenterology* **2016**, *150*, 1778–1785. [CrossRef] [PubMed]
289. Lourens, S.; Sunjaya, D.B.; Singal, A.; Liangpunsakul, S.; Puri, P.; Sanyal, A.; Ren, X.; Gores, G.J.; Radaeva, S.; Chalasani, N. Acute Alcoholic Hepatitis: Natural History and Predictors of Mortality Using a Multicenter Prospective Study. *Mayo Clin. Proc. Innov. Qual. Outcomes* **2017**, *1*, 37–48. [CrossRef] [PubMed]
290. Szabo, G. Clinical Trial Design for Alcoholic Hepatitis. *Semin. Liver Dis.* **2017**, *37*, 332–342. [CrossRef] [PubMed]
291. Singal, A.K.; Shah, V.H. Current trials and novel therapeutic targets for alcoholic hepatitis. *J. Hepatol.* **2019**, *70*, 305–313. [CrossRef]

antioxidants

MDPI

Article

Quercetin Ameliorates Insulin Resistance and Restores Gut Microbiome in Mice on High-Fat Diets

Yuqing Tan [1,2], Christina C. Tam [3], Matt Rolston [4], Priscila Alves [2], Ling Chen [2,5], Shi Meng [6,7,*], Hui Hong [1,*], Sam K. C. Chang [8] and Wallace Yokoyama [2]

1 Beijing Laboratory for Food Quality and Safety, College of Food Science and Nutritional Engineering, China Agriculture University, Beijing 100083, China; yuqingtan@cau.edu.cn
2 Healthy Processed Foods Research Unit, Agricultural Research Service, United States Department of Agriculture, Albany, CA 94710, USA; priscila.alves@usda.gov (P.A.); lingchen@jiangnan.edu.cn (L.C.); wally.yokoyama@usda.gov (W.Y.)
3 Foodborne Toxins Detection and Prevention Research Unit, Agricultural Research Service, United States Department of Agriculture, Albany, CA 94710, USA; christina.tam@usda.gov
4 Host Microbe Systems Biology Core, University of California, One Shields Avenue, Davis, CA 95616, USA; mrrolston@ucdavis.edu
5 School of Food Science and Technology, Jiangnan University, Wuxi 214122, China
6 Nestlé R & D (China) Ltd., Beijing 100015, China
7 Key Research Laboratory of Agro-Products Processing, Institute of Food Science and Technology, Chinese Academy of Agricultural Sciences, Beijing 100193, China
8 Experimental Seafood Processing Laboratory, Costal Research and Extension Center, Mississippi State University, Biloxi, MS 39579, USA; sc1690@msstate.edu
* Correspondence: mengshi@caas.cn (S.M.); hhong@cau.edu.cn (H.H.)

Abstract: Quercetin is a flavonoid that has been shown to have health-promoting capacities due to its potent antioxidant activity. However, the effect of chronic intake of quercetin on the gut microbiome and diabetes-related biomarkers remains unclear. Male C57BL/6J mice were fed HF or HF supplemented with 0.05% quercetin (HFQ) for 6 weeks. Diabetes-related biomarkers in blood were determined in mice fed high-fat (HF) diets supplemented with quercetin. Mice fed the HFQ diet gained less body, liver, and adipose weight, while liver lipid and blood glucose levels were also lowered. Diabetes-related plasma biomarkers insulin, leptin, resistin, and glucagon were significantly reduced by quercetin supplementation. In feces, quercetin supplementation significantly increased the relative abundance of *Akkermansia* and decreased the Firmicutes/Bacteroidetes ratio. The expression of genes *Srebf1*, *Ppara*, *Cyp51*, *Scd1*, and *Fasn* was downregulated by quercetin supplementation. These results indicated that diabetes biomarkers are associated with early metabolic changes accompanying obesity, and quercetin may ameliorate insulin resistance.

Keywords: quercetin; high-fat diet; insulin resistance; gut microbiome

Citation: Tan, Y.; Tam, C.C.; Rolston, M.; Alves, P.; Chen, L.; Meng, S.; Hong, H.; Chang, S.K.C.; Yokoyama, W. Quercetin Ameliorates Insulin Resistance and Restores Gut Microbiome in Mice on High-Fat Diets. *Antioxidants* **2021**, *10*, 1251. https://doi.org/10.3390/antiox10081251

Academic Editor: Baojun Xu

Received: 14 July 2021
Accepted: 3 August 2021
Published: 5 August 2021

1. Introduction

Obesity is recognized as a major global public health crisis. In 2016, over 1.9 billion adults globally were overweight and 650 million were obese globally [1]. Obesity, cardiovascular disease, and type II diabetes are considered inflammatory diseases. Anti-inflammatory phytochemicals such as phenolics and polyphenolics are extremely potent against metabolic diseases. They are concentrated in leaves, peels, and seeds where they protect the plant against environmental pathogens. Their bioactivity against plant pathogens is broad and may be the basis for their demonstrated beneficial health properties in humans.

Quercetin is one of the most abundant and common flavonoids in plant foods [2]. It is well known as a potent antioxidant and scavenger of reactive oxygen species (ROS) and reactive nitrogen species (RNS). Quercetin has been shown to have beneficial effects in human studies [3]. However, the effect of chronic intake of quercetin on the gut microbiome and diabetes-related biomarkers remains unclear. However, quercetin content in typical meals for humans does not reach the high levels used in some animal studies. Therefore, understanding the effect of a chronic intake of a low concentration (0.05%) of quercetin on the gut microbiome in mice on a high-fat diet is vital. Quercetin is found as a glycoside in foods, and studies suggest that the glycoside must be hydrolyzed to the aglycone for efficient absorption [4]. The pharmacological effects of quercetin may also be partly due to its physicochemical properties. The solubility of quercetin is low, about 2.6 mg/L at 25 °C. Despite its low solubility, carbon-14 (^{14}C) studies in humans reported that oral absorption was 36–53%. However, the combined urinary and fecal excretion was less than 10% [5]. The investigators found that as much as 23–81% was excreted as CO_2, suggesting that microbial action may contribute to its degradation. A recent study suggested that the metabolism of the natural product asperuloside altered intestinal metabolites levels and composition via modulation of gut microbiota [6].

Quercetin has been shown to be highly effective in preventing obesity-related metabolic syndrome characteristics in animals. For example, quercetin (0.05%) reduced body weight, visceral fat, blood glucose, insulin, and TNF-α in C57BL/6J mice fed a high-fat diet for 20 weeks but not 8 weeks. However, this study did not investigate the gut microbiome [7,8]. In terms of a low-fat diet, no significant effects on body weight, visceral fat, or blood glucose and lipids were observed in C57BL/6J mice fed 0.05% or 1% quercetin in an AIN93G diet (7% fat by weight) for 20 weeks [9]. However, type 2 diabetic mice (db/db mice) fed 0.04% or 0.08% quercetin on a AIN93G diet for 6 weeks had lower fasting glucose but not insulin, as well as lower adiponectin at 0.08% but not 0.04% quercetin [10]. These results suggest that quercetin reduces biomarkers of metabolic dysfunction when mice have excessive body and visceral adipose weight gain via consumption of a high ft diet or in an obesogenic animal model.

Quercetin also changes the distribution of gut bacteria. In overweight humans and animals, the intake of probiotics and dietary fibers changed the patterns of gut microbiota in comparison to those of normal-weight animals or humans [11]. Gut bacteria transform quercetin to metabolites such as homoprocatechuic acid, protocatechuic acid, 4-hydroxybenzoic acid, and propionic acid [12]. These metabolites can be detected in blood and urine to assess bioactivity in human trials [13]. Quercetin (1% of diet) was reported to reshape the fecal microbiota composition of rats fed a high-fat, sucrose diet, but anti-obesity and anti-inflammation effects were not reported [14]. However, 1% quercetin supplementation is hard to achieve in typical meals for humans (usual consumption is 10–100 mg quercetin a day). While many studies have investigated the impact of quercetin intake on physiological effects including changes in gene expression and, levels of lipid- and carbohydrate-metabolizing enzymes, few have reported alterations in biomarkers associated with insulin resistance (IR) and appetite. This is the first study to investigate chronic intake and its effect on gut microbiota profiles in diet-induced insulin-resistant mice. We hypothesize that the prevention of insulin resistance and obesity, as well as improvements in the gut microbiome, due to quercetin intake in diet-induced obese mice is accompanied by measurable changes in blood biomarkers for diabetes.

2. Materials and Methods

2.1. Animal and Diets

Male C57BL/6J mice (22.3 ± 1.5 g), 5 weeks old, were purchased from Jackson Laboratories (Sacramento, CA, USA). Mice were housed individually in a temperature-controlled room (20–22 °C, 60% relative humidity, 12 h alternating light/dark cycle). Mice were acclimated and given access to drinking water and chow diet (LabDiet #5001, PMI International, Redwood, CA, USA) *ad libitum* for 1 week before feeding of the experimental diet. Mice were sorted by weight, and each weight range was randomly divided into three groups with eight mice each and fed, ad libitum, semi-synthetic diets based on an AIN-93G formulation consisting of a high-fat (HF, 53% fat calories) diet supplemented with 0.05% quercetin (HFQ), an HF control diet, and a low-fat (LF) reference diet (Table 1). Quercetin was dissolved in 5 mL of ethanol dispersed with the dry ingredients, followed by evaporation of ethanol to obtain 1 kg of food. Body weights were recorded once a week, and food intake was monitored twice a week. The study was reviewed and approved by the Institutional Animal Care and Use Committee, Western Regional Research Center, USDA, Albany, CA, USA (Protocol No. 18-4).

Table 1. Diet composition (grams).

Ingredient	High Fat Diet (HF)	High Fat Diet with Quercetin (HFQ)	Low Fat Diet (LF)
Lard fat	225.0	225.0	63.0
Soybean oil	25.0	25.0	7.0
Cholesterol	0.8	0.8	0.8
Cellulose	50.0	49.5	50.0
0.05% quercetin	-	0.5	-
Casein	200.0	200.0	200.0
Corn starch	148.2	148.2	528.2
Sucrose	300.0	300.0	100.0
Cystine	3.0	3.0	3.0
Choline bitartrate	3.0	3.0	3.0
Mineral mix	35.0	35.0	35.0
Vitamin mix	10.0	10.0	10.0
Total weight	1000.0	1000.0	1000.0
Calories/kg	4850.0	4850.0	3950.0

2.2. Plasma and Tissue Collection

After 6 weeks of feeding, mice were fasted for 16 h and then anesthetized with isoflurane (Phoenix Pharmaceutical, St. Joseph, MO, USA). Blood was collected by cardiac puncture into EDTA-rinsed syringes. Plasma was separated by centrifugation at 2000× g for 15 min at 4 °C, and the samples were stored at −80 °C for further analysis. Liver and epididymal adipose were collected, weighed, and frozen in liquid nitrogen for further analysis. Five livers were used for hepatic lipid content analysis, and the remaining livers (3–4) were used for PCR analysis.

2.3. Blood Glucose, Plasma, and Hepatic Lipid Analysis

Tail-vein blood glucose level was determined using a OneTouch Ultrameter (Life Scan Inc., Milpitas, CA, USA). Plasma lipoprotein cholesterol was determined according to our previous method [15,16]. Plasma triglyceride (TG) was determined using an enzyme colorimetric assay kit (Sekisui Diagnostics PEI Inc., Charlottetown, PE, Canada) according to the manufacturer's instructions, and the absorbance was measured at 505 nm (Nanodrop 2000 C spectrophotometer, Thermo Scientific, Pleasanton, CA, USA). Liver lipids were determined as described previously [15]. Liver tissue from five mice from each group was used for hepatic lipid analysis and the residual liver tissue (n = 4) were saved for PCR analysis.

2.4. Glucose Tolerance Test (GTT)

The glucose tolerance test was administered after 6 weeks of diet treatment. After 5 h of fasting, mice were orally administered a 20% glucose solution (10 mL/kg body weight). Tail-vein blood glucose levels were determined at 0, 15, 30, 60, and 120 min using a OneTouch Ultrameter (Lifescan Inc., Milpitas, CA, USA). GTT curves were obtained by plotting glucose concentration versus time, and integrated glucose concentration over 120 min was calculated as the area under the curve (AUC).

2.5. Plasma Levels of Metabolic Biomarkers Relevant to Diabetes and Obesity

Plasma ghrelin, gastric inhibitory polypeptide (GIP), glucagon-like peptide-1 (GLP-1), insulin, leptin, resistin, and glucagon levels were analyzed using a mouse diabetes multiplex antibody assay kit (Bio-Plex Pro Mouse Diabetes Assay, Bio-Rad, Hercules, CA, USA) on the Bio-Plex 200 system (Bio-Rad, Hercules, CA, USA) according to the manufacturer's instructions.

2.6. Plasma Levels of Inflammation Cytokines

Plasma interleukin-2 (IL-2), interleukin-4 (IL-4), interleukin-5 (IL-5), interleukin-10 (IL-10), interleukin-12 (IL-12), granulocyte-macrophage colony-stimulating factor (GM-CSF), interferon gamma (IFN-γ), and tumor necrosis factor alpha (TNF-α) levels were analyzed using a multiplex antibody kit following the manufacturer's instructions (Bio-Plex Pro Mouse Cytokine Assay 8-plex, Bio-Rad) with the Bio-Plex 200 system (Bio-Rad).

2.7. Homeostatic Model Assessment of Insulin Resistance (HOMA-IR) Index Calculation

HOMA-IR was calculated from fasting blood glucose and plasma insulin concentrations. HOMA-IR index was calculated according to Equation (1) [17]. Insulin and plasma glucose concentrations after 16 h fasting were used to calculate the HOMA-IR index.

$$\text{HOMA} - \text{IR} = \frac{\text{Fasting plasma insulin (mU/L)} \times \text{Fasting plasma glucose (mmol/L)}}{22.5} \quad (1)$$

2.8. Fecal Microbiome Analysis

Mice were placed in paper cups fecal pellets were immediately collected and, stored at −80 °C. DNA from feces was extracted using Qiagen DNeasy PowerSoil kits (Qiagen, Valencia, CA, USA) following the standard protocol. The V3–V4 domains of the 16S rRNA were amplified using primers 319F/806R (TCGTCGGCAGCGTCAGATGTGTATAAGAGACAG (spacer) GTACTCCTACGGGAGGCAGCAGT and [GTCTCGTGGGCTCGGAGATGTGTATAAGAGACAG (spacer) CCGGACTACNVGGGTWTCTAAT, respectively) containing an Illumina tag sequence, a variable-length spacer, a linker sequence, and the 16S target sequence. Each sample was barcoded with an Illumina P5 adapter sequence, a unique eight nucleotide (nt) barcode, and a partial matching sequence of the forward primer, as well as reverse primers with an Illumina P7 adapter sequence, unique 8 nt barcode, and a partial matching sequence of the reverse adapter. The final product was quantified on a Qubit 4.0 instrument using the dsDNA Broad Range DNA kit (Invitrogen, Carlsbad, CA, USA), and individual amplicons were pooled in equal amounts. The pooled library was cleaned with Ampure XP beads (Beckman Coulter, Brea, CA, USA), and bands of interest were further isolated by gel electrophoresis (Sage Science, Beverly, MA, USA). The library was quantified via qPCR then sequenced with 300 bp dual end sequencing with an Illumina MiSeq at the Genome Center DNA Technologies Core, UC Davis. The names of the repository/repositories and the accession numbers can be found at https://www.ncbi.nlm.nih.gov/ (accessed on 4 August 2021), PRJNA722496.

The Raw FASTQ files and adapter trimmings were demultiplexed with dbcAmplicons version 0.8.5 (https://github.com/msettles/dbcAmplicons (accessed on 4 August 2021)). Forward and reverse unmerged reads were imported into QIIME2 version 2020.2 (https://qiime2.org (accessed on 4 August 2021)), and sequence variants were determined by utilizing the DADA2 analysis pipeline. Singletons and chimeras were removed as part of the quality filtering process, and the remaining sequences were clustered into amplicon sequence variants (ASVs). The clustered sequences were then compared against the Silva 132 reference database which was used for taxonomic assignment, meeting 99% identity.

Shannon's index was calculated and displayed using the R program through rarefactions to indicate alpha-diversity. Beta-diversity was used to evaluate differences by both weighted and unweighted UniFrac methods. Subsequently, a principal coordinate analysis (PCoA) based on Bray–Curtis distance was performed with an iterative algorithm. An online LEfSe analysis was adopted to search for the biomarkers of different groups (http://huttenhower.sph.harvard.edu/galaxy (accessed on 4 August 2021)). Based on the LEfSe analysis, bacteria with p-values < 0.05 in LDA scores of 3.0 was plotted.

2.9. RT-PCR

RNA was extracted from livers and adipose tissues by TRIzol and an RNA purification kit (Invitrogen, Life Technologies, Carlsbad, CA, USA). All primers and probes for ddPCR were designed by Invitrogen (Invitrogen, Life Technologies, Carlsbad, CA, USA) as per MIQE guidelines [18]. cDNA was synthesized using a GeneAmp RNA PCR kit (Applied Biosystems, Foster City, CA, USA). Synthesized cDNA was diluted 10 times with dH2O, and 1 μL of diluted cDNA was used in each real-time RT-PCR using SYBR green supermix (Bio-Rad, Hercules, CA, USA) with an Mx3000P instrument (Agilent, Cedar Creek, TX, USA). Cycle conditions were as follows: 5 min at 95 °C and 94 °C for 30 s, followed by 60 °C for 1 min, and then 72 °C for 30 s. Primers were validated by PCR product sizes, and no primer dimers were observed in gel electrophoresis of PCR products. Primer amplification efficiency was over 90% for every RT-PCR assay. Differences in mRNA expression in liver and adipose tissues were calculated after normalization to β-actin or 36B4 mRNA expression using the $2^{-\Delta\Delta Ct}$ method [19]. The genes used in this study were Srebf1 (NCBI gene ID:78968), Cyp7a1 (NCBI gene ID:13122), Ppara (NCBI gene ID:19013), Cyp51 (NCBI gene ID:13121), Scd1 (NCBI gene ID:20249), Fasn (NCBI gene ID:14104), Slc2a4 (NCBI gene ID:20525), Adipoq (NCBI gene ID:11450), 36b4 (NCBI gene ID:11837), and β-actin (NCBI gene ID: 11461). Primers are shown in Supplementary Table S1.

2.10. Statistical Analysis

Results were expressed as the mean ± SEM. The significance of differences between treatments was analyzed by ANOVA, followed by Tukey–Kramer HSD tests, using 2016 SAS (version 9.4, SAS Inc., Cary, NV, USA). The significance level was set at $p < 0.05$.

3. Results and Discussion

3.1. Animal Metrics

Mice fed an HF diet had almost three times higher weight gain compared to mice on the LF diet, confirming HF diet-induced obesity (Figure 1A). Published reports of weight gain of diet-induced obese (DIO) mice fed quercetin have not been consistent. In our study, mice fed the HFQ diet gained 69.7% less weight ($p < 0.05$) and had a 66.7% lower feed efficiency ratio (g gain/calories intake, $p < 0.05$) than mice on the HF diet. Weight gain and feed efficiency (Figure 1B) of mice fed HFQ diet were similar to the LF group ($p > 0.05$). The food intake data are shown in Figure 1C. The total food intake of mice on the HFQ diet is less than the HF group. However, the feed efficiency ratio (g gain/g feed) was decreased significantly with quercetin supplementation. Quercetin supplementation might affect appetite. Porras and coworkers [20] reported a similar lower weight gain (77% of the control) and lower feed intake in C57BL/6J mice fed a 60% fat calorie diet containing 0.05% quercetin for 12 weeks. However, others have reported that quercetin supplementation of the HF diet fed to C57BL/6J mice did not result in differences in body weight. Kuipers and coworkers [21] reported no differences in body weight or feed consumption in the same mice model fed a 45% fat diet supplemented with 0.1% quercetin for 12 weeks. The differences in outcomes may be due to the lower percentage of fat calories, 46% vs. 53%, or the low solubility of quercetin. A previous study of isorhamnetin, an O-methylated quercetin glycoside abundant in onions, suggested that the glucose component in the quercetin glucoside affects its bioavailability [22]. The application of quercetin is also limited due to stability and solubility issues [23]. In this study, quercetin solubility and bioavailability were optimized by dissolving in alcohol to a molecular form before dispersing on the dry ingredients. The fat content of diet might influence the bioavailability and metabolism of quercetin [24].

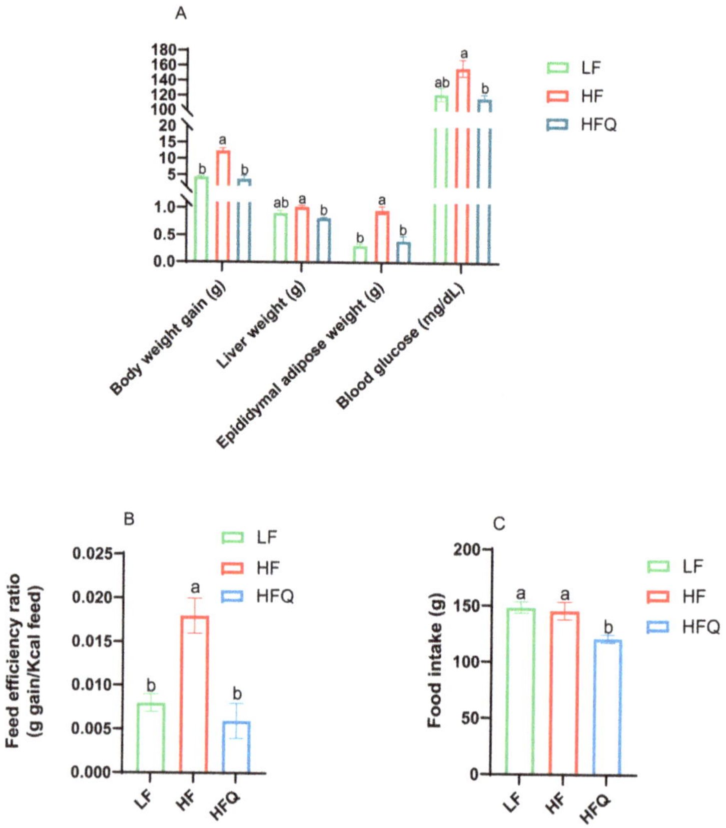

Figure 1. (**A**). Anthropometrics in mice fed HF, HFQ, and LF diets for 6 weeks. HF: high-fat control diet (46% kcal from fat, 16.5% kcal from protein, and 37.5% kcal from carbohydrate); HFQ: 0.05% quercetin in high-fat diet; LF: low-fat control diet (16% kcal from fat, 20% kcal from protein, and 64% kcal from carbohydrate) (**B**). Feed efficiency ratio of mice fed different diets. (**C**). Food intake of mice fed different diets. Values are means ± SEMs, $n = 8$/group. Bars with different letters were significantly different ($p < 0.05$).

Mice on the HFQ diet for 6 weeks had significantly ($p < 0.05$) lower liver and adipose weight, 19.6% and 58.3%, respectively, than those on the HF diet. The fasting blood glucose level of the mice on the HFQ diet was 25.4% lower than that of those on the HF diet. Final body weight, liver weight, adipose weight, and blood glucose levels of mice fed the HFQ and LF diets were not significantly different. This suggests that supplementation of quercetin in the HF diet has health-promoting effects in terms of preventing increases in in body, liver, and epididymal adipose tissue weights associated with mice on the HF diet. Previously, researchers reported that C57BL/6J mice fed a high-fat diet (39.9% energy from fat) supplemented with 0.05% quercetin had reduced body weight, blood glucose, insulin, cholesterol, TNF-α, and other markers of metabolic syndrome after 20 weeks of feeding but not after 8 weeks [7]. Since the present study did not first induce metabolic syndrome in mice followed by treatment with quercetin, this indicated that the supplementation with quercetin delays the development of obesity.

3.2. Plasma/Hepatic Lipid Content and Triglyceride (TG) Levels

Quercetin supplementation of the HF diet lowered low-density lipoprotein (LDL) and plasma TG concentration by 37.6% and 62.9%, respectively, compared with the HF diet ($p < 0.05$) (Figure 2A). However, plasma very-low-density lipoprotein (VLDL) and high-density lipoprotein (HDL) cholesterol concentrations of mice fed the HFQ diet were not different from those on the HF diet (Figure 2A). The hepatic total lipid level of mice fed the HFQ diet was 35.5% lower than that of those on the HF diet (Figure 2B). The reduction in plasma TG and hepatic lipid levels was similar to that seen by Kobori and coworkers [7], who reported 27% lower plasma TG and about 30% lower hepatic lipids in C57BL/6J mice fed a Western diet (39.9% energy from fat) supplemented with 0.05% quercetin after 20 weeks but not at 4 or 8 weeks. The present study indicates that the HFQ diet may be responsible for the lowered total lipid content in liver after 6 weeks.

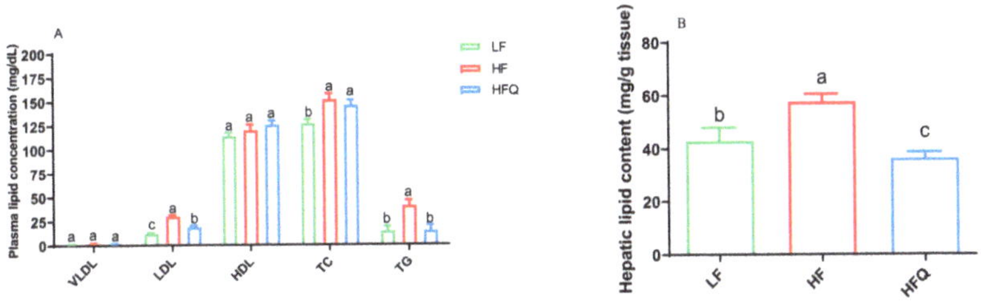

(A) (B)

Figure 2. (**A**). Plasma lipoprotein cholesterol concentration and triglyceride level in mice fed a high-fat (HF) diet, high-fat diet containing 0.05% quercetin (HFQ), and low-fat (LF) diet for 6 weeks. VLDL: very-low-density lipoprotein; LDL: low-density lipoprotein; HDL: high-density lipoprotein; TC: total cholesterol; TG: triglyceride. Data are expressed as means ± SEMs, $n = 8$/group. Bars with different letters within the same plasma lipoprotein were significantly different ($p < 0.05$). (**B**). Hepatic total lipid content in mice fed with high-fat (HF) diet, high-fat diet containing 0.05% quercetin (HFQ), and low-fat (LF) diet for 6 weeks. Data are expressed as means ± SEMs, $n = 5$/group. Bars with different letters were significantly different ($p < 0.05$).

3.3. Glucose Tolerance Test (GTT) and Insulin Resistance

The oral glucose tolerance test (GTT) curve, the area under the curve (AUC), and the HOMA-IR index are shown in Figure 3A–C, respectively. Compared to the HF diet, HFQ significantly lowered the fasting blood glucose level ($p < 0.05$, Figure 1A). The blood glucose levels of the GTT for the HFQ-fed mice were lower at 15, 30, and 60 minutes but were not significantly different from HF. The HFQ diet significantly improved glucose metabolism ($p < 0.05$), lowered AUC by 10.9% compared to the HF diet. The HOMA-IR index of mice fed the HF diet was 82.3% higher than that of mice on the HFQ diet, suggesting that quercetin supplementation contributed to improved HF diet-induced IR. In the Figure 3C, HOMA-IR index of LF groups was 0, but not included in Figure 3C. In B6 mice fed an HF diet containing 0.4% quercetin for 26 weeks, it was reported that fasting blood glucose levels was lower but not significant [25]. Vessal and coworkers [26] reported that the glucose tolerance of STZ-induced diabetic rats returned to normal levels after intraperitoneal (ip) administration of quercetin for 10 days due to the regeneration of pancreatic islets. This experiment suggests that quercetin itself is bioactive as opposed to metabolites of gut bacterial metabolism. However, quercetin or its glycosides may be excreted through the bile, recirculated into the intestinal lumen, and made accessible to gut bacteria for metabolism into phenolic acids [27]. Kobori and coworkers [7] suggested that, in mice fed high-fat Western type diets, quercetin enables the recovery of cell functions in the liver and pancreas by reducing oxidative stress.

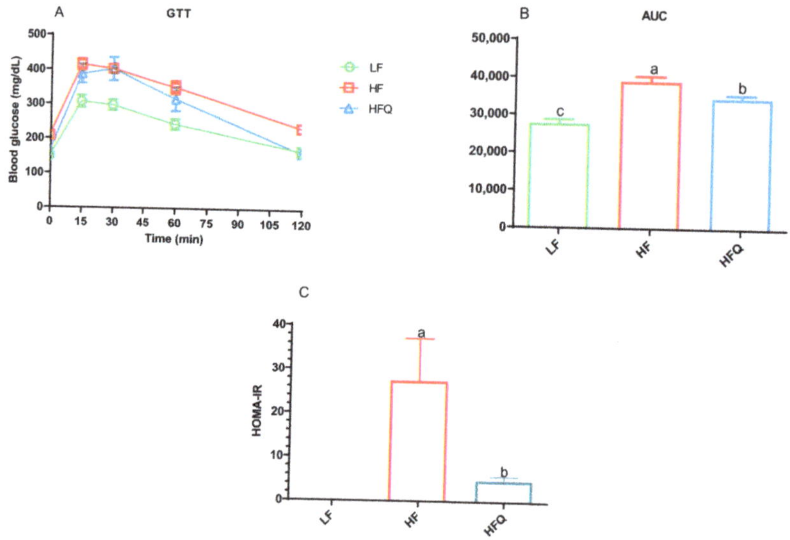

Figure 3. (**A**). Glucose tolerance in mice fed a high-fat (HF) diet, high-fat diet containing 0.05% quercetin (HFQ), and low-fat (LF) diet for 6 weeks. (**B**). Area under glucose tolerance test (GTT) curve values. (**C**). Homeostatic model assessment of insulin resistance (HOMA-IR) index of HF and HFQ diet-fed mice; the HOMA-IR index for the LF group was 0. Data are expressed as means ± SEMs, $n = 8$/group. Bars with different letters were significantly different ($p < 0.05$).

3.4. Plasma Biomarkers of Diabetes and Obesity

Plasma ghrelin, GIP, GLP-1, insulin, leptin, resistin, and glucagon levels were analyzed using a multiplex immunoassay method (Figure 4). Mice fed the HFQ diet had a 34.5% higher ghrelin concentrations and 88%, 92%, 27%, and 97% lower insulin, leptin, resistin, and glucagon levels, respectively, than those on the HF diet. Plasma GIP and GLP-1 levels of HFQ diet fed mice were not significantly different from those on the HF ($p > 0.05$).

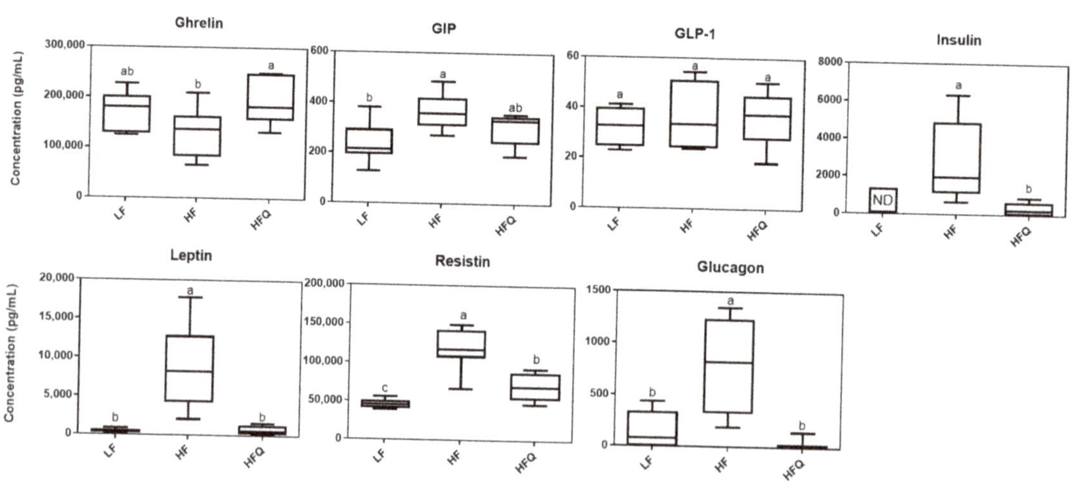

Figure 4. Boxplot of plasma biomarker concentrations related to diabetes and obesity, $n = 8$/group. ND = not detected. Top edge of the box, 75th percentile; bottom edge, 25th percentile; horizontal bar within box, median; top horizontal bar outside box, maximum concentration; bottom horizontal bar outside box, minimum concentration. Boxes with different letters were significantly different ($p < 0.05$). ND represents not detected.

Ghrelin is an orexigenic hormone secreted mainly by the stomach preprandially and is responsible for regulating appetite and energy hemostasis. Ghrelin stimulates appetite; thus, our finding that the ghrelin level in mice fed the HFQ diet was higher but the feed intake and feed efficiency ratio were lower than in the HF diet-fed mice was unexpected. Ghrelin levels are higher before meals and lower between meals. In the current study, mice were sacrificed after a 16 h fast; therefore, high ghrelin levels in mice were expected. However, Moesgaard and coworkers [28] reported that the feeding status (fasting or non-fasting) does not affect ghrelin levels, and the reduced expression of ghrelin in obese C57BL/6J mice was due to decreasing numbers of ghrelin-producing cells. After gastric bypass, DIO mice had higher levels of ghrelin than mice without bypass, suggesting that lower feed intake may be associated with ghrelin levels [29]. In this study, ghrelin levels in mice fed LF diet tended to be higher, supporting the hypothesis that a HF diet reduces ghrelin-producing cells. Previous studies have not reported the effect of quercetin feeding with HF diets on the secretion of ghrelin.

Plasma insulin was lower in mice fed a HFQ diet compared to HF diet-fed mice. A high fasting plasma insulin level is an indicator of IR, and IR often precedes T2D. Plasma insulin and blood glucose levels of mice fed the HF diet were 83.3% and 25.4% higher ($p < 0.05$), respectively, than those on the 0.05% HFQ diet after 6 weeks. Kobori et al. [7] reported higher plasma insulin concentrations of C57BL/6J mice fed a Western diet (39.9% energy from fat) compared to the Western diet containing 0.05% quercetin after a 20 week feeding study but not at 8 weeks. However, the same researchers reported that, in STZ-induced diabetic BALB/c mice, low-fat (AIN93) diets containing 0.5% quercetin restored plasma insulin compared to STZ diabetic controls ($p < 0.05$) but did not reach the levels of the normal nondiabetic mice [30]. These results indicated that quercetin supplementation lowers insulin secretion compared to the HF diet.

Leptin is an adipokine that regulates energy balance by inhibiting hunger. In humans, obesity is also associated with higher serum leptin levels, indicating leptin resistance. In this study, serum leptin levels of mice fed the HFQ and LF diets were significantly ($p < 0.05$) lower compared to HF, suggesting normal leptin sensitivity (Figure 4). Few studies have reported the effect of quercetin intake on plasma leptin level. Hoek-van den Hil and coworkers [31] reported a lower body weight gain in C57bl/6JOlaHsd mice fed a HF diet containing 0.33% quercetin for 12 weeks. They reported that serum leptin and its adipose gene expression were lower compared to HF controls. In rats, Wein and coworkers [32] reported no differences in leptin levels between HF or 0.03% HFQ diets for 4 weeks, possibly due to the lower level of quercetin and the shorter feeding period.

Resistin is a proinflammatory adipokine related to insulin resistance and obesity. Zhang and coworkers [33] reported that quercetin (75 mg/kg/day) decreased serum resistin in a rat model of nonalcoholic fatty liver disease induced by a HF diet for 8 weeks. Studies of quercetin on plasma resistin are scarce, however, related flavonoids from herbs have shown effects on resistin concentration. The infusion of quercetin and *Ruta graveolens*, a traditional medicinal plant native to Europe, inhibited resistin expression in adipose tissue through downregulation of the resistin-encoding gene [34].

Glucagon is a hormone produced by the α cells of the pancreas that promotes glucose synthesis and fatty-acid oxidation in the liver. Glucagon's action opposes that of insulin, and higher levels of glucagon in humans have been associated with insulin resistance [35]. The lower level of plasma glucagon in HFQ fed mice in this study supports the HOMA-IR index for increased insulin sensitivity [36]. To the best of our knowledge, the present study is the first to report the inhibitory effect of a HFQ diet on plasma resistin and improvement of HOMA-IR by reducing glucagon secretion.

3.5. Plasma Levels of Inflammatory Cytokines

There were no differences in all plasma cytokines, IL-2, IL-4, IL-5, IL-10, IL-12, GM-CSF, IFN-g, and TNF-α, between HFQ and HF or LF treatments analyzed using the multiplex immunoassay, as shown in the Supplementary Figure S1. However, despite this lack of difference in cytokine levels, several studies have found that quercetin decreases TNF-α. C57BL/6 mice fed quercetin at the same concentration (0.05% of diet) as this study but in a diet with higher fat content (60% energy from fat) for 9 weeks had a lower TNF-α level [37]. Kobori and coworkers [7] reported a decrease in TNF-α at 20 weeks but not at 8 weeks in C57BL/6J mice fed a diet with a similar fat content to the present study. The lack of a difference between treatments in our study may be due to a need for longer high-fat feeding time required to develop inflammation and the fat level of the diet.

3.6. Fecal Microbiota Analysis

Quercetin supplementation of the HF diet resulted in beneficial changes in the microbiome compared with HF. Quercetin lowered the relative abundance of the phyla Actinobacteria and Firmicutes (87.4% and 14.9%, respectively) but increased the abundance of Bacteroidetes (49.3%) (Figure 5A). The ratio of Firmicutes/Bacteroidetes (F/B) (Figure 5B) was significantly reduced by 65.6% by the HFQ diet ($p < 0.05$), and it was not different from LF diet. The ratio of Firmicutes to Bacteroidetes, F/B, is often cited as a marker of microbiota-associated obesity [38]. Firmicutes and Bacteroidetes account for 90% of the phyla in humans, and other phyla such as Actinobacteria are present at about 5–6%. Porras and coworkers [22] found that an HF diet induced nonalcoholic fatty liver disease in mice, which had an increased relative abundance of Bacteroidetes. However, Firmicutes were not significantly affected by 0.05% quercetin supplementation in the HF diet ($p > 0.05$). Turnbaugh and coworkers [39] claimed that the gut microbiota in ob/ob mice were more efficient at releasing calories from the diet than their lean siblings. The present study is the first to report that mice fed HF diets supplemented with quercetin significantly lowered the relative abundance of Firmicutes and increased the relative abundance of Bacteroidetes ($p < 0.05$). The alpha-diversity (Figure 5C) of the gut microbiota in the HFQ group was similar to that of the HF and LF groups, suggesting that quercetin supplementation did not affect the diversity of the gut microbiota. The results of the beta-diversity analysis illustrated that the taxonomic composition was distinctly different between the HF and LF groups (Figure 5D). To further characterize the changes to the gut microbiota, an LEfSe analysis was used to gain insight into the differences among the groups (Figure 5E,F). One significantly different class and one family were identified; Bacilli was higher in the LF group, and Peptostreptococcaceae was higher in the HFQ group. According to the LEfSe analysis, these abundant taxa could be considered as potential biomarkers (LDA score > 3.0, $p < 0.05$).

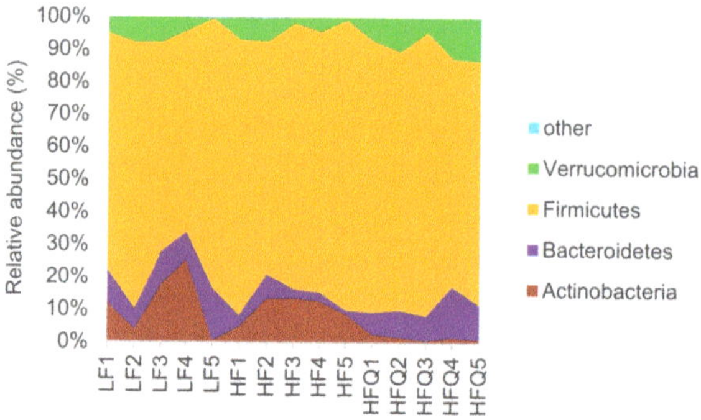

(A)

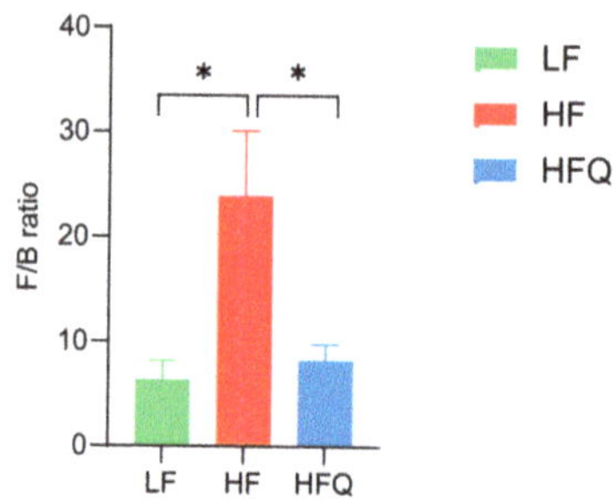

(B)

Figure 5. *Cont.*

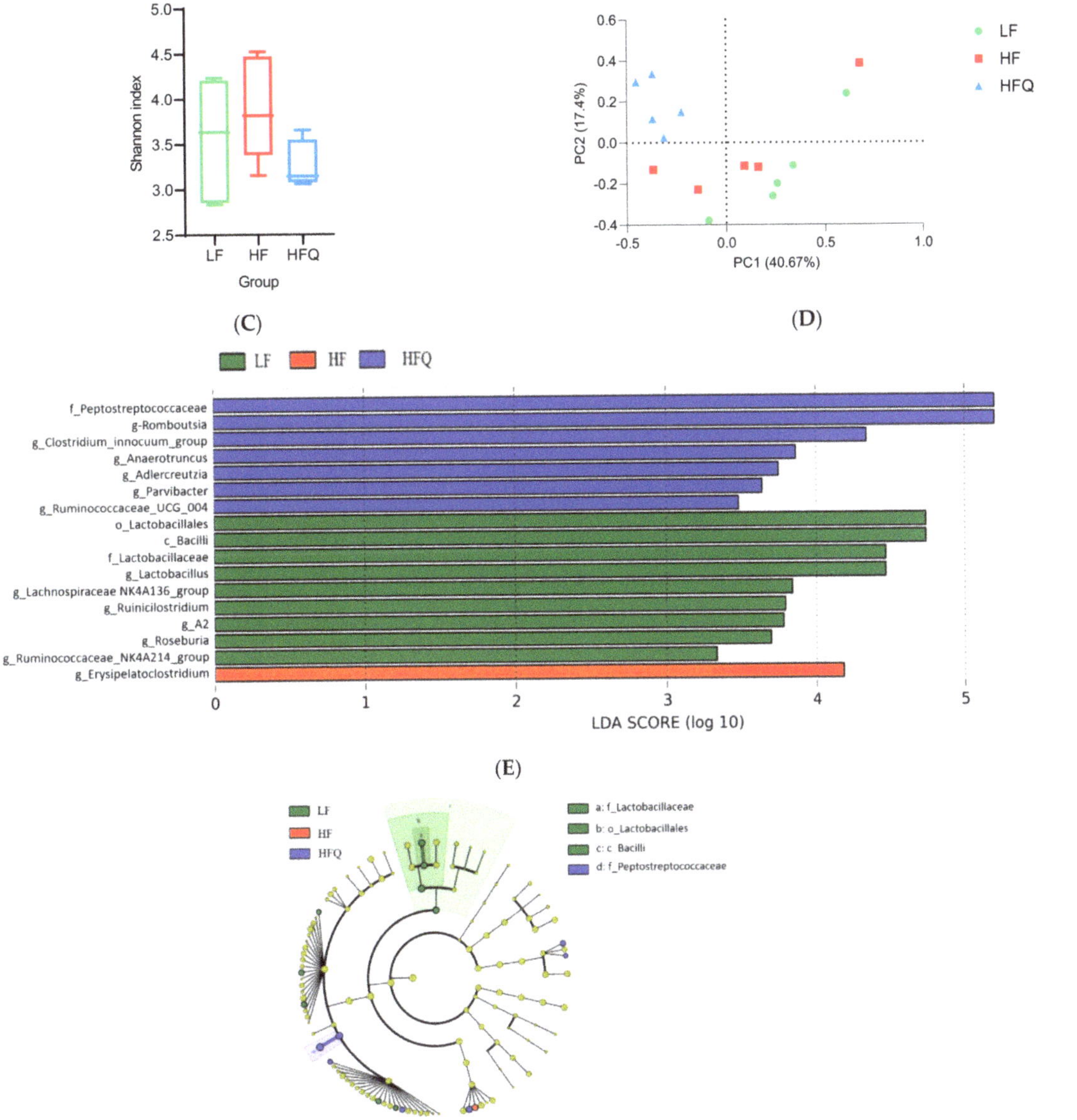

(C) (D) (E) (F)

Figure 5. (**A**) Effect of 0.05% quercetin on the relative abundance of the four most abundant bacterial phyla (n = 5) in LF (low−fat), HF (high−fat), and HFQ (high−fat with 0.05% quercetin) diets. (**B**) Firmicutes/Bacteroidetes ratio of HF, HFQ, and LF diets for 6 weeks. (**C**). The alpha−diversity was assessed by calculating the Shannon index. (**D**) The beta−diversity was calculated by principal coordinate analysis (PCoA) for the visualization of pairwise community dissimilarity (Bray−Curtis index) of the microbial community. (**E**) Results of linear discriminative analysis (LDA). (**F**) Effect size (LefSe) analysis among three groups. Cardiogram showing differentially abundant taxonomic clades with an LDA score of 3.0 among groups with a p−value of 0.05; n = 5/group. An asterisk (*) indicates a significant difference.

At the family level, the relative abundance of Akkermansiaceae, Bacteroidaceae, Eggerthellaceae, and Peptostreptococcaceae in the HFQ diet-fed mice was greater than that of HF diet alone. The HFQ diet significantly reduced the relative abundance of Atopobiaceae and Erysipelotrichaceae compared to the HF diet ($p < 0.05$). Erysipelotrichaceae has associated with diet-induced obesity [14]. Rabot [40] reported that diet-induced obese mice had higher levels of the Erysipelotrichaceae family. The Pearson correlation analysis between leptin and Erysipelotrichaceae suggested a strong correlation $r = 0.99$, $p < 0.02$. In contrast to Erysipelotrichaceae, the relative abundance of the Bacteroidaceae family in HFQ diet-fed mice was higher in the present study and supports the work of Etxeberria et al. who reported an increasing relative abundance of the Bacteroidaceae family induced by the HFQ diet, although the changes were not statistically significant ($p > 0.05$) [14]. The enrichment of the Bacteroidaceae family was found to be negatively associated with HF diets in mice. Thus, the result from the current investigation suggested that the HFQ diet enrichment of the Bacteroidaceae family may have contributed to a gut microbiome with a higher level of Bacteroidaceae family. Tan et al. (2018) reported that quercetin fed for 12 weeks had a similar effect on body weight, metabolic features, and the gut microbiome, particularly when given with a soluble fiber [41]. It was reported that quercetin was metabolized in the gut after administration, and its methylated metabolite isorhamnetin was the dominant form in the serum; however, the metabolites of quercetin in the gut remained unclear [42]. Moreover, the authors also evaluated the short-chain fatty acids (SCFAs) in the fecal samples, and results suggested that quercetin-treated mice exhibited significantly ($p < 0.05$) higher levels of butyrate in their feces compared with the antibiotic-treated mice [42]. Moreover, quercetin supplementation did not seem to modify the intestinal barrier permeability-associated markers TJP-1, TJP-2, and Ocln gene [12]. Jin et al. reported that quercetin increased the expression of ZO-1 in rats treated with quercetin [43]. These two studies indicated that quercetin might affect the tight junction protein level of ZO-1but not Ocln. Weight gain was reported to be accompanied by the release of inflammatory cytokines from adipose tissues in response to lipopolysaccharides (LPS), cell-wall fragments of Gram-negative bacteria that pass from the intestinal lumen through the intestinal wall [44]. While Gram-negative bacteria are the sources of LPS and initiator of inflammation, other commensal and probiotic bacteria were shown to prevent or reduce the severity of diabetes and other metabolic diseases [45]. Correlation between glucagon and Peptostreptococcaceae was strong ($r = 0.99$, $p < 0.04$). The present study is the first to report that the relative abundance of Atopobiaceae, Eggerthellaceae, and Peptostreptococcaceae families was significantly affected by the HFQ diet compared to the HF fed mice.

At the genus level, the relative abundances of *Akkermansia*, *Bacteroides*, *Marvinbryantia*, and *Romboutsia* genera were significantly increased (64.2%, 67.9%, 65.1%, and 75.2%, respectively) by the HFQ diet compared to the HF alone. There have been few reports of the presence of *Romboutsia* in feces, but a recent report suggested that this genus was indicative of a healthy status of patients [46]. The relative abundance of *Blautia*, *Clostridium sensu stricto 1*, *Erysipelatoclostridium*, *Lactobacillus*, and *Turicibacter* genera was lowered by the HFQ diet compared to the HF diet, but this did not reach statistical significance. *Akkermansia muciniphila* is one of the most abundant species in the intestine, which has been found to be lower in obesity and lowered in individuals treated with metformin [47]. *Akkermansia* enrichment was inversely correlated with obesity [14]. We observed an increase by 64.2% in the relative abundance of *Akkermansia* in the feces of HFQ-fed mice, as well as a significant increase by 68.0% in the relative abundances of *Bacteroides*, *Marvinbryantia*, and *Romboutsia* genera in feces of the HFQ-fed mice compared to the HF diet. More propionate was produced by *Bacteroides*, and the lipid synthesis from acetate was inhibited by propionate [48]. The increased abundance of *Bacteroides* may contribute to weight loss via propionate inhibition of lipid synthesis. In addition, a significantly greater relative abundance of *Marvinbryantia* genus was observed in the feces of HFQ-fed mice. The relative abundance of *Marvinbryantia* was positively correlated with body weight [49]. However, the mechanism is unknown. The current results show that a HFQ diet reshapes the gut microbiome of mice in the HF diet group withtaxa associated with a lean phenotype and less metabolic dysfunction.

3.7. RT-PCR Analysis

Plasma TG, LDL and total hepatic lipid contents, fasting glucose, GTT, and HOMA-IR were significantly lower in mice fed HFQ diets compared to HF diet-fed mice. We, therefore, compared the expression of selected genes for fat and glucose metabolic pathways in liver and adipose tissue (Figure 6A,B, respectively). The mRNA levels of hepatic genes for major enzymes for cholesterol and bile acid synthesis, *Cyp51* (cytochrome P450, family 51, encoding lanosterol 14α-demethylase) was 0.35-fold lower than the HF diet, but the expression of *Cyp7a1* (cytochrome P450, family 7, subfamily a, polypeptide 1, encoding cholesterol 7-alpha-monooxygenase) was not changed (0.93 compared to HF). Lanosterol demethylase is considered to be the first committed step for the synthesis of sterols from lanosterol. *Cyp7a1* encodes the gene for cholesterol 7-alpha hydroxylase, the enzyme responsible for the rate-limiting step of bile acid synthesis. These results suggest that the observed lower LDL cholesterol was due to reduced hepatic cholesterol synthesis rather than excretion of bile acids. Fatty-acid metabolism-related genes, *Scd1* (stearoyl-coenzyme A desaturase 1) and *Srebf1* (sterol regulatory element binding transcription factor 1), were reduced by 0.52 and 0.35-fold, respectively. *Srebf1* is a gene regulated by insulin and codes for transcription factor that increase glycolysis and lipogenesis. Its reduced expression relative to HF suggests that the effect may have been due to lower plasma insulin. *Scd1* catalyzes the synthesis of monounsaturated fatty acids that serve as substrates for the synthesis and storage of triglycerides. The relative expression of *Ppara* (peroxisome proliferator activated receptor alpha), a nuclear transcription factor regulating hepatic fat metabolism, was 0.75 of the control. The lower expression of *Pparα* and *Scd1* suggests a lower uptake and oxidation of fatty acids and may be related to the lower liver TG levels (Figure 2B).

In adipose, fatty-acid synthase (Fasn) is a multienzyme protein that metabolizes acetyl- and malonyl-CoA derived from glucose into fatty acids. The downregulation of the *Fasn* gene may be related to the lower liver lipid content (Figures 2B and 6B). *Slc2a4* (solute carrier family 2, facilitated glucose transporter member 4, also known as *Glut4*) is an insulin-regulated glucose transporter, and downregulation of Slc2a4 in adipose tissue indicated less glucose intake and, therefore, less fat storage in adipocytes in mice fed an LF diet. However, overexpression of Slc2a4 increases insulin sensitivity and glucose tolerance in obese mice [50]. Slc2a4 was significantly increased (Figure 6B) and supports the lower HOMA-IR and improved insulin sensitivity by the HFQ diet compared to the HF diet alone. Adiponectin is a protein hormone coded by the *Adipoq* gene by adipocytes which regulates glucose level and fatty-acid oxidation. *Adipoq* expression was almost three times higher than HF control (Figure 6B), confirming the plasma multiplex immunoassay results. Higher adiponectin levels are also associated with improved insulin sensitivity, as indicated by lower HOMA-IR. Decreased *Adipoq* is related to the development of IR [51]. The proposed mechanism responsible for ameliorating insulin resistance in mice fed an HF diet supplemented with quercetin is shown in Figure 6C. The increased level of *Akkermansia* was related to reduced insulin resistance [52]. The *Lactobacillus* level was reported to be positively associated with weight loss [53]. Gut microbiome changes are associated with insulin resistance; however, the mechanism behind the association remains unclear.

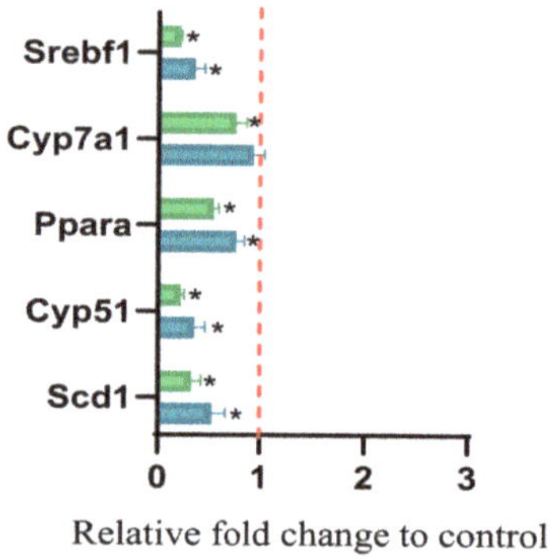

(**A**)

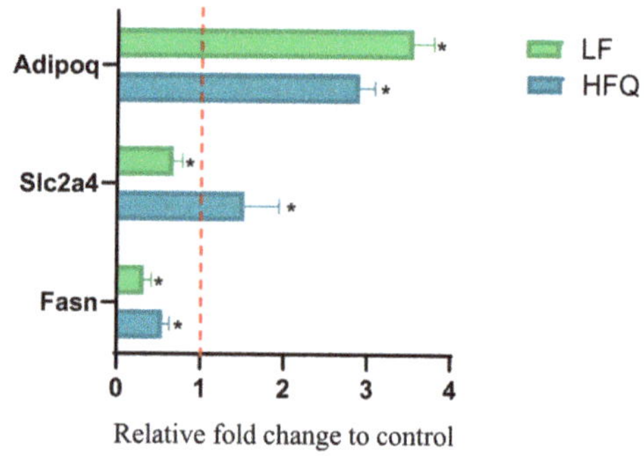

(**B**)

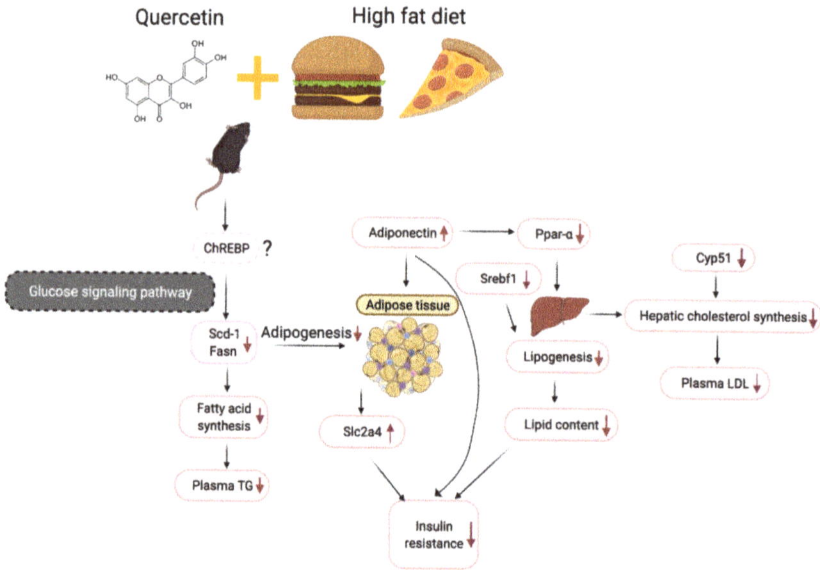

(**C**)

Figure 6. (**A**) Relative hepatic gene expression of *Srebf1*, *Cyp7a1*, *Ppara*, *Cyp51*, and *Scd1* in mice fed an HFQ diet compared to the HF diet (red dotted line), with the LF diet group shown as a reference. (**B**) Relative expression of *Fasn*, *Slc2a4*, and *Adipoq* genes in epididymal adipose tissue of mice fed an HFQ diet compared to the HF diet, with the LF diet group shown as a reference. Data are expressed as the mean ± SE, $n = 4$/group. Differences in mRNA expression for livers and adipose tissue were calculated after normalizing to 35B4 mRNA expression. An asterisk (*) indicates a significant difference ($p < 0.05$) compared to the HF diet. The dotted line (x = 1) represents the expression of control gene. (**C**) Proposed mechanism responsible for ameliorating insulin resistance in mice fed an HF diet supplemented with quercetin, involving the glucose signaling pathway, and ChREBP (dashes outline), as extrapolated from the literature [54].

4. Summary

The present study demonstrated that chronic quercetin supplementation for 6 weeks significantly lowered plasma triglyceride level, decreased body weight gain, reduced liver fat accumulation, ameliorated insulin resistance, and decreased F/B ratio (associated with lean phenotype) of HF diet-induced obese mice. Plasma ghrelin, leptin, resistin, and glucagon changes were associated with improvements in metabolic health. The HFQ diet attenuated the increase in Firmicutes/Bacteroidetes ratio and modulated gut bacteria composition at both family and genus levels. Furthermore, chronic supplementation of quercetin resulted in a significant alternation in genera related to obesity (*Bacteroides* and *Akkermansia*). The present study suggested that chronic quercetin supplementation prevented the increase of many biomarkers of obesity-related metabolic dysfunction, and microbiota associated with lean phenotypes and healthy metabolic profiles were restored by quercetin supplementation.

Supplementary Materials: The following are available online at https://www.mdpi.com/article/10.3390/antiox10081251/s1, Figure S1: Concentrations of plasma inflammatory cytokines, $n = 8$/group. Top edge of the box, 75th percentile; bottom edge, 25th percentile; horizontal bar within box, median; top horizontal bar outside box, maximum concentration; bottom horizontal bar outside box, minimum concentration. Boxes with different letters were significantly different ($p < 0.05$)., Table S1: gene primers used in the PCR assay.

Author Contributions: Conceptualization, Y.T., W.Y. and H.H.; methodology, Y.T. and W.Y.; software, Y.T., H.H. and M.R.; validation, C.C.T., P.A. and L.C.; formal analysis, Y.T.; investigation, Y.T. and C.C.T.; resources, W.Y.; writing—original draft preparation, Y.T.; writing—review and editing, Y.T., W.Y., S.K.C.C. and S.M.; visualization, Y.T. and H.H.; supervision, W.Y., S.M. and H.H.; project administration, W.Y. All authors have read and agreed to the published version of the manuscript.

Funding: This research received no external funding.

Institutional Review Board Statement: This study was approved (13 August 2018) by the Institutional Animal Care and Use Committee, Western Regional Research Center, USDA, Albany, CA, USA (protocol No. 18-4).

Informed Consent Statement: Not applicable.

Data Availability Statement: Data are contained within the article.

Acknowledgments: The authors thank Jackie Miller and Ezekial Martinez for their diligent attention to animal care, and James Pan for helping with liver fat extraction. My daughter Zhiyan Meng (juanjuan) accompanied me revising, editing and proofread the manuscript.

Conflicts of Interest: The authors declare no conflict of interest.

References

1. Chu, D.T.; Nguyet, N.T.M.; Dinh, T.C.; Lien, N.V.T.; Nguyen, K.H.; Ngoc, V.T.N.; Tao, F.; Son, L.H.; Le, D.H.; Nga, V.B.; et al. An update on physical health and economic consequences of overweight and obesity. *Diabetes Metab. Syndr.* **2018**, *12*, 1095–1100. [CrossRef]
2. Hollman, P.C.H.; Arts, I.C.W. Flavonols, flavones and flavanols—nature, occurrence and dietary burden. *J. Sci. Food Agric.* **2000**, *80*, 1081–1093. [CrossRef]
3. Nishimura, M.; Muro, T.; Kobori, M.; Nishihira, J. Effect of daily ingestion of quercetin-rich onion powder for 12 weeks on visceral fat: A randomised, double-blind, placebo-controlled, parallel-group study. *Nutrients* **2020**, *12*, 91. [CrossRef]
4. Ho, T.Y.; Lo, H.Y.; Liu, I.C.; Lin, K.A.; Liao, Y.F.; Lo, Y.C.; Hsiang, C.Y. The protective effect of quercetin on retinal inflammation in mice: The involvement of tumor necrosis factor/nuclear factor-κB signaling pathways. *Food Funct.* **2020**, *11*, 8150–8160. [CrossRef]
5. Walle, T.; Walle, U.K.; Halushka, P.V. Carbon dioxide is the major metabolite of quercetin in humans. *J. Nutr.* **2001**, *131*, 2648–2652. [CrossRef]
6. Nakamura, A.; Yokoyama, Y.; Tanaka, K.; Benegiamo, G.; Hirayama, A.; Zhu, Q.; Kitamura, N.; Suguzaki, T.; Morimoto, K.; Itoh, H.; et al. Asperuloside improves obesity and type 2 diabetes through modulation of gut microbiota and metabolic signaling. *Iscience* **2020**, *23*, 101522. [CrossRef]
7. Kobori, M.; Masumoto, S.; Akimoto, Y.; Oike, H. Chronic dietary intake of quercetin alleviates hepatic fat accumulation associated with consumption of a Western-style diet in C57/BL6J mice. *Mol. Nutr. Food Res.* **2011**, *55*, 530–540. [CrossRef]

8. Kobori, M.; Takahashi, Y.; Akimoto, Y.; Sakurai, M.; Matsunaga, I.; Nishimuro, H.; Ippoushi, K.; Oike, H.; Ohnishi-Kameyama, M. Chronic high intake of quercetin reduces oxidative stress and induces expression of the antioxidant enzymes in the liver and visceral adipose tissues in mice. *J. Funct. Food* **2015**, *15*, 551–560. [CrossRef]
9. Porras, D.; Nistal, E.; Martínez-Flórez, S.; Olcoz, J.L.; Jover, R.; Jorquera, F.; González-Gallego, J.; García-Mediavilla, M.V.; Sánchez-Campos, S. Functional Interactions between Gut Microbiota Transplantation, Quercetin, and High-Fat Diet Determine Non-Alcoholic Fatty Liver Disease Development in Germ-Free Mice. *Mol. Nutr. Food Res.* **2019**, *63*, 1800930. [CrossRef]
10. Jeong, S.-M.; Kang, M.-J.; Choi, H.-N.; Kim, J.-H.; Kim, J.-I. Quercetin ameliorates hyperglycemia and dyslipidemia and improves antioxidant status in type 2 diabetic db/db mice. *Nutr. Res. Pract.* **2012**, *6*, 201–207. [CrossRef] [PubMed]
11. Ferrarese, R.; Ceresola, E.R.; Preti, A.; Canducci, F. Probiotics, prebiotics and synbiotics for weight loss and metabolic syndrome in the microbiome era. *Eur. Rev. Med. Pharmacol. Sci.* **2018**, *22*, 7588–7605.
12. Shabbir, U.; Rubab, M.; Daliri, E.B.M.; Chelliah, R.; Javed, A.; Oh, D.H. Curcumin, quercetin, catechins and metabolic diseases: The role of gut microbiota. *Nutrients* **2021**, *13*, 206. [CrossRef]
13. Dabeek, W.M.; Marra, M.V. Dietary quercetin and kaempferol: Bioavailability and potential cardiovascular-related bioactivity in humans. *Nutrients* **2019**, *11*, 2288. [CrossRef]
14. Etxeberria, U.; Arias, N.; Boqué, N.; Macarulla, M.T.; Portillo, M.P.; Martínez, J.A.; Milagro, F.I. Reshaping faecal gut microbiota composition by the intake of trans-resveratrol and quercetin in high-fat sucrose diet-fed rats. *J. Nutr. Biochem.* **2015**, *26*, 651–660. [CrossRef]
15. Kim, H.; Bartley, G.E.; Arvik, T.; Lipson, R.; Nah, S.Y.; Seo, K.; Yokoyama, W. Dietary supplementation of chardonnay grape seed flour reduces plasma cholesterol concentration, hepatic steatosis, and abdominal fat content in high-fat diet-induced obese hamsters. *J. Agric. Food Chem.* **2014**, *62*, 1919–1925. [CrossRef] [PubMed]
16. German, J.B.; Xu, R.; Walzem, R.; Kinsella, J.E.; Knuckles, B.; Nakamura, M.; Yokoyama, W.H. Effect of dietary fats and barley fiber on total cholesterol and lipoprotein cholesterol distribution in plasma of hamsters. *Nutr. Res.* **1996**, *16*, 1239–1249. [CrossRef]
17. Bowe, J.E.; Franklin, Z.J.; Hauge-Evans, A.C.; King, A.J.; Persaud, S.J.; Jones, P.M. Metabolic phenotyping guidelines: Assessing glucose homeostasis in rodent models. *J. Endocrinol.* **2014**, *222*, G13–G25. [CrossRef] [PubMed]
18. Bustin, S.A.; Benes, V.; Garson, J.A.; Hellemans, J.; Huggett, J.; Kubista, M.; Mueller, R.; Nolan, T.; Pfaffl, M.; Shipley, G.; et al. The MIQE Guidelines: Minimum Information for Publication of Quantitative Real-Time PCR Experiments. *Clin. Chem.* **2009**, *55*, 611–622. [CrossRef] [PubMed]
19. Livak, K.J.; Schmittgen, T.D. Analysis of relative gene expression data using real-time quantitative PCR and the $2-\Delta\Delta CT$ method. *Methods* **2001**, *25*, 402–408. [CrossRef]
20. Porras, D.; Nistal, E.; Martínez-Flórez, S.; Pisonero-Vaquero, S.; Olcoz, J.L.; Jover, R.; González-Gallego, J.; García-Mediavilla, M.V.; Sánchez-Campos, S. Protective effect of quercetin on high-fat diet-induced non-alcoholic fatty liver disease in mice is mediated by modulating intestinal microbiota imbalance and related gut-liver axis activation. *Free Radic. Bio. Med.* **2017**, *102*, 188–202. [CrossRef]
21. Kuipers, E.N.; Dam, A.D.V.; Held, N.M.; Mol, I.M.; Houtkooper, R.H.; Rensen, P.C.; Boon, M.R. Quercetin lowers plasma triglycerides accompanied by white adipose tissue Browning in diet-induced obese mice. *Int. J. Mol. Sci.* **2018**, *19*, 1786. [CrossRef] [PubMed]
22. Gašić, U.; Ćirić, I.; Pejčić, T.; Radenković, D.; Djordjević, V.; Radulović, S.; Tešić, Ž. Polyphenols as possible agents for pancreatic diseases. *Antioxidants* **2020**, *9*, 547. [CrossRef]
23. Ozkan, G.; Kostka, T.; Esatbeyoglu, T.; Capanoglu, E. Effects of Lipid-Based Encapsulation on the Bioaccessibility and Bioavailability of Phenolic Compounds. *Molecules* **2020**, *25*, 5545. [CrossRef] [PubMed]
24. Guo, X.D.; Zhang, D.Y.; Gao, X.J.; Parry, J.; Liu, K.; Liu, B.L.; Wang, M. Quercetin and quercetin-3-O-glucuronide are equally effective in ameliorating endothelial insulin resistance through inhibition of reactive oxygen species-associated inflammation. *Mol. Nutr. Food Res.* **2013**, *57*, 1037–1045. [CrossRef]
25. Zhou, M.; Wang, S.; Zhao, A.; Wang, K.; Fan, Z.; Yang, H.; Liao, W.; Bao, S.; Zhao, L.; Zhang, Y.; et al. Transcriptomic and metabonomic profiling reveal synergistic effects of quercetin and resveratrol supplementation in high fat diet fed mice. *J. Proteome Res.* **2012**, *11*, 4961–4971. [CrossRef]
26. Vessal, M.; Hemmati, M.; Vasei, M. Antidiabetic effects of quercetin in streptozocin-induced diabetic rats. *Comp. Biochem. Physiol. Part C* **2003**, *135*, 357–364. [CrossRef]
27. Matsukawa, N.; Matsumoto, M.; Shinoki, A.; Hagio, M.; Inoue, R.; Hara, H. Nondigestible saccharides suppress the bacterial degradation of quercetin aglycone in the large intestine and enhance the bioavailability of quercetin glucoside in rats. *J. Agric. Food Chem.* **2009**, *57*, 9462–9468. [CrossRef]
28. Moesgaard, S.G.; Ahrén, B.; Carr, R.D.; Gram, D.X.; Brand, C.L.; Sundler, F. Effects of high-fat feeding and fasting on ghrelin expression in the mouse stomach. *Regul. Pept.* **2004**, *120*, 261–267. [CrossRef]
29. Uchida, A.; Zechner, J.F.; Mani, B.K.; Park, W.M.; Aguirre, V.; Zigman, J.M. Altered ghrelin secretion in mice in response to diet-induced obesity and Roux-en-Y gastric bypass. *Mol. Metab.* **2014**, *3*, 717–730. [CrossRef]
30. Kobori, M.; Masumoto, S.; Akimoto, Y.; Takahashi, Y. Dietary quercetin alleviates diabetic symptoms and reduces streptozotocin-induced disturbance of hepatic gene expression in mice. *Mol. Nutr. Food Res.* **2009**, *53*, 859–868. [CrossRef]

31. Hoek-van den Hil, E.F.; van Schothorst, E.M.; van der Stelt, I.; Swarts, H.J.; van Vliet, M.; Amolo, T.; Veroort, J.J.M.; Venema, D.; Hollman, P.C.H.; Rietjens, I.; et al. Direct comparison of metabolic health effects of the flavonoids quercetin, hesperetin, epicatechin, apigenin and anthocyanins in high-fat-diet-fed mice. *Genes Nutr.* **2015**, *10*, 23. [CrossRef] [PubMed]
32. Wein, S.; Behm, N.; Petersen, R.K.; Kristiansen, K.; Wolffram, S. Quercetin enhances adiponectin secretion by a PPAR-γ independent mechanism. *Eur. J. Pharm. Sci.* **2010**, *41*, 16–22. [CrossRef] [PubMed]
33. Zhang, M.H.; Liang, Z.Q.; Qin, Q.; Li, S.L.; Zhou, D.S.; Tang, L. Effects of quercetin on serum levels of resistin and IL-18 and on insulin resistance in nonalcoholic fatty liver disease rats. *Chin. J. Hepatol.* **2013**, *21*, 66–70.
34. Ahmed, O.M.; Moneim, A.A.; Yazid, I.A.; Mahmoud, A.M. Antihyperglycemic, antihyperlipidemic and antioxidant effects and the probable mechanisms of action of Ruta graveolens infusion and rutin in nicotinamide-streptozotocin-induced diabetic rats. *Diabetol. Croat.* **2010**, *39*, 15–35.
35. Kawamori, D.; Katakami, N.; Takahara, M.; Miyashita, K.; Sakamoto, F.; Yasuda, T.; Matsuoka, T.; Shimomura, I. Dysregulated plasma glucagon levels in Japanese young adult type 1 diabetes patients. *J. Diabetes Investig.* **2019**, *10*, 62–66. [CrossRef]
36. Ferrannini, E.; Muscelli, E.; Natali, A.; Gabriel, R.; Mitrakou, A.; Flyvbjerg, A.; Golay, A.; Hojlund, K. Association of fasting glucagon and proinsulin concentrations with insulin resistance. *Diabetologia* **2007**, *50*, 2342–2347. [CrossRef] [PubMed]
37. Kim, C.-S.; Choi, H.-S.; Joe, Y.; Chung, H.T.; Yu, R. Induction of heme oxygenase-1 with dietary quercetin reduces obesity-induced hepatic inflammation through macrophage phenotype switching. *Nutr. Res. Pract.* **2016**, *10*, 623–628. [CrossRef] [PubMed]
38. Ley, R.E.; Turnbaugh, P.J.; Klein, S.; Gordon, J.I. Human gut microbes associated with obesity. *Nature* **2006**, *444*, 1022–1023. [CrossRef] [PubMed]
39. Turnbaugh, P.J.; Ley, R.E.; Mahowald, M.A.; Magrini, V.; Mardis, E.R.; Gordon, J.I. An obesity-associated gut microbiome with increased capacity for energy harvest. *Nature* **2006**, *444*, 1027. [CrossRef] [PubMed]
40. Rabot, S.; Membrez, M.; Blancher, F.; Berger, B.; Moine, D.; Krause, L.; Bibiloli, R.; Bruneau, A.; Gerard, P.; Siddharth, J.; et al. High fat diet drives obesity regardless the composition of gut microbiota in mice. *Sci. Rep.* **2016**, *6*, 32484. [CrossRef]
41. Tan, S.; Caparros-Martin, J.A.; Matthews, V.B.; Koch, H.; O'Gara, F.; Croft, K.D.; Ward, N.C. Isoquercetin and inulin synergistically modulate the gut microbiome to prevent development of the metabolic syndrome in mice fed a high fat diet. *Sci. Rep.* **2018**, *8*, 1–13. [CrossRef]
42. Shi, T.; Bian, X.; Yao, Z.; Wang, Y.; Gao, W.; Guo, C. Quercetin improves gut dysbiosis in antibiotic-treated mice. *Food Funct.* **2020**, *11*, 8003–8013. [CrossRef]
43. Jin, Y.; Huang, Z.L.; Li, L.; Yang, Y.; Wang, C.H.; Wang, Z.T.; Ji, L.L. Quercetin attenuates toosendanin-induced hepatotoxicity through inducing the Nrf2/GCL/GSH antioxidant signaling pathway. *Acta Pharmacol. Sin.* **2019**, *40*, 75–85. [CrossRef] [PubMed]
44. Hersoug, L.G.; Møller, P.; Loft, S. Gut microbiota-derived lipopolysaccharide uptake and trafficking to adipose tissue: Implications for inflammation and obesity. *Obes. Rev.* **2016**, *17*, 297–312. [CrossRef]
45. Kootte, R.S.; Vrieze, A.; Holleman, F.; Dallinga-Thie, G.M.; Zoetendal, E.G.; de Vos, W.M.; Groen, A.K.; Hekstra, J.B.L.; Stroes, E.S.; Nieuwdorp, M. The therapeutic potential of manipulating gut microbiota in obesity and type 2 diabetes mellitus. *Diabetes Obes. Metab.* **2012**, *14*, 112–120. [CrossRef] [PubMed]
46. Mangifesta, M.; Mancabelli, L.; Milani, C.; Gaiani, F.; de'Angelis, N.; de'Angelis, G.L.; de'Angelis, G.L.; Sinderen, D.S.; Ventura, M.; Turroni, F. Mucosal microbiota of intestinal polyps reveals putative biomarkers of colorectal cancer. *Sci. Rep.* **2018**, *8*, 13974. [CrossRef]
47. Cani, P.D.; de Vos, W.M. Next-generation beneficial microbes: The case of Akkermansia muciniphila. *Front. Microbiol.* **2017**, *8*, 1765. [CrossRef]
48. Pouteau, E.; Nguyen, P.; Ballèvre, O.; Krempf, M. Production rates and metabolism of short-chain fatty acids in the colon and whole body using stable isotopes. *Proc. Nutr. Soc.* **2003**, *62*, 87–93. [CrossRef] [PubMed]
49. Henning, S.M.; Yang, J.; Hsu, M.; Lee, R.-P.; Grojean, E.M.; Ly, A.; Tseng, C.H.; Heber, D.; Li, Z. Decaffeinated green and black tea polyphenols decrease weight gain and alter microbiome populations and function in diet-induced obese mice. *Eur. J. Nutr.* **2018**, *57*, 2759–2769. [CrossRef]
50. Shepherd, P.R.; Kahn, B.B. Glucose transporters and insulin action—implications for insulin resistance and diabetes mellitus. *N. Engl. J. Med.* **1999**, *341*, 248–257. [CrossRef]
51. Yamauchi, T.; Kamon, J.; Waki, H.; Terauchi, Y.; Kubota, N.; Hara, K.; Mori, Y.; Murakami, I.; Tsuboyama-Kasaoka, N.; Akanuma, Y.; et al. The fat-derived hormone adiponectin reverses insulin resistance associated with both lipoatrophy and obesity. *Nat. Med.* **2001**, *7*, 941. [CrossRef] [PubMed]
52. Anhê, F.F.; Roy, D.; Pilon, G.; Dudonné, S.; Matamoros, S.; Varin, T.V.; Garofalo, C.; Moine, Q.; Desjardins, Y.; Levy, E.; et al. A polyphenol-rich cranberry extract protects from diet-induced obesity, insulin resistance and intestinal inflammation in association with increased Akkermansia spp. population in the gut microbiota of mice. *Gut* **2015**, *64*, 872–883. [CrossRef] [PubMed]
53. Tilg, H.; Moschen, A.R. Microbiota and diabetes: An evolving relationship. *Gut* **2014**, *63*, 1513–1521. [CrossRef] [PubMed]
54. Postic, C.; Girard, J. Contribution of de novo fatty acid synthesis to hepatic steatosis and insulin resistance: Lessons from genetically engineered mice. *J. Clin. Investig.* **2008**, *118*, 829–838. [CrossRef] [PubMed]

antioxidants

Article

Bacillus amyloliquefaciens SC06 Induced AKT–FOXO Signaling Pathway-Mediated Autophagy to Alleviate Oxidative Stress in IPEC-J2 Cells

Li Tang [1,2], Zihan Zeng [1,2], Yuanhao Zhou [1,2], Baikui Wang [1,2], Peng Zou [1,2], Qi Wang [1,2], Jiafu Ying [1,2], Fei Wang [1,2], Xiang Li [1,2], Shujie Xu [1,2], Pengwei Zhao [3,*] and Weifen Li [1,2,*]

1 Key Laboratory of Molecular Animal Nutrition of the Ministry of Education, Institute of Feed Science, College of Animal Sciences, Zhejiang University, Hangzhou 310058, China; litang2017@zju.edu.cn (L.T.); 21817016@zju.ecu.cn (Z.Z.); 12017005@zju.edu.cn (Y.Z.); wangbaikui@zju.edu.cn (B.W.); 21917085@zju.edu.cn (P.Z.); 21917012@zju.edu.cn (Q.W.); 21817401@zju.edu.cn (J.Y.); 22017007@zju.edu.cn (F.W.); 22017014@zju.edu.cn (X.L.); 22017075@zju.edu.cn (S.X.)

2 Key Laboratory of Animal Nutrition and Feed Science (Eastern of China), Ministry of Agriculture and Rural Affairs, Hangzhou 310058, China

3 Department of Biochemistry, Department of Cardiology of Second Affiliated Hospital, Zhejiang University School of Medicine, Hangzhou 310058, China

* Correspondence: zhaopengwei@zju.edu.cn (P.Z.); wfli@zju.edu.cn (W.L.)

Citation: Tang, L.; Zeng, Z.; Zhou, Y.; Wang, B.; Zou, P.; Wang, Q.; Ying, J.; Wang, F.; Li, X.; Xu, S.; et al. *Bacillus amyloliquefaciens SC06* Induced AKT–FOXO Signaling Pathway-Mediated Autophagy to Alleviate Oxidative Stress in IPEC-J2 Cells. *Antioxidants* **2021**, *10*, 1545. https://doi.org/10.3390/antiox10101545

Academic Editors: Baojun Xu and Stanley Omaye

Received: 19 July 2021
Accepted: 25 September 2021
Published: 28 September 2021

Abstract: Autophagy is a conserved proteolytic mechanism, which degrades and recycles damaged organs and proteins in cells to resist external stress. Probiotics could induce autophagy; however, its underlying molecular mechanisms remain elusive. Our previous study has found that *BaSC06* could alleviate oxidative stress by inducing autophagy in rats. This research aimed to verify whether *Bacillus amyloliquefaciens SC06* can induce autophagy to alleviate oxidative stress in IPEC-J2 cells, as well as explore its mechanisms. IPEC-J2 cells were first pretreated with 10^8 CFU/mL *BaSC06*, and then were induced to oxidative stress by the optimal dose of diquat. The results showed that *BaSC06* significantly triggered autophagy, indicated by the up-regulation of LC3 and Beclin1 along with downregulation of p62 in IPEC-J2 cells. Further analysis revealed that *BaSC06* inhibited the AKT–FOXO signaling pathway by inhibiting the expression of p-AKT and p-FOXO and inducing the expression of SIRT1, resulting in increasing the transcriptional activity of FOXO3 and gene expression of the ATG5–ATG12 complex to induce autophagy, which alleviated oxidative stress and apoptosis. Taken together, *BaSC06* can induce AKT–FOXO-mediated autophagy to alleviate oxidative stress-induced apoptosis and cell damage, thus providing novel theoretical support for probiotics in the prevention and treatment of oxidative damage.

Keywords: *Bacillus amyloliquefaciens SC06*; IPEC-J2; oxidative stress; autophagy; apoptosis; AKT–FOXO

1. Introduction

The gastrointestinal tract is an important primary digestive organ, whilst intestinal health directly affects the health and growth of animals. While exposed to exogenous stimulators, the gastrointestinal tract is susceptible to oxidative stress [1,2]. The mucosal barrier, composed of intestinal epithelium, the mucus layer, and cells involved in local immune responses [3], plays a key role in maintaining intestinal homeostasis. The epithelium is composed of a single layer of columnar intraepithelial cells. IPEC-J2 cells, derived from the columnar epithelial cells of a piglet's jejunum, were first isolated from the middle jejunum of neonatal piglets by A.J. Brosnahan et al. at the University of North Carolina [4]. IPEC-J2 cells were initially used as a porcine small intestine model to study the pathogenic bacteria that induce porcine hyperplastic bowel disease [5], and then gradually applied to various studies, including oxidative stress [6], intestinal microorganism [7], and intestinal immune response [8]. IPEC-J2 cells are highly similar to normal intestinal epithelial cells due to their tumor-free genes, and are derived from piglets. The high similarity between IPEC-J2 cells and the real pig and human small intestine has gradually attracted researchers' attention in recent years.

Autophagy is an evolutionary conserved cell procedure that participates in the lysosomal degradation of proteins and organelles, contributing significantly to cell growth maintenance, differentiation, and homeostasis in some adverse conditions, such as hypoxia and starvation [9]. It plays an integral role in maintaining cellular homeostasis and promoting cell survival, while apoptosis exerts a defensive role, selectively removing cells and renewing cells and tissues [10,11]. Therefore, an equilibrium of autophagy and apoptosis in intestinal epithelial cells affects intestinal health. During the process of autophagy, portions of the cytoplasm or the entire organelle are sequestered in double-membranous vesicles, called autophagy vacuoles (AV) or autophagosomes. The autophagosome then fuses with the lysosome to form a monofilm autophagosome that degrades its contents [12].

In vivo and in vitro models, increasing evidence reveals that probiotics have the potential to alleviate oxidative stress in livestock and prevent them from oxidative stress. Some probiotics, such as *Lactobacillus and Bifidobacterium*, exhibit a good therapeutic effect on alleviating intestinal oxidative stress [13–15]. Probiotic *Bacillus* spp. are widely used to improve animal growth performance and prevent gastrointestinal disorders [16,17] and significantly alleviated intestinal oxidative stress in aquatic products and piglets [18–20]. In addition, *Bacillus* spp. could prevent oxidative stress and LPS-induced inflammatory responses in Raw 264.7 macrophages [21]. Furthermore, in chickens, *Bacillus subtilis* increased their antioxidant capacity and oxidative stability [22]. Studies have elucidated that probiotic can protect intestinal epithelial cells from damage and necrotizing apoptosis by regulating the autophagy and apoptosis signaling pathways, recruiting immune cells, and anti-inflammatory factors [23,24]. Some recent studies have shown that probiotics can regulate autophagy to alleviate oxidative stress. For example, metronidazole and *L. reuteri* combination treatment could decrease oxidative stress and inflammatory and autophagic pathways to prevent NAFLD progression [25]. *Lactobacillus reuteri ZJ617* and *Lactobacillus rhamnosus GG* supplementation suppressed lipopolysaccharide-induced oxidative stress by attenuating apoptosis and autophagy via the mTOR signaling pathway [26].

Currently, studies on the effect of probiotics on oxidative stress by regulating autophagy are still rare. In our previous experiments involving rats, it was uncovered that *Bacillus amyloliquefaciens SC06* (*BaSC06*) alleviated oxidative stress through autophagy via the p38 signaling pathway, but the other specific signaling pathway was inadequately investigated [27]. This study therefore sought to verify whether *BaSC06* can induce autophagy to alleviate oxidative stress in IPEC-J2 cells, as well as explore the other related signaling pathways.

2. Materials and Methods

2.1. *BaSC06 Bacterial Strain Preparation*

For this study, the probiotic *BaSC06* (CCTCC No: M2012280) was isolated from soil by the Laboratory of Microbiology, Institute of Feed Sciences, Zhejiang University, and preserved at the China Center for Type Culture Collection Afterward, the *BaSC06* strains were cultured at 37 °C in Luria-Bertani (LB) broth overnight, and then gathered by centrifugation (8000 rpm for 5 min). After that, the *BaSC06* strains were washed twice with PBS (pH = 7.4) and suspended at 10^8 CFU/mL in the cell culture media. The fresh bacteria suspensions were prepared for cell incubation.

2.2. *IPEC-J2 Cell Culture*

IPEC-J2 cells were provided by Northwest Sci-Tech University of Agriculture and Forestry, which was then incubated at 37 °C in a humidified 5% CO_2 with DMEM/F12 (HyClone, Logan, UT, USA) media, containing 10% FBS (Gibco, Grand Island, NE, USA) and 1% antibiotics (100 mg/mL of streptomycin and 100 U/mL of penicillin G).

2.3. Establishing Oxidative Stress Model in IPEC-J2 Cells

The diquat (DQ)-induced oxidative stress model was evaluated utilizing an MTT cell assay kit for cell proliferation and cytotoxicity (Nanjing Jiancheng Bioengineering Institute, Nanjing, China). According to instructions, 10^4 cells per well were seeded in 96-well plates and cultured for 12 h, followed by DQ treatment at various concentrations (0, 250, 500, 750, 1000, and 1250 μmol/L) for 6 h, with nine parallel holes in each group. Thereafter, to each well was added 50 μL of MTT assay solution, and then incubated for 4 h. Afterward, a Spectra Max M5 microplate reader (Sunnyvale, CA, USA) was used to determine the absorbance of the plate at 570 nm. In order to set up the oxidative stress model for IPEC-J2 cells, the optimal DQ concentration was selected according to the IC_{50}, calculated using a probability unit based on the MTT assay. The optimal concentration of *BaSC06* was determined using a Cell Counting Kit-8 (CCK-8 kit, Nanjing Jiancheng Bioengineering Institute, Nanjing, China) as per the instructions of the manufacturer. IPEC-J2 cells were treated with *BaSC06* at various concentrations (0, 10^5, 10^6, 10^7, 10^8, and 10^9 CFU/mL) using the same seeding method as MTT, with nine parallel holes in each group, for 6 h. A total of 10 μL of CCK-8 solutions were added to every well, and the plate incubated for 1 h. The optimal *BaSC06* concentration was selected and calculated according to the viability of the cells based on the CCK-8 assay. The Spectra Max M5 microplate reader (Sunnyvale, CA, USA) was used to determine the absorbance of the plate at 450 nm.

IPEC-J2 cells were further divided into four groups: CK (PBS treatment only), DQ (diquat (DQ) treatment only), Ba (*BaSC06* treatment only), and Ba+DQ (*BaSC06* combined with diquat treatment) groups. Notably, IPEC-J2 cells in the CK group were treated with PBS, while those in the DQ and Ba groups were treated with 950 μmol/mL diquat and 10^8 CFU/mL *BaSC06*, respectively. In the Ba+DQ group, IPEC-J2 cells was pretreated with 10^8 CFU/mL *BaSC06* for 6 h and then treated with 950 μmol/mL diquat for the same period. Before *BaSC06* or DQ treatment, each group was washed twice with PBS at the same time.

Furthermore, 25 μM AKT phosphorylation inhibitor Perifosine (Biotime Biotechnology) was used to pre-incubate with IPEC-J2 cells for 48 h according to the instruction, and we then repeated the former *BaSC06* and DQ treatment methods.

2.4. ROS Generation Analysis

A reactive oxygen species (ROS) assay kit (Nanjing Jiancheng Bioengineering Institute, Nanjing, China) was utilized to detect the ROS level in treated IPEC-J2 cells in 96-well plates, 10^4 cells in one well, with nine parallel holes in each group. The 2′,7′-dichlorohydrofluorescein diacetate (DCFH-DA) is the most sensitive and commonly used probe for detecting intracellular ROS. Specifically, all cell samples were treated with 10 μM DCFH-DA solution, after pretreated with *BaSC06* and DQ at 37 °C for 30 min and subsequently washed with FBS-free media and PBS. DCFH is oxidized into a strong green fluorescence substance DCF (dichlorofluorescein) in the presence of ROS in cells, and its emission wavelength is 502 nm, while the fluorescence peaks at the wavelength of 530 nm. Its fluorescence intensity is proportional to the ROS levels in cells. The fluorescence signals were monitored using the microplate reader SpectraMax M5 (Sunnyvale, CA, USA) and an Olympus BX61W1-FV1000 laser scanning confocal microscope (Tokyo, Japan).

2.5. Detection of Antioxidant Capacities

IPEC-J2 cell lysates were gathered to determine oxidative stress indexes including the level of methane dicarboxylic aldehyde (MDA), and the activities of superoxide dismutase (SOD), catalase (CAT), as well as the glutathione peroxidase (GSH-Px), each group with 6 replications. All assays were performed following the manufacturer's guidelines (Jiancheng Bioengineering Institute, Nanjing, China) [28].

2.6. Apoptosis Cell Analysis by TUNEL Assay

For this experiment, an Apoptosis Detection Kit named TUNEL BrightGreen (Vazyme, Nanjing, China) was used to detect the apoptosis levels in IPEC-J2 cells (n = 6). The 3′-hydroxyl terminus of the broken DNA can bind to FITC-12-DUTP, which is activated by Terminal Deoxynudleotidyl Transferase (TdT). Based on the manufacturer's instruction, 4% paraformaldehyde was used to fix the slides. Then all the slides were incubated with proteinase K solution at a concentration of 20 μg/mL, followed by 50 μL BrightGreen labeling mix as well as 50 μL recombinant TDT enzyme. Then, the stained IPEC-J2 cells were instantly examined under an Olympus BX61W1-FV1000 laser scanning confocal microscope (Tokyo, Japan). Furthermore, IPEC-J2 cell were collected to be measured by a FC500 flow cytometer (Beckman Coulter, Fullerton, CA, USA).

2.7. Annexin V-FITC/PI Apoptosis Assay

Annexin V has been used as a sensitive indicator of early apoptosis because it binds to the membrane of early apoptotic cells via phosphatidylserine exposed externally. Propidium iodide (PI) is a nucleic acid dye that cannot penetrate the entire cell membrane, but PI can penetrate the cell membrane and make the nucleus red due to increased membrane permeability in the middle and late apoptotic cells as well as dead cells. Therefore, annexin V matched with PI could be used to distinguish cells at different stages of apoptosis. IPEC-J2 cells were collected and washed in PBS. Then they were suspended in a 1× annexin binding buffer, mixed, and incubated with 5 μL annexin V-FITC V and 5 μL PI for 10 min. After that, a FC500 flow cytometer (Beckman Coulter, Fullerton, CA, USA) was utilized to measure the stained cells.

2.8. Detection of Caspase-3 Activity

The activity of caspase-3 was measured by the Caspase-3 Activity Kit (Beyotime, Shanghai, China). After treatment, IPEC-J2 cells were collected after rinsed with cold PBS and centrifugation. Caspase-3 activity in the supernatant were assayed using the kit following the instruction. The caspase activity was expressed as the percentage of enzyme activity compared to the control.

2.9. FOXO3a siRNA and Transfection

Three small interfering RNAs targeting pig FOXO3a and negative control siRNA were synthesized by Sangon Biotech, Shanghai, China, the sequences are summarized in Table 1. IPEC-J2 cells were cultured in antibiotic-free medium, then treated with Lipo-2000 and siRNA mixture. All the cells were collected for western blotting.

Table 1. List of siRNA.

Gene Name.	Access No.	Sequences	
		Sense(5′-3′)	Antisense(5′-3′)
sscFOXO3a-608	NM_001135959.1	CCGGCUGGAAGAACUCUAUTT	AUAGAGUUCUUCCAGCCGGTT
sscFOXO3a-1115	NM_001135959.1	CCGGAACCAUGAAUCUCAATT	UUGAGAUUCAUGGUUCCGGTT
sscFOXO3a-1928	NM_001135959.1	CCCUCAUCUCCACACAGAATT	UUCUGUGUGGAGAUGAGGGTT
Negative Control	NM_001135959.1	UUCUCCGAACGUGUCACGUTT	ACGUGACACGUUCGGAGAATT

2.10. Western Blotting

We refer to the protocol of Xiao et al. for our Western blot assay [29]. The cell lysis buffer produced by Western and American Psychological Association (RIPA, Biotime Biotechnology, Xiamen, China) was used to prepare the total IPEC-J2 cell lysates (n = 3), while the nuclear protein extraction kit (Biotime Biotechnology, Xiamen, China) was utilized to gather nuclear proteins (n = 3). In accordance with the manufacturer's protocol, a BCA Protein assay kit (Biotime Biotechnology, Xiamen, China) was used to measure protein concentrations. After the SDS-PAGE, proteins were electrophoretically transferred to nitrocellulose membranes (Sangon Biotech, Shanghai, China). Subsequently, the membranes were incubated with first antibodies at 4 °C overnight, and thereafter incubated with secondary antibodies goat anti-mouse IgG-HRP and goat anti-rabbit IgG-HRP after washing by TBST. The blots were then developed with an ECL detection system (Tanon 5200, Shanghai, China). Anti-Bcl2 and anti-LC3II primary antibodies were purchased from Abcam (Cambridge, MA, USA), while primary antibodies such as anti-Bax, anti-Beclin1, anti-SQSTM/p62, anti-phosphor-mTOR (Ser2448), anti-mTOR, anti-phosphor-AKT (Ser473), and anti-AKT used in this study were obtained from CST (Danvers, MA, USA). Additionally, anti-FOXO3, anti-phosphor-FOXO3 (Ser253), and anti-Histone1 were acquired from Huabio (Hangzhou, China). Mouse anti-β-actin monoclonal antibody was obtained from Biotime Biotechnology, Xiamen, China. For all the proteins mentioned above, the relative density was analyzed with ImageJ software.

2.11. Immunofluorescence Analysis

IPEC–J2 cells were seeded and cultured in 12-well plates for 12 h to reach 70% confluence. Cell samples (n = 6) were fixed with cold methanol for 5 min, and after that, all the samples were blocked with 2.5% BSA at room temperature for 2 h, followed by incubating with anti-LC3II antibody (Abcam, Cambridge, MA, USA) for 12 h at 4 °C. Thereafter, Alexa Fluor 488 conjugated antibody (Biotime Biotechnology, Xiamen, China) and DAPI solution (Biotime Biotechnology, Xiamen, China) were used to stain cells and nucleus. An Olympus BX61W1-FV1000 laser scanning confocal microscope (Tokyo, Japan) was utilized to obtain images.

2.12. RNA Extractions and Quantitative Real-Time PCR (qPCR) Analysis

Total RNA was extracted utilizing RNAiso plus (Takara, Japan) from IPEC-J2 cells. The Nanodrop was used to examine the concentration of RNA. We utilized the PrimeScript II 1st Strand cDNA Synthesis Kit (Vazyme, Nanjing, China) to synthesize cDNA based on the manufacturer's manual. The qRT-PCR was conducted with the use of the HiScript II One Step qRT-PCR SYBR Green Kit (Vazyme, Nanjing, China) based on the manufacturer's manual. Glyceraldehyde-3-phosphate dehydrogenase (GAPDH) was used to normalize the amount of total RNA as an endogenous control. The Primer 5.0 as well as the Oligo 7.0 software were utilized to design and validate the primers. The primer sequences are summarized in Table 2. We estimated the abundance of the mRNA using the $2^{-\Delta\Delta Ct}$ method.

Table 2. List of qPCR primers.

Gene	Access No.	Primers Sequence	Length
GAPDH	NM_001206359.1	F: CGGAGTGAACGGATTTGGC R: CACCCCATTTGATGTTGGCG	248
ATG5	NM_001037152.2	F: AGTCAACCCTCCAATACCCAG R: TGTGGCCCTCTCTAGGTTTCT	299
ATG12	NM_001190282.1	F: AGGTTGGATACCCGCCTACT R: ACTTGGTTGGAGCAATCT	111
ATG16L1	NM_001190272.1	F: CTGCCAGTCGAACAGGATGA R: AGCGCCTCCCAAAGATATTAGT	166
ATG8	NM_001190288.1	F: CCACCTTCCCACTCAGCTTT R: GTGTATCCTACCTTCCCCGC	187
ATG14	XM_001924990.5	F: GCTTACACTGGACACCGCTA R: CCCCACGGCTTAACCTCTTT	125
Caspase-8	NM_001031779.2	F: CCAGGATTTGCCTCCGGTTA R: GCCAGGTCATCACTGTCCAA	108

2.13. RNA Extraction and RNA-Seq Analysis

After the total RNA extraction, mRNA was purified from total RNA using poly-T oligo-attached magnetic beads. First-strand cDNA was synthesized using random primers. Second-strand cDNA was synthesized by DNA polymerase I, Rnase H, dNTP, and buffer. Then, the cDNA fragments were purified with an AMPure XP system (Beckman Coulter,

Beverly, GA, USA). Afterwards, PCR was performed with Phusion High-Fidelity DNA polymerase, Universal PCR primers, and Index (X) Primer. Lastly, the PCR products were purified (AMPure XP system) and the library quality assessed on an Agilent Bioanalyzer 2100 system. The ligated products were size-selected by agarose gel electrophoresis and PCR, and then sequenced using Illumina HiSeqTM 4000. For the bioinformatics analysis, the original reading containing the adapter or low quality (Q value $\leq$ 10) was removed and mapped to the pig reference genome (assembly SSCROFA 11–1). We further analyzed the differentially expressed genes (DEGs) between different samples or groups with the DeSeq2 R package (1.16.1). Genes with a fold change of $\geq$1, at a false discovery rate of $p \leq 0.05$, were considered significantly differentially expressed. The enrichment of DEGs was performed using the KEGG pathway database.

2.14. Statistical Analysis

All data are showed in the form of the mean $\pm$ standard deviation (SD). Statistically significant differences between means were calculated with a one-way ANOVA with a Tukey test and analyzed in SPSS statistical software, version 23.0 (IBM®, Chicago, IL, USA). The IC_{50} of the DQ to IPEC-J2 cells was calculated utilizing the probit method in SPSS statistical software, version 23.0 (Chicago, IL, USA). $p < 0.05$ was considered statistically significant. Figures were prepared utilizing Prism 9.0 software (GraphPad Software Inc., San Diego, CA, USA) and Origin 8.0 software (Origin Lab Corporation, Northampton, MA, USA).

3. Results

3.1. Establishment of Oxidative Stress Model Induced by Diquat in IPEC-J2 Cells

IC_{50} represents 50% of the inhibitor concentration required for inhibition of enzymes, cells, cell receptors, or microorganisms. DQ, which is widely utilized as an herbicide in agriculture, was used to establish an oxidative stress model. Notably, DQ reduced the IPEC-J2 cells' viability in a dose-dependent manner (Figure 1a). Utilizing the probit method, the IC_{50} for DQ in IPEC-J2 cells was 932.2 µmol/mL. Therefore, 950 µmol/mL DQ was used in the following experiment. IPEC-J2 cells were exposed to *BaSC06* at different concentrations (0, 10^5, 10^6, 10^7, 10^8, and 10^9 CFU/mL) for 6 h, and their cell viability was further detected. As shown in Figure 1b, the cell viability at the 10^8 CFU/mL treatment was the closest to 100% (105.384%), showing no significant change compared to the cells in the CK group ($p > 0.05$). Hence, 10^8 CFU/mL *BaSC06* was used in further experiments.

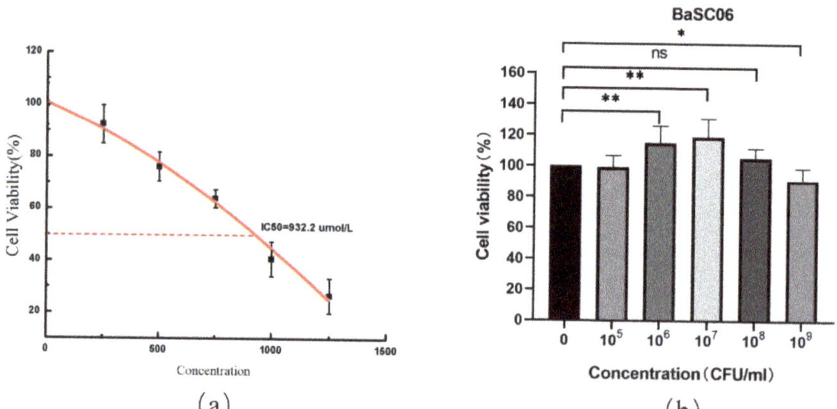

Figure 1. The establishment of a diquat-induced oxidative stress model. (**a**) IPEC-J2 cells were treated with DQ at various concentrations (0, 250, 500, 750, 1000, 1250 µmol/L) for 6 h. The IC_{50} was calculated using the probit method. (**b**) IPEC-J2 cells were treated with *BaSC06* for 6 h at different concentrations (0, 10^5, 10^6, 10^7, 10^8, and 10^9 CFU/mL). Cell viability was calculated using an CCK-8 kit. Data are exhibited in the form of the mean $\pm$ SD (n = 9), and significance was measured by one-way ANOVA with a Tukey test: * $p < 0.05$, ** $p < 0.01$, and ns = no significance ($p > 0.05$).

3.2. BaSC06 Alleviated Oxidative Stress Induced by DQ in IPEC-J2 Cells

Consequent ROS generation and redox cycling are believed to be the key causes of oxidative stress induced by DQ [30]. The production of ROS in IPEC-J2 cells was examined using a DCFH-DA fluorescence assay and Microplate Reader. Compared with the CK group, DQ treatment significantly increased the generation of ROS (32.73 ± 4.755%, $p < 0.001$), which was markedly inhibited by *BaSC06* pretreatment (10.70 ± 1.588%, $p < 0.01$) (Figure 2a,b). A similar result detected by microplate reader was obtained (Figure 2c). In addition, MDA content was significantly increased in the DQ group (14.12 ± 1.07 nmol/mg protein, $p < 0.01$) (Figure 2d), but markedly decreased in the Ba+DQ group (4.03 ± 1.07 nmol/mg protein, $p < 0.01$). However, no significant difference was observed in the GSH-Px activity between CK and DQ groups, which was significantly increased in *BaSC06*-treated cells ($p < 0.05$) (Figure 2e). At the same time, compared with the CK group, the activity of T-SOD decreased by 48.2% (p <0.05) in the DQ group, which significantly increased by 26.3% in *BaSC06* pretreatment group ($p < 0.05$) (Figure 2f). These results suggest that *BaSC06* could enhance the antioxidant capacity of IPEC-J2 cells by decreasing the production of ROS and increasing the activities of antioxidant enzymes.

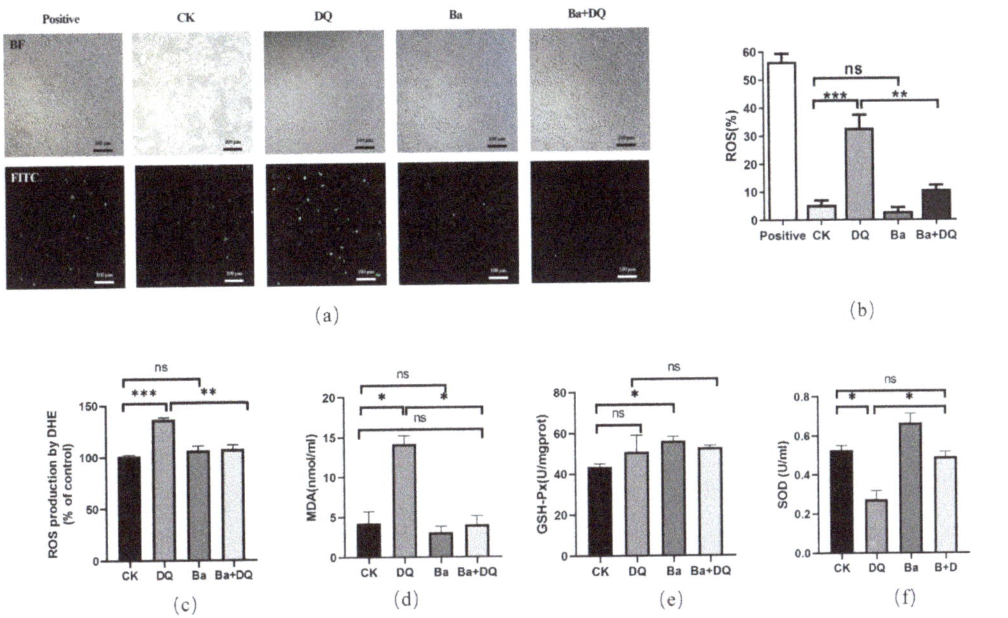

Figure 2. *BaSC06* alleviated oxidative stress induced by DQ in IPEC-J2 cells. (**a–c**) Levels of ROS measured by DCFH-DA fluorescence assay, and the data are presented as the ratio of green fluorescence and the percentage of ROS production by DHE. Scale bar: 100 μm; $n = 9$ in each group. (**d–f**) Antioxidant capacity in cell lysates indicated by MDA, T-SOD, and GSH-Px; $n = 6$ in each group. All data were analyzed utilizing one-way ANOVA with Tukey test: * $p < 0.05$, ** $p < 0.01$, *** $p < 0.001$, and ns = no significance ($p > 0.05$).

3.3. BaSC06 Alleviated Oxidative Stress-Induced Apoptosis in IPEC-J2 Cells

The results of the expression of apoptosis-related proteins showed that *BaSC06* markedly upregulated the protein level of Bcl2 ($p < 0.05$), while that in the DQ treatment group indicated no significant changes compared with untreated cells ($p > 0.05$) (Figure 3a,c). The expression of Bax and the ratio of Bax/Bcl2 was significantly increased in cells treated with DQ, but *BaSC06* pre-treatment altered this trend significantly ($p < 0.05$) (Figure 3a,b,d). DQ also increased the activity of caspases-3 (Figure 3f), the relative intensity of cleaved caspase-3 and the mRNA levels of caspase-8 considerably ($p < 0.01$), which were significantly reversed by BaSC06 pre-treatment ($p < 0.01$, Figure 3e,g). In addition, compared with the CK group, the number of apoptotic bodies (green puncta) in the DQ group, detected using a TUNEL kit, was considerably increased by 31.40 ± 7.19% ($p < 0.01$), but no significant changes and a significant decrease (18.25 ± 5.843%) were observed in the Ba group and Ba+DQ group ($p < 0.01$), respectively (Figure 3h,i). Furthermore, the similar results were obtained using TUNEL assay by flow cytometry (Figure 3j,k). *BaSC06* pretreatment dramatically downregulated DQ-triggered apoptosis ($p < 0.01$). The percentages of early and late apoptotic cells in the CK, DQ, Ba, and Ba+DQ groups were 5.05 ± 1.31%, 12.95 ± 0.64%, 5.67 ± 0.45%, and 6.48 ± 0.40%, respectively (Figure 3l,m). Overall, these findings demonstrate that pre-treatment with *BaSC06* might significantly alleviate apoptosis induced by DQ in IPEC-J2 cells.

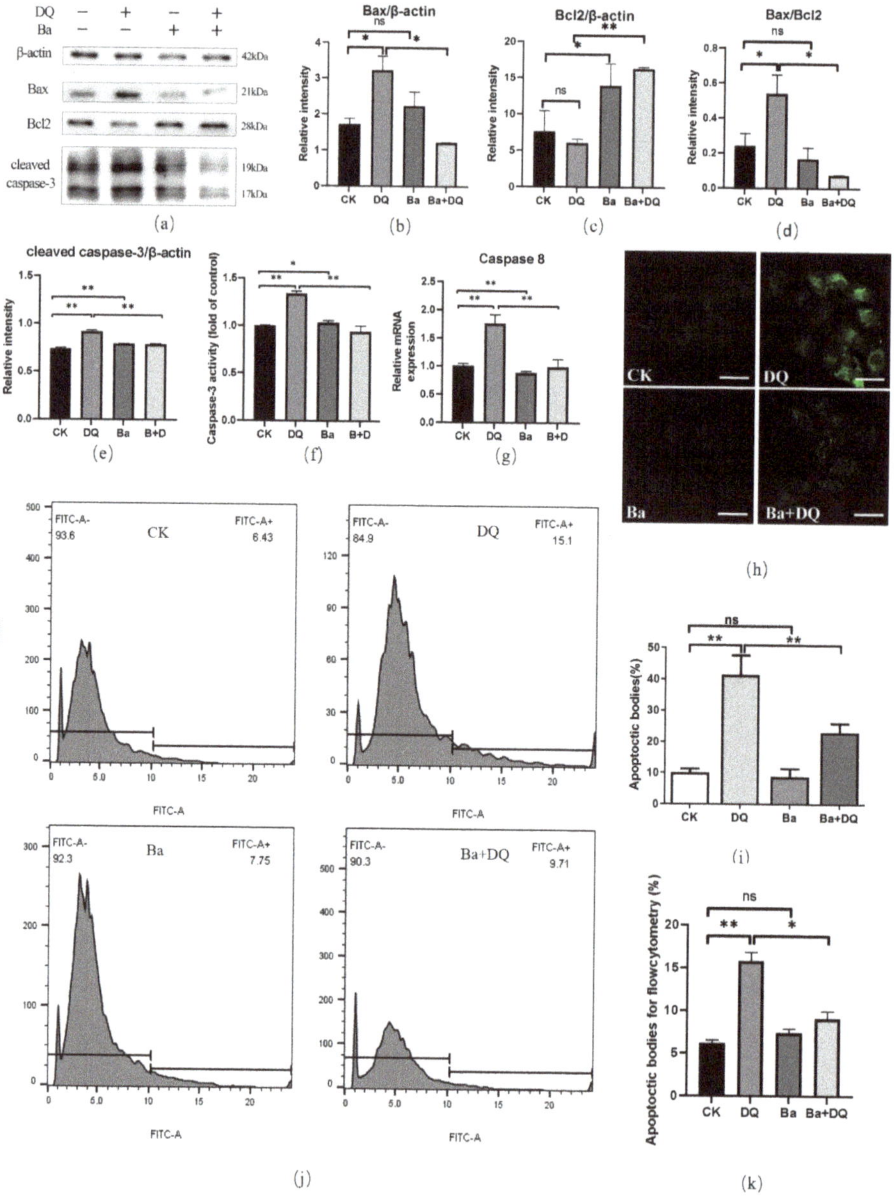

Figure 3. *Cont.*

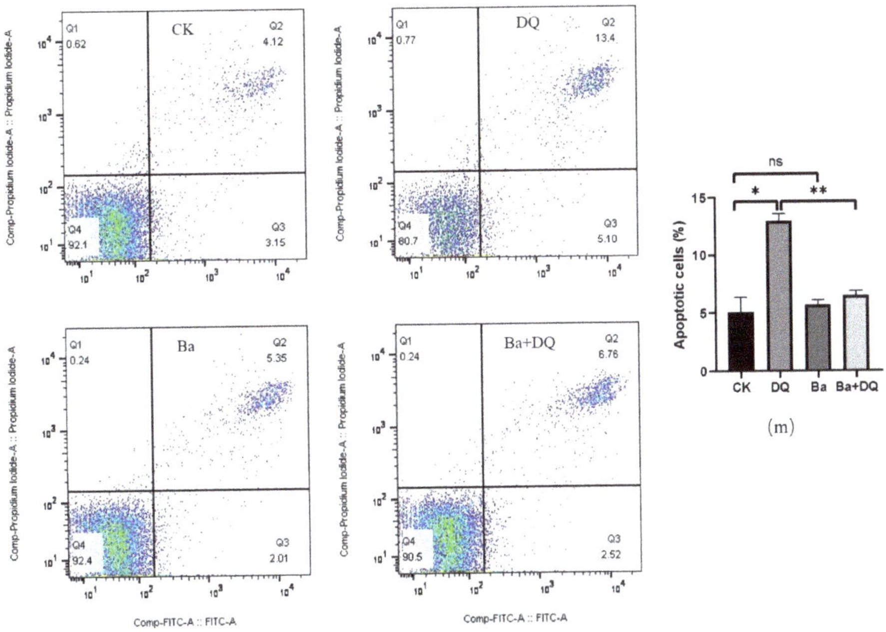

(l)

Figure 3. *BaSC06* pretreatment inhibited apoptosis in IPEC-J2 cells during diquat exposure. (**a–e**) The ratio of Bcl2/β-actin, Bax/β-actin and Bax/Bal2, and cleaved caspase-3/β-actin were analyzed using ImageJ software. Data were analyzed utilizing one-way ANOVA with a Tukey test; $n = 3$ in each group; * $p < 0.05$, ** $p < 0.01$, and ns = no significance ($p > 0.05$). (**f**) Caspase-3 activity was evaluated using a caspase-3 activity assay kit. (**g**) The mRNA expression levels of caspase-8 were determined using quantitative real-time PCR. (**h**,**i**) TUNEL assay. IPEC-J2 cells were stained with a BrightGreen apoptosis detection kit, after being treated with 10^8 CFU/mL BaSC06 and 950 µmol/mL diquat. Scale bar: 5 µm. (**j**,**k**) TUNEL assay by flow cytometry. (**l**,**m**) Annexin V-FITC/PI apoptosis assay. Apoptotic cell rates were detected with a FITC annexin V-FITC/PI apoptosis kit, and then analyzed by flow cytometry. All data were analyzed using one-way ANOVA with a Tukey test; $n = 6$ in each group; * $p < 0.05$, ** $p < 0.01$, and ns = no significance ($p > 0.05$).

3.4. *BaSC06* Triggered Autophagy during Oxidative Stress in IPEC-J2 Cells

We further evaluated whether *BaSC06* can induce autophagy in IPEC-J2 cells under oxidative stress. As expected, 10^8 CFU/mL *BaSC06* upregulated the LC3II/LC3I ratio dramatically in a time-dependent manner, particularly from 6 h to 10 h ($p < 0.05$) (Figure 4a,b), but decreased p62 expression (Figure 4a,c). A significant decreased expression of LC3-II in the DQ and Ba+DQ group was found ($p < 0.05$) (Figure 4d,e), and no significant change was observed between the two groups ($p > 0.05$). Further, DQ treatment substantially inhibited the degradation of p62 ($p < 0.05$); however, *BaSC06* pretreatment dramatically blocked this trend ($p < 0.05$) (Figure 4d,f). The ratio of LC3II/LC3I markedly increased in the *BaSC06* group compared to the DQ group ($p < 0.01$) (Figure 4d,e). Additionally, compared with the CK group, *BaSC06* pretreatment markedly upregulated the expression of Beclin1 ($p < 0.05$), while that in the DQ group exhibited no significant change ($p > 0.05$) (Figure 4d,g). The results of autophagy based on the percentage of LC3-positive cells demonstrated that, compared with the CK group, DQ significantly downregulated LC3-positive cells (45.92 ± 3.23%, $p < 0.01$), while *BaSC06* significantly increased LC3 puncta (62.81 ± 4.90%, $p < 0.001$) (Figure 4h,i). To sum up, these findings suggest that *BaSC06* treatment can induce autophagy in IPEC-J2 cells.

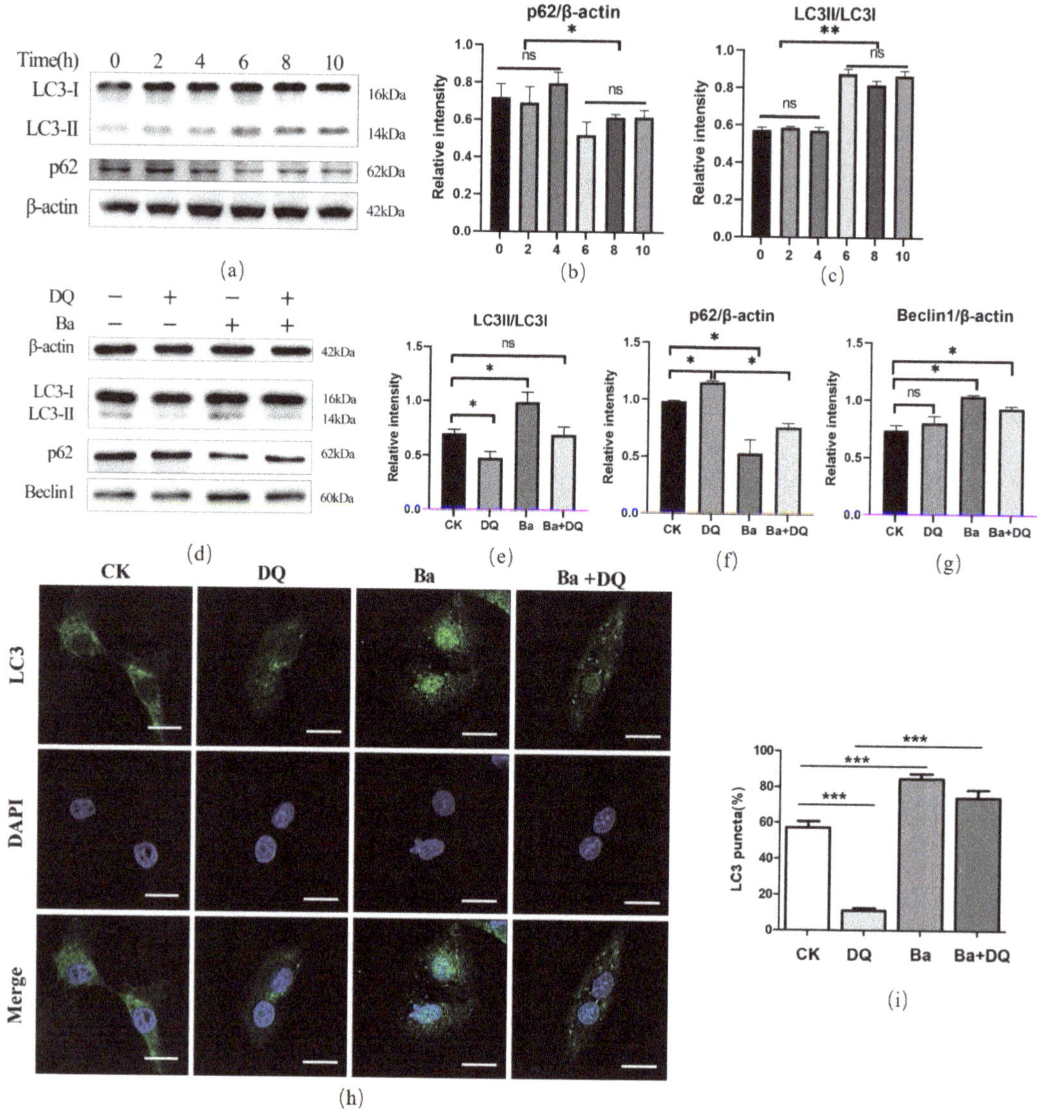

Figure 4. Effects of *BaSC06* on autophagy during oxidative stress in IPEC-J2 cells. (**a–c**) *BaSC06* triggered autophagy in IPECJ-J2 cells in a time-dependent manner. Cell lysates were collected to detect the protein levels of LC3-II/LC3-I and p62/β-actin, and data were analyzed using ImageJ software. (**d–g**) IPEC-J2 cells pretreated with *BaSC06* in the concentration of 10^8 CFU/mL for 6 h, subsequently treated with DQ in the concentration of 950 μmol/mL for another 6 h. The LC3II/LC3I, p62/β-actin, or Beclin1/β-actin ratio was analyzed by ImageJ software. Data were analyzed using one-way ANOVA with a Tukey test; $n = 3$ in each group; * $p < 0.05$, ** $p < 0.01$, and ns = no significance ($p > 0.05$). (**h**,**i**) *BaSC06* increased LC3 puncta in IPEC-J2 cells. After the same treatment with (**d**), IPEC-J2 cells were stained and visualized under confocal microscopy for immunofluorescence analysis. Scale bar: 5 μm; $n = 6$ in each group. The number of LC3-positive cells was statistically analyzed: * $p < 0.05$, ** $p < 0.01$, *** $p < 0.001$, and ns = no significance ($p > 0.05$).

3.5. KEGG Pathway Analysis of DEGs

The DEGs' numerical analysis is depicted in Figure 5a,c,e. It was uncovered that compared to the CK group, *BaSC06* induced the upregulation of 115 and downregulation of 148 genes, while DQ induced the upregulation of 546 and downregulation of 500 genes. In addition, *BaSC06* upregulated 215 and downregulated 309 genes compared to the Ba+DQ group (fold change > 1 and $p < 0.05$). Then, a KEGG enrichment pathway analysis was performed and obtained the autophagy-related pathways, including MAPK, AMPK, PI3K-AKT, P53, FOXO, JAK-STAT, NF-kappa B, TNF, TGF-beta, and mTOR signaling pathways (Figure 5b,d,f). Among them, the FOXO signaling pathway was enriched with the most significant differential genes, which implies that this key pathway may regulate *BaSC06* or diquat-induced autophagy in IPEC-J2 cells. In addition, *BaSC06* was also found to have an effect on Akt in rats in our previous study [27], so we speculated that autophagy may be regulated by the Akt–FOXO signaling pathway.

Next, differential genes in the FOXO signaling pathway were examined, and 83 genes with significant differences were identified (Figure 5g). To verify the identified DEGs by qRT-PCR, we randomly selected 5 genes, SGK1, Beclin1, Raf1, MDM2, and STAT3, from the significant DEGs. The results revealed that the sequencing results were consistent with the RT-qPCR data, which confirmed the reliability of the sequencing data (Figure 5h–l).

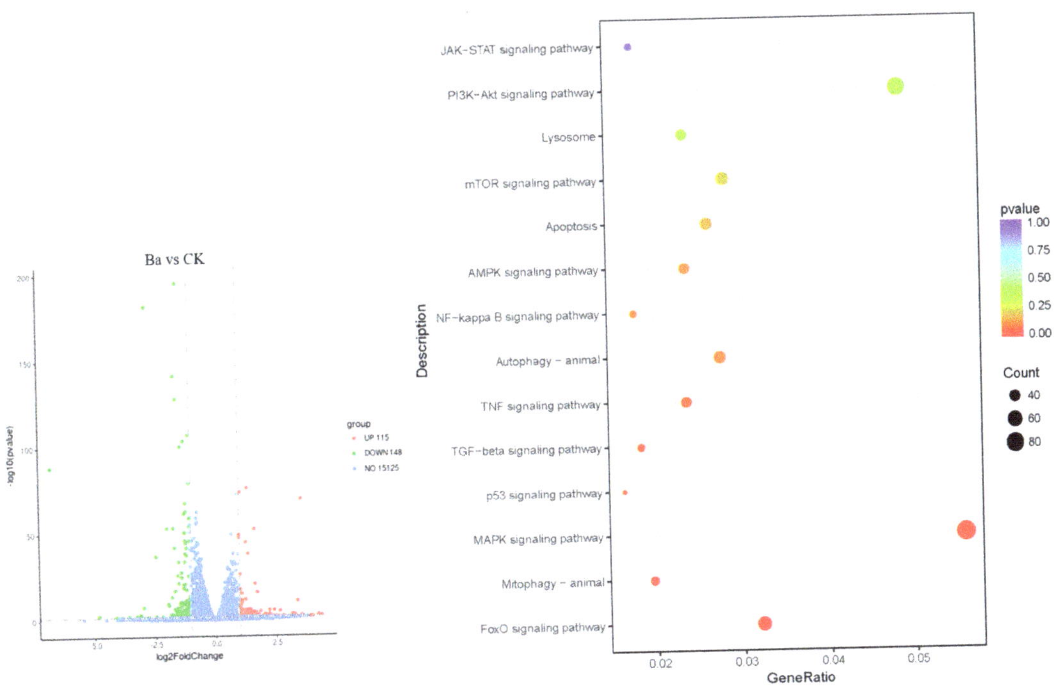

(a) (b)

Figure 5. *Cont.*

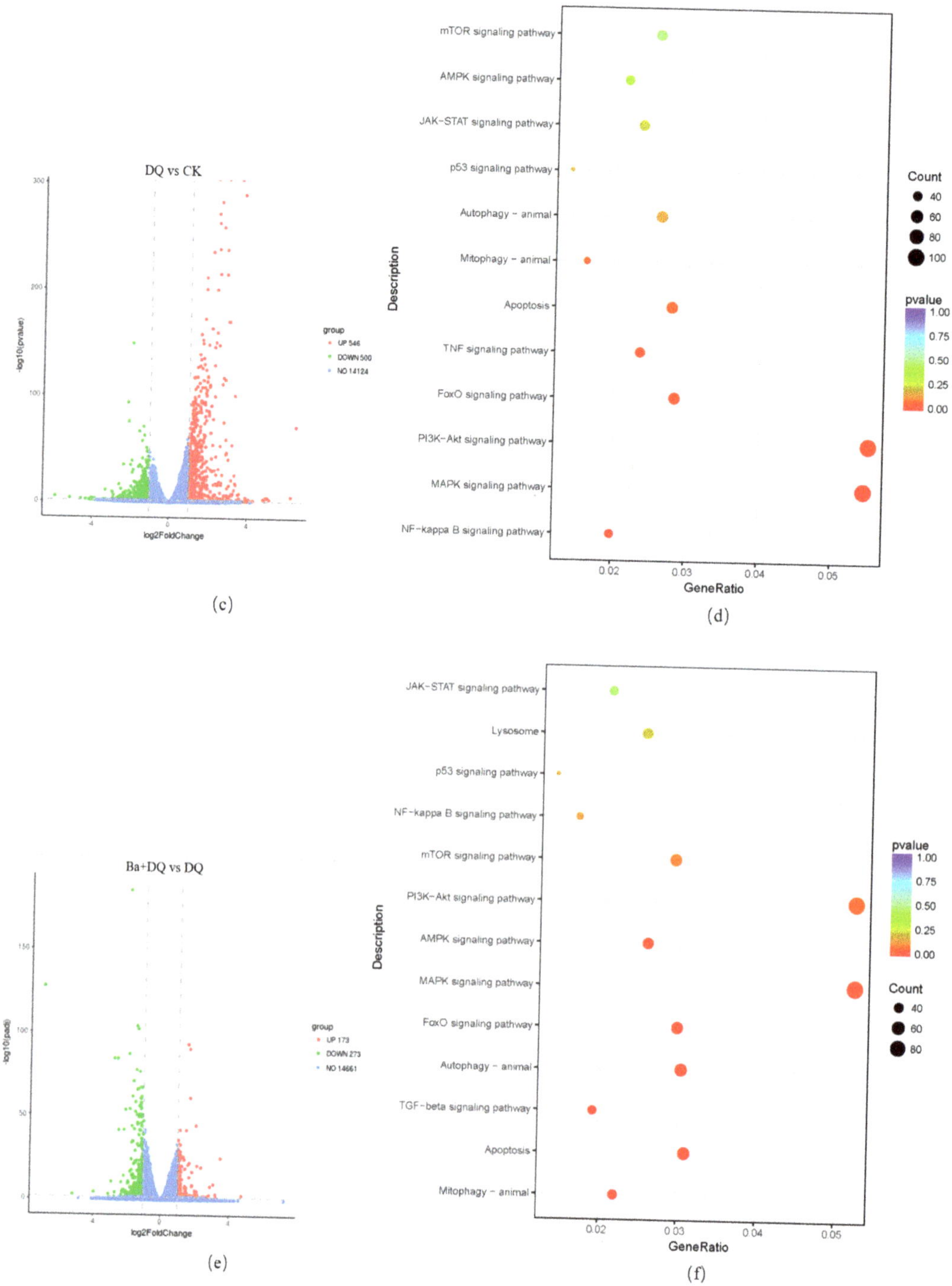

Figure 5. *Cont.*

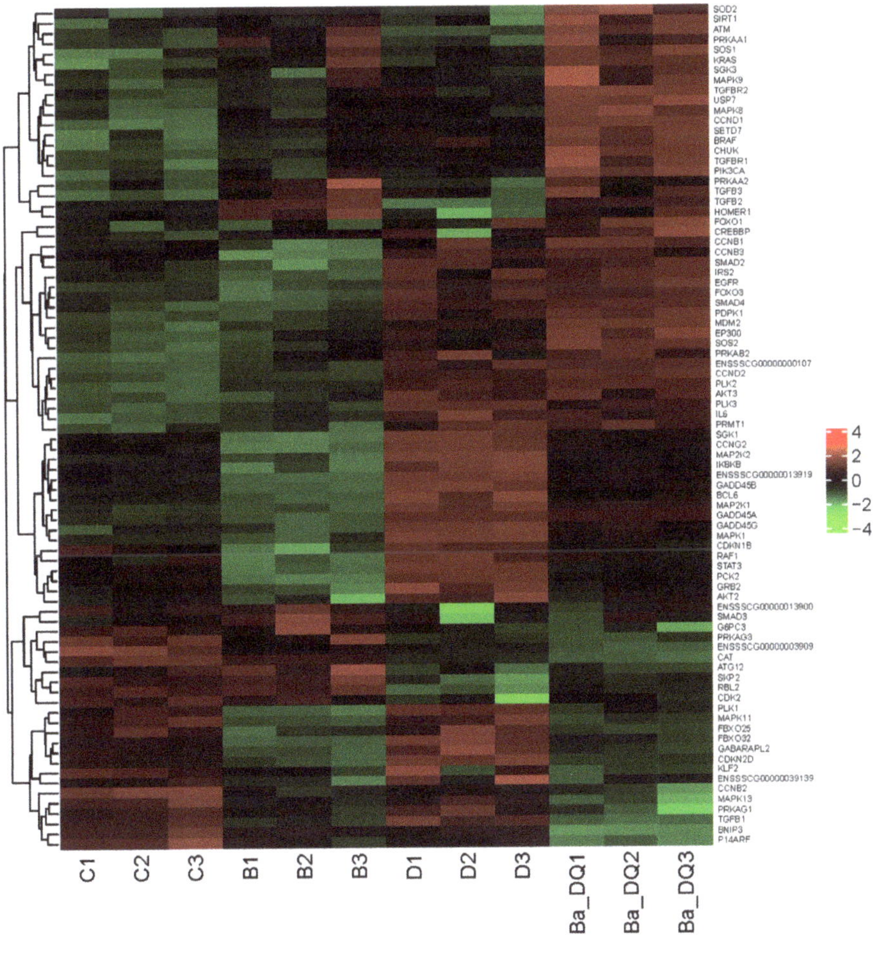

(g)

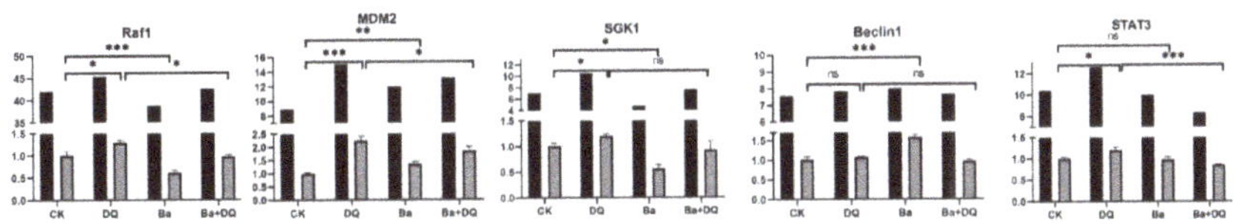

(h) (i) (j) (k) (l)

Figure 5. Upregulation and downregulation genes, and differential genes in the FOXO signaling pathway and validation of the DEGs data by RT-qPCR. (**a–f**) Volcano plots of the DEGs. The x-axis indicates the difference in expression level on a log2 scale (fold change), while the y-axis represents the *p*-value. Red represents the upregulation gene whereas green denotes downregulation genes. Bubbles of KEGG pathways for differential gene enrichment. The circle presents the gene number. The color of the circles indicates the *p*-value. (**g**) Heatmaps of all differential genes in the FOXO signaling pathway. (**h–l**) Validation of the RNA-Seq expression profiles of genes randomly selected from DEGG by RT-qPCR; $n = 9$ in each group. Black bars represent FPKM (Fragments Per Kilobase Million), while grey bars represent the fold change. FPKM normalized values as gene expression in RNA-seq. * $p < 0.05$, ** $p < 0.01$, *** $p < 0.001$, and ns = no significance ($p > 0.05$). Values are the mean ± SD.

3.6. *BaSC06* Can Regulate the Transcriptional Activity of FOXO3 Transcription Factor in IPEC-J2 Cells

It is well known that the subcellular localization and transcriptional activity of FOXO proteins are mainly regulated by posttranslational modifications, including phosphorylation, acetylation, and ubiquitination [31]. Phosphorylation of FOXO3 can transfer it from the nucleus to the cytoplasm, thereby inactivating FOXO3. Currently, the expression of the p-FOXO3 protein in the DQ group was significantly increased ($p < 0.01$), which was reversed by the pretreatment of *BaSC06* ($p < 0.01$) (Figure 6a,b). More importantly, the expression of Sirt1 is crucial due to its deacetylation function. Noticeably, *BaSC06* significantly elevate the reduced expression of SIRT1 induced by DQ ($p < 0.05$) (Figure 6a,c), which implied that *BaSC06* may decrease acetylation of FOXO3. The protein expression FOXO3 in the nucleus decreased significantly in the DQ group ($p < 0.05$), while pretreatment with *BaSC06* inhibited this change in IPEC-J2 cells (Figure 6a,d). In summary, these outcomes indicate that *BaSC06* can increase the FOXO3 expression in nuclear by inhibiting its phosphorylation and increasing its deacetylation, thereby increasing the transcriptional activity of the FOXO3.

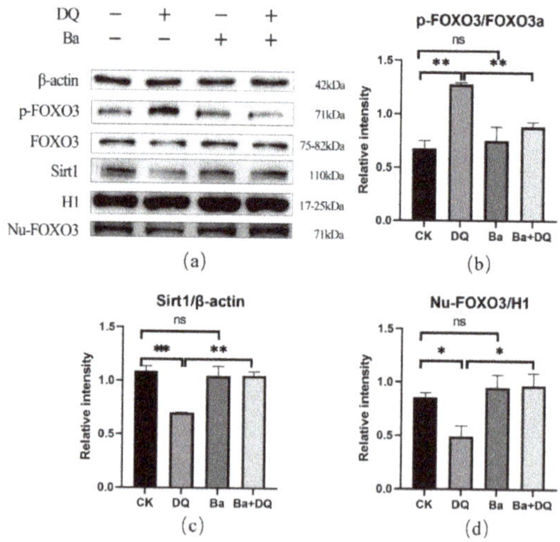

Figure 6. Effects of *BaSC06* on FOXO3 transcription factor in IPEC-J2 cells. (**a–d**) The expression of phosphate-FOXO3, FOXO3, and Sirt1 protein in IPEC-J2 cells. Cell lysates were collected to detect the protein levels of p-FOXO3/FOXO3, Nu-FOXO3/H1, and SIRT1/β-actin, and data analyses were executed using ImageJ software. Data were analyzed using one-way ANOVA with a Tukey test; $n = 3$ in each group; * $p < 0.05$, ** $p < 0.01$, *** $p < 0.001$, and ns = no significance ($p > 0.05$).

3.7. *BaSC06* Mediated Autophagy by the AKT–FOXO Signaling Pathway Independent of mTOR

The ratio of p-AKT/AKT exhibited significantly reduced in the *BaSC06* pretreatment groups compared to the DQ group ($p < 0.05$) (Figure 7a, b). However, no significant change for the ratio of p-mTOR/mTOR was observed in all groups ($p > 0.05$) (Figure 7a,c). The results showed that the addition of perifosine (the inhibitor of AKT phosphorylation) significantly reduced the phosphorylation of FOXO3a ($p < 0.05$) (Figure 7d,e), confirming that the transcriptional activity of FOXO3a was upregulated. Then we used siRNA to reduce FOXO3 expression, and further detected the expression of autophagy related proteins in IPEC-J2 cells. SscFOXO3a-608 had the best silence effect (34.53 ± 0.16%), the protein expression level of FOXO3a was significantly reduced, so we used it in subsequent experiments (Figure 7h,i). The FOXO3/β-actin and the LC3II/LC3I ratio decreased significantly in sscFOXO3a-608-added groups ($p < 0.05$) compared to the NC group, but no significant difference was observed among all sscFOXO3a-608-added groups ($p > 0.05$) (Figure 7j–l). Although the expression level of P62 showed the increasing tend when Ba was added, there was no significant difference compared with NC group ($p > 0.05$). These results demonstrated that the decrease of FOXO3 expression significantly inhibited autophagy in IPEC-J2 cells, and that *BaSC06* induced autophagy by the AKT-FOXO signaling pathway, not by the AKT/mTOR signaling pathway.

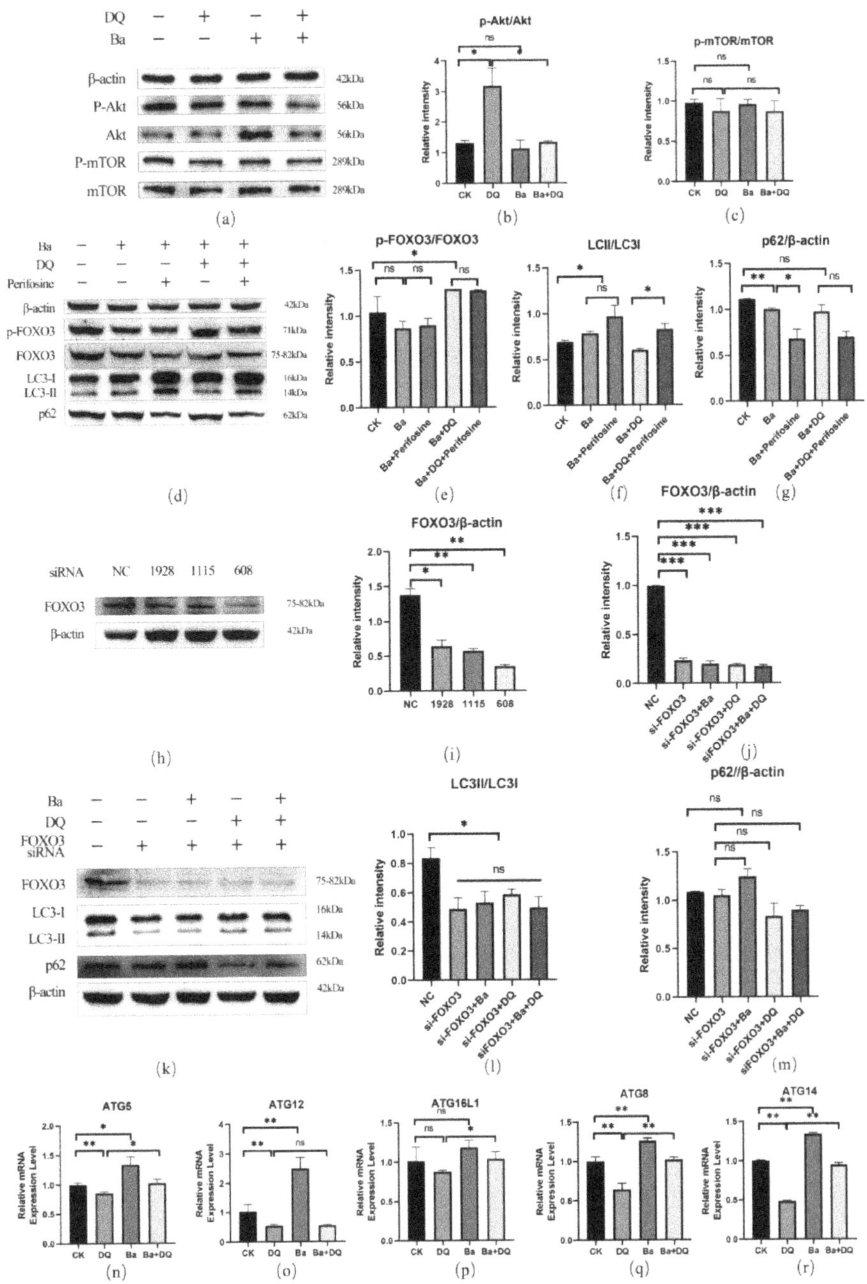

Figure 7. *BaSC06* mediated autophagy was triggered by the AKT–FOXO signaling pathway independent of mTOR, and the effect of *BaSC06* on FOXO downstream autophagy-related target genes. (**a**–**c**) The expression of p-AKT, AKT, p-mTOR, and mTOR protein in IPEC-J2 cells. Cell lysates were gathered to detect the ratio of *p*-AKT/AKT and p-mTOR/mTOR.

(**d–g**) We added 25 μM perifosine into the *BaSC06* pretreated groups and collected the cell lysates to detect the protein levels of p-FOXO3/ FOXO3, LC3II/LC3I, and p62/β-actin. (**h,i**) The efficacy of all siRNAs. (**j–m**) The expression of FOXO3//β-actin, LC3II/LC3I, and p62 /β-actin in siRNA treatment groups. All data analyses were implemented using ImageJ software. Data were analyzed using one-way ANOVA with a Tukey test; $n = 3$ in each group; * $p < 0.05$, ** $p < 0.01$, *** $p < 0.001$, and ns = no significance ($p > 0.05$). (**n–r**) The relative mRNA expression levels of ATG5, ATG12, ATG8, ATG14, and ATG16L1. Data were evaluated using one-way ANOVA with a Tukey test; $n = 9$ in each group; * $p < 0.05$, ** $p < 0.01$, and ns = no significance ($p > 0.05$).

We further explored the expression of downstream autophagy-related target genes of FOXO. Our results demonstrated that compared with the CK group, the mRNA expressions level of ATG5, ATG12, ATG8, as well as ATG14 of the DQ group were substantially decreased, whereas these genes expression level significantly up-regulated in *BaSC06* treatment group ($p < 0.05$) (Figure 7n,o,q,r). Although treatments of DQ and *BaSC06* alone insignificantly affected the mRNA expression level of ATG16L1 ($p > 0.05$), pretreatment with *BaSC06* still markedly increased the mRNA expression level of ATG16L1 when it was compared with the DQ group ($p < 0.05$) (Figure 7p). These results verified that *BaSC06* could activate the AKT-FOXO signaling pathway to induce autophagy.

4. Discussion

Diquat, which is widely used as an herbicide in agriculture, can be absorbed by green plants that turn it into peroxide-free radicals. Toxicological studies have demonstrated that diquat can cause damage to the digestive system as well as to the lung, liver, and other organs, eventually leading to death [30] and can induce oxidative stress and injuries in intestinal epithelial cells [32]. Therefore, Diquat has been largely applied to induce oxidative stress in vivo, because it can significantly increase the serum MDA level as well as inhibit the activities of antioxidant enzymes (including SOD and GSH-Px) [33–35]. But different cells have different sensitivities to it. 80 μmol/L DQ was used in our previous study in IEC-6 cells [27]. Current study found that the IC_{50} for DQ in IPEC-J2 cells was 932.2 μmol/ml, which is consistent with the results by Xu C et al. (2018), who used 1 mmol/L diquat to successfully establish an oxidative stress model on IPEC cells [36].

Elevated ROS production and low antioxidant capacity has been identified to induce oxidative stress in cells. MDA is an important biomarker of lipid peroxidation. Importantly, GSH-Px, CAT, and T-SOD are the main antioxidant enzymes in cells [37]. SOD terminates lipid peroxidation and eliminates ROS; however, its activity might be defected during acute injuries, leading to DNA damage, aggregated lipid peroxidation, and cell dysfunction [38]. The results of the present study showed that DQ exposure significantly increased MDA level in IPEC-J2 cells, while *BaSC06* pretreatment dramatically decreased the levels of ROS and MDA by reversing the decreased activities of SOD and GSH-Px induced by DQ, suggesting the potential role of *BaSC06* on alleviating oxidative stress, which corresponded with previous reports [27,36,39,40].

Evidence suggests that apoptosis can be caused by excessive oxidative stress, while excessive ROS production results in cell injury and death. Mitochondria are the major intracellular source of ROS [41]. The Bcl-2 protein family consists of Bcl-2, Bax, and Bak, among others. Bcl-2 is primarily located in the inner mitochondria membrane, the endoplasmic reticulum, and the perinuclear membrane and its function is regulated by its protein products, Bax and Bcl-xl [42]. What's more, the Caspase family is a key protein of apoptosis [43]. Our results also revealed that DQ can induce cell apoptosis and damage in IPEC-J2 cells, which was indicated by the increased caspase-3 activity and the expression level of cleaved caspase-3 protein and the mRNA expression of caspase-8 gene, which is consistent with the results of our previous research on rats, *BaSC06* also decreased apoptosis of rats caused by diquat both in vivo and in vitro, which was verified by the reversal of the upregulated caspase-3 and Bax expression, and the decreased expression of Bcl2 [27]. Besides, DNA fragmentation has been found to occur during cell apoptosis. A study has reported that ROS reacts with cellular macromolecules through oxidation causing the cells to undergo an active process of cell death, which is set in motion by a high-conserved genetic program and culminated in DNA fragmentation and the formation of apoptotic bodies [44]. In this study, we verify that *BaSC06* can decrease apoptosis caused by diquat-induced oxidative stress via regulating the expression of apoptotic proteins and reducing the production of apoptotic bodies.

Autophagy is a catabolic pathway that is activated in response to different cellular stressors, such as damaged organelles, accumulation of misfolded or unfolded proteins, ER stress, accumulation of ROS, and DNA damage [45]. Several studies have reported that autophagy activation could alleviate oxidative stress. For instance, spermidine provides neuroprotection against oxidative stress and apoptosis by activating autophagy in aging male rats [46]. Tetrahedral framework nucleic acid inhibits chondrocyte apoptosis and oxidative stress through activation of autophagy [47]. In addition, quercetin helps the retina external barrier avoid oxidative stress injury by promotion of autophagy [48]. However, little information is available on the effect of probiotics on oxidative stress by regulating autophagy. Our previous study demonstrated that the autophagy induced by *BaSC06* was involved in decreasing oxidative stress in the rat [27]. Interestingly, we herein obtained similar results that *BaSC06* can upregulate the expression of LC3 and Beclin1, degrade p62, and increase LC3 puncta during diquat-induced oxidative stress in IPEC-J2 cells. Thus, it was suggested that the probiotic *BaSC06* contributes to alleviating oxidative stress by inducing autophagy in IPEC-J2 cells.

The results of KEGG enrichment pathway analysis revealed that DEGs and enriched major autophagy-related pathways contained P53, FOXO, JAK-STAT, NF-kappa B, TNF, TGF-beta, MAPK, AMPK, PI3K-AKT, and mTOR signaling pathways. In the Ba vs. CK, DQ vs. CK, and Ba+DQ vs. CK comparisons, FOXO was the primary signaling pathway that enriched the most significant DEGs regulating the autophagy process. The main DEGs in the FOXO signaling pathway are depicted in the heatmap included FOXO3, among others. Transcriptional factor FOXO3 has been uncovered to be extensively involved in autophagy and apoptosis, regulating the cell cycle, and participating in antioxidant stress response [49]. Moreover, FOXO3 activates autophagy by upregulating autophagy regulatory genes, including LC3, ATG12, γ-GABA receptor-associated protein 1 gene, yeast ATG8, and BNIP3 [50,51]. The acetylation, phosphorylation, and other post-transcriptional modification sites in FOXO significantly affect its DNA-binding activity, as well as its subcellular localization. The acetylation/deacetylation of FOXO regulates its transcriptional activity. Oxidative stress increase FOXO acetylation, and then acetylated-FOXO (Ac-FOXO) accumulates in the nucleus and binds to nucleosomes to block its transcriptional activity [52,53]. Furthermore, SIRT1 indirectly regulates autophagy by deacetylation of FOXO3, leading to increased expression of autophagy-related genes, including Bnip3, which are critical for autophagy induction [54]. The signal crossovers of SIRT1 and ROS can cause the decrease of autophagy and reduce the occurrence of the inflammatory response [55]. SIRT1 might deacetylate FOXO3 for oxidative stress response, thus improving the anti-stress ability of cells [56]. Overwhelming evidence suggests that SIRT1 deacetylates the FOXO factors, including FOXO1, FOXO3a, and FOXO4, and subsequently stimulate the expression of antioxidants, such as CAT, MnSOD, and Trx. Besides, an automatic feedback loop also potentiates SIRT1 expression [56–60]. In this work, *BaSC06* pretreatment significantly slowed the decline of SIRT1 induced by DQ, implying the increased deacetylation of FOXO3.

Phosphorylation/dephosphorylation of FOXO determines its subcellular localization. Phosphorylation of FOXO is mainly affected by protein kinase B (PKB or AKT). Studies have shown that p-AKT can phosphorylate FOXO3 at Thr32 /Ser315 /Ser253, thus preventing FOXO3 from entering the nucleus and inhibiting autophagy gene transcription, thus down-regulating autophagy level [61]. Hence, AKT-FOXO3 primarily exerts a negative regulatory role on autophagy regulation. Reduced nucleation of FOXO3a leads to decreased expression of reactive oxygen scavenging enzymes (superoxide dismutase and catalase), which results in increased intracellular ROS. However, some studies have suggested that PI3K-AKT-FOXO3 can promote the level of autophagy flow. The underlying mechanism is that FOXO3, activated by phosphorylation, promotes the synthesis of glutamine synthase, and also prevents mTOR translocation to the lysosomal membrane in a glutamine synthase-dependent manner, causing mTOR inhibition, thus promoting autophagy [62]. Our results revealed that in the DQ group, the significant increased expression of p-AKT, p-FOXO3 and the significant decreased nuclear FOXO expression were observed, although no significant changes were found in the expression of mTOR. Nevertheless, pretreatment with *BaSC06* reduced the expression of p-FOXO3 and restored it to the same level as in the CK group. We therefore hypothesized that *BaSC06* could increase the expression of FOXO3 by inhibition of p-AKT, thereby inducing the autophagy of IPEC-J2 cells. To further verify this result, we treated IPEC-J2 cells with perifosine, an AKT inhibitor. Perifosine can significantly reduce the phosphorylation of AKT and then reduce the extracellular signal-regulated kinase (ERK) 1/2, inducing cell cycle stagnation in G1 and G2 [63]. Prolong treatment of breast cancer cells with AKT inhibitors (which inhibit AKT phosphorylation) induces dephosphorylation of FOXO3a, nuclear translocation, and destruction of its binding to SIRT6, resulting in FOXO3a acetylation and BRD4 recognition [64]. Our results showed that *BaSC06* and the inhibitor perifosine exhibited almost the same effect. This suggests that both *BaSC06* and the inhibitor perifosine might significantly reduce p-FOXO3 expression, increase LC3 expression, and promote the degradation of p62. In addition, our results showed that *BaSC06* could not upregulate the ratio of LC3II/LC3I after FOXO3 knockdown, suggesting that FOXO3 is necessary for *BaSC06* to induce autophagy in IPEC-J2 cells. Overall, these outcomes indicate that *BaSC06* can promote autophagy and antioxidant enzyme production in IPEC-J2 cells by inhibiting the AKT-FOXO signaling pathway, thus alleviating oxidative stress.

The extension of autophagic vesicles leads to the formation of autophagosomes, usually bilayer organelles. During the process of autophagy, there are two ubiquitination processes that occur in the ATG5-ATG12 and LC3 systems, which are essential for the extension of autophagic vesicles and the maturation of autophagosomes [65]. In particular, ATG5 is a key protein involved in phagocytic membrane elongation in autophagic vesicular, which forms a constituent complex with ATG12. In this process, ATG7 activates ATG12 as an E1-like ubiquitin activase. Then, ATG12 is delivered to the E2-like ubiquitin transferase ATG10. Eventually, ATG12 binds to ATG5 for forming a complex [66]. Afterward, the ATG12-ATG5 complex and ATG16L further form the ATG12-ATG5-ATG16L complex, which is found to be located on the outer membrane of the autophagosome. Of note, the ATG5-ATG12-ATG16L complex exhibits E3 ligase-like activity, essentially by activating ATG3 enzyme activity, promoting LC3 (namely ATG8) transfer from ATG3 to the bottom phosphatidylethanolamine (PE) [65]. The ATG5-ATG12-ATG16L complex is disintegrated from the membrane, once the autophagosome formed. Our results suggest that *BaSC06* may induce autophagy in IPEC-J2 cells by promoting the formation of the ATG12-ATG5 complex, while the increase of ATG12 may be activated by FOXO deacetylation. These outcomes indicated that *BaSC06* can promote the expression of autophagy-related genes by increasing FOXO activity.

5. Conclusions

Collectively, the possible molecular mechanisms underlying *BaSC06*-induced autophagy to attenuate oxidative stress are summarized as follows: *BaSC06* inhibited the AKT–FOXO signaling pathway by decreasing the expression of p-AKT and p-FOXO and increasing the expression of SIRT1, thereby increasing the transcriptional activity of FOXO3 and gene expression of the ATG5–ATG12 complex, which induce autophagy to alleviate oxidative stress in IPEC-J2 cells. Besides, *BaSC06* attenuates apoptosis by modulating the activities of antioxidant enzymes and the expression of the apoptotic proteins Bcl2, Bax, Caspase 3 and Caspase 8. All the above findings provide a valuable theoretical basis for the application of probiotics in preventing and treating diseases caused by oxidative stress and improving human and animal health (Figure 8).

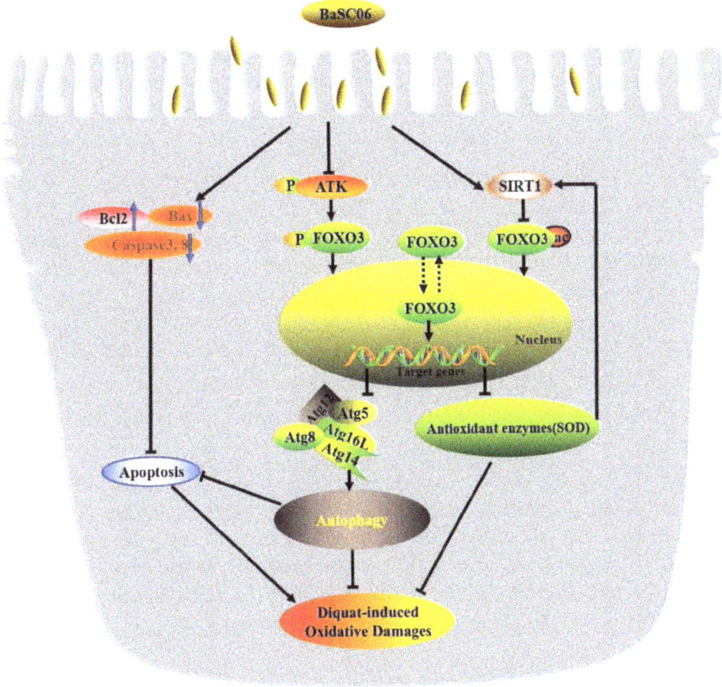

Figure 8. Proposed model of the protective effect of *BaSC06* in DQ-induced oxidative stress and apoptosis. (1) *BaSC06* inhibited the AKT-FOXO signaling pathway by inhibiting the expression of p-AKT, p-FOXO and increasing the expression of SIRT1, thereby increasing the gene expression of the ATG5-ATG12 complex to induce autophagy. (2) *BaSC06* attenuate apoptosis by modulating the activities of antioxidant enzymes, the expression of the apoptotic proteins Bcl2, Bax, Caspase 3 and Caspase 8 to alleviate oxidative stress.

Author Contributions: Conceptualization, W.L. and L.T.; methodology, L.T.; software, L.T. and B.W.; validation, L.T., Z.Z., Y.Z. and P.Z. (Pengwei Zhao); formal analysis, L.T.; resources, Q.W. and P.Z. (Peng Zou); data curation, L.T. and B.W.; writing—original draft preparation, L.T.; writing—review and editing, L.T. and Z.Z.; visualization, S.X.; supervision, F.W. and J.Y.; project administration, X.L.; funding acquisition, W.L. and P.Z. (Pengwei Zhao). All authors have read and agreed to the published version of the manuscript.

Funding: This study is supported by the National Natural Science Foundation of China (No. 31672460, 32072766 and 31472128), the Natural Science Foundation of Zhejiang province (No. LZ20C170002), National High-Tech R&D Program (863) of China (No. 2013AA102803D), the Major Science and Technology Project of Zhejiang Province (No. 2006C12086), PRC.

Institutional Review Board Statement: Not applicable.

Informed Consent Statement: Not applicable.

Data Availability Statement: All data that support the findings of this study are available in the submitted article except the RNA-seq, and the RNA-seq data are not publicly available due to privacy. The data are available from the corresponding author upon reasonable request.

Conflicts of Interest: The authors declare that the research was conducted in the absence of any commercial or financial relationships that could be construed as a potential conflict of interest.

References

1. Qiao, Y.; Sun, J.; Ding, Y.; Le, G.; Shi, Y. Alterations of the gut microbiota in high-fat diet mice is strongly linked to oxidative stress. *Appl. Microbiol. Biotechnol.* **2012**, *97*, 1689–1697. [CrossRef] [PubMed]

2. Diaz-Ochoa, V.E.; Lam, D.; Lee, C.S.; Klaus, S.; Behnsen, J.; Liu, J.Z.; Chim, N.; Nuccio, S.-P.; Rathi, S.G.; Mastroianni, J.R.; et al. Salmonella Mitigates Oxidative Stress and Thrives in the Inflamed Gut by Evading Calprotectin-Mediated Manganese Sequestration. *Cell Host Microbe* **2016**, *19*, 814–825. [CrossRef] [PubMed]
3. Artis, D. Epithelial-cell recognition of commensal bacteria and maintenance of immune homeostasis in the gut. *Nat. Rev. Immunol.* **2008**, *8*, 411–420. [CrossRef] [PubMed]
4. Brosnahan, A.J.; Brown, D.R. Porcine IPEC-J2 intestinal epithelial cells in microbiological investigations. *Vet. Microbiol.* **2012**, *156*, 229–237. [CrossRef] [PubMed]
5. McOrist, S.; Jasni, S.; Mackie, R.; Berschneider, H.; Rowland, A.; Lawson, G. Entry of the bacterium ileal symbiont intracellularis into cultured enterocytes and its subsequent release. *Res. Vet. Sci.* **1995**, *59*, 255–260. [CrossRef]
6. Cai, X.; Chen, X.; Wang, X.; Xu, C.; Guo, Q.; Zhu, L.; Zhu, S.; Xu, J. Pre-protective effect of lipoic acid on injury induced by H2O2 in IPEC-J2 cells. *Mol. Cell. Biochem.* **2013**, *378*, 73–81. [CrossRef]
7. Zhu, J.; Yin, X.; Yu, H.; Zhao, L.; Sabour, P.; Gong, J. Involvement of Quorum Sensing and Heat-Stable Enterotoxin a in Cell Damage Caused by a Porcine Enterotoxigenic Escherichia coli Strain. *Infect. Immun.* **2011**, *79*, 1688–1695. [CrossRef]
8. Paszti-Gere, E.; Csibrik-Nemeth, E.; Szeker, K.; Csizinszky, R.; Jakab, C.; Galfi, P. Acute Oxidative Stress Affects IL-8 and TNF-α Expression in IPEC-J2 Porcine Epithelial Cells. *Inflammation* **2011**, *35*, 994–1004. [CrossRef]
9. Scherz-Shouval, R.; Elazar, Z. ROS, mitochondria and the regulation of autophagy. *Trends Cell Biol.* **2007**, *17*, 422–427. [CrossRef]
10. Das, G.; Shravage, B.; Baehrecke, E.H. Regulation and Function of Autophagy during Cell Survival and Cell Death. *Cold Spring Harb. Perspect. Biol.* **2012**, *4*, a008813. [CrossRef]
11. Chowdhury, I.; Tharakan, B.; Bhat, G.K. Current concepts in apoptosis: The physiological suicide program revisited. *Cell. Mol. Biol. Lett.* **2006**, *11*, 506–525. [CrossRef]
12. Yoshimori, T. Autophagy: A regulated bulk degradation process inside cells. *Biochem. Biophys. Res. Commun.* **2003**, *313*, 453–458. [CrossRef] [PubMed]
13. Jiang, X.; Gu, S.; Liu, D.; Zhao, L.; Xia, S.; He, X.; Chen, H.; Ge, J. Lactobacillus brevis 23017 Relieves Mercury Toxicity in the Colon by Modulation of Oxidative Stress and Inflammation Through the Interplay of MAPK and NF-κB Signaling Cascades. *Front. Microbiol.* **2018**, *9*, 2425. [CrossRef] [PubMed]
14. Wang, Y.; Guo, Y.; Chen, H.; Wei, H.; Wan, C. Potential of Lactobacillus plantarum ZDY2013 and Bifidobacterium bifidum WBIN03 in relieving colitis by gut microbiota, immune, and anti-oxidative stress. *Can. J. Microbiol.* **2018**, *64*, 327–337. [CrossRef] [PubMed]
15. de Souza, M.; Baptista, A.A.S.; Valdiviezo, M.J.; Justino, L.; Menck-Costa, M.F.; Ferraz, C.R.; da Gloria, E.M.; Verri, W.A.; Bracarense, A.P.F. Lactobacillus spp. reduces morphological changes and oxidative stress induced by deoxynivalenol on the intestine and liver of broilers. *Toxicon* **2020**, *185*, 203–212. [CrossRef]
16. Ai, Q.; Xu, H.; Mai, K.; Xu, W.; Wang, J.; Zhang, W. Effects of dietary supplementation of Bacillus subtilis and fructooligosaccharide on growth performance, survival, non-specific immune response and disease resistance of juvenile large yellow croaker, Larimichthys crocea. *Aquaculture* **2011**, *317*, 155–161. [CrossRef]
17. Li, X.; Qiang, L.; Xu, C.; Liu. Effects of Supplementation of Fructooligosaccharide and/or Bacillus Subtilis to Diets on Performance and on Intestinal Microflora in Broilers. *Arch. Anim. Breed.* **2008**, *51*, 64–70. [CrossRef]
18. Shang, X.; Yu, P.; Yin, Y.; Zhang, Y.; Lu, Y.; Mao, Q.; Li, Y. Effect of selenium-rich Bacillus subtilis against mercury-induced intestinal damage repair and oxidative stress in common carp. *Comp. Biochem. Physiol. Part C Toxicol. Pharmacol.* **2020**, *239*, 108851. [CrossRef]
19. Lin, Y.-S.; Saputra, F.; Chen, Y.-C.; Hu, S.-Y. Dietary administration of Bacillus amyloliquefaciens R8 reduces hepatic oxidative stress and enhances nutrient metabolism and immunity against Aeromonas hydrophila and Streptococcus agalactiae in zebrafish (Danio rerio). *Fish Shellfish Immunol.* **2018**, *86*, 410–419. [CrossRef]
20. Jia, R.; Sadiq, F.A.; Liu, W.; Cao, L.; Shen, Z. Protective effects of Bacillus subtilis ASAG 216 on growth performance, antioxidant capacity, gut microbiota and tissues residues of weaned piglets fed deoxynivalenol contaminated diets. *Food Chem. Toxicol.* **2021**, *148*, 111962. [CrossRef]
21. Diao, Y.; Xin, Y.; Zhou, Y.; Li, N.; Pan, X.; Qi, S.; Qi, Z.; Xu, Y.; Luo, L.; Wan, H.; et al. Extracellular polysaccharide from Bacillus sp. strain LBP32 prevents LPS-induced inflammation in RAW 264.7 macrophages by inhibiting NF-κB and MAPKs activation and ROS production. *Int. Immunopharmacol.* **2013**, *18*, 12–19. [CrossRef] [PubMed]
22. Bai, K.; Huang, Q.; Zhang, J.; He, J.; Zhang, L.; Wang, T. Supplemental effects of probiotic Bacillus subtilis fmbJ on growth performance, antioxidant capacity, and meat quality of broiler chickens. *Poult. Sci.* **2017**, *96*, 74–82. [CrossRef]
23. Liang, Z.; Yuan, Z.; Guo, J.; Wu, J.; Yi, J.; Deng, J.; Shan, Y. Ganoderma lucidum Polysaccharides Prevent Palmitic Acid-Evoked Apoptosis and Autophagy in Intestinal Porcine Epithelial Cell Line via Restoration of Mitochondrial Function and Regulation of MAPK and AMPK/Akt/mTOR Signaling Pathway. *Int. J. Mol. Sci.* **2019**, *20*, 478. [CrossRef]
24. Nunes, T.; Bernardazzi, C.; De Souza, H.S. Cell Death and Inflammatory Bowel Diseases: Apoptosis, Necrosis, and Autophagy in the Intestinal Epithelium. *BioMed Res. Int.* **2014**, *2014*. [CrossRef]
25. El-Din, S.H.S.; Salem, M.; El-Lakkany, N.; Hammam, O.; Nasr, S.; Okasha, H.; Ahmed, L.; Saleh, S.; Botros, S. Early intervention with probiotics and metformin alleviates liver injury in NAFLD rats via targeting gut microbiota dysbiosis and p-AKT/mTOR/LC-3II pathways. *Hum. Exp. Toxicol.* **2021**, *40*, 1496–1509. [CrossRef]

26. Cui, Y.; Liu, L.; Dou, X.; Wang, C.; Zhang, W.; Gao, K.; Liu, J.; Wang, H. Lactobacillus reuteri ZJ617 maintains intestinal integrity via regulating tight junction, autophagy and apoptosis in mice challenged with lipopolysaccharide. *Oncotarget* **2017**, *8*, 77489–77499. [CrossRef]
27. Wu, Y.; Wang, B.; Xu, H.; Tang, L.; Li, Y.; Gong, L.; Wang, Y.; Li, W. Probiotic Bacillus Attenuates Oxidative Stress- Induced Intestinal Injury via p38-Mediated Autophagy. *Front. Microbiol.* **2019**, *10*, 2185. [CrossRef]
28. Chalubinski, M.; Wojdan, K.; Gorzelak-Pabiś, P.; Borowiec, M.; Broncel, M. The effect of oxidized cholesterol on barrier functions and IL-10 mRNA expression in human intestinal epithelium co-cultured with dendritic cells in the transwell system. *Food Chem. Toxicol.* **2014**, *69*, 289–293. [CrossRef]
29. Xiao, K.; Jiao, L.; Cao, S.; Song, Z.; Hu, C.; Han, X. Whey protein concentrate enhances intestinal integrity and influences transforming growth factor-β1 and mitogen-activated protein kinase signalling pathways in piglets after lipopolysaccharide challenge. *Br. J. Nutr.* **2016**, *115*, 984–993. [CrossRef] [PubMed]
30. Jones, G.M.; Vale, J.A. Mechanisms of Toxicity, Clinical Features, and Management of Diquat Poisoning: A Review. *J. Toxicol. Clin. Toxicol.* **2000**, *38*, 123–128. [CrossRef] [PubMed]
31. Daitoku, H.; Sakamaki, J.-I.; Fukamizu, A. Regulation of FoxO transcription factors by acetylation and protein–protein interactions. *Biochim. et Biophys. Acta (BBA)-Bioenerg.* **2011**, *1813*, 1954–1960. [CrossRef]
32. Abdollahi, M.; Ranjbar, A.; Shadnia, S.; Nikfar, S.; Rezaie, A. Pesticides and oxidative stress: A review. *Med. Sci. Monit.* **2004**, *10*, RA141–RA147.
33. Lv, M.; Yu, B.; Mao, X.B.; Zheng, P.; He, J.; Chen, D.W. Responses of growth performance and tryptophan metabolism to oxidative stress induced by diquat in weaned pigs. *Animal* **2012**, *6*, 928–934. [CrossRef]
34. Lu, T.; Piao, X.; Zhang, Q.; Wang, D.; Kim, S.W. Protective effects of Forsythia suspensa extract against oxidative stress induced by diquat in rats. *Food Chem. Toxicol.* **2010**, *48*, 764–770. [CrossRef]
35. Wu, K.C.; Zhang, Y.; Klaassen, C.D. Nrf2 protects against diquat-induced liver and lung injury. *Free Radic. Res.* **2012**, *46*, 1220–1229. [CrossRef]
36. Xu, C.; Guo, Y.; Qiao, L.; Ma, L.; Cheng, Y.; Roman, A. Biogenic Synthesis of Novel Functionalized Selenium Nanoparticles by Lactobacillus casei ATCC 393 and Its Protective Effects on Intestinal Barrier Dysfunction Caused by Enterotoxigenic Escherichia coli K88. *Front. Microbiol.* **2018**, *9*, 1129. [CrossRef]
37. Kang, R.; Li, R.; Dai, P.; Li, Z.; Li, Y.; Li, C. Deoxynivalenol induced apoptosis and inflammation of IPEC-J2 cells by promoting ROS production. *Environ. Pollut.* **2019**, *251*, 689–698. [CrossRef] [PubMed]
38. McCord, J.M.; Edeas, M. SOD, oxidative stress and human pathologies: A brief history and a future vision. *Biomed. Pharmacother.* **2005**, *59*, 139–142. [CrossRef] [PubMed]
39. Yin, J.; Liu, M.; Ren, W.; Duan, J.; Yang, G.; Zhao, Y.; Fang, R.; Chen, L.; Li, T.; Yin, Y. Effects of Dietary Supplementation with Glutamate and Aspartate on Diquat-Induced Oxidative Stress in Piglets. *PLoS ONE* **2015**, *10*, e0122893. [CrossRef] [PubMed]
40. Cao, S.; Wu, H.; Wang, C.; Zhang, Q.; Jiao, L.; Lin, F.; Hu, C.H. Diquat-induced oxidative stress increases intestinal permeability, impairs mitochondrial function, and triggers mitophagy in piglets1. *J. Anim. Sci.* **2018**, *96*, 1795–1805. [CrossRef]
41. Chong, S.J.F.; Low, I.C.C.; Pervaiz, S. Mitochondrial ROS and involvement of Bcl-2 as a mitochondrial ROS regulator. *Mitochondrion* **2014**, *19*, 39–48. [CrossRef] [PubMed]
42. Wu, C.; Fujihara, H.; Yao, J.; Qi, S.; Li, H.; Shimoji, K.; Baba, H. Different Expression Patterns of Bcl-2, Bcl-xl, and Bax Proteins After Sublethal Forebrain Ischemia in C57Black/Crj6 Mouse Striatum. *Stroke* **2003**, *34*, 1803–1808. [CrossRef] [PubMed]
43. Boulares, H.; Yakovlev, A.; Ivanova, V.; Stoica, B.; Hasan, S.K.; Iyer, S.; Smulson, M. Caspase-3 resistant PARP mutant increases rates of apoptosis in transfected osteosarcoma cells. *Faseb J.* **1999**, *13*, A518.
44. Loh, K.P.; Huang, S.H.; De Silva, R.; Tan, B.K.H.; Zhu, Y.Z. Oxidative Stress: Apoptosis in Neuronal Injury. *Curr. Alzheimer Res.* **2006**, *3*, 327–337. [CrossRef] [PubMed]
45. Galati, S.; Boni, C.; Gerra, M.C.; Lazzaretti, M.; Buschini, A. Autophagy: A Player in response to Oxidative Stress and DNA Damage. *Oxid. Med. Cell. Longev.* **2019**, *2019*. [CrossRef]
46. Singh, S.; Kumar, R.; Garg, G.; Singh, A.K.; Verma, A.K.; Bissoyi, A.; Rizvi, S.I. Spermidine, a caloric restriction mimetic, provides neuroprotection against normal and d-galactose-induced oxidative stress and apoptosis through activation of autophagy in male rats during aging. *Biogerontology* **2020**, *22*, 35–47. [CrossRef]
47. Shi, S.; Tian, T.; Li, Y.; Xiao, D.; Zhang, T.; Gong, P.; Lin, Y. Tetrahedral Framework Nucleic Acid Inhibits Chondrocyte Apoptosis and Oxidative Stress through Activation of Autophagy. *ACS Appl. Mater. Interfaces* **2020**, *12*, 56782–56791. [CrossRef]
48. Li, D.L.; Mao, L.; Gu, Q.; Wei, F.; Gong, Y.-Y. Quercetin protects retina external barrier from oxidative stress injury by promoting autophagy. *Cutan. Ocul. Toxicol.* **2020**, *40*, 7–13. [CrossRef]
49. Mammucari, C.; Milan, G.; Romanello, V.; Masiero, E.; Rudolf, R.; Del Piccolo, P.; Burden, S.J.; Di Lisi, R.; Sandri, C.; Zhao, J.; et al. FoxO3 Controls Autophagy in Skeletal Muscle In Vivo. *Cell Metab.* **2007**, *6*, 458–471. [CrossRef]
50. Cao, D.J.; Jiang, N.; Blagg, A.; Johnstone, J.L.; Gondalia, R.; Oh, M.; Luo, X.; Yang, K.; Shelton, J.M.; Rothermel, B.A.; et al. Mechanical Unloading Activates FoxO3 to Trigger Bnip3-Dependent Cardiomyocyte Atrophy. *J. Am. Heart Assoc.* **2013**, *2*, e000016. [CrossRef]
51. Sengupta, A.; Molkentin, J.; Yutzey, K.E. FoxO Transcription Factors Promote Autophagy in Cardiomyocytes. *J. Biol. Chem.* **2009**, *284*, 28319–28331. [CrossRef]

52. Dey, G.; Bharti, R.; Dhanarajan, G.; Das, S.; Dey, K.K.; Kumar, B.N.P.; Sen, R.; Mandal, M. Marine lipopeptide Iturin A inhibits Akt mediated GSK3β and FoxO3a signaling and triggers apoptosis in breast cancer. *Sci. Rep.* **2015**, *5*, 10316. [CrossRef]
53. Hagenbuchner, J.; Ausserlechner, M.J. Mitochondria and FOXO3: Breath or die. *Front. Physiol.* **2013**, *4*, 147. [CrossRef]
54. Bánréti, Á.; Sass, M.; Graba, Y. The emerging role of acetylation in the regulation of autophagy. *Autophagy* **2013**, *9*, 819–829. [CrossRef] [PubMed]
55. Salminen, A.; Kaarniranta, K.; Kauppinen, A. Crosstalk between Oxidative Stress and SIRT1: Impact on the Aging Process. *Int. J. Mol. Sci.* **2013**, *14*, 3834–3859. [CrossRef] [PubMed]
56. Brunet, A.; Sweeney, L.B.; Sturgill, J.F.; Chua, K.F.; Greer, P.L.; Lin, Y.; Tran, H.; Ross, S.E.; Mostoslavsky, R.; Cohen, H.Y.; et al. Stress-Dependent Regulation of FOXO Transcription Factors by the SIRT1 Deacetylase. *Science* **2004**, *303*, 2011–2015. [CrossRef] [PubMed]
57. van der Horst, A.; Tertoolen, L.G.J.; de Vries-Smits, L.M.M.; Frye, R.A.; Medema, R.; Burgering, B.M.T. FOXO4 Is Acetylated upon Peroxide Stress and Deacetylated by the Longevity Protein hSir2. *J. Biol. Chem.* **2004**, *279*, 28873–28879. [CrossRef]
58. Sengupta, A.; Molkentin, J.; Paik, J.-H.; DePinho, R.; Yutzey, K.E. FoxO Transcription Factors Promote Cardiomyocyte Survival upon Induction of Oxidative Stress. *J. Biol. Chem.* **2011**, *286*, 7468–7478. [CrossRef]
59. Xiong, S.; Salazar, G.; Patrushev, N.; Alexander, R.W. FoxO1 Mediates an Autofeedback Loop Regulating SIRT1 Expression. *J. Biol. Chem.* **2011**, *286*, 5289–5299. [CrossRef]
60. Yamamoto, T.; Sadoshima, J. Protection of the Heart Against Ischemia/Reperfusion by Silent Information Regulator 1. *Trends Cardiovasc. Med.* **2011**, *21*, 27–32. [CrossRef]
61. Shin, H.R.; Kim, H.; Kim, K.I.; Baek, S.H. Epigenetic and transcriptional regulation of autophagy. *Autophagy* **2016**, *12*, 2248–2249. [CrossRef]
62. Van Der Vos, K.E.; Eliasson, P.; Proikas-Cezanne, T.; Vervoort, S.J.; van Boxtel, R.; Putker, M.; Van Zutphen, I.J.; Mauthe, M.; Zellmer, S.; Pals, C.; et al. Modulation of glutamine metabolism by the PI(3)K–PKB–FOXO network regulates autophagy. *Nature* **2012**, *14*, 829–837. [CrossRef]
63. Momota, H.; Nerio, E.; Holland, E.C. Perifosine Inhibits Multiple Signaling Pathways in Glial Progenitors and Cooperates With Temozolomide to Arrest Cell Proliferation in Gliomas In vivo. *Cancer Res.* **2005**, *65*, 7429–7435. [CrossRef] [PubMed]
64. Liu, J.Y.; Duan, Z.B.; Guo, W.J.; Zeng, L.; Wu, Y.D.; Chen, Y.L.; Tai, F.; Wang, Y.F.; Lin, Y.W.; Zhang, Q.; et al. Targeting the BRD4/FOXO3a/CDK6 axis sensitizes AKT inhibition in luminal breast cancer. *Nat. Commun.* **2018**, *9*, 5200. [CrossRef]
65. Nakatogawa, H. Two ubiquitin-like conjugation systems that mediate membrane formation during autophagy. *Essays Biochem.* **2013**, *55*, 39–50. [CrossRef] [PubMed]
66. Romanov, J.; Walczak, M.; Ibiricu, I.; Schüchner, S.; Ogris, E.; Kraft, C.; Martens, S. Mechanism and functions of membrane binding by the Atg5-Atg12/Atg16 complex during autophagosome formation. *EMBO J.* **2012**, *31*, 4304–4317. [CrossRef]

MDPI
St. Alban-Anlage 66
4052 Basel
Switzerland
www.mdpi.com

Antioxidants Editorial Office
E-mail: antioxidants@mdpi.com
www.mdpi.com/journal/antioxidants

www.ingramcontent.com/pod-product-compliance
Lightning Source LLC
Chambersburg PA
CBHW061800210926
51850CB00006B/606
9783036585321